Activated Sludge – 100 Years and Counting

Activated Sludge – 100 Years and Counting

Edited by

David Jenkins and Jiří Wanner

PUBLISHING

Published by **IWA Publishing**
Alliance House
12 Caxton Street
London SW1H 0QS, UK
Telephone: +44 (0)20 7654 5500
Fax: +44 (0)20 7654 5555
Email: publications@iwap.co.uk
Web: www.iwapublishing.com

First published 2014, reprinted 2016

British Library Cataloguing in Publication Data
A CIP catalogue record for this book is available from the British Library

ISBN 9781780404936 (Hardback)
ISBN 9781780404943 (eBook)

Cover images:
Row 1 Image 1 and 2: Photography by Ken Osborn. Photographs of East Bay Municipal Utility District, Oakland, CA, USA by permission of Ben Horenstein, Manager Wastewater Division.
Row 1 Image 3: Courtesy of Dr. Iveta Ruzickova, Institute of Chemical Technology, Prague
Row 2 Image 1: Photograph of secondary clarifier of the WWTP Plzen by permission of Miloslav Vostry, Manager General, Vodarna a.s., Plzen. Courtesy of Jiri Wanner
Row 2 Image 2: Photograph of regeneration zone of the WWTP Liberec by permission of Milan Kuchar, Manager General, Severoceske Vodovody a Kanalizace, a.s., Teplice Coutes. Courtesy of Jiri Wanner
Row 2 Image 3: Photography by Ken Osborn. Photographs of East Bay Municipal Utility District, Oakland, CA, USA by permission of Ben Horenstein, Manager Wastewater Division.
Row 3 Image 1: Fluorescence *in situ* hybridization. Appearing in Chapter 4 as Figure 4.3a. Courtesy of Simon McIlroy
Row 3 Image 2: Photography by Ken Osborn. Photographs of East Bay Municipal Utility District, Oakland, CA, USA by permission of Ben Horenstein, Manager Wastewater Division.

Contents

Abbreviations

AGS	aerobic granular sludge
AOX	absorbable organic halogens
AS	activated sludge
ASM	activated sludge model
BNR	biological nutrient removal
BOD	biochemical oxygen demand
CAS	conventional activated sludge
CEPT	chemically enhanced primary treatment
CFD	computational fluid dynamics
CIP	clean-in-place
COD	chemical oxygen demand
DGGE	denaturing gradient gel electrophoresis
DNA	deoxyribonucleic acid
DO	dissolved oxygen
DOC	dissolved organic carbon
DSVI	diluted sludge volume index
EBPR	enhanced biological phosphorus removal
EDC	endocrine disrupting compound
EDI	energy dissipating inlet
EPS	exocellular polymer
ESS	effluent suspended solids
F/M	food to mass ratio

FISH	fluorescent in situ hybridization
FOG	fat, oil and grease
GAC	granular activated carbon
GAO	glycogen accumulating organism
GHG	greenhouse gas
HPO	high purity oxygen
HRT	hydraulic residence time
ICZ	initial contact zone
IFAS	integrated fixed-film activated sludge system
K_S	half saturation constant
MAR	microautoradiography
MBBR	moving bed biofilm reactor
MBR	membrane bioreactor
MLSS	mixed liquor suspended solids
MUCT	modified University of Capetown (process)
N	nitrogen
N/DN	nitrification/ denitrification
NGS	next generation sequencing
OHO	ordinary heteroptrophic organism
OUR	oxygen uptake rate
P	phosphorus
PAC	powdered activated carbon
PAO	polyphosphate accumulating organism
PE	population equivalents
PHA	polyhydroxyalkanoate
PhAC	pharmacologically active compound
polyP	polyphosphate
POTW	publically owned treatment works
PPCP	pharmaceuticals and personal care products
PWWF	peak wet weather flow
QA/QC	quality assurance /quality control
q-PCR	quantitative polymerase chain reaction
RAS	return activated sludge
RBC	rotating biological contactor
rbCOD	readily biodegradable chemical oxygen demand

rDON	refractory dissolved organic nitrogen
rRNA	ribosomal ribonucleic acid
SAE	standard aeration efficiency
SBR	sequencing batch reactor
SLR	solids loading rate
SND	simultaneous nitrification/denitrification
SOR	surface overflow rate
SOTR	standard oxygen transfer rate
SOUR	specific oxygen uptake rate
SRT	solids residence time, mean cell residence time, sludge age
SS	suspended solids
SSVI	stirred sludge volume index
SVI	sludge volume index
SWD	side wall depth
TEFL	total extended filament length
TF	trickling filter
TMDL	total maximum daily load
TMP	transmembrane pressure
TN	total nitrogen
TOC	total organic carbon
TP	total phosphorus
TSS	total suspended solids
UASB	upflow anaerobic solids basin
UCT	University of Capetown (process)
UF	ultrafiltration
VFA	volatile fatty acid
VOC	volatile organic compound
VSS	volatile suspended solids
WAS	waste activated sludge
WLA	waste load allocation
WWTP	wastewater treatment plant
Y	yield
ZSV	zone settling velocity
μ_{max}	maximum growth rate

About the authors

TECHNICAL EDITORS

David Jenkins
University of California at Berkeley
11 Yale Circle
Kensington CA 94708
USA
Email: flocdr@yahoo.com

Jiří Wanner
Institute of Chemical Technology, Prague
Technická 5
CZ-166 28 Prague 6
Czech Republic
Email: jiri.wanner@vscht.cz

AUTHORS AND CO-AUTHORS

Chapter 1

Glen T. Daigger
CH2M Hill
9191 South Jamaica Street
Englewood CO 80112-5946
USA
Email: gdaigger@ch2m.com

Chapter 2

Hermann H. Hahn
Institut für Wasser und Gewässerentwicklung
Bereich Siedlungswasserwirtschaft und Wassergütewirtschaft
Karlsruher Institut für Technologie
Adenauerring 20
76131 Karlsruhe
Germany
Email: hermann@hahn-family.net

Chapter 3

H. David Stensel
University of Washington
303 More Hall Box 352700
Seattle WA 98195-270
USA
Email: stensel@u.washington.edu

Jacek Makinia (Co-author)
Gdansk University of Technology
11/12 Gabriela Narutowicza St
80-233 Gdansk-Wrzeszcz
Poland
Email: jmakinia@pg.gda.pl

Chapter 4

Per Halkjær Nielsen
Center for Microbial Communities
Department of Biotechnology
Chemistry and Environmental Engineering
Aalborg University
Sohngaardsholmsvej 49
9000 Aalborg
Denmark
Email: phn@bio.aau.dk

Katherine D. McMahon (Co-author)
Civil & Environmental Engineering
University of Wisconsin
3204 Engineering Hall
1415 Engineering Drive
Madison WI 53706-1691

USA
Email: tmcmahon@engr.wisc.edu

Chapter 5

Wendell O. Khunjar
Hazen & Sawyer
4035 Ridge Top Road, Suite 400
Fairfax VA 22030
USA
Email: wkhunjar@hazenandsawyer.com

Paul A. Pitt (Co-author)
Hazen & Sawyer
2121 North California Blvd, Suite 290
Walnut Creek CA 94596
USA
Email: ppitt@hazenandsawyer.com

Charles B. Bott (Co-author)
Hampton Roads Sanitation District
1434 Air Rail Avenue
Virginia Beach, VA 23455
USA
Email: cbott@hrsd.com

Kartik Chandran (Co-author)
Earth and Environmental Engineering
Columbia University
500 West 120th St
New York NY 10027
USA
Email: kc2288@columbia.edu

Chapter 6

James Barnard
Black and Veatch
12869 Cambridge Terrace
Leawood KS 66209
USA
Email: barnardjl@bv.com

Yves Comeau (Co-author)
Department of Civil, Geological and Mining Engineering (CGM)
Université de Montreal campus
PO Box 6079
Station Centre-Ville
Montreal
Quebec
Canada H3C 3A7
Email: cgm@polymtl.ca

Chapter 7

Hansruedi Siegrist (Switzerland)
Department of Process Engineering
EAWAG
PO Box 611
Ch-8600 Duebendorf
Switzerland
Email: hansruedi.siegrist@eawag.ch

Adriano Joss (Co-author)
Department of Process Engineering
EAWAG
PO Box 611Ch-8600 Duebendorf
Switzerland
Email: adriano.joss@eawag.ch

Thomas A. Ternes (Co-author)
Bundesanstalt fuer Gewasserkunde
Kaiserin-Augusta-Anlagen 15-17 D-56068
Koblenz
Germany
Email: ternes@bafg.de

Chapter 8

Martin Wagner
Technische Universitaet Darmstadt
Institut IWAR – Facharbeit Wastewater
Technology Work
Franziska-Braun-Straße 7

64287 Darmstadt
Germany
Email: m.wagner@iwar.tu-darmstadt.de

Michael K. Stenstrom (Co-author)
University of California Los Angeles
5714 Boelter Hall
Los Angeles CA 90095-1593
USA
Email: stenstro@seas.ucla.edu

Chapter 9

Jay R. Witherspoon
CH2M Hill
155 Grand Avenue Suite 1000
Oakland CA 94612
USA
Email: jay.witherspoon@ch2m.com

Michael D. Short (Co-author)
UNSW Water Research Centre
School of Civil and Environmental Engineering
University of New South Wales
Sydney, New South Wales 2052
Australia
Email: m.short@unsw.edu.au

Kate Simmonds (Co-author)
CH2MHill
Beca House 21 Pitt Street
Auckland 1010
New Zealand
Email: kate.simmonds@ch2m.com.au

Ben van den Akker (Co-author)
SA Water Centre for Water Management and Reuse School of Natural and Built Environments
University of South Australia
Adelaide, South Australia 5095
Australia
Email: ben.vandenakker@sawater.com.au

Ewa Madon (Co-author)
CH2M Hill
Level 40
385 Bourke Street
Melbourne
Australia
Email: ewa.madon@ch2m.com.au

Richard M. Stuetz (Co-author)
UNSW Water Research Centre
School of Civil and Environmental Engineering
The University of New South Wales
Sydney NSW 2052
Australia
Email: r.stuetz@unsw.edu.au

Chapter 10

Jiří Wanner
Institute of Chemical Technology, Prague
Technická 5
CZ-166 28 Prague 6
Czech Republic
Email: jiri.wanner@vscht.cz

Andrea Jobbágy (Co-author)
Budapest University of Technology and Economics
Department of Applied Biotechnology and Food Sciences
Szt. Gellért tér 4
H-1111 Budapest
Hungary
Email: ajobbagy@mail.bme.hu

Chapter 11

Denny S. Parker
Brown and Caldwell
201 North Civic Drive, Suite 115
Walnut Creek, CA 94596
USA
Email: DParker@brwncald.com

Wolfgang Günthert (Co-author)
Universität der Bundeswehr München
Institut für Wasserwesen
Werner-Heisenberg-Weg 39
85577 Neubiberg
Germany
Email: wolfgang.guenthert@unibw.de

Britt-Marie Wilén (Co-author)
Chalmers University of Technology
Department of Civil and Environmental Engineering
Sven Hultins gata 8
412 96 Göteborg
Sweden
Email: britt-marie.wilen@chalmers.se

Chapter 12

Helmut Kroiss
Vienna University of Technology
Institute for Water Quality, Resource & Waste
Vienna 1040
Austria
Email: hkroiss@iwag.tuwien.ac.at

Yeshi Cao (Co-author)
Water Reclamation (Plants) Department
PUB
40 Scotts Road
#15-01 Environment Building
Singapore 228231
E-mail: cao_yeshi@pub.gov.sg

Chapter 13

Gustaf Olsson
Dept. of Industrial Electrical Engineering and Automation (IEA)
Lund University Box 118
SE-22100 Lund
Sweden
E-mail: gustaf.olsson@iea.lth.se

Zhiguo Yuan (Co-author)
Advanced Water Management Centre
The University of Queensland
St Lucia Brisbane QLD 4072
Australia
Email: zhiguo@awmc.uq.edu.au

Changwon Kim (Co-author)
Department of Environmental Engineering
Pusan National University
Keumjungkoo Changjundong San 30
Busan
Korea 609-735
Email: cwkim100@gmail.com

Chapter 14

George A. Ekama
Department of Civil Engineering
University of Capetown
Rondebosch CAPE 7701
South Africa
Email: george.ekama@uct.ac.za

Imre Takács (Co-author)
Dynamita
66 bis Avenue du Parc d'Espagne
Pessac 6600
France
Email: imre@dynamita.com

Chapter 15

Hallvard Ødegaard
Norwegian University of Science and Technology (NTNU)
Department of Hydraulic and Environmental Engineering
SP Andersens vei 5
N-7491 Trondheim
Norway
E-mail: hallvard.odegaard@ntnu.no

Magnus Christensson (Co-author)
AnoxKaldnes AB
Klosterängsvägen 11A
Lund

Sweden
Email: magnus.christensson@anoxkaldnes.com

Kim Helleshoj Sorensen (Co-author)
Wastewater Processes Technical Department, Veolia Water
Immeuble Giovanni Battista B 1 rue Giovanni Battista Pirelli
94410 SAINT-MAURICE Cedex
France
E-mail: kim.sorensen@veoliaeau.fr

Chapter 16

George V. Crawford
CH2MHill
245 Consumers Road Suite 400
Toronto ON M2J 1R3
Canada
Email: george.crawford@ch2m.com

Simon Judd (Co-author)
Cranfield Water Science Institute
Cranfield University
College Road
Cranfield
Bedfordshire MK43 0AL
United Kingdom
Email: s.j.judd@cranfield.ac.uk

Tamas Zsirai (Co-author)
Cranfield Water Science Institute
Cranfield University
College Road
Cranfield
Bedfordshire MK43 0AL
United Kingdom
Email: t.zsirai@cranfield.ac.uk

Chapter 17

Karl-Heinz Rosenwinkel
Institute for Sanitary Engineering and Waste Management
Leibniz Universität Hannover
Welfengarten 1

30167 Hannover
Germany
Email: rosenwinkel@isah.uni-hannover.de

Willy Verstraete (Co-author)
LabMET, Ghent University
Dasstraat 49032
Wondelgem
Belgium
Email: willy.verstraete@gmail.com

Siegfried E. Vlaeminck (Co-author)
Ghent University Faculty Bioscience Engineering
LabMET Coupure Links 653
9000 Ghent
Belgium
Email: siegfried.vlaeminck@ugent.be

Martin Wagner (Co-author)
Technische Universitaet Darmstadt
Institut IWAR – Wastewater Technology
Franziska-Braun-Straße 7
64287 Darmstadt
Germany
Email: m.wagner@iwar.tu-darmstadt.de

Sabrina Kipp (Co-author)
Institute for Sanitary Engineering and Waste Management
Leibniz Universität Hannover
Welfengarten 1
30167 Hannover
Germany
Email: kipp@isah.uni-hannover.de

Nina Manig (Co-author)
Institute for Sanitary Engineering and Waste Management
Leibniz Universität Hannover
Welfengarten 1
30167 Hannover
Germany
Email: manig@isah.uni-hannover.de

Chapter 18

Burkhard Teichgräber
Emschergenossenschaft
Postfach 101161
45011 Essen
Germany
Email: teichgraeber.burkhard@eglv.de

Chapter 19

Norbert Jardin
Planning Department
Ruhrverband
Kronprinzenstraße 37
45128 Essen
Germany
Email: nja@ruhrverband.de

Julian Sandino (Co-author)
CH2M HILL
13113 Melrose St
Overland Park KS 66213
USA
Email: julian.sandino@ch2m.com

Chapter 20

Mark van Loosdrecht
Delft University of Technology Julianalaan 67
Delft 2628 BC
The Netherlands
Email: m.c.m.vanloosdrecht@tudelft.nl

Harry Seah (Co-author)
PUB Headquarters
40 Scotts Road #22-01
Environment Building
Singapore 228231
Tel: + 65 6284 6600
Email: harry_seah@pub.gov.sg

Yuen Long Wah (Co-author)
Water Reclamation (Plants) Department
PUB Headquarters
40 Scotts Road #15-01 Environment Building
Singapore 228231
Singapore
E-mail: WAH_Yuen_Long@pub.gov.sg

Yeshi Cao (Co-author)
Water Reclamation (Plants) Department
PUB Headquarters
40 Scotts Road #15-01 Environment Building
Singapore 228231
Singapore
E-mail: cao_yeshi@pub.gov.sg

Preface

Even though the saying 'May you live in interesting times' is neither ancient nor Chinese, I think that it is an appropriate descriptor for the professional life of those of us who, like us, have lived during the latter half of the 20th century. It was early in that century when Edward Ardern and William Lockett, two chemists working at the Davyhulme sewage treatment plant in Manchester, England, first coined the term 'activated sludge' to describe some particulate material that accumulated during their sewage aeration experiments. We have seen the process that developed from these experiments come to fruition, to find uses all over the world and to expand its capabilities far beyond those for which it was originally conceived. Our ability to turn a treatment method, with what now seem rather mundane objectives, into a wide range of sophisticated processes has been the result of a remarkable, though not always completely congenial, partnership between engineers and scientists of every stripe.

This book and the conference for which it serves as proceedings was the result of discussions several years ago between the editors and Dr Glen T. Daigger. We resolved to invite the world's leading specialists on the various aspects of the activated sludge process to gather for a special conference at which they would deliver a paper and to prepare this paper for publication in a book of proceedings that would be available at the conference and on the market following that.

We asked the authors to present their personal assessment of the specific topic within the short space of 15 pages and to make it a summary of 'Where did we come from?' 'Where are we now?' and 'What does the future hold?' When you read this book, we are sure that you will agree with us that the authors have done a remarkable job ... and we thank them. The book starts with a remembrance of Ardern and Lockett, and summaries of the development of water quality requirements and of the development of the activated sludge process itself. Next, microbiology and macro- and micro-nutrient removals are discussed.

Following this, the subjects of aeration, mixing and air emissions are addressed. Then chapters on solids separation and secondary clarifiers are presented. Energy, automation, control and modeling form the next group of papers. These are followed by chapters on hybrid systems, membrane systems and industrial wastewater treatment. The book concludes with chapters on planning, design and economics and finally a look into what we might expect in the next 100 years.

We chose Essen, Germany because the first activated sludge plant in continental Europe was built between 1924 and 1926 in Rellinghausen, which is now part of Essen.

We extend our thanks to: the conference sponsors, Ruhrverband and Emschergenossenschaft/Lippeverband; the conference organizers, Norbert Jardin and Burkhard Teichgräber; to the all of the authors for their hard work and fine product; and last, but certainly not least to Joan Jenkins for her invaluable assistance during the preparation of the manuscripts for publication in this book.

David Jenkins
Berkeley, CA, USA

Jiří Wanner
Prague, Czech Republic
December 2013

Chapter 1

Ardern and Lockett remembrance

Glen T. Daigger (USA)

1.1 INTRODUCTION

In this volume we recount the development of the AS process over the past 100 years. In so doing we also celebrate its contributions to enhancing human life and environmental protection and speculate about its role in the future. This is an ambitious task to say the least, but one which we hope will inform both the current generation about the future and future generations about the past. To do this it is necessary to go back to 'the beginning' and understand the circumstances which resulted in the initial development of the AS process. The reasons that it has been so transformational to the sanitary engineering profession, both throughout the past 100 years and especially as the sanitary engineering profession evolved into the environmental engineering profession in the late 1960s and 1970s, are also addressed.

Edward Ardern and William T. Lockett, pictured in Figure 1.1, are widely recognized as the 'discoverers' and 'inventors' of the AS process. Their paper before the Society of Chemical Industry on April 3, 1914 in Manchester, England is broadly recognized as the 'official' presentation of the AS process to the professional community. How did this come to pass and why was this such a momentous event in the history of the sanitary/environmental engineering profession? The answers to these questions require an understanding of the historical context for the development of the AS process – the circumstances which existed in society in general as the process was being developed. In English there is a common expression that states 'necessity is the mother of invention', so we might ask 'what was the necessity that drove this invention?' Addressing this question is essential for setting the stage for all that follows, both in the following chapters of this book and as we collectively look at the future of the process. Why is this the case? It is because, as in the past, it is 'necessity' that will drive the further practical advancement of the process.

Figure 1.1 Ardern and Lockett. (Photo from J. Wanner, courtesy of V. Madera.)

To accomplish this task I will:

- Review the social, economic, scientific, and environmental context of the late 19th and early 20th century when the AS process was developed.
- Briefly summarize the principal events which led to development of the process.
- Discuss the role that Ardern and Lockett specifically played in development of the process.
- Discuss the subsequent implementation of the process, which set it on the path to become the dominant technology that it is today.
- Identify the key characteristics of the process which have allowed it to remain so dominant for such a long time.

It is clear that the AS process has had a major impact on the sanitary/environmental engineering profession in its approach to addressing human health and environmental protection over the past 100 years. Each quarter century or so of AS process's existence has brought with it a review of its historical development (Mohlman, 1938; Sawyer, 1965; Alleman & Prakasam, 1983). It will be interesting to see if yet another major review and assessment is prepared in about 2040.

1.2 INVENTION OF AS

1.2.1 The context

The late 19th and early 20th century was a time of significant change. Roughly 100 years following the initiation of the first industrial revolution in the UK and the USA, and at the conclusion of the second industrial revolution in which technological, economic, and population growth gained increased momentum, the living standards of a rapidly increasing population were, likewise, rapidly increasing. Millions of people were literally being lifted out of poverty into a standard of living never before experienced by so many (a situation not unlike what is occurring in developing countries such as China today). The urban population was also increasing exponentially (a trend which continues today) as the economic wealth created by the industrial revolution was concentrated there.

While a tendency exists to romanticize this period, the actual situation was far from idyllic (Bettmann, 1974). Indeed, the increased urban population suffered from unsanitary conditions that adversely impacted their health, comfort, and standard of living, and negative environmental impacts were broadly felt. The average life span during this time period was roughly 45 to 50 years, compared to the 75 to 80 years that it is today. Infant mortality was a principal contributor to reducing the lifespan in the late 19th and early 20th century. It is thought by many that two of the three decades of additional life span now enjoyed in the developed world are due to the water and sanitation systems installed during the 20th century. Indeed, a survey of public health researchers and professionals in the *British Medical Journal* indicated that the implementation of modern water and wastewater systems was the most significant step that improved public health over the past 150 years (Anon., 2007). Observations such as these led the US National Academy of Engineering to list modern water and sanitation as one of the significant engineering accomplishments of the 20th century (Constable & Sommerville, 2003).

As with all transformations of the magnitude of the Industrial Revolution, there were both positive and negative effects. The principal negative impact was the absence of acceptable methods of managing human and industrial wastewaters. These circumstances resulted in intense studies by numerous researchers on both sides of the Atlantic. Schneider (2011) provides a thorough and interesting summary of developments in the late 19th and early 20th century in the USA and the UK. The problem of sewage arose in the 19th century in cities throughout the industrializing world as the modern potable water service was extended, thereby addressing one aspect of public health protection – the provision of safe drinking water. This led to the development of waterborne sewage, with the associated dramatic increase in volume which overwhelmed traditional waste management approaches. Identifying solutions to this problem was the subject in the UK of a number of Royal Commission studies, one of which in 1865 established land treatment as the only acceptable sewage treatment system (Congress on the Sewage of Towns, 1866). This finding led to the development of sewage farms to serve major cities throughout Europe. Sewage farms proved highly effective in some locations, with notable examples developing to serve Paris and Berlin (Schneider, 2011). However, in other locations, with poorly drained soils and rainy climates, they were less than successful. Moreover, the large land requirements became a practical constraint, especially as industrialization resulted in the rapid growth of urban areas. Thus, research into alternative sewage treatment methods was intense on both sides of the Atlantic. Schneider (2011) also discusses the ideological debate which developed concerning the 'nature' of an appropriate treatment system. Land-based systems were viewed as 'natural' and therefore appropriate by some proponents, in contrast to 'artificial' systems such as anaerobic septic and Imhoff tanks or contact and infiltration beds. Schneider (2011) discusses the development of modern sewage treatment processes as a leading example of the industrialization of biological process technology, which occurred during this same time frame. While this ideological debate continued through the second half of the 19th century, especially in the UK, it was largely resolved in the first decades of the 20th century (although, as we shall see, it reappears in subsequent decades).

By later in the 19th century various forms of anaerobic treatment, such as septic tanks and the Imhoff tank, were in practice, along with physical–chemical methods. Biofilm-based biological systems, such as intermittent filters, contact beds, and trickling filters, were also developed and transferred into practice in the later portion of the 19th century. At the same time, some researchers hypothesized that the provision of aerobic conditions would address the unpleasant and odorous conditions associated with anaerobic treatment, leading to experiments on blowing air through sewage. Experiments on the aeration of sewage were initiated as early as in 1882 by Dr Angus Smith, followed by several other researchers (Alleman & Prakasam, 1983). While obnoxious odors were avoided, effective sewage treatment was not achieved. Experiments providing direct aeration to biological filter systems conducted at the Lawrence Experimental Station, Massachusetts, USA were more successful. We understand today that this occurred because the system incorporated the provision of oxygen with a sufficient quantity of biomass to metabolize the biodegradable organic matter in the wastewater. It was the biomass which was missing from the earlier experiments on the aeration of the sewage. Based on our knowledge today, this sequence of events foreshadowed development of the AS process. In fact, during the later 1890s and early 1900s various researchers hypothesized the need for accumulated 'humus' to accelerate treatment. However, some such as Gilbert John Fowler thought that solids (particulates) must be fully solubilized and, consequently, considered accumulated humus to be contrary to the objective of sewage purification. Then, in the early 1910s, researchers coupled aeration with the accumulation of biomass on wooden slats in tanks – essentially an early embodiment of the current moving bed biofilm reactor (MBBR) process. Black and Phelps (1914) conducted successful full-scale tests of this concept in New York, while laboratory studies of the concept were conducted at the Lawrence Experimental Station. Of course, at this time the biological nature of wastewater treatment was not established, so the importance of maintaining a sufficient biomass was not recognized. Researchers were beginning, however, to document the importance of both aeration and the retention of solids required for treatment.

1.2.2 The discovery

At this point Fowler and the City of Manchester in the UK enter the picture. Fowler was both an academic and a researcher as well as a practitioner. While at the University of Manchester, he also served as Superintendent Chemist at the Manchester Sewage Works. In 1912 Fowler was enlisted to consult on the pollution problems in the New York Harbor. During this visit to the USA he also visited the Lawrence Experimental Station and observed the experiments on aeration of filters being conducted there. Fowler was well familiar with the previous work, including his own, on aeration of sewage. He later credited this visit with an 'illuminating idea' and referred to the Lawrence Experimental Station as the 'Mecca of sewage purification' (Alleman & Prakasam, 1983). The 'illuminating idea', was the concept of using a suspended biomass culture in an aerated bioreactor. What he observed in Lawrence were the results of experiments by Clark and Gage (1912) who inoculated bottles of sewage with algal suspensions and aerated them. Upon return to the UK, Fowler initiated studies using a mixture of iron salts and a selected bacterial seed added into an aerated tank followed by a clarifier. Unfortunately, he did not recycle the solids which settled in the clarifier back to the aerated tank. Thus, while purification occurred, the rate was not sufficient. Alleman and Prakasam (1983) point out that, at this point, thirty-one years had passed since the first experiments on sewage aeration. Researchers had also empirically identified the need for a sufficient mass of treatment solids, such as those that were retained by biofilms developing on media provided in contact beds, and so on. or through their direct addition. However, no one had disclosed the idea of retaining the solids produced through treatment by sedimentation.

Then, on April 3, 1914 in Manchester, two of Fowler's students, Edward Ardern and William T. Lockett (Ardern & Lockett, 1914a), presented what Fowler himself characterized as the 'bombshell'. Employed at the Manchester Sewage Works, they presented the results of their experiments using aerated batch fill-and-draw reactors where they retained the settled solids following sewage purification. In their laboratory they aerated sewage in glass bottles that were covered with paper to prevent the growth of algae. Unlike previous researchers, they retained the sediment formed following aeration and fed subsequent batches of sewage. They found that the rate of purification (at that time defined as the removal of biodegradable organics and full nitrification) increased as they retained the settled solids over the course of several treatment cycles. The time required for purification was eventually reduced to less than 24 h, which made the process economically feasible. The solids were said to become 'activated' by retention and recycling. Today we understand that retention of settled solids allowed the development of the biomass needed for effective wastewater purification, but at the time, this was a novel concept. In this, their first paper, and in their two subsequent papers (Ardern & Lockett, 1914b, 1915) they provided a comprehensive treatment of many practical factors, such as energy considerations, sludge handling, sensitivity of nitrifiers to temperature and pH, continuous flow versus batch fill-and-draw operation, the effects of industrial wastes, aeration methods and levels, and the need to acclimate the 'activated sludge' to the subject wastewater.

While Ardern and Lockett are almost universally referred to as the inventors of the AS process, it appears obvious that they were able to do what they did because of the work of many others. They simply added the final concept, the last 'piece of the puzzle'. It would also appear obvious that their association with Fowler and his wide access to, and knowledge of the work of others, positioned them to make this highly significant contribution. In fact, Fowler suggested to Ardern and Lockett that they duplicate the lab-scale research that he had observed Clark and Gage conducting at the Lawrence Experimental Station in 1912. It is interesting that, at the same meeting in 1914 where Ardern and Lockett presented their second paper, Melling (1914) announced that he had already successfully applied the AS process in a 302 m^3/d facility in Salford, England. This must have required some previous knowledge of the essential elements of the AS process, suggesting that others in the UK were considering similar concepts. Perhaps it was both the presentation of this last 'piece of the puzzle', as well as the fact that their three publications provided a comprehensive treatment of many of the relevant factors, that resulted in the widespread attribution of the invention of the process to them. In any event, the critical importance of the results presented by Ardern and Lockett were immediately recognized and acted on.

1.3 AFTERMATH OF THE INVENTION

1.3.1 Accelerated implementation

The substance of the 'bombshell' announcement by Ardern and Lockett was immediately adopted by practitioners who began implementing it. Just think, it was not necessary to buy and install all of the media needed in aerated biofiltration systems! Rather, just keep the solids produced as a result of treatment and the treatment rate accelerates. Elegantly simple! Engineers immediately began to explore and develop the equipment needed to implement this new process:

- *Batch fill-and-draw, as used by Ardern and Lockett, or continuous flow?* The practical answer in the early part of the 20th century was continuous flow due to the absence of automation for the valves and level controls necessary to implement batch processing. It would not be until the early

1970s, when programmable logic controllers became widely available, that the fill-and-draw operation needed for the now well-known sequencing batch reactor process was possible.

- *Aeration devices to efficiently transfer the needed oxygen?* Both mechanical and diffused air systems were developed and used. Practice quickly evolved to what we currently do today – mechanical aeration in smaller plants and diffused air in larger ones. However, practice in this regard differed somewhat between the USA and UK, with mechanical aeration becoming more popular in the UK in larger plants.
- *Clarifier configurations?* This is still a controversial aspect that will be addressed in Chapter 11 of this book.
- *Sludge pumping (RAS and WAS) and treatment of excess sludge?*

These are all topics which continue to be addressed today and, perhaps, will continue to be in the future. The need for practical answers to these questions and for the manufacture and provision of the required equipment led to whole new industries to develop and provide it. The intensity of interest in the process is illustrated by the fact that in a review published by Porter (1917) just 3 years after presentation of the Ardern and Lockett paper there was a bibliography containing 200 papers which increased to 800 papers in a subsequent review published in 1921 (Porter, 1921).

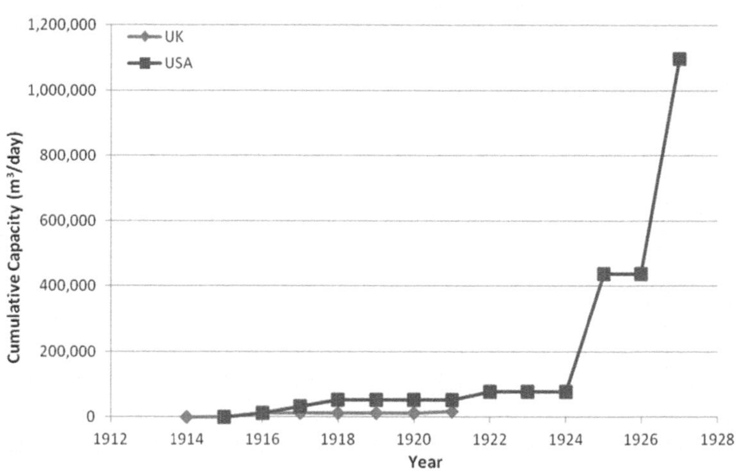

Figure 1.2 AS Treatment capacity expanded rapidly in the UK, and especially the USA, following announcement of the process by Ardern and Lockett (1914). (Based on data from Alleman & Prakasm, 1983.)

What amazes is the immediate adoption and implementation of the process. Alleman and Prakasam provide a list of facilities that had been constructed in the UK and the USA immediately following announcement of the process. Figure 1.2

provides a graphical presentation of the data showing the immediate acceptance of the process in the UK (which was subsequently interrupted by World War I) and especially in the USA. The US data include major facilities in Milwaukee (170,000 m^3/d), Indianapolis (190,000 m^3/d) and Chicago (662,000 m^3/d). Nowadays in our field, new technology adoption is often a 20-year process, perhaps because new technologies are not sufficiently transformational or their need is not initially recognized. Neither of these factors were the case in the early 20th century since full-scale adoption of AS occurred almost immediately. This may be because the process met an urgent need and because it proved to be transformational in terms of both the human health and environmental protection that it provided.

While the engineers were busy implementing AS systems (and learning as they went), researchers were debating whether the process was biological in nature or physical–chemical. Ardern and Lockett's results, while transformational, provided no clues to the mechanism. They demonstrated that retention of the solids produced by treatment resulted in increasing purification activity (they became 'activated'), but by what mechanism? The debate over mechanism raged on for about a decade, but by the early 1920s it was generally acknowledged that biological mechanisms were predominant, although of course there was some dissent. Today, with all of the scientific tools at our disposal, we know this to be the case. But, in the early 20th century, science itself was still quite primitive and was not able to provide an unequivocal answer to this fundamental question. In fact, the discovery of microorganisms and the recognition of their ubiquitous role in nature were relatively new. As indicated above, however, engineers were not constrained by the absence of an answer to this question and simply moved forward to implementation.

1.3.2 The patent

Then the patent came along. The firm of Jones and Attwood, Ltd successfully filed and obtained four patents for the AS process, first in the UK and subsequently in the USA. One may ask how they were able to do this when Ardern and Lockett invented the process. The simple answer is that the rights that Fowler (remember him) claimed were conveyed to Jones and Attwood, Ltd and formed the basis for their successful acquisition of the subject patents. How was Fowler able to do this? Well, quite frankly this is somewhat cloaked in mystery and, while we might speculate in this volume, the facts are quite straightforward. Jones and Attwood, Ltd successfully applied for and received the patents (prior to Ardern and Lockett's public presentation of their data), and they subsequently went about systematically pursuing payments from those who implemented the process. Table 1.1 summarizes these four patents. Jones and Attwood, Ltd must also be credited with accelerating the early adoption of AS in the UK, because they financed the construction of some of the initial installations, including those at Worcester and Stamford.

Table 1.1 Jones and Attwood, Ltd AS patents.

British patent number	Filing date	Subject
19,915	October 11, 1913	Wastewater aeration device based on an airlift piping system.
22,952	October 11, 1913	Wastewater aeration based on a porous diffuser.
729	January 10, 1914	Wastewater treatment system based on a looped-channel reactor with fluid rotation induced through a diffused aeration unit mounted on a recessed, dip chamber. Intermittent operation was used to settle solids and decant clear top fluid, or a separate clarifier unit was used for solids recycle during continuous-flow operation.
19,916	April 11, 1914	Wastewater aeration system based on a bottom-mounted diffuser array.

Source: From Alleman & Prakasm (1983).

In the USA, a patent was issued to Leslie Frank in 1915 which substantially covered the same substance as the Jones and Attwood, Ltd patents. Leslie subsequently assigned the 'activated slude' (his misspelling) process into the public domain. Clark, from the Lawrence Experimental Station, also challenged Fowler's claim as the inventor of AS. In spite of this, in late 1914 Jones and Attwood, Ltd began pressing claims in the USA concerning use of the AS process as well as their specific equipment innovations. These claims were largely ignored until the late 1920s when Activated Sludge Ltd (the US licensee of Jones & Attwood) sued first the city of Chicago, then Milwaukee, Cleveland, Indianapolis, and several other smaller cities over use of the AS process. Legal actions continued until 1933 when a district judge ruled against the city of Milwaukee and found that they had violated the patents held by Activated Sludge Ltd. In all, the cities of Milwaukee and Chicago were each assessed fines in excess of $1 million US (which was a lot of money back then!). The results were quite chilling as some communities shut down their plants to avoid paying license fees, while others paid the royalty fee of $0.25 US per capita served. For example 150 out of 203 AS facilities identified in the USA in 1938 had paid the requested license fee to Activated Sludge Ltd (*Engineering News Record*, 1939). Many other communities either selected alternative technology (such as the trickling filter process) or postponed installing biological treatment until the patents expired.

A bit of background relevant to the AS patent controversy is best presented by Schneider (2011). In 1896 Donald Cameron, a British engineer then with the city of Exeter in England, patented in the UK his design for a septic tank that he had designed for the city. To minimize obnoxious conditions, previous approaches had focused on killing the resident organisms. In contrast, Cameron allowed the anaerobic biological process to develop under controlled conditions and, consequently, he referred to the process as 'septic' to contrast it with the *antiseptic* approach previously used. He also applied for a US patent, which was granted in 1899. While the British patent was on the tank configuration, the US patent was broader and was on the septic tank process itself. This is where the difficulty began.

As discussed by Schneider (2011), debate over wastewater treatment patents, especially the Cameron and the AS patents, were crucial in addressing what was and was not patentable. Patents were initially allowed on inventions of new apparatus – new physical means for accomplishing tasks. However, over time, the issue of whether 'methods' could be patented surfaced. One can classify knowledge into three categories: (1) knowledge of, and understanding of, the principals of nature, (2) methods for using knowledge of the principles of nature to produce useful results, and (3) apparatus for accomplishing useful tasks (often needed to implement category 2). Treatment processes fall into category 2. Category 1 may be thought of as discovering the 'secrets' of nature, while category 2 may be thought of as identifying how to create conditions which apply these secrets for beneficial uses. This distinction was not clear in the late 19th and early 20th centuries, and only became clear to the legal profession as various patents on biological processes, including those associated with wastewater treatment, were addressed in the first half of the 20th century. It is now recognized that discoveries in category 2 are patentable. Thus, the key to determining whether a process is patentable is distinguishing it from what already occurs in nature. Knowledge about how to harness and make useful this knowledge is patentable. The resolution of these issues provided the basis for adjudicating numerous contemporary patent issues, including those associated with modern biotechnology, patenting genes, and patenting business processes – all of immense economic value.

Another issue with patents is that they must identify the 'inventors' – those individuals solely responsible for the critical insights that lead to the subject invention. However, in an era of collaboration and rapid exchange of information and knowledge (such as within the sanitary engineering research and practice areas in the late 19th and early 20th centuries), it may not be clear who is solely responsible since, in reality, many may be responsible.

Engineers in the USA quickly recognized the potential for the septic tank process but began to experiment with tank configurations of their own design rather than duplicating the configuration of the tanks at Exeter. In response, the Cameron Septic Tank Co. began approaching cities and engineers for fees set at 5% of the construction cost for the treatment plant, along with an additional 3% for every year the plant operated without a license. The reaction of the US engineering community was to organize and oppose the patent. While the reasons for such opposition may be complex, Schneider (2011) highlights three. First, sewage treatment was viewed as a public good and the supporting technology was considered, by many, to not be proprietary but should be free for all to use. Second, much of the research and development resulting in the development of sewage treatment technology was funded by the public and, consequently, the results of this research should not be conveyed into private hands. In fact, significant financial investments were made by cities such as Milwaukee and Chicago to develop their applications of the process. Third, engineers were focused on development of the engineering profession to achieve a status similar to that then developing for the medical and legal professions. It was judged that engineers should remain separate from commercial considerations to maintain their professionalism. This latter view varied depending on the employment of the engineer, whether in private practice, public employ, or employed by private concerns.

Opposition to the septic tank patent by the Cameron Septic Tank Co. was initially articulated by Leonard Metcalf of Boston but led to organized opposition and legal challenges. In 1907, a trial judge sided with the engineers that it was not possible to patent a 'natural' process, only the specific means to implement it, but this was subsequently reversed by an appeals court. The reaction of the engineering community was swift. Organizations such as the 'Association for the Defense of Septic Process Suits', the League of California Municipalities, and subsequently the 'National Septic Process Protective League' developed to oppose the septic tank patent. In resisting efforts by the Cameron Septic Tank Co. in court, these Associations were successful in exhausting the financial resources of the Cameron Septic Tank Co., leading in 1919 to an offer by Cameron to settle all remaining claims for the nominal sum of $5000US. The engineers had won!

Thus, when Activated Sludge Ltd began pressing its claims on the AS process it is not surprising that US engineers presumed that such efforts would not be successful. As Schneider (2011) further describes, the initial communication by Fowler to US engineers was ambiguous at best concerning the issue of patents, which perhaps created expectations about the nature of any necessary license and payment requirements. For example, Fowler was initially a consultant to Milwaukee, leading Milwaukee to believe that they had discharged their legal requirements before being approached by Activated Sludge Ltd.

Two schools of thought exist relative to the role of the patents in advancing the AS process. Activated Sludge Ltd. took the financial risk and was actually the developer of some of the initial AS facilities in the UK. It can be argued that these initial installations helped to 'jump start' practical application of the AS process, thereby accelerating its early use and the associated learning that facilitated subsequent adoption. It is also certainly true that the trickling filter process was one of the beneficiaries of the AS patent controversy of the first half of the 20th century as trickling filters were the popular alternative to AS during this time. There is little doubt, however, that the patent and associated license fees subsequently slowed adoption as communities selected other options to avoid the license fees. One may consider that the truth concerning the role of patents in accelerating or restraining development of AS is somewhere in between.

Given his central role in the development and initial commercialization of AS, one might wonder what eventually happened to Fowler? In 1916 Fowler became Professor at the Indian Institute of Science in Bangalore and continued in India as an influential academic and researcher, and especially as a proponent of nitrogen management in agriculture. He lived a long and productive life, dying in India in 1953 at the age of 85. Why relocate to India when the AS process was just taking off? Who knows?

1.4 SUBSEQUENT DEVELOPMENTS

Details of the subsequent development of the AS process and the performance requirements for the process over the years will be addressed in detail in Chapters 2 and 3. The intervening 100 years, since the presentation by Ardern and Lockett, have brought about significant experimentation, development, innovation, and learning. Some may argue that the AS process, as practiced today, is not fundamentally different than what was initially advanced in 1914, while others will argue that our knowledge and experience base has extended in many ways and that our ability to predict and control outcomes far exceeds that which existed in the 'early days' of the process. Again, the truth is probably somewhere in between. It is interesting to recognize that history may be repeating itself in recent times. While the AS process controversy is behind us, patent issues still arise on occasion, along with the historic opposition of the user community to their imposition. The controversy over patents on the EBPR process is a recent, well known example. It is clear that the contrasting views of proprietary technology and opposition of the environmental engineering professional community to patents as characterized by Schneider (2011) still exist. Schneider's fundamental hypothesis was that the debate over biological sewage treatment in the latter part of the 19th and early 20th century framed and initially addressed patent issues which resulted in societal changes and legal precedents that subsequently affect a broad range of biotechnological, technological, and business areas.

Speaking of EBPR, many of us remember debates over its basis – whether biological or chemical. We know today that it is principally biological in nature, although biologically mediated chemical P removal can occur. As discussed above, the basis for the AS process, whether biological or chemical in nature, was widely debated at its advent.

The controversy between 'natural' and 'artificial' processes, and their advancement based not only on technical merits but on ideological grounds, has also continued to the present time. The popularity of land treatment in the 1970s and 1980s represents this debate, along with the emergence of 'green infrastructure'. The debate about centralized versus decentralized infrastructure often includes ideological along with technological elements.

The emergence of partial nitritation (conversion of about half of the ammonia-N in a wastewater stream to nitrite-N) coupled with deammonification based on the Anaerobic Ammonium Oxidation, or Anammox, metabolism, and its rapid adoption, demonstrates that the wastewater engineering profession is capable of rapid change when the right idea comes along at the right time. Some express concern that the engineering profession is not able to change rapidly. Initial adoption and rapid development of the AS process, but also the current rapid adoption of partial nitritation–deammonification, demonstrates that this is not always the case. We can hope that these positive trends for rapid adoption of new and beneficial technologies continue as the need for increased protection of the water environment is great.

One can also argue that the AS process played a key role in transformation of the sanitary engineering profession into the environmental engineering profession during the 1960s and early 1970s. Inspired by developments in the biological and chemical sciences, a number of now notable sanitary engineers engaged much more deeply in the biological and chemical sciences. They brought these sciences into sanitary engineering research, bringing about a transformation to a much more science-based engineering practice that is now the foundation for environmental engineering. AS formed a key element of this research, as illustrated by the classical publication by Lawrence and McCarty on the AS process (Lawrence & McCarty, 1970).

1.5 FUTURE PROSPECTS

The final chapter of this volume will speculate about the future of the AS process. It is not my purpose here to intrude on this topic. However, in looking to the future, it is worth considering the past and speculating on the features of the AS process which have allowed it to play such a dominant role in water pollution control over the past 100 years. While many hypotheses exist, some conspiratorial in nature, I would suggest that it is the adaptability of the process which is the core reason for its long-term success. The process consists essentially of an aerated (or partially aerated) suspended growth bioreactor with downstream liquid–solids separation unit (usually gravity separation using clarifiers but increasingly using membranes), with recirculation of the separated solids back to the bioreactor. Solids separation, recycle, and retention was the key concept introduced by Ardern and Lockett. This simple concept can be implemented in any number of ways. Many bioreactor configurations exist, along with numerous approaches to liquid–solids separation and recycle. The process is further defined by the guiding principles for suspended growth processes, as first articulated by my mentor and colleague Professor C. Leslie Grady, Jr, and summarized in Table 1.2. Essentially the same bioreactor and liquid–solids separation system with solids recycle can be configured and operated to accomplish an extremely wide range of treatment objectives by simply varying the size of the bioreactor relative to the liquid–solids separation system and the internal configuration of the bioreactor (mixed versus aerated zones, internal recirculation). The system can be designed as a high-rate system for partial carbon removal, for full carbon removal, to nitrify, or as a low-rate system which produces a stabilized sludge. The macronutrients N, P, or both can also be removed. A range of trace constituents can be removed. All of these treatment objectives can be achieved with essentially the same basic system components. It will be interesting to see whether this feature allows the AS process to continue to evolve over the next 100 years to meet the challenges of the future.

Table 1.2 Suspended growth biological process guiding principles.

1. The biochemical environment determines the nature of the microbial community that develops in a bioreactor and the character of the reactions they perform.
2. SRT is the most important design and control parameter.
3. A COD balance across the bioreactor provides valuable information about the amount of electron acceptor required and the amount of excess biomass produced.
4. The excess biomass production rate is essentially the same for all suspended growth systems with the same SRT and biochemical environment, regardless of the bioreactor configuration.

Source: Adapted from Grady *et al.* (2011).

The question is whether this flexibility and adaptability will position the AS process to meet future needs. The situation facing human society today is very different to that in the late 19th and early 20th centuries when the process was developed and through the 20th century as the process became so widely applied. In the 21st century the human population is continuing to grow from the just over 1 billion present at the beginning of the 20th century and is likely to either stabilize at about 10 billion in the second half of this century or perhaps to continue to grow somewhat. The industrialization process which was such a driving force at the beginning of the 20th century is running its course, and in many developed countries a post-industrial economy is established. While cities were growing exponentially at the beginning of the 20th century, this was from a small base so that human habitation was still largely rural. But, exponential growth over a century can change a lot, and the human population has now become an urban one. More than half of humanity now lives in urban areas, and this is expected to increase to more than 70% by the middle of this century. Most significantly, we are transitioning from our historic situation where resources were available in excess of our needs to one of resource scarcity. This is causing an increasing number of water professionals to conclude that water management in the future must be fundamentally different from that in the past. While this is neither a topic of this volume nor of this chapter, it is the evolving and future needs of society, and the ability of the AS process to adapt to those needs, which will determine its future.

1.6 ACKNOWLEDGEMENTS

The excellent review of Alleman and Prakasam (1983) was relied on for significant parts of this chapter. The interested reader is referred to this invaluable article for further detail and references.

1.7 REFERENCES

Alleman J. E. and Prakasam T. B. S. (1983). Reflections on seven decades of activated sludge history. *J. Water Pollut. Control Fedn*, **55**(5), 436–443.

Anon (1939). Inventory of sewage disposal facilities – 1938. *Eng. News Record*, **122**, 122.

Anon (2007). Medical milestones. *Brit. Med. J.*, **334**, s1–s20.

Ardern E. and Lockett W. T. (1914a). Experiments on the oxidation of sewage without the aid of filters. *J. Soc. Chem. Ind.*, **33**, 523–539.

Arden E. and Lockett W. T. (1914b). Experiments on the oxidation of sewage without the aid of filters, Part II. *J. Soc. Chem. Ind.*, **33**, 1122–1124.

Ardern E. and Lockett W. T. (1915). Experiments on the oxidation of sewage without the aid of filters, Part III, *J. Soc. Chem. Ind.*, **34**, 937–943.

Black E. B. and Phelps E. B. (1914). Brooklyn sewage experiment station. *Eng. News Record*, **74**, 826.

Bettmann O. L. (1974). The Good Old Days – They Were Terrible! Random House, NY.

Clark H. W. and Gage S. D. M. (1912). Experiments on the purification of sewage and water at the Lawrence experiment station, Massachusets State Board of Health. Annual report, 1912.

Constable G. and Sommerville B. (2003). A Century of Innovation: Twenty Engineering Achievements That Transformed Our Lives. Joseph Henry Press, Washington, DC.

Grady C. P. L. Jr., Daigger G. T., Love N. G. and Filipe C. D. M. (2011). Biological Wastewater Treatment, 3rd edn, CRC Press, Boca Raton, FL.

Lawrence A. W. and McCarty P. L. (1970). Unified basis for biological treatment design and operation. *J. San. Eng. Div., Amer. Soc. Civil Eng.*, **96**, 757–778.

Melling S. E. (1914). The purification of Salford sewage along the lines of the Manchester experiments. *J. Soc. Chem. Ind.*, **33**, 1124.

Mohlman F. W. (1938). Twenty-five years of activated sludge. *Modern Sewage Disposal*, Federation of Sewage Works Association.

Porter J. E. (1917). The Activated Sludge Process of Sewage Treatment: A bibliography of the subject. General Filtration Company, Rochester, NY.

Porter J. E. (1921). The Activated Sludge Process of Sewage Treatment: A bibliography of the subject, 2nd edn, General Filtration Company, Rochester, NY.

Sawyer C. N. (1965). Milestones in the development of the activated sludge process. *J. Water Pollut. Control Fedn*, **37**(2), 151–162.

Schneider D. (2011). Hybrid Nature: Sewage Treatment and the Contradictions of the Industrial Ecosystem. The MIT Press, Cambridge, MA.

Chapter 2

Wastewater treatment requirements through the years (exemplified by the development in Germany)

Hermann H. Hahn (Germany)

2.1 INTRODUCTION – THE EMERGENCE OF SYSTEMATIC WASTEWATER TREATMENT (IN GERMANY)

The development of wastewater treatment requirements in Germany had a relatively late start compared to Great Britain. With intensified industrialization and urbanization in Germany initial efforts to install and operate wastewater treatment, in part profited from insights and developments elsewhere. Today, German standards for wastewater treatment are part of the European requirements, which were frequently developed by adjusting the existing German standards to newly formulated European directives (*ec.europa.eu/environment/water/water-urbanwaste/* (1997); *www.bmu.de/fileadmin/bmu-import/files/pdfs/...*, (2000)). German standards are not at the limits of technological feasibility so the method and setting of their development can provide an example for other regions.

An analysis of the historic development of treatment requirements in Germany, as is attempted here, will be foremost concentrated on treatment of municipal wastewaters (i.e. a combination of domestic wastewaters and a mixture of various industrial wastewater streams). The analysis will illustrate the triggers of emerging concepts of wastewater treatment here and elsewhere, both in terms of the possibilities and limitations of pollution control. It seems that quasi-aesthetic considerations at the point of wastewater discharge were the starting point for developing technologies to curb water pollution by such wastewater discharges. As these efforts became more successful and industrialization/urbanization intensified, the analysis of the ambient water situation turned to other aspects and extended to areas more distant from the discharge point. Today one can identify an additional concept for the development of wastewater treatment in which the maintenance of suitable water quality beyond just the near future has become an objective. Pollution control at one specific point includes the integral analysis of an entire water basin as well as of long(er) time periods of discharge and discharge control. The (German) development of wastewater treatment standards could be described as one:

- from aesthetic desiderata to chemical, biological and health considerations;
- from emissions control (at the discharge point) to immissions control analysis (within the receiving water);
- from corrections that aim for immediate improvements to those that guarantee future quality.

These developments were accompanied and facilitated by changes in the attitude of the public to wastewater treatment or even to the public demanding these changes. They were also influenced (furthered and hindered) by societal developments such as national or international economic (or even military) crises.

A historic and possibly original 'description' by von der Emde (1999) of the development of the AS process (p. 83) reads more like a description of a technological process rather than a discussion of wastewater treatment efficiency data or the emergence of treatment requirements: von der Emde writes that a COD reduction (most likely permanganate-COD) of around 90% is possible *if* the following operating conditions are met and maintained in a continuous flow-through operation in a channel-like reactor with a width of 1.2 m and a depth of 1.7 m with 3 h aeration (fine bubbles pressurized air) or 5 h aeration if nitrification is intended and subsequently 2 h sedimentation in circular deep funnel-type tanks. Other AS treatment process variations, such as surface-aeration or rotating brush-aeration were described as being similarly effective in reducing the COD and led to some nitrification if the aeration time was long enough.

Design of AS plants in Germany in the years after the discovery and description of the process was aimed primarily not at achieving a particular effluent quality but predominantly at achieving stable and high-level operation, that is operations optimization. In later years, with the construction of a larger number of plants, the possibilities for optimizing design and operation even more, particularly with respect to minimizing construction and operational efforts, were sought (for instance higher AS concentrations, higher reactor loadings, minimizing aeration expenditures, etc.) An orientation towards an effluent-requirement-related process design and operation began (in Germany) only after Worl War II with the differentiation of 'full' biological treatment versus 'partly biologically' treated.

Other historic books and analyses of the development of requirements for wastewater treatment similarly show that the aim of protecting or rehabilitating receiving waters was formulated such that a treatment process was proposed and designed/operated optimally but not to specific or performance requirements. According to Strell (1913) the dominant aspect was to keep settlements clean and free of unwanted or health-endangering materials and to, as quickly as possible, route wastewater (German: 'Ab-Wasser' – literally translated as 'ab-transportiert', i.e. water that transports materials away) to a receiving (and hopefully also transporting) river or body of water. However, the *"Sanitätspolizei", the sanitary police (soon) saw it as their task in Cologne (1877) and Frankfurt (1876) to keep all significant pollution of rivers by sewers, industrial waste and similar substances away from rivers and where this is not possible, to discharge these substances after purification or treatment'* (translated freely from Strell, pp. 202/203). Historically, this seems to be the first mention of 'requirements for wastewater' discharge.

König (1899) not only found that river pollution was increasing all over Germany but he also attempted a user-oriented or use-oriented protection of waters. Most ambitious were the partly quantified surface water quality requirements for drinking water. Concepts for the protection of fishing waters followed. König also described the relevant German water legislation (based on experiences in other countries, such as Great Britain) that in typical German fashion, differed from state to state. For example in the Grand-Dukedom of Baden (the region of Karlsruhe) in 1884 the limits for the discharge of 'Abwasser' into fishery-relevant surface waters were no more than 10% of suspended or dissolved substances, no specified materials such as acids, salts, and so on higher in concentration than 1000 times the river concentrations, no wastewaters containing degradable organic materials (no concentration given), chlorine and petroleum compounds and last but not least, no wastewaters with temperatures above 40°C. This is a rather definitive and quantitative directive for wastewater discharge, however, it was still independent of the wastewater treatment processes employed.

The ever growing insights into treatment processes and the formation of national and international committees aimed at collecting and standardizing process information led to the definition of design and operating conditions that enabled operators (mostly engineers) to have the plant perform according to the demands of the receiving body of water (e.g. no nitrification or full N removal). In parallel with the growing understanding of process capabilities, the water authorities demanded 'guarantees' of performance before discharge permits were issued. These performance guarantees were developed more and more toward quantified discharge concentrations and/or discharge loads of wastewater constituents that were to be controlled. And the list of substances of concern that had to be partly removed or removed with maximum efficiency grew longer and longer, that is the authorities' aims became more and more ambitious.

2.2 DEVELOPING WASTEWATER TREATMENT CHARACTERISTICS – FROM QUASI-AESTHETIC CONSIDERATIONS TO CHEMICAL, BIOLOGICAL AND HEALTH CONSIDERATIONS

Dunbar (1912) presented the first systematic text in Germany on wastewater treatment and related issues. He attempted an overview of the achievements of various wastewater treatment processes. Such analyses later on served as a basis for the setting of requirements for wastewater treatment. His view of water pollution which serves directly or indirectly as a basis for setting standards for the receiving water as well as the discharged streams were as follows: (*'Eine Verschmutzung kann nur zustande kommen...'*) *'Pollution can only result when the organisms that are responsible for the self-purification of the rivers, are no longer successfully combating these pollutional loads.'* (p. 588). In addition the intended or existing uses of the body of water under consideration should always be taken into account (p. 590).

The view that the capacity of a river for self-purification can (or should) be taken into consideration in determining the measures of pollution control were then held by every responsible specialist. Similarly the differentiation of use or user affected to a large degree the answer to the question as to whether a natural body of water would be seen as polluted or not and whether consequently some control measures were to be taken. In terms of formulating requirements for pollution control for specific bodies of water, Dunbar stated that *('Es ist nicht zu bezweifeln...') 'There is no doubt, that the responsible authorities follow closely the advances in wastewater treatment technology and that they will act against existing violations once the correct moment is reached.... No doubt, difficulties in wastewater treatment (technology) have often been used, not only by industry but also by several municipalities to hinder over years the fulfillment of authoritative requirements.... A basis for hope (for less expensive and safely operating) has been laid some fifteen years ago with the emergence of artificial biological processes.... That one must always be careful in installing newly developed wastewater treatment processes shows the history of the development of such processes.'* (pp. 44–45).

Dunbar formulated clearly, that the definition of requirements for wastewater treatment was only possible on the basis of a thorough understanding and practical testing of treatment processes. For Germany, he reported on the basis of extensive travel and/or literature study, the possibilities and limitations of various wastewater treatment processes (Table 2.1). The data represent average treatment efficiencies (% reduction) of specified material, excerpted from Dunbar's reports. His findings are interesting in two respects: first they represent a list of available and proven treatment processes, and second, they give a hint of the efficiency of each process with respect to the reduction of some wastewater constituents.

These German experiences correspond to some degree with data that Dunbar found reported for US plants (Table 2.2). Thus, they might be taken as a description of the historic state and capacity of wastewater technology in general, that is best available technology, at that time. Whether or not this became the basis for subsequent definitions of treatment requirements is not clear. For a later period in Germany it can be stated that hearings by the authorities with technology specialists on the highest efficiency delivered by specific treatment processes led then to the clear formulation of requirements.

Table 2.1 Average treatment efficiency of wastewater treatment plants around the beginning of the 20th century in Germany as % reduction of specified materials.

Treatment process	Non-dissolved substances, % removal	Organic material, % removal	Bacteria, % removal
Grit chamber/fine screens	20–25	–	–
Sedimentation	60–70	nr	nr
Fermentation process	60–70	30–50	nr
Chemical treatment	75–86	20–30	nr
Land application	>90	85–89	99.7
Submerged filters (Füllkörper)	nr	70–80	nr
Artificial biological processes (trickling filters)	nr	≈75	nr

nr: not reported.
Source: Excerpted from Dunbar (1912 pp. 62ff and 594ff).

Table 2.2 Treatment efficiency of US American (treatment) plants as % reduction.

Method	Suspended Solids, % removal	Organic Material, % removal	Bacteria, % removal
Fine screens	15	10	15
Sedimentation	65	30	65
Fermenting processes	65	30	65
Chemical treatment	85	50	85
Submerged filters	85–90	65–70	80–85
Trickling filters	85–90	65–70	90–95
Intermittent soil filtration	95–99	90–98	98–99

Source: Dunbar (1912, p. 607).

2.3 FROM CONSIDERATION OF ONE SPECIFIC POINT OF DISCHARGE TO INTEGRAL ANALYSIS OF AN ENTIRE WATER BASIN

The development of wastewater treatment and the efficiency of various processes in Germany were systematically assessed between 1907 and 2009 by Imhoff (initially K. Imhoff, then later K. R. Imhoff) in the *Handbook of Wastewater Collection and Treatment*. These books are mirrors of technical and administrative developments. The authors also include in the step-by-step account the development of emissions-oriented requirements. Table 2.3 presents only those

Table 2.3 Indirect or direct requirements for wastewater treatment.

Year of publication	Item	Text/Wording	Comment
1925 (4th Edn)	The treatment plant ('Berechnung der Kläranlagen').	For instance: Aeration time of activated sludge plant 4–6 h depending upon the concentration of the wastewater.	Oxidative degradation of wastewater constituents to the highest known extent. No direct requirements mentioned.
1932 (6th Edn)	Necessary degree of treatment ('Erforderlicher Reinheitsgrad').	Extent of river self-purification determines the necessary degree of (pre)treatment of wastewater.	Requirements depending upon river characteristics. Dilution-based calculations explicitly excluded.
1939 (8th Edn)	Efficiency of wastewater treatment processes ('die Verfahren der Abwasserreinigung und ihre Leistung') Demands of the receiving water ('Ansprüche des Gewässers').	AS plant calculated according to a given aeration time. The latter is irrespective of type of wastewater or receiving water characteristics.	In parallel it is suggested that an oxygen balance calculation be done. Indirect mention of influence of discharge characteristics (i.e. wastewater treatment requirements).
1941(9th Edn)	Under the heading Demands ('Ansprüche des Gewässers') definite requirements formulated.	Depending upon the state of the receiving water and the intended use of this water body, types of wastewater treatment (implicitly their efficiency) and emergency measures indicated.	No numbers either for the efficiency of treatment processes or the concentrations or loads discharged.

1951 (14th Edn)	Data on the efficiency of wastewater treatment processes ('Verfahren der Abwasserreinigung und ihre Leistung').	For so-called mechanical, chemical and biological (unit) processes the efficiency as % reduction of wastewater constituents are given (BOD, SS and bacteria) – see Table 2.4.	The quantitative information is given with a warning that plants (processes) have been sized on a risk-oriented basis and may not perform as postulated.
1972 (23th Edn)	Standard values for effluent concentrations from municipal wastewater treatment plants (including some industrial wastewater ('Normalanforderungen').	Settleable matter in all treatment options: 0.3 mL/L $KMnO_4$ in mg/L: 0 (mech.); 150 (part biol.); 100 (fully biol.) BOD in mg/L: 0 (mech.); 80 (part biol.); 25 (fully biol.).	Requirements as defined by the administration as aid for the designing engineer and/or the municipality.
1976 (24th Edn)	Indicating the possible need for additional treatment stages implying intensified requirements ('Dritte Reinigungsstufe').	For chemical precipitation stages a standard dosage of precipitant is given without explicit effluent guarantee.	This unit process was not yet fully included in (German) options for wastewater treatment.
1979 (25th Edn)	Defining minimum requirements ('Mindestanforderungen') instead of normal standards (see above) – Wastewater may legally not be discharged if those minimum requirements are not met. Furthermore the existing or future quality of the (receiving) water body is quantified.	Introducing the significant aspect of size of plant (load concept) and formulating increasingly stricter values for settleable material, COD and BOD). Surface water quality is defined in biological terms (saprobic index) and organic loading as well as resulting oxygen concentration.	At the time of publication by the administration, new plants had to perform accordingly – older plants were allowed 8 years to achieve the limits. An attempt was made to reconcile chemical and biological parameters.

(*Continued*)

Table 2.3 Indirect or direct requirements for wastewater treatment (*Continued*).

Year of publication	Item	Text/Wording	Comment
1990 (27th Edn)	Adjusting minimum requirements (enlarging the list of parameters and deleting one insignificant parameter).	No more requirements for settleable material. Newly included requirements for NH_4-N and P.	Adjustment was made to a large degree on pollution evidence in more removed areas and in view of concentration time.
1993 (28th Edn)	Under the heading of general requirements for surface water quality, requirements are formulated in one specific German State for a large number of water constituents.	Adding TOC, NH_4-N, P_{tot} heavy metals and adsorbable organic carbon enlarges the existing list of parameters to be considered (of originally minimum requirements for wastewater treatment).	Wastewater treatment requirements can/must now be re-routed from desired quality of surface waters (through water quality modeling).
2007 (30th Edn)	Dimensioning a treatment plant ('Berechnung der Klärwerke') such as to satisfy the minimum (effluent) standards. Requirements of wastewater treatment in view of pollution control ('Gewässerschutz') Standards ('Normung, DIN und EN').	Calculation according to the effluent requirements for a plant of a certain size and therefore predetermined German standards. Requirements are indirectly related to receiving body of water through size of plant Standards are monitored.	(German) Emission standards determine the requirements. Advanced EU requirements for instance through WFD not yet included. Immissions standards implicitly referred to through suggested water quality modeling.
2009 (31st Edn)	Under the heading 'Leistung der Reinigungsverfahren' requirements for municipal wastewater at a specified point of discharge are given (quasi: a combined approach of emission and immissions analysis).	Five classes of treatment plant requirements depending upon the size of the treatment plant – the larger the plant, the more stringent the treatment requirements (see Table 2.5 for parameters and respective numbers).	EU Requirements included (municipal wastewater treatment directive and water framework directive). Receiving water characteristics might cause the authorities to formulate specific requirements.

Source: Excerpts from Imhoff/Imhoff (1925 to 2009).

editions that report a change or development that the authors thought worthwhile to present to the practitioner. In examining Table 2.3 it becomes apparent, that in the 1970s the term 'requirements' appears in that precise wording for the first time. At the end of that decade it is reported that such requirements were mandatory for all newly constructed plants (in Germany) while existing plants had to follow this ruling within eight years.

At this point it is necessary to point out, that Germany has developed its (water) pollution control concepts in terms of emission controls. The reasons for this are manifold and begin with the slowly emerging understanding of the processes and their efficiencies. They also include administrative aspects, such as assignability of pollution acts and quantification of pollution effects. For many wastewater constituents, such a focus on effluent quality was logical, useful and readily attainable. However, there are also pollution control aspects that cannot be interpreted or linked to specific discharges, that do not occur at discharge points and that need to be taken into account to guarantee satisfactory water quality throughout the entire water body. This immissions orientation is at the basis of pollution control concepts in other regions and countries. The European pollution control concept accepts both a control of emissions and an analysis of the immissions situation following a polluting discharge.

Table 2.4 Efficiency of wastewater treatment processes observed for Germany for the period after World War II.

Efficiency of treatment process	BOD reduction (%)	Suspended solids reduction (%)	Bacteria reduction (%)
Fine sieves	5–10	5–20	10–20
Chlorination of raw sewage or settled sewage	15–30	–	90–95
Sedimentation (tanks)	25–40	0–70	25–75
Coagulation/ Flocculation	40–50	50–70	–
Chemical coagulation	50–85	70–90	40–80
High-load trickling filters	65–90	65–92	70–90
Low load trickling filters	80–95	70–92	90–95
High load AS	50–75	80	70–90
Low load AS	75–95	85–95	90–98
Soil filters	90–95	85–95	95–98
Chlorination (biol. pretreated)	–	–	98–99

Source: Imhoff (1951, 14th Edn).

The historic developments to be reported here for Germany are predominantly emissions oriented. Emissions standards originally were directly linked to the (maximum) capacity of a specific treatment process. They were then assigned to each discharger, irrespective of size. Only later was it accepted and then regulated that the discharge of smaller loads need not necessarily require the maximum performance of a given process with all its complications of safe and consistent operation. Similarly, with the improved understanding of both receiving water demands and treatment process characteristics, the authorities required that the dischargers of high volumes and loads must guarantee the highest possible process efficiency. The twelve selected editions of the Imhoff handbook series from various periods show how process understanding developed and how the formulation of requirements occurred.

At the end of World War II the process technology available was what might be called the first generation of wastewater treatment tools. The following unit processes and their efficiencies were identified by Imhoff (Table 2.4). However these efficiency data were not yet translated by water authorities into requirements for wastewater treatment.

2.4 FROM CORRECTIONS OF TODAY'S WATER POLLUTION PROBLEMS TO ACHIEVING WHOLESOMENESS OF WATER FOR FUTURE GENERATIONS

German development up until about the last decade of the twentieth century as described above, was focused predominantly on analyzing the effects of wastewater discharge at the point of discharge or in the neighborhood of that location. The requirements were tightened systematically by including an increasing number of parameters and mostly lower maximum allowable concentrations (Figure 2.1). With the dramatic die-off of some maritime fauna before the turn of the century in the North Sea ('Robbensterben'- dying seals) it became clear to the German authorities (and even more to the German public) that river pollution occurring far away from the North Sea might lead to water pollution effects in that shallow estuarine/maritime body of water, in particular if aquatic nutrients were concerned.

Contrary to the oxygen-consuming wastewater constituent requirements for free-flowing rivers (more than 50% of all German receiving waters), the stipulations for treatment plants now also include standards for eutrophying constituents, such as N and P compounds.

This line of thinking was extended not only to the current impairment of geographically distant water quality (such as eutrophication in the North Sea and Baltic due to discharge of N and P compounds into inland rivers) but also to the future anticipated deleterious effects from continuing discharges. Wastewater treatment requirements were revised, such that discharged substances would not be anticipated to endanger water quality in the foreseeable future. Such requirements were formulated foremost for heavy metals and complex non-degradable, refractory organic compounds. Some of these treatment process requirements were not necessarily formulated directly on the treatment plant effluent but they influenced the treatment plant design and operation through immissions standards. The EU water framework directive (WFD) of 2000 clearly shows an orientation towards protecting the total aqueous environment and attempting, or at least suggesting, a return to its natural quality for future generations.

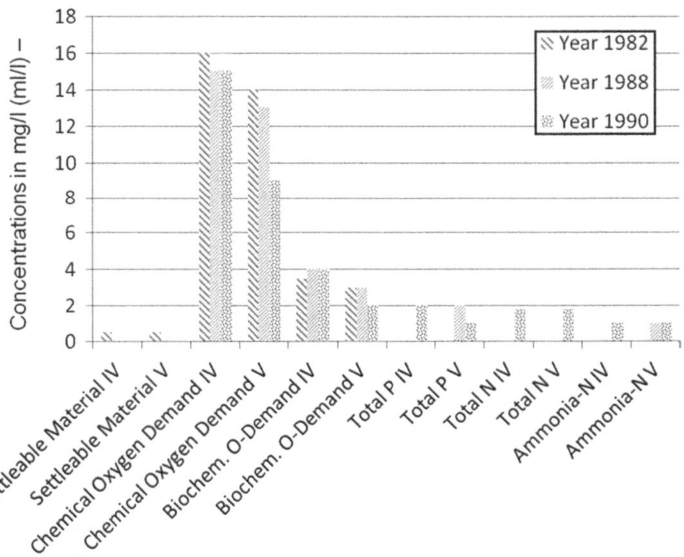

Figure 2.1 Development of wastewater treatment requirements in Germany for two different treatment plant sizes (IV: less that 100,000 PE, and V: more than 100,000 PE) between 1982 and 1990.

In summary then it is useful to repeat that in Germany the list of requirements for wastewater treatment plants has become longer and longer with respect to water constituents or groups thereof; furthermore a general tendency to tighten those standards with time has also occurred and may be expected to continue in the future (Figure 2.1 and Table 2.5). In only one instance has a standard been lowered or abandoned. This parameter is 'settlable material' which appeared to be a relic from the times when aesthetic concerns were dominant and treatment options focused on the removal of such substances. With the beginning of more intensive wastewater treatment and the use of more effective unit processes for wastewater treatment this parameter was no longer of concern. Abandoning it also allowed the plant regulatory authorities to concentrate on more urgent matters.

This development, which was in part supported by the European Union legislation, also took place in Switzerland, a country that distinguishes itself as being quite protective of its waters and rather ambitious in its legislation (Table 2.6).

Table 2.5 Relevant (German) requirements for wastewater treatment plant effluents (BGBl, 2004; PE – population equivalent) as listed in Imhoff (2009).

Size of treatment plant	COD (mg/L)	BOD (mg/L)	NH_4-N (mg/L)	N_{total} (mg/L)	P_{total} (mg/L)
<1000 PE (Class I)	150	40	–	–	–
<5000 PE (II)	110	25	–	–	–
<10,000 PE (III)	90	20	10	–	–
<100,000 PE (IV)	90	20	10	18	2
>100,000 PE (V)	75	15	10	13	1

Table 2.6 Swiss discharge regulations and their development with time.

Year	1966	1976	Since 1980	Minimum
TSS (mg/L)	20	5	5	20
BOD_5 (mg/L)	20	10	5–10	20
DOC (mg/L)	–	10	10	10
Ammonia-N (mg/L)	–	2	1–2	–
Nitrite-N (mg/L)	0.3	0.3	0.3	0.3
Nitrate-N (mg/L)	–	–	10	–
Phosphate-P (mg/L)	0.7	–	–	–
Total P (mg/L)	–	1	0.2–0.8	0.8

Source: According to Gujer (1999, p. 284).

As a result of (earlier) German developments and harmonization with corresponding EU directives, all values refer to analyses of a specific grab sample or of a 2 h composite sample. The Swiss requirements, cited by Gujer (1999), are indicative or representative for 'advanced' wastewater treatment. These minimum requirements must be met by practically all plants, irrespective of size. The requirements are considered fulfilled, if four of five load-averaged daily samples taken under dry weather flow conditions do not exceed the regulated values.

2.5 HOW TO GUARANTEE THAT STANDARDS ARE MET (OPERATIVE AND ADMINISTRATIVE INSTRUMENTS)

To develop requirements for wastewater treatment for pollution control is one task. To assure that these standards are continuously met is another. There are all too many instances where well formulated goals or requirements remain unfulfilled. In a certain sense, it should be part of the strategy of developing such standards that they actually can be technically and administratively obeyed, and likewise that it is possible to monitor this.

In general there are two different reasons for monitoring treatment plant operation and two corresponding types of monitoring. First, the water authorities must make sure that the requirements stipulated for the discharge permit are met: state supervision ('amtliche Überwachung'). Second, the plant operator is encouraged to monitor the performance of his plant to a much larger extent than the authorities do, for optimizing its operation and efficiency: operator monitoring ('Eigenüberwachung' in the 'Eigenkontrollverordnung', that is the water authority directive that the plant operator conduct self-monitoring for operation control).

Depending upon the various formulations of the respective state regulations in Germany, comparisons between both series of monitoring data can be made. Operator monitoring is clearly more extensive in frequency and parameters are measured by either real-time continuous methods or laboratory methods.

The quasi-legal formulation that describes the present German wastewater treatment requirement is the so-called 'appendix 1 to the general administrative ruling on wastewater treatment' (Table 2.7). Other appendices regulate specific wastewater streams from industries. For a long time, German pollution control concepts favored joint collection and treatment of domestic and industrial wastewater streams. First, this seemed to allow better control over collection and treatment. Second, there can be process benefits such as providing nutrients for a nutrient-deficient industrial wastewater from the excess nutrients present in the domestic wastewater. Nevertheless the current requirements for wastewater treatment (in Germany), in addition to those in appendix 1 (municipal wastewater treatment requirements) also contain numerous regulations for specific industrial streams in a rather long list of appendices to the general wastewater directive.

Table 2.7 Appendix 1 to the German wastewater directive.

Plant size	COD (mg/L)	BOD_5 (mg/L)	Ammonia-N (mg/L)	Total P (mg/L)	Total-N (mg/L)
I <1000 PE	150	40	–	–	–
II <5000 PE	110	25	–	–	–
III <20,000 PE	90	20	10	–	18 ($WW_{temp} > 12°C$)
IV <100,000 PE	90	20	10	2	18 ($WW_{temp} > 12°C$)
V >100,000 PE	75	15	10	1	18 ($WW_{temp} > 12°C$)

Source: BGBl (2004).

As indicated before, these minimum requirements must be met based on samples taken and analyzed by the authorities. The samples are either a specifically defined ('qualified') grab sample or a 2-h composite-sample.

For the plant operator, it is important to realize that average plant operation efficiency must be set much higher than these standards in order to 'guarantee' that these minimum requirements are met throughout most of the time (statistically formulated in the Swiss regulations as 4 out of 5 samples – in the German regulations as the 80th%-tile of all analyses in grab samples or 2-h composite samples.

2.6 CONCLUDING REMARKS – ISSUES NOT CONSIDERED

The above discussion focused on the development of standards for municipal wastewater treatment, that is the treatment of wastewaters from domestic sources and various industrial enterprises. The reason is that, in Germany, combined treatment often has advantages for operation and supervision over separate systems. When the industrial complex is large and the wastewater has a unique composition, separate treatment can be advantageous. In this case the treatment process performance and the wastewater discharge quality must meet the requirements laid out in the appropriate industrial appendix to the general wastewater directive. The specific treatment of undiluted and 'one-dimensionally polluted' streams may make recycling of materials technically feasible. Such concepts are followed today in Germany and in the European Union (under the heading of integrated environmental production). Numerous working groups are sharing information and experiences on collection and treatment of liquid waste streams with the object of recycling both, the aqueous phase and the materials present in the wastewater stream.

These aspects of materials recycling are not covered in the discussion presented above. First of all, it requires a rather different form for collecting wastewater streams, that is mostly avoiding mixing streams, which creates a multitude of constituents that require a very large number of different unit processes for treatment. Second, the objective of pollution control is no longer dominant in setting treatment efficiency and in supervising operations. The term 'closed' water cycles, often applied in this industrial water management indicates the intent not to create waste streams at all. – However, some of the experiences in industrial water management, and also in the area of solid waste management, might be transferred into the area of classical collection and treatment of municipal wastewaters. These approaches might have to involve changing the wastewater collection scheme (a complex and rather future oriented task) to take full advantage of taking a new look at treatment process requirements.

2.7 REFERENCES

BGBl (2004). *AbwV, Ausfertigungsdatum: 21.03.1997, Abwasserverordnung in der Fassung der Bekanntmachung vom 17. June 2004* (Wastewater Directive in the form of the June 2004 Publication) IS. 1108, 2625. www.gesetze-im-internet.de/bundesrecht/abwv/ (Last accessed 2005)

Dunbar W. P. (1912). *Leitfaden für die Abwasserreinigungsfrage* (Guideline for wastewater treatment) Oldenbourg, München und Berlin.

ec.europa.eu/environment/water/water-urbanwaste/ (1997) Urban Waste Water Directive.

von der Emde W. (1999). *Geschichte der Abwasserentsorgung* (The history of wastewater collection and discharge), ATV Publishing, Hennef, p. 83.

Gujer W. (1999). *Siedlungwirtschaft* (Sanitary engineering), Springer Publishing Co., Berlin, Heidelberg, p. 284.

Imhoff K. and Imhoff K. R. (1925 to 2009). *Taschenbuch der Stadtentwässerung* (Handbook of wastewater engineering), 4th to 31st edns, Oldenbourg, München and Berlin.

König J. (1899). *Die Verunreinigung der Gewässer* (The pollution of waters), Springer Verlag, Berlin.

Strell M. (1913). *Die Abwasserfrage* (The wastewater challenge), Ledinweber, Leipzig, p. 202.

www.bmu.de/fileadmin/bmu-import/files/pdfs/... (2000). *Richtlinie 2000/60/EG des Europäischen* Rates (Directive of the European Council), BMU.

Chapter 3

Activated sludge process development

H. David Stensel (USA), Jacek Makinia (Poland)

3.1 INTRODUCTION

AS is the most common biological treatment process in the world. It is used for municipal and industrial wastewater treatment with applications ranging from small package plants for single homes to huge facilities serving metropolitan areas. The AS process was first announced on April 3, 1914 to the Society for Chemical Industry at the Grand Hotel, Manchester, England by Ardern and Lockett (Ardern & Lockett, 1914) and resulted from investigations on sewage treatment in the USA and UK at the end of the 19th, and the beginning of the 20th centuries.

The AS process has five basic functions: (1) suspension of a flocculent slurry of microorganisms by mixing and/or aeration, (2) oxidation of soluble and particulate organic matter with oxygen or nitrate/nitrite to produce gaseous products and additional biomass, (3) liquid–solids separation to produce a treated effluent with a low TSS concentration, (4) return of solids from the liquid–solids separation zone to the suspended growth treatment reactor or the retention of the solids in the treatment reactor during effluent separation and removal, and (5) wasting of excess sludge to maintain a desired mass or SRT. AS process developments have been motivated by wastewater treatment needs and goals; from BOD and TSS removals in its infancy in 1914 to N removal in the 1960s, and P removal in the 1970s. During this period and into the 1980s, process development continued to overcome common problems of poor settling and foaming. During the 1980s, 1990s, and early in the 21st century, process development continued with the objectives of reducing AS reactor volume and site footprint and saving energy. This paper summarizes the most significant steps in AS process development and recognizes key contributors.

3.2 THE BEGINNING – 1882–1914

Recognition of the roles of oxygen and living organisms was the first step in AS process development. Schloesing and Müntz (1877) found that putrescible matter reduction in sewage purification was related to living organisms consuming oxygen and occurred with nitrate production. According to Martin (1927), Angus Smith, in 1882 reported that blowing air into sewage tanks led to nitrification and the reduction of putrescible matter. In 1884, Dupre (1884) advocated the need for oxygen in sewage purification and tested the effect of bubbling air and aeration by a cascading water flow. At about the same time Mason and Hine (1892) in the USA conducted similar studies. A natural draft aeration step was included in the first installed intermittent filters in 1871 (Bailey-Denton, 1882), in contact beds by Dibdin in 1892 (Dibdin, 1903) in England, and in the first intermittent filters evaluated in the USA in 1888 at the Lawrence Experimental Station (Mills, 1890). Waring (1894) designed a percolating gravel filter system with forced aeration and was issued a USA patent in 1895 (Waring, 1895). Clogging problems occurred with these systems, and after 1910, aeration was evaluated with other media.

Black and Phelps (1911) studied sewage aeration in tanks containing inclined wooden gratings and for periods up to 24 h found only slight organic matter oxidation but no N oxidation. Clark and Gage (1912), at the Lawrence Experimental Station, Massachusetts aerated sewage in sand filters, in bottles inoculated with algae, and in a tank containing slabs of slate. After several months, nitrification occurred in the aerated filter containing algae. For the tank containing slate slabs they reported that the slate slabs and tank sides became covered with a compact brown growth, effluent clarity greatly improved, and nitrification was accomplished within 24 h. At the end of each dosing cycle the tank was emptied and refilled with sewage.

These results inspired further work by Fowler at Manchester University, UK – who was active in research on using microorganisms for industrial wastewater treatment. Fowler was also a consulting chemist to the Manchester Corporation Rivers Division and had directed their research in the Davyhulme laboratory since 1904. In 1912, Fowler visited New York City to consult on harbor pollution. He also visited Lawrence Experimental Station and observed their work on sewage aeration with algae inoculation and fixed media. On returning to Manchester he supervised an experiment in which an iron-bacteria culture (M-7) was added to raw sewage and about 17 mg/L of an iron salt and aerated for 6 h. This was followed by clarification. They did not recycle the settled solids from the clarifier but continued to add M-7 to the aeration vessel with each sewage dose (Fowler & Mumford, 1913).

Fowler also suggested work on aeration to E. Ardern and W. T. Lockett, at the Davyhulme, Manchester, Sewage Works. They aerated sewage in batch treatment using 1-gal bottles that, in contrast to previous work were: (1) covered to exclude light and algal growth and, (2) not emptied completely after each aeration period so that the accumulated solids were mixed and aerated with new raw sewage batches. After about 5 weeks, complete nitrification occurred in 24 h. Next they investigated the effect of the remaining solids volume to feed volume and reported that with a ratio of one volume of solids to four volumes of sewage a well-oxidized effluent could be obtained in 6–9 h. Ardern and Lockett (1914) named the accumulated solids 'activated sludge' because they consumed organic material in the raw sewage feed. They noted that it was a dark brown, flocculent mass that rapidly separated from the treated water.

3.3 RAPID ACCEPTANCE OF AS – 1914–1930

The AS process was rapidly accepted and applied. Even before the 1914, experimental work was under way for the treating of 303 m^3/d at Salford, UK (Melling, 1914). Further testing and development followed rapidly in the USA, including work at Mt Vernon, New York, the University of Illinois and Milwaukee, Wisconsin. Professor Edward Bartow at the University of Illinois visited Fowler's group in August 1914 and then carried out AS experiments at the Illinois State Water Survey laboratories in 3-gal batch-fed test bottles and with 4 batch-fed 8-ft 5-in deep, 3-ft square aeration tanks (Bartow & Mohlman, 1915). This was followed by a 750 m^3/d continuous-flow test facility with bar screens, grit removal, separate aeration and settling tanks, and sludge drying beds at Champaign IL (Bartow, 1917). These tests showed the effect of increasing AS concentration on nitrification rate, the superiority of filter plate over wooden air diffusers, and the ratio of diffuser area to tank area. The larger continuous flow plant testing was done after following a visit by Barstow to the Mt Vernon, NY test facility operated by C. L. Peck of the Dorr Company. The Dorr–Peck process used an airlift for returning RAS to the settling tank.

USA designs encouraged using a separate AS clarifier and continuous flow treatment. Only 18 months after Ardern and Lockett's presentation, Milwaukee, Wisconsin, USA started a 7570 m^3/d continuous flow AS plant and then, in 1925, a 170,325 m^3/d system. The *plug flow* aeration system used eight tanks in series and a circular clarifier; two tanks in series were used for RAS reaeration. The Milwaukee studies provided design guidance on several AS process parameters including aeration HRT (4–8 h), air application rates (5–15 m^3 air/m^3 sewage treated), and clarifier application rates (1.4 m/h) (Metcalf & Eddy, 1922).

Both the Dorr Company and Professor Barstow were interested in the WAS fertilizer value, particularly its N content. Barstow and Mohlman (1915) noted the high AS N level and showed pictures of plants grown with and without WAS to illustrate its attractiveness as a fertilizer. The fertilizer value of WAS also interested Fowler and Milwaukee, and in 1926 Milwaukee began sale of the dried AS product Milorganite (Hatton, 1926). In 1919, Fowler and Mumford (1919) received a USA patent describing heat sterilization of WAS at 60–70°C. Fowler noted that this was an opportunity for chemists to return N to soil instead of wasting it as was the current practice.

Over 20 full-scale facilities were operating in the UK and USA within 12 years of the AS process introduction (Table 3.1). By 1917, continuous flow treatment with separate settling tanks was the common design approach. By 1927 the largest plants were in the USA at Houston, Texas, Milwaukee, Wisconsin, Indianapolis, Indian and Chicago, Illinois.

Table 3.1 First full-scale AS plants in the UK and USA.

Year	Location		Flow, m^3/d	Operating mode	Aeration type
1914	Salford	UK	303	Fill-and-draw	Diffused
1914	Salford	UK	45	Continuous	Diffused
1915	Davyhulme	UK	378	Fill-and-draw	Diffused
1916	Worcester	UK	7570	Continuous	Diffused
1916	San Marcos, TX	USA	454	Continuous	Diffused
1916	Milwaukee, WI[1]	USA	7570	Continuous	Diffused
1916	Cleveland, OH[1]	USA	3785	Continuous	Diffused
1917	Houston (North), TX	USA	20,817	Continuous	Diffused
1917	Sheffield	UK	3028	Fill-and-draw	Mechanical
1917	Withington	UK	946	Continuous	Diffused
1918	Houston (South), TX	USA	18,925	Continuous	Diffused
1920	Stamford	UK	378	Continuous	Diffused
1920	Tunstall	UK	3104	Continuous	Mechanical
1921	Sheffield	UK	1340	Continuous	Mechanical
1921	Davyhulme	UK	2509	Continuous	Diffused
1921	Bury	UK	1363	Continuous	Mechanical
1922	Des Plaines, IL	USA	20,817	Continuous	Diffused
1922	Calumet, IN	USA	5677	Continuous	Mechanical
1925	Milwaukee, WI	USA	170,325	Continuous	Diffused
1925	Indianapolis, IN	USA	189,250	Continuous	Diffused
1927	Chicago (North) IL	USA	662,375	Continuous	Diffused

[1]Experimental.

The AS process was also being installed in other countries. By 1924, Ontario, Canada had seven municipal plants (Wolman, 1924). In 1925, K. Imhoff in Essen managed the first experimental plant in Germany. The first German full-scale facility at Essen-Recklinghausen (1925) was the first to use anaerobic digester gas as a fuel (Seeger, 1999). In 1927, an abattoir effluent at Apeldoorn (Netherlands) used a brush aerator developed by Kessener (Cooper, 2001). In 1936, New York City started an AS plant at Wards Island with a capacity of 681,300 m^3/d. This plant was eventually expanded to 1.5×10^6 m^3/d and recently was converted to step feed biological nutrient removal by modifying the existing tanks. By 1938, there were 203 AS sludge plants in the USA and the process was being used on five continents (Alleman & Prakasam, 1983; Cooper, 2001).

Though many early AS designs accomplished nitrification, treatment goals were close to the 1972 USA definition of secondary treatment. The 8th British Royal Commission Report on Sewage Disposal included the following constituent levels as a typical AS effluent standard: $BOD_5 = 20$ mg/L, TSS = 20 mg/L and total organic N = 3 mg/L (Royal Commission on Sewage Disposal, 1912).

3.4 THE BEGINNING OF AS PATENTS

Jones and Attwood, Ltd obtained a number of Great Britain (GB) patents during the formative years of the AS process. Their initial interest was in aeration equipment and the first patent (GB 22,952, Oct. 11, 1913) was on aeration and an airlift method for mixing aeration tanks. A second patent (GB 19,915) was also related to aeration and was issued on the day of Ardern and Lockett's AS presentation. The third patent (GB 19,916), also issued on that day, described a treatment process with a repeated cyclic sequence of sewage feed, aeration, settling, and effluent withdrawal, much like the process described by Ardern and Lockett and what today is called a sequencing batch reactor SBR (Table 3.2). A fourth patent (GB 21,976), issued on Nov. 4, 1914 described a similar process with details on the aeration and settling times. In all, Jones and Atwood obtained seven patents between 1913 and 1914 that addressed aeration and mixing methods and reactor configurations. Between 1914 and 1918 they obtained eleven more patents, one of which (GB 132,826) described an aeration tank design with a sidewall inner overhang and diffuser location along the length of one side to create what has become known as 'spiral roll' aeration. The term, *activated sludge* was used for the first time in one of their patents (GB 122,428) on August 24, 1917. This patent described continuous-flow operation with aeration tanks in series and liquid–solids separation. Many of these patents were issued in the USA about 4 years after the GB patents. GB 19,916 patent describing SBR operation was issued in the USA on Oct. 22, 1918 (USA 1,282,587). On May 11, 1915 a USA patent (1,139,024) describing continuous flow separate aeration and liquid/solids separation zones was issued to L. C. Frank, a USA Public Health Officer who then granted the patent to the public (Porter, 1921; Alleman & Prakasam, 1983).

In Jul. 1919, all the patents of Jones and Attwood Ltd were given to Activated Sludge Ltd, which was formed by J. A. Coombs, (Managing Director), G. Fowler and W. Jones (Owner). In the late 1920s, Activated Sludge Ltd filed a patent infringement suit against Chicago, then suits against Milwaukee, Cleveland, Indianapolis and several others. In 1933 the courts ruled against Milwaukee. Some cities shut down their AS operations, including San Marcos, Texas, to avoid the monetary payments but many continued using them for a royalty fee of 25 c/cap. By 1938, 105 of the 203 USA AS facilities were licensed by Activated Sludge Ltd (Kappe, 1938).

3.5 FURTHER PROCESS UNDERSTANDING AND INNOVATION – 1930–1970

New process designs developed steadily in response to changing treatment needs, advances in process understanding, and site-specific issues. Increased understanding of how the oxygen demand rate decreased with time (or length of the aeration tank) led Kessler *et al.* (1936) to design the first 'Tapered Aeration' system (Table 3.2). 'Step Feed' AS (also referred to as step aeration) was used by Gould (1942) for New York City AS designs. Besides providing a more uniform

Table 3.2 AS process developments.

Year	Process	Schematic	Comments
1914	Sequencing batch reactor (SBR)	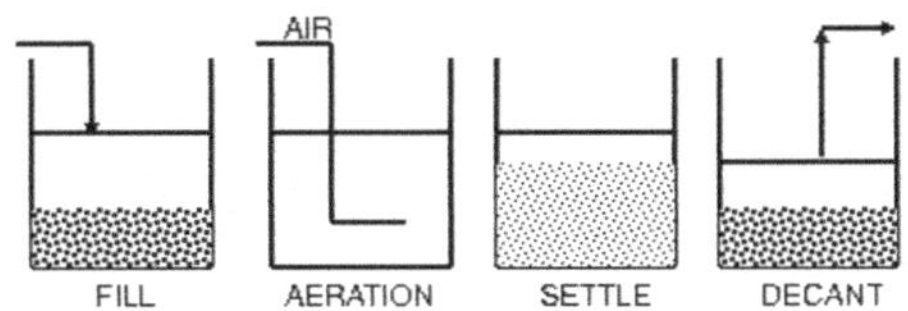	AS process described by Ardern and Lockett (1914) following testing at Davyhulme, England.
1916	Plug flow AS	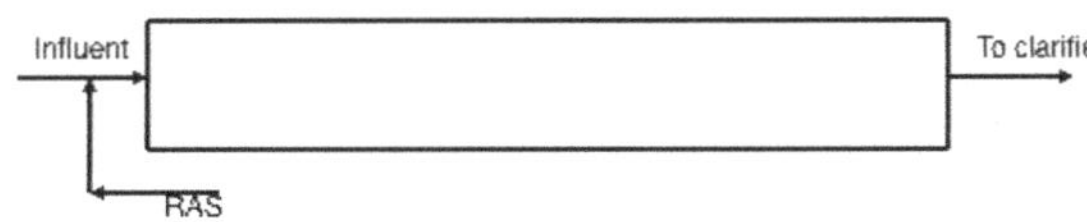	Long narrow tanks used for continuous flow AS after studies in Milwaukee (Metcalf & Eddy, 1922).
1936	Tapered aeration	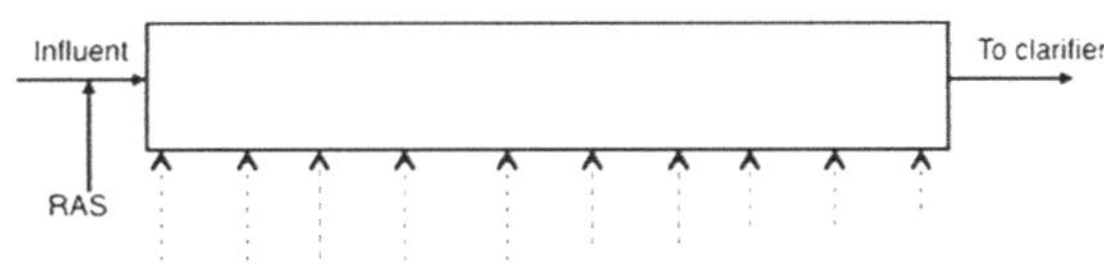	More efficient designs account for changing oxygen demand along length of tank. First applied in Monroe, WI (Kessler *et al.*, 1936).
1942	Step feed (step aeration)	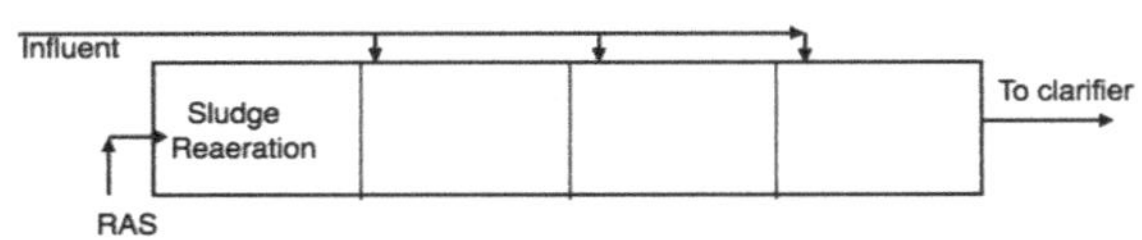	Gould (1942) introduces step feed in New York City to reduce the amount of nitrification and required air supply.
1944	Kraus process	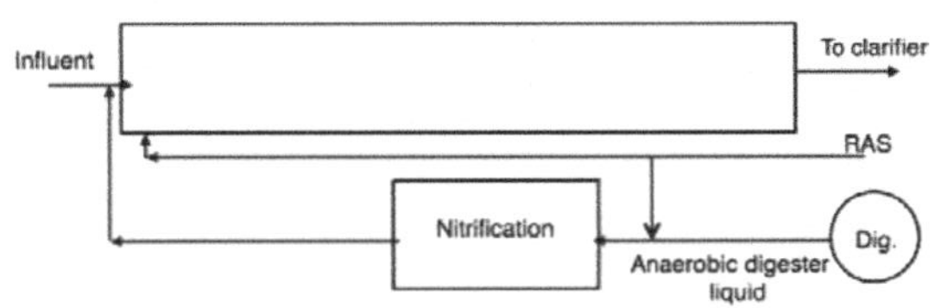	Sidestream nitrification in Peoria, IL to provide nitrate to AS tank inlet zone for treating high strength wastes and help prevent sludge bulking (Kraus, 1945).

1944	Contact stabilization	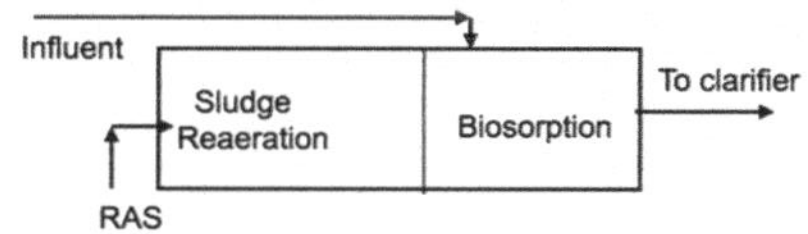	Short contact initial period for biosorption and reaeration period for RAS results in smaller tank volume. Used in 1951 at Austin, TX (Ulrich & Smith, 1957).
1956	Complete mixing AS	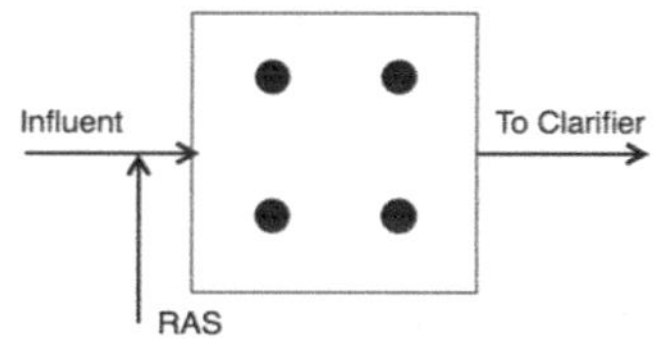	Use of complete-mixed aeration tanks to buffer toxic and shock loads and maintain constant substrate conditions for microbes (McKinney, 1957).
1957	High purity oxygen	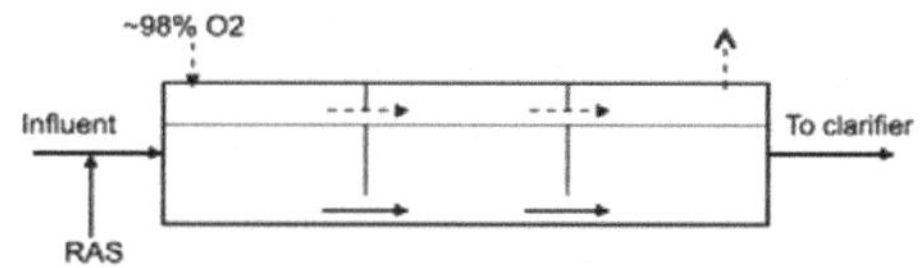	Pilot testing of high purity oxygen AS system in Baltimore, MD (Budd & Lambeth, 1957). Commercialized in USA in 1970s by Union Carbide Corp. First full-scale plant, Detroit, MI, 1975.
1959	Oxidation ditch	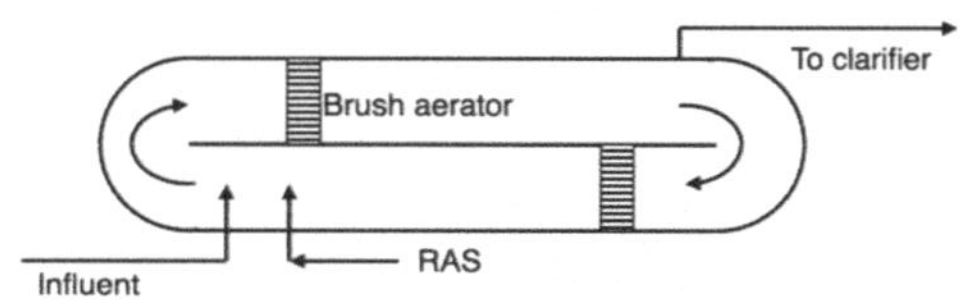	Intermittent feeding and effluent withdrawal in a ditch configuration for simple, extended aeration process with minimal equipment (Pasveer, 1959). In 1960s continuous flow designs with different aerator designs.
1962	Postanoxic denitrification	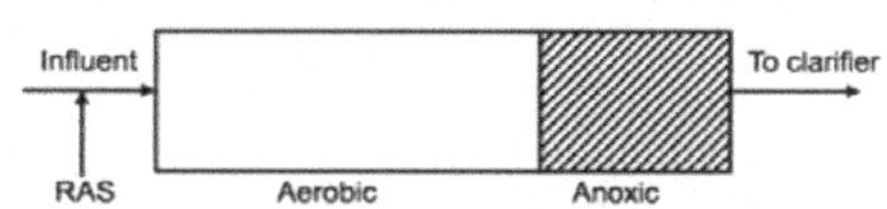	Use of a mixed unaerated zone to remove nitrate/nitrite after aerobic nitrification (Wuhrmann, 1964).

(*Continued*)

Table 3.2 AS process developments (*Continued*).

Year	Process	Schematic	Comments
1962	Ludzack–Ettinger preanoxic denitrification	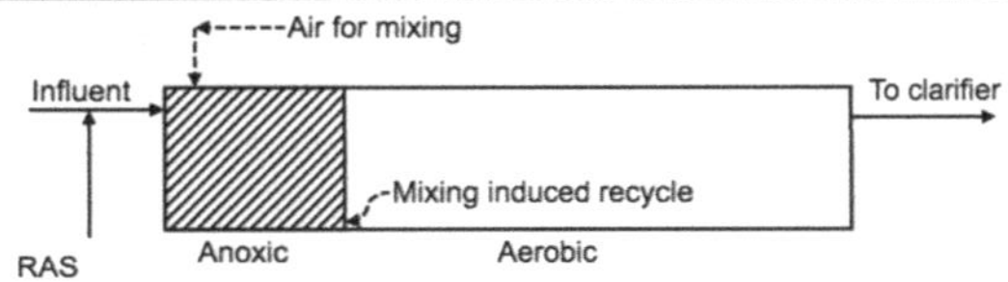	Lab experiments with recycling of AS from the aerobic zone upstream to an anoxic zone for nitrate removal with biological oxidation of influent BOD (Ludzack & Ettinger, 1962).
1973	High F/M selector	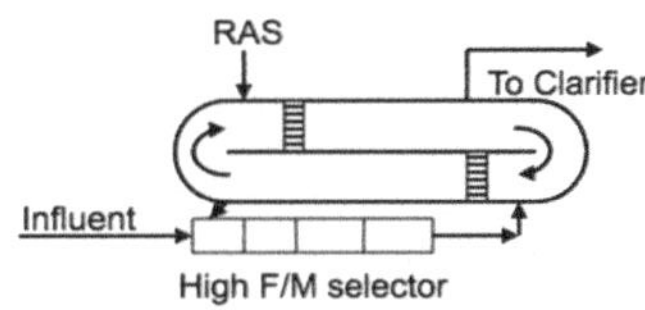	Use of high a F/M staged selector before AS aeration tank to control filamentous bulking (Heide & Pasveer, 1973).
1974	Modified Bardenpho (illustrates anaerobic and anoxic selectors)	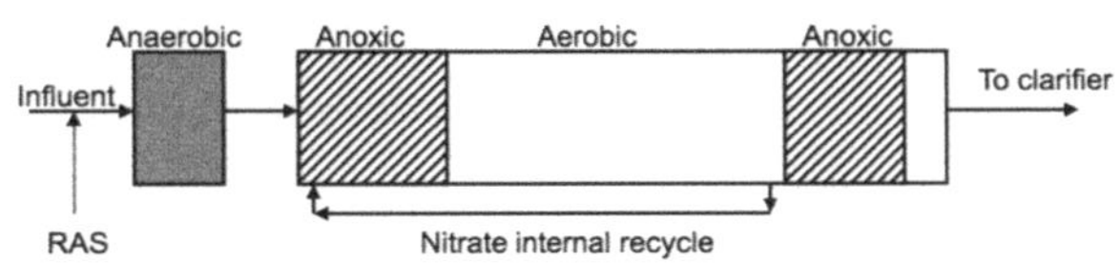	First full-scale Modified Bardenpho process for Meyerton WWTP in 1974 (Barnard, 1998) and full-scale testing at Alexandra WWTP (Nicholls, 1975), both near Johannesburg, RSA. First system in North America at Palmetto, FL in 1979 (Burdick *et al.*, 1982).
2002	Integrated fixed film AS (IFAS)	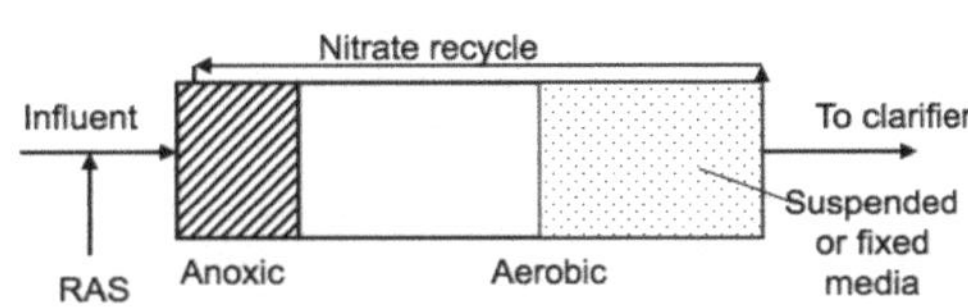	Anoxic/aerobic IFAS process design used at Broomfield, CO with plastic carrier (Rutt *et al.*, 2006). First installation with sponge carrier at Wyk auf Fohr, Germany in 1988 (Morper & Wildmoser, 1990).

1990	Membrane bioreactor (MBR)	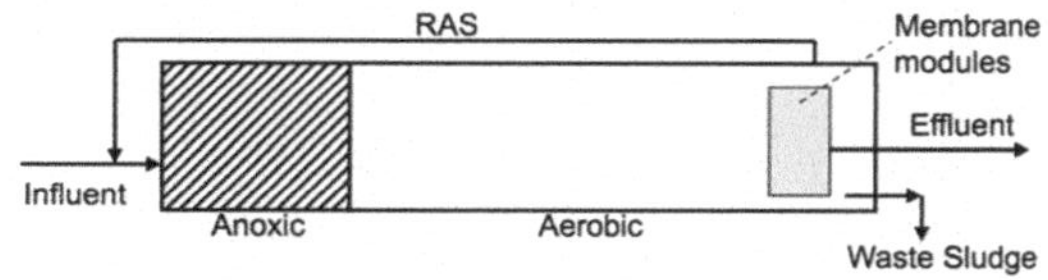	First municipal MBR installations in Japan and Canada in 1990 and 1993, respectively (Judd, 2011).
2004	A/B Process	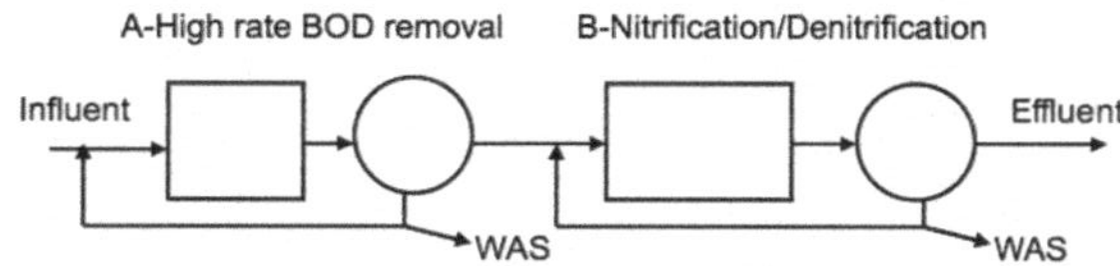	Primary clarification and high rate carbonaceous removal in A-stage increases methane production in anaerobic digestion to approach energy self-sufficiency at Strass WWTP Austria.
2009	SBR with granular AS	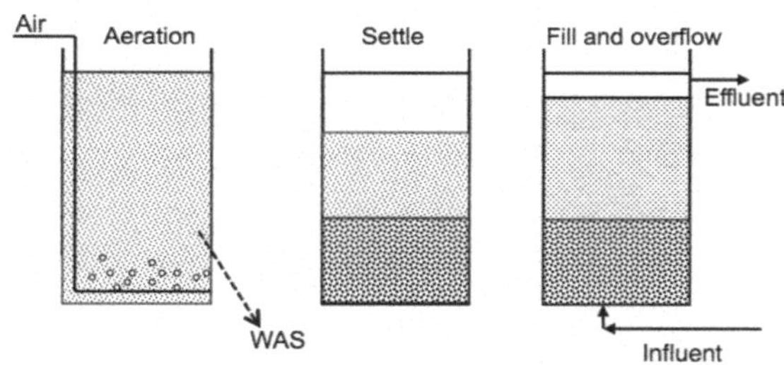	First full-scale facility in 2009 at Gansbaai WWTP, Western Cape, RSA. New facility at Epe, The Netherlands (2012) for treating domestic and slaughterhouse wastewater.

air demand, Gould felt that it would lead to higher effluent ammonia and thus a lower air demand. The use of these designs in multiple-pass aeration tanks at New York City plants fortuitously provided for a relatively easy conversion of existing facilities to step feed BNR to meet the 21st century treatment needs (Carrio *et al.*, 2000). The Kraus process (1945) was also developed to address oxygen demand issues at the head end of the aeration basin, especially when treating industrial or combined industrial/municipal wastewaters. The Kraus process flow sheet (Table 3.2) has a side reactor for nitrification of anaerobic digester return flow. The nitrate/nitrite produced provided an additional electron acceptor at the head end of the aeration tank, where the oxygen demand rate could not be met by diffused aeration. Although not used widely, this process had two similarities to AS developments that followed 30–50 years later. The process helped to control sludge bulking as in the anoxic/aerobic selector processes to be discussed below (Jenkins *et al.*, 2004). Also, many sidestream nitrification processes have been developed in the past fifteen years to bioaugment nitrification in the AS mainstream process or to remove N from the return flow (Tchobanoglous *et al.*, 2014).

The development of the 'Contact Stabilization' (also called a biosorption process; Table 3.2) provided an AS alternative requiring less space that could meet required BOD and TSS removals. For BOD and TSS removals of 65–75%, Setter and Edwards (1943) promoted the *Modified Aeration* or high rate AS process. This process required about 1/3 of the volume of conventional AS and was applied in other large US cities (Sawyer, 1965). Much later this low SRT process became part of the A/B process (Table 3.2). At the other end of the scale, long SRT *Extended Aeration* AS processes gained favor for smaller flows because of their simplicity (typically no primary settling), more stabilized WAS, and lower WAS production. This approach was used in package plants for small and decentralized facilities in the 1950s and 1960s. Extended aeration was also the design basis for many oxidation ditch systems.

The oxidation ditch process (Pasveer, 1959) had a major impact on AS process design and selection. The first facility was in Voorschoten, The Netherlands (1954); now there are thousands of these installations worldwide and in the USA alone more than 9200 municipal facilities use them (WEF, 1998). Pasveer's first designs were intermittently fed with batch settling of sludge; he returned to the SBR concept abandoned by 1920 in favor of continuous-flow systems. Various modifications of the oxidation ditch process configuration and operation have been used for efficient N removal (Tchobanoglous *et al.*, 2014).

Other activities in the 1970s set the stage for new AS processes and applications that had major impacts. Budd and Lambreth (1957) investigated a high purity oxygen AS (HPOAS) process (Table 3.2) in pilot plants at Baltimore, MD and Stamford, CT. These preceded the widely successful commercialization of HPOAS by Union Carbide and Air Products and Chemicals in the 1970s. There was controversy about the sludge production and substrate utilization kinetics of HPOAS (Chapman *et al.*, 1976; Kalinske, 1976); however a careful review

(Parker & Merrill, 1976) showed that its main advantages were the ability to provide high oxygen transfer rates per unit volume and to greatly reduce the space required for AS treatment. This led to the construction of many HPOAS systems in large cities with limited available land area.

Biological denitrification was known by the 1920s. Buswell and Long (1923) reported that AS could use nitrate in the absence of DO. Rising sludge in secondary clarifiers following nitrification in the aeration tank was attributed to denitrification (Sawyer & Bradney, 1945). The use of discrete anoxic zones in the AS process for N removal was introduced in the 1960s and preceded the development of BNR processes in the 1970s. Wuhrmann (1964) recommended a mixed, unaerated tank after the aeration tank (Table 3.2) for nitrate removal and noted an abundance of denitrifying bacteria in AS. Preanoxic zone and internal recycle concepts were introduced in lab studies (Ludzack & Ettinger, 1962). AS from the nitrification zone was recycled to a low aerated preanoxic zone to allow use of the wastewater BOD for nitrate reduction. In 1969, Balakrishnan and Eckenfelder (1969) showed a N removal flow scheme with a preanoxic zone receiving influent wastewater and nitrified recycled flow from a final clarifier.

During this period, significant advances were made in understanding key process kinetics, developing design equations, recognizing the importance of SRT and understanding the fundamental parameters for oxygen demand, sludge production, and nitrification kinetics. These provided a basis for less empirical process analysis and further process developments (McKinney, 1962; Downing *et al.*, 1964; Eckenfelder, 1966; Jenkins & Garrison, 1968; Lawrence & McCarty, 1970). Nevertheless, the age-old problem of AS settling continued but there were solutions waiting to be uncovered in the 1970s.

3.6 THE AGE OF THE SELECTOR AND BNR – 1970–1990

The most dramatic changes in AS process understanding and treatment application occurred during the 1970s and 1980s. During this time treatment needs were shifting from secondary treatment for BOD and TSS removal to requirements for nutrient limits. Just as the pioneers in the 1900s influenced AS process designs, investigations by Chudoba *et al.* (1973) and Barnard (1973) in the early 1970s had far-reaching, long-lasting impacts on the future of AS. Chudoba *et al.* (1973) showed that filamentous bulking of AS could be controlled by introducing a *selector*, which influenced both the microbial population of the AS and its treatment performance. They showed that certain AS configurations could select for good settling floc-forming bacteria with little if any of the filamentous microorganisms that caused sludge bulking. Their *kinetic selector* designs consisted of several short HRT reactors in series in which influent mixed with RAS prior to the main reactor. Tomlinson (1976) showed that SVI decreased as the size of the initial contact zone (ICZ) decreased and its F/M increased. Rensink (1974) confirmed the benefits of a staged selector for preventing sludge bulking and Heide and Pasveer (1973) showed improved SVI for oxidation ditch treatment after a staged selector zone with a high ICZ F/M ratio (Table 3.2). Albertson (1991) recognized the importance of multiple selector stages and recommended an ICZ F/M of >2.0 kg BOD_5/kg MLVSS-d followed by stages with an F/M gradient. Selectors designed in this fashion produced AS with SVIs below 100 mL/g.

Chiesa and Irvine, 1985 noted that an SBR could be manipulated to serve as a selector that produced good settling AS. The SBR treatment scheme of the early 1900s was revived in the 1970s because of advances in programmable logic controllers, on-line instrumentation, and the design and technology transfer efforts of Irvine (Irvine & Davis, 1971; Irvine & Busch, 1979).

Although this paper is not intended to review the extensive history of the development of BNR processes, Barnard's findings in the early 1970s (Barnard, 1973) are noted here for their broad impact on selector designs that discouraged filamentous bacterial growth and promoted EBPR and/or biological N removal. Barnard first observed EBPR and biological N removal in 1972 in a 100 m^3/d anoxic-aerobic-anoxic-aerobic pilot plant at Daspoort WWTP, Pretoria, RSA. He later separated an anaerobic contact zone ahead of the anoxic zone in recognition of its importance in selecting for EBPR (Barnard, 1974). Many BNR processes followed this work (Tchobanoglous *et al.*, 2014) The Modified Bardenpho process, which incorporates EBPR by having the initial anaerobic contact zone and biological N removal in the subsequent anoxic-aerobic-anoxic-aerobic treatment zones is shown in Table 3.2. Removing the anoxic zones results in an anaerobic selector-aerobic process that produces low SVI AS and EBPR. Removing the anaerobic zone results in biological N removal and an anoxic selector for low SVI. These types of *metabolic selectors* have been described by Jenkins *et al.* (2004) and used in some facilities for achieving good settling sludge even if BNR is not required. Rosso and Stenstrom (2005) showed that for nitrifying systems it was more economical on a life cycle cost basis to include a preanoxic zone. The process stability provided by a lower and more reliable SVI was also found to be a major benefit. Numerous well-performing BNR facilities and/or selector designs were applied worldwide in the 1980s and are of great importance today in view of increasing discharge permit requirements for nutrient removal.

Patents on the anaerobic/aerobic process for achieving good sludge settling were issued well before Barnard's findings; to Davidson (1957) and later Spector of Air Products and Chemicals Inc. (1977). Barnard's USA patent (1976) filed initially in 1973 showed a Modified Bardenpho process schematic with a focus on N removal rather than AS settling properties. Both Davidson and Spector showed the importance of an initial anaerobic contact zone for producing low SVIs. Davidson's work was mainly in the industrial wastewater area; he reported SVIs as low as 34 mL/g for a distillery wastewater AS plant (Albertson, 1991). However his process never gained much attention, since there was not much interest in P removal in the municipal sector at that time and there was a strong fixation on providing only aerobic conditions for AS treatment.

3.7 SMALLER FOOTPRINT, HIGHER EFFLUENT QUALITY – 1990–THE PRESENT

During this period wastewater treatment goals became more complex and variable, depending on local conditions and needs such as: (1) much lower effluent N and P concentrations, (2) higher quality effluent for water reuse, (3) methods for upgrading existing facilities without additional tank volume, (4) designs that maximize micropollutant removal, (5) sustainable engineering methods that have lower energy demand, increased energy production, and nutrient recovery rather than just wastewater treatment. New sidestream nutrient removal technology was developed in the late 1990s and early 2000s. These include sidestream nitrification methods such as InNitri® for bioaugmenting the mainstream AS process with nitrifying bacteria (something that Kraus likely accomplished but did not recognize in 1945) and the SHARON® process for N removal. Since 2000, there has been a rapid acceptance of the Anammox process for sidestream N removal with many full-scale facilities in Europe; current efforts are being made to incorporate Anammox into the mainstream AS process for N removal without carbon addition. During this period advances were made in the application of IFAS and other hybrid AS processes, MBRs, and granular AS (Table 3.2).

The placement of media for biofilm growth into the suspended growth AS process goes back to the 1930s with processes such as the Hays process in which asbestos-concrete sheets were placed in the aeration tank (Lackey & Dixon, 1943). During the 1980s a number of proprietary media designs and processes were developed that incorporated fixed webs or strands, sponges, and low-density plastic media. The sponge-type media was developed by Reimann and Wildmoser in Germany, and the first commercial installation was at Wyk auf Fohr, Germany in 1988 (Morper & Wildmoser, 1990); several plants are now operating in southern Germany (Muller, 1998) and there is a USA installation in Providence, RI. Small plastic carrier media has been used mainly in moving bed bioreactors (MBBRs) in Europe as well as in IFAS systems in the USA with the first installation at Broomfield, CO in 2002 (Rutt *et al.*, 2006). Randall and Sen (1996) have described full-scale evaluations of fixed-media IFAS systems in the USA. The ability to increase capacity and/or upgrade existing facilities for N removal (Table 3.2) without constructing new tank volume is a major advantage of IFAS processes.

After almost 80 years of using gravity settling to separate the treated effluent from AS, liquid-solids separation by synthetic membranes was used in MBRs in the early 1990s (Table 3.2). This produced dramatic changes in AS treatment. Plant footprints were less than 1/3rd of a gravity clarifier AS plant, operation was simpler along with less concern about filamentous bacteria growth, and a reclaimed water effluent quality containing unmeasurable TSS was obtained. In the first USA patent for AS treatment with membrane separation (Budd & Okey, 1969) the membrane separation was external to the AS tank. Dorr Oliver installed the first system at Pikes Peak, CO, in 1974. Improvements in membrane materials, lower fabrication costs, and the placement of the membranes in the AS tank with coarse bubble aeration to control membrane fouling by Yamamoto *et al.* (1989) made possible the modern day MBRs. The first commercialized flat plate membranes by Kubota were used in Japan in 1990 and by 1996 there were 60 MBR installations (Judd, 2011). Membrane technology interested Andrew Benedek at McMasters University in Hamilton, Canada, and in 1980 he started Zenon Environmental Inc. to develop and commercialize hollow fiber membranes for water and wastewater treatment. The first Zenon hollow fiber, ZeeWeed®, system for wastewater treatment was installed at Stoney Creek, Ontario, Canada in 1993, and the first USA installation was in 1998 at the Arapohoe County, CO, Lone Tree Creek WWTP. The first European MBR installations for Kubota and Zenon were respectively at Porlock, England in 1997 and at Perthes en Gatinais, France in 1999; both with design flows of about 2000 m^3/d. As of 2009, there were about 4400 MBR installations worldwide, with 85–90% supplied by Kubota, Zenon (now GE) and Mitsubishi Rayon (Judd, 2011).

One of the sustainability goals in wastewater treatment in the first decade of the 21st century was to become energy self-sufficient. Efforts were made to increase methane production for cogeneration and minimize electrical consumption. This revived the A/B process (Table 3.2) developed in Germany in the early 1970s at the Aachen University of Technology (Böhnke, 1977). Since early 1980 the A/B process has been commonly used in Germany. The 1st or A stage is designed for high-rate BOD removal to route more influent carbon to anaerobic digestion. The 2nd or B-stage system is designed for nitrification and denitrification at low DO. Because of the limited amount of BOD remaining after the 1st stage, research is currently being carried out to determine how to incorporate the Anammox process for N removal into the 2nd stage.

Research led by van Loosdrecht at Delft University of Technology has developed an AS process that produces dense granular sludge in a modified SBR operation (Table 3.2; de Kreuk *et al.*, 2007). A key to this process is microbial selection based on high upflow velocity during a short 'wastewater feed/effluent withdrawal' period. The dense granules provide a high biomass concentration that allows the reactor volume to be reduced. The granular AS settles in 5 min. to about the same extent that would take 30 min for conventional SBR AS. The process can achieve EBPR by using an anaerobic feed followed by aerobic operation. The dense granular AS allows for simultaneous nitrification/denitrification (SND) with nitrification in the outer area and denitrification in the inner region. The substrate removal in the granules is analogous to the removal mechanisms in a biofilm even though the process fits the definition of a suspended growth reactor.

SND using low DO or cyclic aeration processes has been used for many years in a number of the systems described above. Low DO SND is of particular interest today because of the aeration energy savings from the larger driving force for oxygen transfer. Oxidation ditches have a long history of excellent N removal; they have cyclically low DO concentrations caused by using fixed aerators that result in a decreasing DO concentration along the length of the channel following the aerator. The long HRT and SRT of these systems provide sufficient time for the growth of nitrifying bacteria at low DO concentrations. Results of recent studies of low DO operation in an MBR system by Giraldo *et al.* (2011) suggest that the effect of low DO on nitrification kinetics is much less than previously thought because of the selection of nitrifying organisms that can function at low DO. In SND AS nitrification to nitrite is desired because this results in the consumption of less carbon for denitrification compared to nitrate reduction. Low DO operation has not always been able to suppress nitrite-oxidizing bacteria. However recent observations (Wett *et al.*, 2013) suggest that cyclic DO operation from anoxic to aerobic conditions over short enough intervals can suppress the nitrite oxidizers.

3.8 THE FUTURE OF AS

Lessons from the last 100 years that may help define the expectations for the future are: (1) treatment needs and goals will continue to result in the use of many types of AS process, (2) changing treatment needs and goals will spur more innovation, (3) the type of AS process employed will vary with plant size and site conditions; perhaps current process developments taking place in the larger plants will be applicable in the future to smaller facilities, (4) the need for retrofitting existing facilities will affect innovation and process selection, and (5) the use of modern molecular biological tools will enable better process analysis and aid innovation.

Our short-term crystal ball tells us that there will be major interest in, and activity with, granular sludge reactors, low DO SDN systems, and EBPR processes with P recovery. We hope that our profession will also show serious interest in, and concern with, the removal of micropollutants and endocrine disruptor compounds in municipal plant discharges. These can be a serious threat to fish and other aquatic life. We also caution against becoming too single-minded about treatment technology or goals. For example, removing more carbon at the front of the WWTP for methane production will result in less heterotrophic growth in the AS process for the biodegradation of endocrine disruptor compounds and other micropollutants and less EBPR. We are going to be faced with competition for the influent BOD, which now appears to be a useful resource, in contrast to how it was viewed when AS began 100 years ago.

3.9 REFERENCES

Albertson O. E. (1991). Bulking sludge control – progress, practice and problems. *Water Sci. Technol.*, **23**(4–6), 835–846.

Alleman J. E. and Prakasam T. B. S. (1983). Reflections on seven decades of activated sludge history. *J. Water Pollut. Control Fed.*, **55**(5), 436–443.

Ardern E. and Lockett W. T. (1914). Experiments in the oxidation of sewage without the aid of filters. *J. Soc. Chem. Industr.*, **33**(10), 523–539.

Bailey-Denton J. (1882). Sewage Disposal: Ten Years Experience, with Notes on Sewage Farming. London (From Kinnicutt *et al.*, 1919).

Balakrishnan S. and Eckenfelder W. W. (1969). Nitrogen relationships in biological treatment processes-III. Denitrification in the modified activated sludge process. *Water Res.*, **3**(3), 177–188.

Barnard J. L. (1973). Biological denitrification. *Water Pollut. Control.*, **72**, 705–720.

Barnard J. L. (1974). Cut P and N without chemicals. *Water Wastes Engr.*, Part 1, **11**(7), 33–36; Part 2, **11**(8), 41–44.

Barnard J. L. (1976). US Patent 3,964,998, Improvements in and Relating to Waste Water Treatment, June 22.

Barnard J. L. (1998). The development of nutrient-removal processes. *Water Environ J.*, **12**, 330–337.

Barstow E. and Mohlman F. W. (1915). Purification of sewage by aeration in the presence of activated sludge, I and II. *J. Ind. Eng. Chem.*, **7**(4), 318–323.

Barstow E. (1917). Purification of sewage by aeration in the presence of activated sludge–III. *J. Ind. Eng. Chem.*, **9**(9), 845–850.

Black W. M. and Phelps E. B. (1911). Report location of sewer outlets and discharge of sewage into New York harbor, 64–78. (From Buswell, 1923).

Böhnke B. (1977) Das Adsorptions-Belebungsverfahren, *Korresp. Abwasser* **24**, 121–127. (From Makinia, 2010).

Budd W. E. and Lambeth G. F. (1957). High-purity oxygen in biological sewage treatment. *Sewage Industr. Wastes*, **29**(3), 237–257.

Budd W. E. and Okey R. W. (1969). US Patent 3,472,765, Membrane Separation in Biological Reactor Systems, October 14.

Burdick C. R., Refling D. R. and Stensel H. D. (1982). Advanced biological treatment to achieve nutrient removal. *J. Water Pollut. Control Fed.*, **48**(11), 2486–2510.

Buswell A. M. (1923). Activated Sludge Studies, 1920–1922; Bulletin No. 18, State of Illinois, Division of the State Water Survey, Springfield, IL, USA.

Buswell A. M. and Long H. L. (1923). Microbiology and theory of activated sludge. *J. Amer. Water Works Assoc.*, **10**(2), 309–321.

Carrio L., Street F., Mahoney K., deBarbadillo C., Anderson J., Abraham K. and Passerelli N. (2000). Practical considerations for design of a step-feed biological nutrient removal system. Proc. 73rd WEFTEC, Anaheim, CA.

Chapman T. D., Match L. C. and Zander E. H. (1976). Effect of high dissolved oxygen concentration in activated sludge systems. *J. Water Pollut. Control Fed.*, **48**(11), 2486–2510.

Chiesa S. C. and Irvine R. L. (1985). Growth and control of filamentous microbes in activated sludge; an integrated hypothesis. *Water Res.*, **19**(4), 471–479.

Chudoba J., Grau P. and Ottova V. (1973). Control of activated sludge filamentous bulking II. Selection of microorganisms by means of a selector. *Water Res.*, **7**(10), 1389–1406.

Clark H. W. and Gage S. DeM. (1912). Experiments upon the purification of sewage and water at the Lawrence Experiment Station, 44th Ann. Rept, State Board of Health of Mass., 275–367.

Cooper P. F. (2001). Historical aspects of wastewater treatment. In: Decentralised Sanitation and Reuse: Concepts, Systems and Implementation, P. Lens, G. Zeeman and G. Lettinga (eds), IWA Publishing, London, pp. 26–29.

Davidson A. B. (1957). US Patent 2,788,127. Apparatus for Treating Sewage, Apr. 9.

de Kreuk M. K., Kishida N and van Loosdrecht M. C. M. (2007). Aerobic granular sludge – state of the art. *Water Sci. Technol.*, **55**(8), 75–81.

Dibdin W. J. (1903). The Purification of Sewage and Water, 3rd edn., Sanitary Publishing Company, New York, D. Van Nostrand Company.

Downing A. L., Painter H. A. and Knowles G. (1964). Nitrification in the activated-sludge process. *J. Inst. Sewage Purif.,* **64**(2), 130–158.

Dupre A. (1884). On the changes in aeration of water as indicating the nature of the impurities present within it. Report of the Metropolitan Board of Works for 1884. London.

Eckenfelder W. W. Jr. (1966). Industrial Water Pollution Control. McGraw Hill Inc., New York.

Fowler G. J. and Mumford E. M. (1913). Preliminary note on the bacterial clarification of sewage, *Jour. Royal Sanit. Inst.,* **34**, 497–500.

Fowler G. J. and Mumford G. (1919). US Patent 1,294,080, Improved Fertilizer, February 11.

Frank L. C. (1915). Process of purifying sewage or other wastes and apparatus therefor. US Patent 1,139,024. May 11.

Giraldo E., Jjemba P., Liu P. and Muthukrishnan S. (2011). Ammonia oxidizing archaea, AOA, population and kinetic changes in a full scale simultaneous nitrogen and phosphorous removal MBR. Proc. 84th WEFTEC. Los Angeles, CA. October 15–19.

Gould R. H. (1942). Operating experiences in New York City. *Sewage Works J.,* **14**(1), 70–80.

Hatton T. C. (1926). Some of the tuning-up difficulties of the Milwaukee sewage plant. *Eng. News Record,* **96**(11), 446.

Heide B. A. and Pasveer A. (1973). Oxidation ditch: prevention and control of filamentous sludge. *Water (Netherlands),* **7**, 373–377.

Irvine R. L. and Davis W. B. (1971). Use of sequencing batch reactors for waste treatment. CPC International, Corpus Christi, TX. In: Proc. 26th Ann. Purdue Industr. Waste Conf., Purdue Univ., West Lafayette, IN, 450–462.

Irvine R. L. and Busch A. W. (1979). Sequencing batch biological reactors – an overview. *J. Water Pollut. Control Fed.,* **51**(2), 235–243.

Jenkins D. and Garrison W. E. (1968). Control of activated sludge by mean cell residence time. *J. Water Pollut. Control Fed.,* **40**(11), 1905–1919.

Jenkins D., Richard M. G. and Daigger G. T. (2004). Manual on the Causes and Control of Activated Sludge Bulking, Foaming, and Other Solids Separation Problems, 3rd edn., CRC Press Lewis Publishers, Boca Raton, FL, USA.

Judd S. (2011). The MBR Book, 2nd edn, Elsevier Ltd., UK.

Kalinske A. A. (1976). A comparison of air and oxygen activated sludge systems. *J. Water Pollut. Control Fed.,* **48** (11), 2472–2485.

Kappe S. E. (1938). Resume of operational experience of mechanical surface aerators. *Sewage Works J.,* **19**(6), 1007–1016.

Kessler L. H., Rohlich G. A. and Smart J. (1936). Tapered aeration of activated sludges. *Municipal Sanit.,* **7**, 268.

Kinnicutt L. P., Winslow C.-E. A. and Winthrop Pratt R. (1919). Sewage Disposal, 2nd edn, John Wiley & Sons, New York.

Kraus L. S. (1945). Dual aeration as a rugged activated sludge process. *Sewage Industr. Wastes,* **27**(12), 1347–1355.

Lackey J. B. and Dixon R. M. (1943). Some biological aspects of the Hays Process of sewage treatment. *Sewage Works J.,* **15**(6), 1139–1152.

Lawrence A. W. and McCarty P. L. (1970). Unified basis for biological treatment design and operation. *J. Sanit. Eng. Div., Amer. Soc. Civil Eng.,* **96**(3), 757–778.

Ludzack F. J. and Ettinger M. B. (1962). Controlling operation to minimize activated sludge effluent nitrogen. *J. Water Pollut. Control Fedn,* **34**(9), 920–931.

Makinia J. (2010). Mathematical Modelling and Computer Simulation of Activated Sludge Systems. IWA Publishing, London.

Martin A. J. (1927). The Activated Sludge Process. MacDonald and Evans, London.

Mason W. P., and Hine S. K. (1892). Note on the direct oxidation of organic matter in water. *J. Amer. Chem. Soc.*, **14**, 233–238.

McKinney R. E. (1957). Activity of microorganisms in organic waste disposal – II. Aerobic processes. *Appl. Microbiol.*, **5**, 167–174.

McKinney R. E. (1962). Mathematics of complete mixing activated sludge. *J. Sanit. Eng. Div. Amer. Soc. Civil Eng.*, **88**(SA3), 87–113.

Melling S. A. (1914). The purification of Salford sewage alongs the lines of the Manchester experiments. *J. Soc. Chem. Industr.,* **33**(23), 1124–1130.

Metcalf L. and Eddy H. P. (1922). Sewerage and Sewage Disposal, A Textbook. McGraw-Hill, New York.

Mills, H. F. (1890). Filtration of sewage and water and chemical precipitation of sewage. Special Report of the State Board of Health of Massachusetts upon the Purification of Sewage and Water. Part II, 1. (From Kinnicutt *et al.*, 1919).

Morper M. and Wildmoser A. (1990). Improvement of existing wastewater treatment plants efficiencies without enlargement of tankage by application of the LinPor-process-case-study. *Water Sci. Technol.,* **22**(7/8), 207–215.

Muller M. (1998). Implementing biofilm carriers into activated sludge process-15 years of experience. *Water Sci. Technol.,* **41**(4/5), 5–12.

Nicholls H. A. (1975). Full-scale experimentation on the new Johannesburg extended-aeration plants. *Water SA*, **1**(3), 121–132.

Parker D. S. and Merrill M. S. (1976) Oxygen and air activated sludge: another view. *J. Water Pollut. Control Fed.,* **48**(11), 2511–2582.

Pasveer A. (1959). A contribution to the development in activated sludge treatment. *J. Proc. Inst. Sewage Purif.,* **4**, 436.

Porter J. E. (1921). The Activated Sludge Process of Sewage Treatment: a bibliography of the subject. 2nd edn, General Filtration Co., Rochester, NY.

Randall C. W. and Sen D. (1996). Full-scale evaluation of an integrated fixed-film activated sludge (IFAS) process for enhanced nitrogen removal. *Water Sci. Technol.,* **33**(12), 155–162.

Rensink J. H. (1974). New approach to preventing bulking sludge. *J. Water Pollut. Control Fed.,* **46**(8), 1888–1901.

Rosso D. and Stenstrom M. K. (2005). Comparative economic analysis of the impacts of mean cell retention time and denitrification on aeration systems. *Water Res.*, **39**(16), 3773–3780.

Royal Commission on Sewage Disposal (1912). 8th Report, Volume 1. Standards and tests for sewage and sewage effluents discharging into rivers and streams. HMSO, London.

Rutt K., Seda J. and Johnson C. H. (2006). Two year case study of integrated fixed film activated sludge (IFAS) at Broomfield, CO, WWTP. Proc. 79th WEFTEC, Washington, DC, Oct. 22, 2006.

Sawyer C. N. and Bradney L. (1945). Rising of activated sludge in final settling tanks. *Sewage Works J.,* **17**(6), 1191–1209.

Sawyer C. N. (1965). Milestones in the development of the activated sludge process. *J. Water Pollut. Control Fedn,* **37**(2), 151–162.

Schloesing T. and Müntz A. (1877). Sur la nitrification par les ferments organisés. *Comptes Rendus hebdomadaires des séances de l'académie des sciences.* LXXXIV, 301, LXXXV, 1018.

Seeger H. (1999). The history of German wastewater treatment. *Europ. Water Management,* **2**(5), 51–56.

Setter L. R. and Edwards G. P. (1943). Modified sewage aeration: Part I. *Sewage Works J.,* **14**(4), 629–641.

Spector M. L. (1977). US Patent 4,056,465, Production of Non-Bulking Activated Sludge, Nov. 1.

Tchobanoglous G., Stensel H. D., Tsuchihashi R. and Burton F. (2014). Wastewater Engineering: Treatment and Resource Recovery, 4th edn, McGraw Hill, New York.

Tomlinson E. J. (1976). Bulking-a survey of activated sludge plants. Water Research Centre Tech. Rept TR35, Stevenage, UK.

Ulrich A. H. and Smith M. W. (1957). Operational experiences with activated sludge-biosorption at Austin, Texas. *Sewage Industr. Wastes,* **29**(4), 400–413.

Waring G. E. (1894). Modern Methods of Sewage Disposal. D. Van Nostrand Company, London.

Waring G. E. (1895). US Patent 535,516, Method of Deodorizing – Sewage Filters, March 12.

Water Environment Federation (1998). Design of Municipal Wastewater Treatment Plants, 4th edn., Manual of Practice No. 8, Vol. 2, Water Environment Federation, Alexandria, VA, USA.

Wett B., Omari A., Podmirseg S. M., Han M., Akintayo O., Gomez Brandon M., Murthy S., Bott C., Hell M., Takacs I., Nyhuis G. and O'Shaughnessy M. (2013). Going for mainstream deammonification from bench to full scale for maximized resource efficiency. *Water Sci. Technol.,* **68**(2), 283–289.

Wolman A. (1924). Sanitary engineering. *Amer. J. Publ. Hlth.,* **14**, 450–452.

Wuhrmann K. (1964). Nitrogen removal in sewage treatment processes. *Proc. Internatl Assoc. Theoret. Appl. Limnol.,* **15**, 580–596.

Yamamoto K., Hiasa M., Mahmood T. and Matsuo T. (1989). Direct solid liquid separation using hollow fiber membrane in an activated sludge aeration tank, *Water Sci. Technol.,* **21**(4–5), 43–54.

Chapter 4

Microbiology and microbial ecology of the activated sludge process

Per Halkjær Nielsen (Denmark), Katherine D. McMahon (USA)

4.1 INTRODUCTION

The Ardern and Lockett breakthrough in 1914 was simple yet ingenious. Other researchers and engineers working around the same time knew that pollutants in sewage would eventually oxidize upon exposure to air, but most were focusing on fixed-film processes. The approach of aerating with forced air to provide both mixing and oxygen transfer, combined with recycling of biomass, accelerated the oxidation process by orders of magnitude. These two hallmark features of the AS process transformed sewage treatment, as evidenced by its rapid adoption worldwide. It is important to note that Ardern and Lockett (and others working on sewage treatment technologies at around the same time) were originally focused on three metrics for assessing treatment process effectiveness: reduction of oxygen demand, ammonia oxidation (i.e. nitrification), and propensity for putrefaction. Other researchers working with fixed-film systems realized that algal growth on surfaces in contact with sewage improved oxidation, but not nearly to the extent observed in Ardern and Lockett's configuration. One of their mentors, G. J. Fowler, was acknowledged as a source of formative ideas leading to Ardern and Lockett's discovery (Ardern & Lockett, 1914). It is worth noting that Fowler and his graduate student Mrs E. Mumford had done extensive work with a pure culture of bacteria known as 'M7' (Fowler & Mumford, 1913). Although Fowler and Mumford were clever enough to consider the controlled use of microorganisms to treat sewage, they were hampered by the need to continuously inoculate the sewage with the pure culture. Thus, although their work inspired Ardern and Lockett, the notion of recycling sludge (i.e. a naturally selected mixed culture of microorganisms) was truly the breakthrough. It is also worth noting that Ardern and Lockett suspected the importance of microorganisms in their AS, but acknowledged that they could not confirm the relative contributions of physical, chemical, and biological processes at the time of their first report.

Ever since Ardern and Lockett's fundamental configuration was adopted throughout the developed world in the early- to mid-1900s, work by both researchers and practitioners focused on two main challenges: (1) control of foaming and bulking, and (2) optimization of N and P removal. Both of these have strong microbiological components, and thus researchers especially have worked hard to identify and characterize the filaments responsible for foam and poor settling, nitrifiers, denitrifiers, and PAOs.

The aim of this chapter is to provide historical perspective on how our understanding of AS microbiology has developed over the past 100 years. We discuss three different eras defined mainly by the methodological tools for studying microbes available during that time (Figure 4.1). Microbial ecology has always been a methods-driven discipline and its application to AS has been no exception.

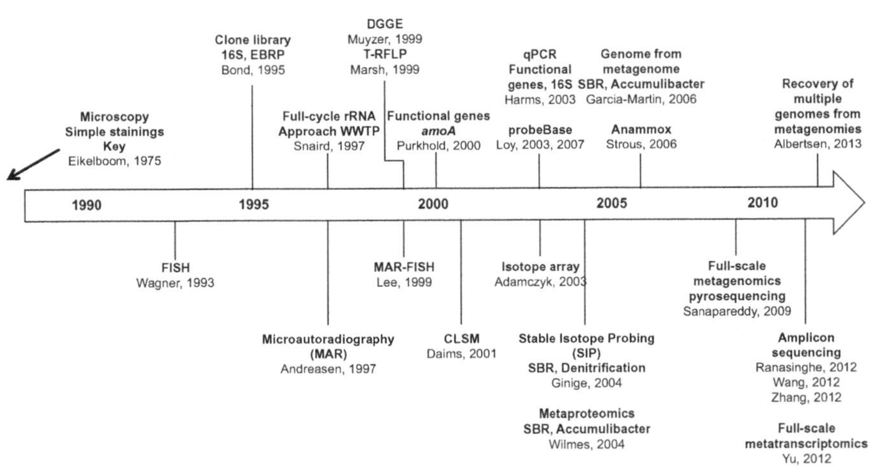

Figure 4.1 Timeline of methods development for detection of identity and function of activated sludge microorganisms.

4.2 WHICH BACTERIA ARE PRESENT? – CULTURING AND LIGHT MICROSCOPY

Soon after the first report from Ardern and Lockett, researchers confirmed that microorganisms in AS were critical to the treatment processes. However, little effort was made initially to identify these microbes and design of treatment systems was highly empirical, as was typical of engineering activities at that time. During this era, researchers were limited to two primary techniques: (1) isolation, cultivation and subsequent taxonomic investigations based on morphological and biochemical characteristics, and (2) light microscopy.

In 1923 *Zoogloea* was already considered important for floc structure (Buswell & Long, 1923) and pure cultures were isolated and studied extensively (Butterfield, 1935) but very few other species were identified as important floc-formers until molecular tools became available. In the 1970s a great deal of attention was paid to the physical structure of flocs and its relationship to effective settling (Li *et al.*, 1986). Implicit throughout this work was the importance of certain floc-forming species. During this period, researchers determined that filamentous bacteria could also be important for developing and maintaining floc structure together with different floc-formers.

Some researchers and practitioners noticed that certain organisms could cause operational problems such as poor settling and foaming. The identity of these problem-causing bacteria was not well established until the late 1960s, and it was assumed the filamentous bacteria were assumed to be primarily *Sphaerotilus,* based on work conducted in the 1920s (Eikelboom, 1975). In the early 1970s, several studies were carried out to isolate and identify the problem-causing filamentous organisms but is was not until Eikelboom in 1975 made an extensive survey of 1100 AS samples by light microscopy that it was accepted that filamentous bacteria are a normal and diverse part of the populations in most treatment plants. He constructed an identification key to classify different kinds of filaments (Figure 4.2) and a number of pure cultures were isolated. Although some of the isolates were given a genus and species name by Eikelboom, many were identified only based on an alpha-numeric code (e.g. Type 021N). Legend states that the codes correspond to page numbers in Eikelboom's laboratory notebooks. Their codes persist in the literature to this day and Eikelboom's work is widely recognized as groundbreaking with respect to our fundamental understanding of AS microbiology, as well as practical tools for diagnosing the causes of bulking sludge. Updated manuals by Eikelboom (1983, 2000, 2006) and Jenkins *et al.* (1984, 1993, 2003) are still very valuable tools today.

In the late 1980s and during the 1990s the groups of Linda Blackall and Robert Seviour in Australia and Valter Tandoi in Italy were able to isolate several bacteria by micromanipulation and grow these in pure culture. Isolates of filamentous *Gordonia, Microthrix parvicella* and several others gave a novel understanding of the physiology of some key bacteria in AS (e.g. Blackall *et al.*, 1996, 1997). Pure culture studies with isolates also formed the basis for the formulation of hypotheses about growth and control of filamentous bacteria in AS based on differences in substrate uptake rate and substrate affinity of filamentous organisms and floc-formers (Chiesa & Irvine, 1985; Jenkins, 1992). From these an understanding was developed to design selectors based on kinetic and metabolic selection criteria (Wanner, 1994) (Chapter 10). These principles are fundamental for most control methods today although with refinements or adjustments.

Ammonia oxidation was one of the original goals for early researchers searching for ways to quickly and cheaply treat sewage. The toxicity of ammonia and nitrite to fish and other aquatic life was widely recognized in the days of Ardern and Lockett. Chemolithotrophic nitrification by soil bacteria was first described by Winogradsky (1890), who first recognized that oxidation to nitrate was a 2-step process carried out by two distinct groups of organisms (i.e. ammonia oxidizers

Identification key 1 of filamentous organisms in activated sludge

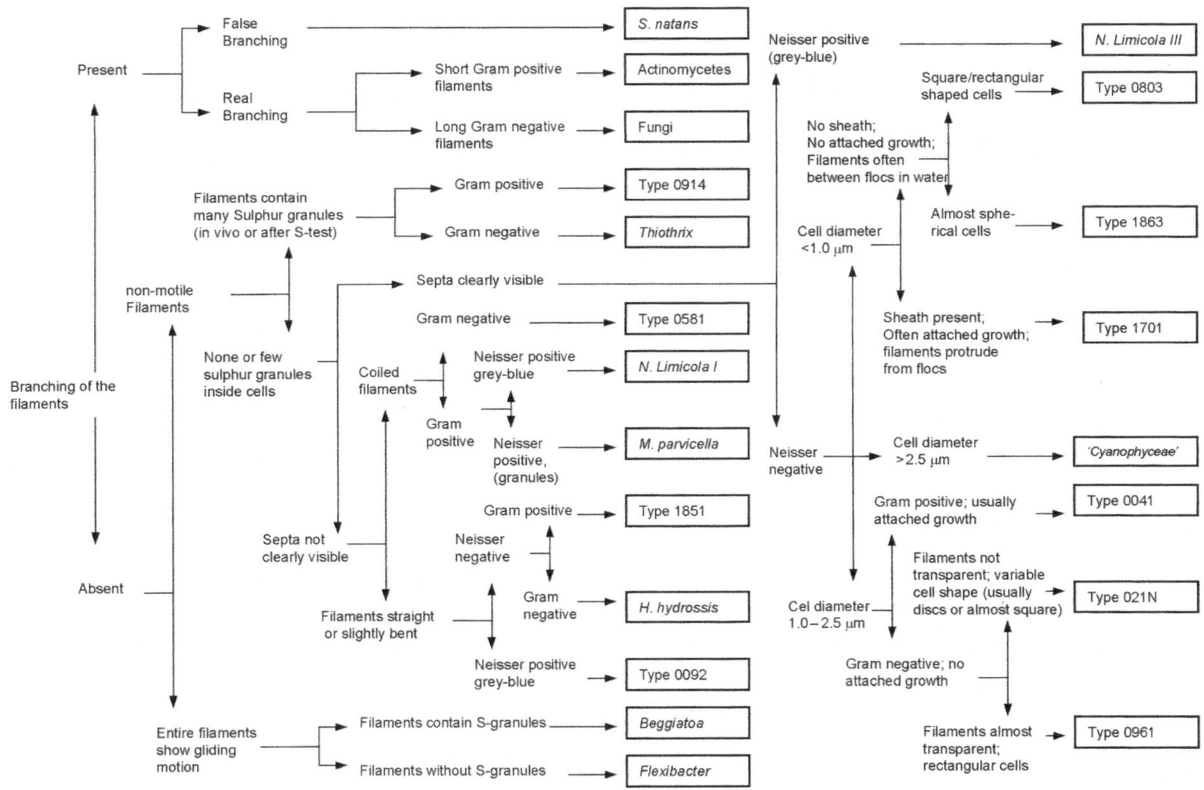

Figure 4.2 Example of identification key (from Eikelboom's manuals, 1983 and 2000).

and nitrite oxidizers). When AS researchers applied cultivation-based methods to recover these two groups from AS, they enriched and isolated members of the genus *Nitrosomonas* (ammonia oxidizer) and the genus *Nitrobacter* (nitrite oxidizer). For many years it was believed that these were the most important and abundant nitrifiers in AS. However, in 1996 it was shown by FISH that *Nitrobacter* (a member of the Alphaproteobacteria class) is not present in significant amounts in AS plants (Wagner *et al.*, 1996). Instead, the majority of nitrite oxidation activity was attributed to members of the genus *Nitrospira* (a member of the phylum *Nitrospirae*) in conventional nitrifying AS plants (Juretschko *et al.*, 1998).

Removal of nitrate by denitrification in AS plants became common during the 1980s and 1990s with new requirements for N-removal to protect sensitive recipient water bodies. It was known that a large number of aerobic bacteria could use nitrate as an alternative electron acceptor and produce nitrite, N_2O or N_2 under anoxic conditions, but details about their identity and physiology were scarce. Many isolates from AS and soil were able to denitrify and it was assumed that they were responsible for the process. Examples are bacteria in the genera *Alcaligenes, Bacillus, Hyphomicrobium, Methylobacterium, Paracoccus, Pseudomonas* and *Zoogloea* (Knowles, 1982; Zumft, 1992). However, it was well recognized that the diversity of denitrifiers was large and poorly described. We know today that the diversity among denitrifiers is large and that only some of the isolates mentioned here are important in full-scale AS plants.

The EBPR process is another textbook case on how culture-based methods can lead researchers astray. Upon its discovery in the late 1950s, many researchers were convinced that the process depended on primarily abiotic precipitation reactions (Jenkins & Menar, 1968). However, addition of a respiratory inhibitor to EBPR sludge prevented P-uptake and the biological nature of the process was confirmed (Levin & Shapiro, 1965). Culture-based approaches combined with light microscopy led early EBPR researchers to believe that members of the genus *Acinetobacter* (a member of the Gammaproteobacteria) were responsible for polyP accumulation in AS (Fuhs & Chen, 1975; Jenkins & Tandoi, 1991; McMahon *et al.*, 2010). As was the case for nitrite oxidizers, molecular techniques showed in the 1990s that *Acinetobacter* was irrelevant to EBPR and eventually two major unrelated groups were confirmed to be important polyP accumulators in AS: the betaproteobacterial candidate genus *Accumulibacter* (Bond *et al.*, 1995; Hesselmann *et al.*, 1999; Crocetti *et al.*, 2000; Zilles *et al.*, 2002) and the actinobacterial genus *Tetrasphaera* (Maszenan *et al.*, 2000; Kong *et al.*, 2005).

4.3 IDENTITY AND FUNCTION REVEALED BY THE MOLECULAR TOOLS – FROM THE EARLY 1990S

Fundamental questions about the identity of important microorganisms in AS plants could finally be answered, at least partly, upon the development of molecular methods targeting nucleic acids. Fluorescence *in situ* hybridization (FISH) uses

DNA oligonucleotides and fluorochromes to probe the abundant rRNA molecules in active cells and thereby allowing visualization of whole microbial cells using epifluorescent or confocal laser scanning microscopy (Figure 4.3). Based on the work of Pace and colleagues (DeLong *et al.*, 1989), Amann *et al.* (1990) optimized the method, making it possible to identify, visualize and quantify bacteria without any isolation procedure. This method completely revolutionized the understanding of the microbiology of AS (and other microbial ecosystems). It was possible to differentiate among different phylogenetic groups of microorganisms in the AS by microscopy.

Studies by Wagner, Amann and colleagues (Wagner *et al.*, 1993, 1994, 1996; Manz *et al.*, 1994; Schuppler *et al.*, 1995) developed and used FISH in full-scale plants to visualize for the first time some of the abundant filamentous bacteria, nitrifiers, and putative PAOs. FISH was an integral part of the so-called 'full-cycle rRNA approach' that was first used in a full-scale AS plant (Snaidr *et al.*, 1997). DNA was extracted from the AS, then 16S genes were amplified and cloned into an *E. coli* host. These clone libraries of 50–100 clones were sequenced and, after phylogenetic analyses, provided a new level of resolution of the community composition (Juretschko *et al.*, 2002). The sequences were also used for gene probe design and with the application of the confocal laser microscope, the 3-D architecture of the flocs and biofilms could be visualized and later on used for quantification of the populations using specialized software (Daims *et al.*, 2001b, 2006). Information about available FISH probes became accessible in the public probeBase (Loy *et al.*, 2003, 2007). Semi-quantitative analyses of the populations were also carried out with fingerprinting techniques after DNA extraction and PCR such as denaturing gradient gel electrophoresis (DGGE) (Muyzer, 1999) and terminal restriction fragment length polymorphism (T-RFLP) (Marsh, 1999).

With a better knowledge of the identity of the various microorganisms in AS, researchers moved on to the more relevant questions surrounding their activity and function. What are the distinct niches of different microorganisms in AS and how do these relate to process performance? How does the composition of the influent wastewater control the growth of specific types of bacteria? It was assumed that pure-culture isolates, where available, represented the activity of species *in situ*. However, the application of microautoradiography (MAR) and radiolabeled substrates showed that this was not always the case and many species were rather specialized in their substrate uptake (Andreasen & Nielsen, 1997). Combined with FISH (FISH–MAR) the approach provided information about the physiology of many probe-defined organisms (Lee *et al.*, 1999; Wagner *et al.*, 2006). The subsequent findings were supported with other methods designed to query function, such as isotope arrays (Wagner *et al.*, 2006; Hesselsoe *et al.*, 2009), stable isotope probing, SIP (Ginige *et al.*, 2004a; Huang *et al.*, 2007), expression of functional genes (Dionisi *et al.*, 2002; Harms *et al.*, 2003; Aoi *et al.*, 2004; Burow *et al.*, 2008; He & McMahon, 2011) and other single cell techniques (Figure 4.4).

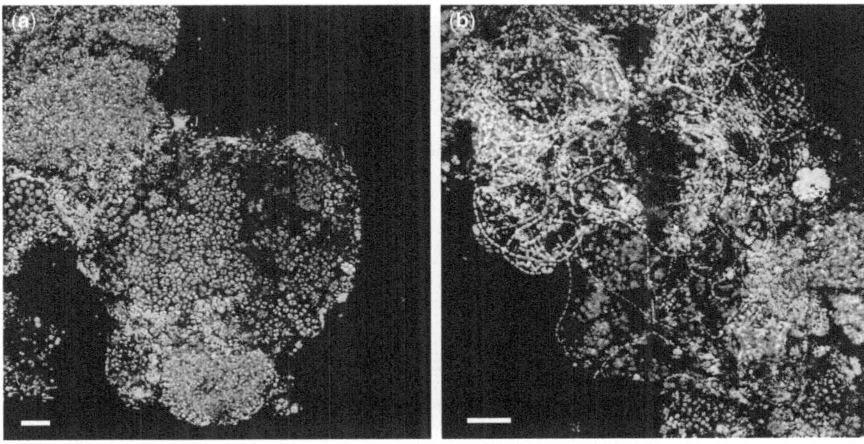

Figure 4.3 Fluorescence *in situ* hybridization. FISH micrograph overlays of AS showing: a. *Defluviicoccus* GAO as yellow (probe DF2mix (Cy3 = red) + EUBmix (FLUOS = green)); b. Filamentous *Kouleothrix* (Eikelboom type 1851) in yellow (probe CHL1851 (Cy3 = red) + EUBmix (FLUOS = green)). In both images *Candidatus* Accumulibacter PAO are light blue (PAOmix (Cy5 = blue) + EUBmix (FLUOS = green)) and cells not targeted with specific probes appear green (EUBmix). Scale bar = 10 μm. (Photo Simon McIlroy.)

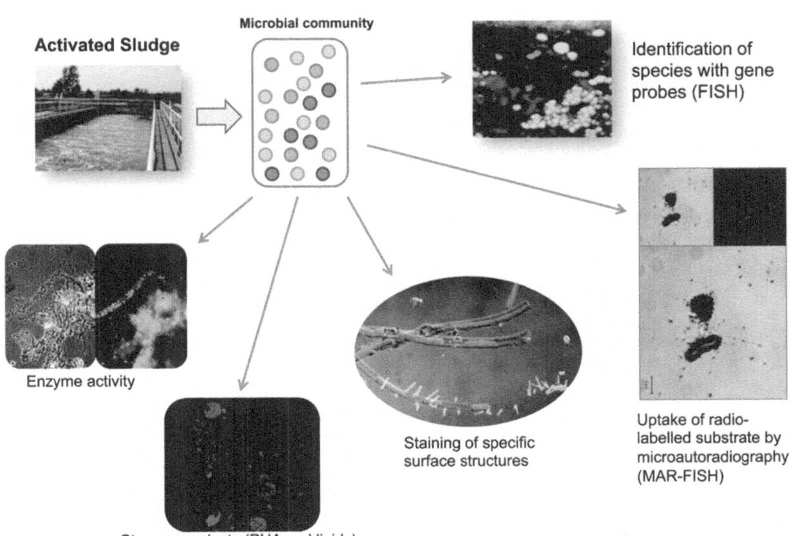

Figure 4.4 Single cell techniques for detection of bacterial ecophysiology under *in situ* conditions.

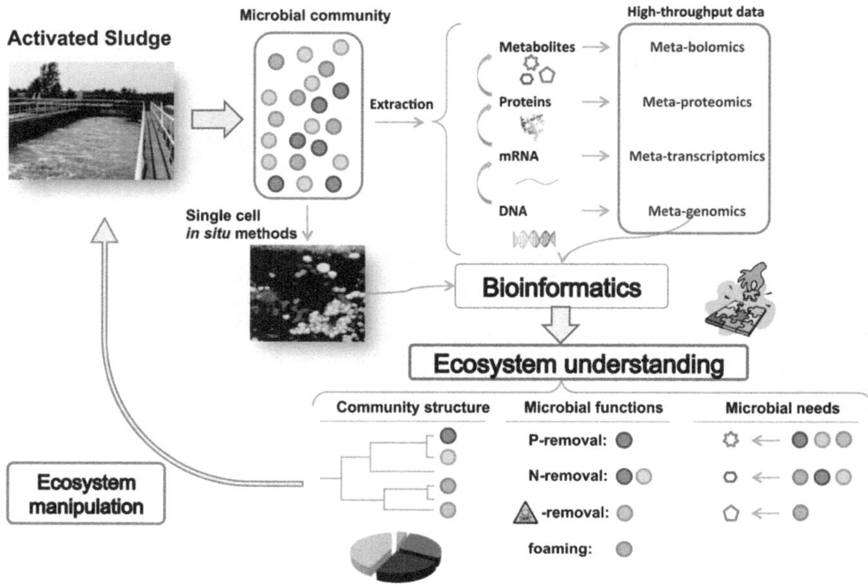

Figure 4.5 Systems microbiology. An overview of the approach to study AS structure and function by integrated 'omics methods, single cell microbiology and bioinformatics.

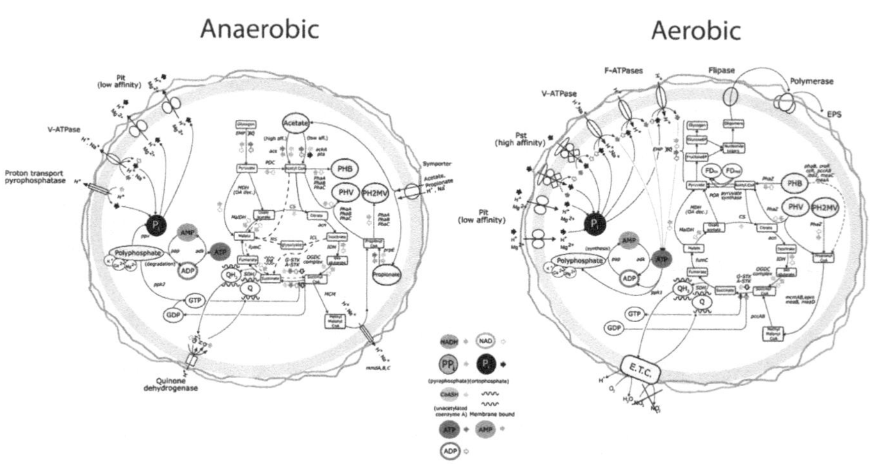

Figure 4.6 Metabolic model of *Candidatus* Accumulibacter.

The identities of nearly all important AS filamentous microorganisms were revealed during this period. Many of Eikelboom's morphotypes were identified by molecular methods and could be detected with rather specific FISH probes such as *Candidatus* Microthrix (Erhart *et al.*, 1997). FISH probes to most filamentous bacteria are now available (Nielsen *et al.*, 2009a, b) however, some morphotypes cover several species. For example, each of the morphotypes 1701, 0041, and 0675 seem to cover several species in the phyla TM7 and Chloroflexi and can be difficult to identify properly without FISH-probes. Together with functional studies (e.g. FISH–MAR) a good understanding is obtained about their key physiologies in terms of substrate utilization and electron acceptor preferences. This knowledge has together with many years of experience with laboratory, pilot and full-scale studies (Eikelboom, 2006; Jenkins *et al.*, 2003; Tandoi *et al.*, 2005) translated into control strategies for most filamentous bacteria, all summarized in several chapters of Seviour and Nielsen (2010). We are now in a situation where most bulking problems can be controlled if the available information about control measures is applied (which it is not always!) – a success story in the microbiology of wastewater treatment.

Effective sludge–liquid separation in AS plants does not just depend on the presence of filamentous bacteria but also on other floc properties. In the late 1980s and beginning of the 1990s floc structure was investigated in great detail by microscopy and physico-chemical methods to describe size, shape, fractal dimension, density, strength, extracellular polymeric substances (EPS) and other properties and components (Li & Ganczarczyk, 1990; Jorand *et al.*, 1995; Frolund *et al.*, 1996). These are important for AS properties such as clarification and settling, drainage and dewatering. Until then AS was often considered 'strange' and 'unique from plant to plant' but it was now realized that flocs were fairly similar among AS plants and were largely composed of the same constituents, just in different ratios, reflecting, often minor, differences in microbial composition and chemistry. With FISH it became possible to identify 'strong' and 'weak' floc-formers, for example, nitrifiers and some Alphaproteobacteria, respectively (Klausen *et al.*, 2004; Larsen *et al.*, 2008b) and it could be shown that they produced specific EPS components, for example, proteinaceous amyloids (Larsen *et al.*, 2008a). Studies also showed that the presence of divalent and trivalent cations were critical for a good floc structure (Novak *et al.*, 1998). Lack of these cations or changes in their concentrations, such as removal of Fe(III) due to FeS precipitation by sulfide under anaerobic conditions, can cause serious deflocculation (Nielsen & Keiding, 1998), which has a major impact on solid/liquid separation, particularly effluent quality and dewaterability.

The N cycle has been investigated in great detail with culture-independent tools. Members of the genus *Nitrosomonas* and *Nitrosospira* are common ammonia oxidizers in full-scale AS plants and two members of the genus *Nitrospira* are common nitrite oxidizers (Juretschko *et al.*, 1998; Purkhold *et al.*, 2000). Interestingly, *Nitrospira* is also able to consume organic substrate but the contribution to its growth is not yet known (Daims *et al.*, 2001a). It has been debated whether ammonium oxidizing archaea could be important contributors to ammonium oxidation in full-scale AS plants (Limpiyakorn *et al.*, 2013), however, the archaea do not seem to be widely distributed (Park *et al.*, 2006; Kayee *et al.*, 2011).

Molecular methods have revealed a number of candidate denitrifiers in AS plants with N-removal, typically by analyzing for genes of nitrite reductases, *nirS* and *nirK*. However, it was difficult to measure *in situ* whether they were capable of denitrification and only with MAR–FISH and SIP, was it possible to identify the abundant and important denitrifiers (Juretschko *et al.*, 2002; Thomsen *et al.*, 2004; Ginige *et al.*, 2004b; Osaka *et al.*, 2006; Thomsen *et al.*, 2007; Morgan-Sagastume *et al.*, 2008). These primarily belong to the genera *Acidovorax, Azoarcus/Thauera* complex, *Curvibacter, Dechloromonas, Rhodobacter, Zoogloea* and some others. In addition, many PAOs and GAOs can denitrify, including some, but not all *Candidatus* Accumulibacter, *Tetrasphaera* and *Candidatus* Competibacter (Kong *et al.*, 2004, 2005; Burow *et al.*, 2007). Thus while it seems that the majority of heterotrophic microbes in N-removing AS plants can denitrify, there is an interesting exception. Very few filamentous bacteria can denitrify fully although some reduce nitrate to nitrite, for example *Candidatus* Microthrix (Hesselsoe *et al.*, 2005; Nielsen *et al.*, 2009b). Potential denitrifying candidates are some Chloroflexi species (e.g. type 0803) (Kragelund *et al.*, 2011) and perhaps some Alphaproteobacteria (Nielsen *et al.*, 2009b). The N cycle has received special attention since 2000 since the AS plants are potential sources for N_2O production. Many full-scale studies have been carried out but the relative importance of nitrifiers and denitrifiers in N_2O production is still not resolved (Kampschreur *et al.*, 2009; Desloover *et al.*, 2012).

Early FISH-based studies of PAOs in full-scale treatment plants showed that Betaproteobacteria and Actinobacteria were likely responsible for P removal, and not the gammaproteobacterial *Acinetobacter* (Wagner *et al.*, 1994). Over the next ten years, 16S rRNA-gene-based evidence mounted that a close relative of the betaproteobacterial genus *Rhodocyclus* was responsible for P-removal in acetate-fed laboratory scale systems (Bond *et al.*, 1995; Hesselmann *et al.*, 1999; Crocetti *et al.*, 2000; McMahon *et al.*, 2002) and in full-scale systems (Zilles *et al.*, 2002; Kong *et al.*, 2004). Similar approaches showed that polyP accumulation is carried out largely by the actinobacterial *Tetrasphaera* (Kong *et al.*, 2005). Full-cycle FISH analysis was also applied extensively to the study of GAOs because of their perceived importance as a competitor for PAOs (Crocetti *et al.*, 2002; Burow *et al.*, 2007), though later work suggests that they do not typically affect EBPR performance (Gu *et al.*, 2008).

4.4 THE MODERN TOOLS – THE NGS ERA – SINCE EARLY 2000

The development of next generation sequencing (NGS) has completely changed our capability to sequence DNA and RNA in unlimited amounts. Microbial community composition can be investigated by amplicon sequencing (e.g. 16S rRNA gene sequencing, sometime called 'tag sequencing'), which identifies the different species based on a molecular barcode. These sequencing technologies are based on 'sequencing by synthesis' and those that have been used include Roche 454 Life Sciences pyrosequencing and Illumina (Genome Analyzer, HiSeq or MiSeq). This approach provides a rapid and relatively precise identification and quantification of the composition of entire microbial communities in AS plants. The amplicon sequencing method can also be applied to any gene of interest that can be amplified from a community (e.g. *amoA, nirK, ppk*).

Several studies have used NGS for amplicon sequencing of AS communities (e.g. Ranasinghe *et al.*, 2012; Wang *et al.*, 2012; Zhang *et al.*, 2012; Ibarbalz *et al.*, 2013, Ye & Zhang, 2013). The 16S rRNA genes and the relative distributions of particular phylogenetic groups within the total pool of sequences recovered using these techniques was generally similar to that obtained using traditional Sanger-based sequencing of nearly full-length 16S rRNA genes, with some differences reported in certain cases (Kwon *et al.*, 2010). Similar findings have been reported for ammonia oxidizing bacteria and archaea (Kim *et al.*, 2011; Zhang *et al.*, 2011). However, most of these studies have been limited to purely descriptive cataloging of community richness and a coarse overview of which phyla and major groups were recovered. More sophisticated exploration of long-term temporal patterns and general community assembly principles originally attempted using 16S rRNA fingerprinting techniques (Wells *et al.*, 2009; Ofiteru *et al.*, 2010) have only recently been attempted using NGS (Kim *et al.*, 2013). Many of these studies, however, suffer from the lack of curated taxonomic databases for AS communities. A new public database, MiDAS, 'A field guide to AS microbes' is attempting to establish a joint taxonomic and functional information resource for all researchers in wastewater microbiology (www.midasfieldguide.com).

NGS has also opened the door to sequencing entire community genomes (i.e. 'metagenomes') from AS (Sanapareddy *et al.*, 2009; Albertsen *et al.*, 2012, 2013a, b). Metagenomics provides comprehensive information about the functional potential of community members by studying all their genes (National Research Council, 2007; Gilbert & Dupont, 2011).

The main obstacle to using metagenomics to study the physiology of microorganisms directly in AS plants (e.g. expressed genes and proteins) is that the reference databases needed to interpret sequences recovered from AS still are insufficient (Albertsen *et al.*, 2012, 2013b). Most of the microorganisms present in wastewater treatment systems are uncultured and not well represented in public databases, although the databases now have more than 3000 finished pure-culture genomes. The solution is to get more genomes from the wastewater treatment systems, but this is a slow process and tedious work. New methods are, however, accelerating this process so it is possible to obtain genomes from individual species from the metagenomes by advanced bioinformatic tools, even if present in lower abundances (Albertsen *et al.*, 2013a; McIlroy *et al.*, 2013a). With reference genomes in hand, the metabolic potential of all the important microbial populations could be established and metabolic models of all key members of the communities developed. A robust collection of such reference genomes will support studies to confirm and refine the metabolic models and add metabolic information to the genomic blueprint, using (meta)transcriptomics and (meta)proteomics. Metatranscriptomics can be conducted by reverse-transcribing RNA extracted from AS samples and proceeding with the same methods used for metagenomics. Metaproteomics is more challenging since it requires contemporary metagenomic analysis for every sample, which is currently often prohibitively expensive.

Only a few NGS-fueled meta'omic studies have been carried out in wastewater systems so far, since they require specific expertise and are subject to many methodological pitfalls. However the preliminary results are promising (Yu & Zhang, 2012; Albertsen *et al.*, 2013a; Flowers *et al.*, 2013). Integration of all these methods is known as 'Systems Microbiology' (Figure 4.5) and is anticipated to revolutionize the study of microbial communities in the coming years. Examples of such investigations are bacteria involved in nitrification, denitrification and P removal along with some filamentous bacteria (Table 4.1). Thus it will be possible to better link community composition with function and overall plant performance.

Table 4.1 Genomes of important activated sludge microorganisms.

Species	Function	Reference
Candidatus Kuenenia	Anammox process	Strous *et al.* (2006)
Nitrosomonas sp	Ammonium oxidizer	Suwa *et al.* (2011)
Candidatus Nitrospira	Nitrite oxidizer	Lucker *et al.* (2010)
Haliscomenobacter hydrossis	Filamentous	Daligault *et al.* (2011)
Candidatus Microthrix parvicella	Filamentous lipid consumer	McIlroy *et al.* (2013b)
Candidatus Accumulibacter	PAO, denitrifier	Garcia Martin *et al.* (2006), Flowers *et al.* (2013)
Tetrasphaera spp.	PAO, denitrifier, fermenter	Kristiansen *et al.* (2013)
Candidatus Competibacter	GAO, denitrifier, fermenter	McIlroy *et al.* (2013a)

The first application of metagenomic sequencing to AS was designed to target *Candidatus* Accumulibacter and was conducted at the US Department of Energy Joint Genome Institute (Garcia Martin *et al.*, 2006). AS from laboratory scale SBRs performing EBPR with acetate as a primary carbon source in Madison, WI, USA, and Brisbane, Queensland, Australia, were shotgun sequenced using traditional Sanger sequencing techniques. The genome of *Candidatus* Accumulibacter phosphatis str. UW-1 (member of clade IIA) was later completed with the aid of NGS approaches and standard genome finishing techniques. A draft genome of a member of Accumulibacter clade IA was then obtained using a combination of traditional Sanger, 454 pyrosequencing, and Illumina HiSeq technologies (Flowers *et al.*, 2013). The original draft strain UW-1 (clade IIA) genomes were used for reconstruction of the proposed EBPR metabolic model (Figure 4.6) through bioinformatic identification of genes necessary for the archetypal carbon and P transformations originally proposed based on AS enrichments only (Comeau *et al.*, 1986; Wentzel *et al.*, 1986; Mino *et al.*, 1987). The most surprising key findings included the absence of genes involved in the Entner Doudoroff (ED) pathway for glycolysis, a potentially novel route for re-oxidation of reduced quinones using a novel cytochrome, pathways enabling CO_2- and N_2-fixation, and absence of a respiratory nitrate reductase (*nar*) that ought to be required for energy-yielding denitrification (Garcia Martin *et al.*, 2006). This initial metagenomic effort spawned several studies of Accumulibacter gene expression based on mRNA (Burow *et al.*, 2008; He *et al.*, 2010; He & McMahon, 2011) and protein (Wilmes & Bond, 2004; Wilmes *et al.*, 2008; Wexler *et al.*, 2009). A metagenomic analysis of phage later recovered from the Madison, Wisconsin, USA bioreactor revealed an exciting potential mechanism for interaction between Accumulibacter and its viral predators, based on heat-stable nucleoid structuring protein (Skennerton *et al.*, 2011). This study was only possible because of the previously published metagenome.

Another example: two complete genomes from *Candidatus* Competibacter were obtained from laboratory-scale enrichment reactors through metagenomics (McIlroy *et al.*, 2013a). Phylogenetic analysis identified the two genomes as sub-groups 1 and 5, respectively. Both have genes for glycogen and PHA cycling and for the metabolism of VFAs. Marked differences were found in their potential for the Embden–Meyerhof–Parnas (EMP) and ED glycolytic pathways, denitrification, N_2 fixation, fermentation, and utilization of glucose and lactate. It was surprising that one of the Competibacter species was able to consume and ferment glucose, and this species was also the most abundant GAO in a full-scale EBPR plant. This observation cast doubt on the standing notion that competition between Accumulibacter and Competibacter is primarily driven by acetate and propionate (Lopez-Vazquez *et al.*, 2009; Oehmen *et al.*, 2010). Furthermore, genetic comparison of P metabolism pathways in Competibacter and sequenced PAOs revealed the absence of the Pit phosphate transporter in the Competibacter-lineage genomes – identifying a key metabolic difference between the two groups. This example show that information obtained by lab-scale reactors may be too simplistic to represent full-scale plants and that NGS can provide much more comprehensive and reliable understanding than the simpler molecular methods.

4.5 COMPREHENSIVE ECOSYSTEM MODEL – WHERE ARE WE TODAY?

The recent application of molecular tools has shown that AS ecosystems in similar treatment processes seem to have a common core community of 'species' shared among the different plants. A recent example is the microbial communities in 25 Danish treatment plants performing EBPR, where most of the bacterial types were present in all plants despite significant differences in plant design, operation and wastewater type (Nielsen *et al.*, 2010). This shows that we only need to deal with a limited number of core microbes for a certain type of treatment process, almost independently of the variations in plant design and operation. For the EBPR plant, approximately 50 genera were above 0.5% abundance. By various culture-independent techniques, many details of their metabolism have been carried out and a conceptual ecosystem model constructed (Nielsen *et al.*, 2010, 2012). It includes microbe identity, abundances, substrate preferences and other ecophysiological data, which are integrated into entire food webs that map the nutrient transformations of the treatment process. Recent studies of AS community compositions from around the world support a universal core population of AS microbes carrying out similar processes (Yang *et al.*, 2011; Zhang *et al.*, 2012). As for the NGS of amplicons, many of these studies are difficult to compare in detail due to methodological differences and lack of curated taxonomic databases for AS communities. The aforementioned public MiDAS database aims to solve this problem.

Stability and resilience of the populations are important for plant function but this is still poorly described. However, the 25 EBPR Danish plants had significant population stability over a 3–4 year period (Mielczarek *et al.*, 2013); other studies show little population stability although the functional stability may be higher (Ofiteru *et al.*, 2010). This may depend on factors such as SRT, temperature and plant design/operation. However, these studies were primarily carried out with quantitative FISH and as some gene probes have a poor resolution and specificity, future amplicon sequencing analyses will better reveal the diversity and may also reveal new, hitherto overlooked core species. Several studies also show the presence of a certain microdiversity in the plants, that is several closely related strains of same species/genus and they may contribute to a functional stability in the plants (Albertsen *et al.*, 2012, 2013a; McIlroy *et al.*, 2013a, b).

The first steps have been taken to form general concepts and theories for management of microbial communities, for example by 'Microbial resource management' (Curtis *et al.*, 2003; McMahon *et al.*, 2007; Verstraete *et al.*, 2007). However, we still need more comprehensive studies of AS communities by the novel 'systems microbiology' approaches to obtain more general concepts and theories for management of microbial communities.

4.6 THE FUTURE

The AS process is applied widely around the world and is extremely successful. In most cases the configuration of such plants will probably not change substantially in the near future, but in many places, we will see a change from treatment to energy production, reuse and recovery of nutrients – the biorefinery concept. This is challenging, as more microbiological processes should be integrated in one large plant: improved pre-settling, biological N and P removal, methane production, N-removal by anammox, P-recovery, and so on. Also, AS will not stand alone but granules and carrier materials will complement and membranes may substitute for clarifiers. If this should be successful, a better management of the microbial communities is mandatory and that can be done by applying systems microbiology approaches. We expect that in a very few years it will be possible to completely deconstruct and mathematically re-assemble wastewater treatment microbial communities using the omics approaches described above combined with systems biology and traditional AS modeling. This should allow for scaling from the level of individual genes/genomes up to whole communities and ecosystem-level processes in a way that no one could have ever imagined in 1914.

4.7 REFERENCES

Albertsen M., Hansen L. B. S., Saunders A. M., Nielsen P. H. and Nielsen K. L. (2012). A metagenome of full-scale microbial community carrying out enhanced biological phosphorus removal. *ISME J.*, **6**, 1094–1106.

Albertsen M., Hugenholtz P., Skarshewski A., Nielsen K. L., Tyson G. W. and Nielsen P. H. (2013a). Genome sequences of rare, uncultured bacteria obtained by differential coverage binning of multiple metagenomes. *Nature Biotechnol.*, **31**, 533–528.

Albertsen M., Saunders A. M., Nielsen K. L. and Nielsen P. H. (2013b). Metagenomes obtained by 'dee sequencing'–what do they tell about the EBPR communities. *Water Sci. Technol.*, **68,** 1959–1968.

Amann R. I., Binder B. J., Olson R. J., Chisholm S. W., Devereux R. and Stahl D. A. (1990). Combination of 16S rRNA-targeted oligonucleotide probes with flow cytometry for analyzing mixed microbial populations. *Appl. Environ. Microbiol.*, **56**, 1919–1925.

Andreasen K. and Nielsen P. H. (1997). Application of microautoradiography to the study of substrate uptake by filamentous microorganisms in activated sludge. *Appl. Environ. Microbiol.*, **63**, 3662–3668.

Arden E. and Lockett W. T. (1914). Experiments on the oxidation of sewage without the aid of filters. *J. Soc. Chem. Industr.*, **33**, 523–539.

Aoi Y., Masaki Y., Tsuneda S. and Hirata, A. (2004). Quantitative analysis of amoA mRNA expression as a new biomarker of ammonia oxidation acivities in a complex microbial community. *Letters Appl. Microbiol.*, **39**, 477–482.

Blackall L. L., Stratton H., Bradford D., DelDot T., Sjorup C., Seviour E. M. and Seviour R. J. (1996). 'Candidatus Microthrix parvicella', a filamentous bacterium from activated sludge sewage treatment plants. *Internatl J. Syst. Bacteriol.*, **46**, 344–346.

Blackall L. L., Rossetti S., Christensson C., Cunningham M., Hartman P., Hugenholtz P. and Tandoi V. (1997). The characterization and description of representatives of 'G' bacteria from activated sludge plants. *Lett. Appl. Microbiol.*, **25**, 63–69.

Bond P. L., Hugenholtz P., Keller J. and Blackall L. L. (1995). Bacterial community structures of phosphate-removing and non-phosphate removing activated sludges from sequencing batch reactors. *Appl. Environ. Microbiol.*, **61**, 1910–1916.

Burow L. C., Kong Y., Nielsen J. L., Blackall L. L. and Nielsen P. H. (2007). Abundance and ecophysiology of *Defluviicoccus* spp., glycogen-accumulating organisms in full-scale wastewater treatmen processes. *Microbiol.-SGM*, **153**, 178–185.

Burow L. C., Mabbett A. N. and Blackall L. L. (2008). Anaerobic glyoxylate cycle activity durin simultaneous utilization of glycogen and acetate in uncultured Accumulibacter enriched in enhanced biological phosphorus removal communities. *ISME J.*, **2**, 1040–1051.

Buswell A. and Long H. (1923). Microbiology and theory of activated sludge. *J. Amer. Water Works Assoc.*, **10**, 309–321.

Butterfield C. T. (1935). A Zoogloea forming bacterium isolated from activated sludge. *Publ. Hlth Repts.*, **50**, 671–684.

Chiesa S. C. and Irvine R. L. (1985). Growth and control of filamentous microrobes in activated sludge: an integrated hypothesis. *Water Res.*, **19**(4), 471–479.

Comeau Y., Hall K. J., Hancock R. E. W. and Oldham W. K. (1986). Biochemical model for enhanced biological phosphorus removal. *Water Res.*, **20**, 1511–1521.

Crocetti G. R., Hugenholtz P., Bond P. L., Schuler A., Keller J., Jenkins D. and Blackall L. L. (2000). Identification of polyphosphate accumulating organisms and the design of 16S rRNA-directed probes for their detection and quantitation. *Appl. Environ. Microbiol.*, **66**, 1175–1182.

Crocetti G. R., Banfield J. F., Keller J., Bond P. L. and Blackall L. L. (2002). Glycogen-accumulating organisms in laboratory-scale and full-scale wastewater treatment processes. *Microbiol.*, **148**, 3353–3364.

Curtis T. P., Head I. M. and Graham D. W. (2003). Theoretical ecology for engineering biology *Environ. Sci. Technol.*, **37**, 64a–70a.

Daims H., Nielsen J. L., Nielsen P. H., Schleifer K.-H. and Wagner M. (2001a). *In situ* characterization of *Nitrospira*-like nitrite-oxidizing bacteria active in wastewater treatment plants. *Appl. Environ. Microbiol.*, **67**, 5273–5284.

Daims H., Ramsing N. B., Schleifer K. H. and Wagner M. (2001b). Cultivation-independent, semiautomatic determination of absolute bacterial cell numbers in environmental samples by fluorescence *in situ* hybridization. *Appl. Environ. Microbiol.*, **67**, 5810–5818.

Daims H., Lucker S. and Wagner M. (2006). daime, a novel image analysis program for microbial ecology and biofilm research. *Environ. Microbiol.*, **8**, 200–213.

Daligault H., Lapidus A., Zeytun A., Nolan M., Lucas S., *et al.* (2011). Complete genome sequence of *Haliscomenobacter hydrossis type strain (O-T) Standards in Genomic Sci.*, **4**, 352–360.

DeLong E. F., Wickham G. S. and Pace N. R. (1989). Phylogenetic stains: ribosomal RNA-based probes for the identificatin of single cells. *Sci.*, **243**, 1360–1363.

Desloover J., Vlaeminck S. E., Clauwaert P., Verstraete W. and Boon N. (2012). Strategies to mitigat N_2O emissions from biological nitrogen removal systems. *Current Opinion Biotechnol.*, **23**, 474–482.

Dionisi H. M., Layton A. C., Harms G., Gregory I. R., Robinson K. G. and Sayler G. S. (2002). Quantification of *Nitrosomonas oligotropha*-like ammonia-oxidizing bacteria and *Nitrospira* spp. from full-scale wastewater treatment plants by competitive PCR. *Appl. Environ. Microbiol.*,**68**, 245–253.

Eikelboom D. (1975). Filamentous organisms observed in activated sludge. *Water Res.*, **9**, 365–388.

Eikelboom D. H. (2000). Process control of activated sludge plants by microscopic investigation. IWA Publishing, London, England.

Eikelboom D. (2006). Identification and Control of Filamentous Micro-organisms Observed in Industrial Waste Water Treatment Plants. IWA Publishing, London, UK.

Eikelboom D. and van Buijsen H. J. J. (1983). Microscopic Sludge Investigation Manual. TNO Research Inst. Environ. Hyg., 2nd edn, Delft.

Erhart R., Bradford D., Seviour R. J., Amann R., and Blackall L. L. (1997). Development and use of fluorescent *in situ* hybridization probes for the detection and identification of 'Microthrix parvicella' in activated sludge. *Sys. Appl. Microbiol.,* **20**, 310–318.

Flowers J. J., He S., Malfatti S., Glavina Del Rio T., Tringe S. G., Hugenholtz P. and McMahon K. D. (2013). Comparative genomics of two 'Candidatus Accumulibacter phosphatis' clades performing biological phosphorus removal. *ISME J.,* In press.

Fowler G. and Mumford E. (1913) Preliminary note on the bacterial clarification of sewage. *J. Royal Sanitary Inst.,* **34**, 497–510.

Frolund B., Palmgren R., Keiding K. and Nielsen P. H. (1996). Extraction of extracellular polymers from activated sludge using a cation exchange resin. *Water Res.,* **30**, 1749–1758.

Fuhs G. W. and Chen M. (1975). Microbiological basis of phosphate removal in the activated sludge process for treatment of wastewater. *Microbial Ecol.,* **2**, 119–138.

Garcia Martin H., Ivanova N., Kunin V., Warnecke F., Barry K., McHardy A. C., Yeates C., He S., Salamov A., Szeto E., Dalin E., Putnam N., Rigoutsos I., Kyrpides N., Blackall L. L., McMahon K. D. and Hugenholtz P. (2006). Metagenomic analysis of phosphorus removing sludge communities. *Nature Biotechnol.,* **24**, 1263–1269.

Gilbert J. A. and Dupont C. L. (2011). Microbial Metagenomics: Beyond the Genome. In, *Ann. Rev. Marine Sci.* **3**, 347–371.

Ginige M. P., Hugenholtz P., Daims H., Wagner M., Keller J. and Blackall L. L. (2004a). Use of stable-isotope probing, full-cycle rRNA analysis, and fluorescence *in situ* hybridization-microautoradiography to study a methanol-fed denitrifying microbial community. *Appl. Environ. Microbiol.,* **70**, 588–596.

Ginige M. P., Hugenholtz P., Daims H., Wagner M., Keller J. and Blackall L. L. (2004b). Use of stable-isotope probing, full-cycle rRNA analysis, and fluorescence *in situ* hybridization-microautoradiography to study a methanol-fed denitrifying microbial community. *Applied and Environ. Microbiol.,* **70**, 588–596.

Gu A. Z., Saunders A., Neethling J. B., Stensel H. D. and Blackall L. L. (2008). Functionally relevant microorganisms to enhanced biological phosphorus removal performance at full-scale wastewater treatment plants in the United States. *Water Environ. Res.,* **80**, 688–698.

Harms G., Layton A. C., Dionisi H. M., Gregory I. R., Garrett V. M., Hawkins S. A., Robinson K. G. and Sayler G. S. (2003). Real-time PCR quantification of nitrifying bacteria in a municipal wastewater treatment plant. *Environ. Sci. Technol.,* **37**, 343–351.

He S., Kunin V., Haynes M., Garcia Martin H., Ivanova N., Kyrpides N., Rohwer F., Hugenholtz P. and McMahon K. D. (2010). Metatranscriptomic analysis of 'Candidatus Accumulibacter'-enriched enhanced biological phosphorus removal sludge. *Environ. Microbiol.,* **12**, 1205–1217.

He S. and McMahon K. D. (2011) 'Candidatus Accumulibacter' gene expression in response to dynamic EBPR conditions. *ISME J.,* **5**, 329–340.

Hesselmann R. P. X., Werlen C., Hahn D., van der Meer J. R. and Zehnder A. J. B. (1999). Enrichment, phylogenetic analysis and detection of a bacterium that performs enhanced biological phosphate removal in activated sludge. *Syst. Appl. Microbiol.,* **22**, 454–465.

Hesselsoe M., Nielsen J. L., Roslev P. and Nielsen P. H. (2005). Isotope labeling and microautoradiography of active heterotrophic bacteria on the basis of assimilation of (CO_2)-C^{14}. *Appl. Environ. Microbiol.,* **71**, 646–655.

Hesselsoe M., Fureder S., Schloter M., Bodrossy L., Iversen N., Roslev P., Nielsen P. H., Wagner M. and Loy A. (2009). Isotope array analysis of Rhodocyclales uncovers functional redundancy and versatility in an activated sludge. *ISME J.*, **3**, 1349–1364.

Huang W. E., Stoecker K., Griffiths R., Newbold L., Daims H., Whiteley A. S. and Wagner M. (2007). Raman-FISH: combining stable-isotope Raman spectroscopy and fluorescence *in situ* hybridization for the single cell analysis of identity and function. *Environ.l Microbiol.*, **9**, 1878–1889.

Ibarbalz F. M., Figuerola E. L. M. and Erijman L. (2013). Industrial activated sludge exhibit unique bacterial community composition at high taxonomic ranks. *Water Res.*, **47**, 3854–3864.

Jenkins D. (1992). Towards a comprehensive model of activated sludge bulking and foaming. *Water Sci. Technol.*, **25**, 215–230.

Jenkins D. and Menar A. B. (1968). Pathways of phosphorus in biological treatment processes. *Water Res.*, **2**, 97–99.

Jenkins D. and Tandoi V. (1991). The applied microbiology of enhanced biological phosphate removal-accomplishments and needs. *Water Res.*, **25**, 1471–1478.

Jenkins D., Richard M. G. and Daigger G. T. (1984). Manual on the Causes and Control of Activated Sludge Bulking and Foaming, 1st edn., Water Research Commission, Pretoria, South Africa.

Jenkins D., Richard M. G. and Daigger G. T. (1993). Manual on the Causes and Control of Activated Sludge Bulking and Foaming, 2nd edn., CRC Press, Lewis Publishers, Boca Raton, FL, USA.

Jenkins D., Richard M. G., and Daigger, G. T. (2003). Manual on the Causes and Control of Activated Sludge Bulking, Foaming and other Solids Separation Problems, 3rd edn., CRC Press, Lewis Publishers, Boca Raton, FL, USA.

Jorand F., Zartarian F., Thomas F., Block J. C., Bottero J. Y., Villemin G., Urbain V. and Manem, J. (1995). Chemical and structural (2d) linkage between bacteria within activated-sludge flocs. *Water Res.*, **29**, 1639–1647.

Juretschko S., Timmermann G., Schmid M., Schleifer K.-H., Pommerening-Roeser A., Koops H.-P. and Wagner M. (1998). Combined molecular and conventional analyses of nitrifying bacterium diversity in activated sludge: *Nitrosococcus mobilis* and *Nitrospira*-like bacteria as dominant populations. *Appl. Environ. Microbiol.*, **64**, 3042–3051.

Juretschko S., Loy A., Lehner A. and Wagner M. (2002). The microbial community composition of a nitrifying-denitrifying activated sludge from an industrial sewage treatment plant analyzed by the full-cycle rRNA approach. *Syst. Appl. Microbiol.*, **25**, 84–99.

Kampschreur M. J., Temmink H., Kleerebezem R., Jetten M. S. M. and van Loosdrecht, M. C. M. (2009). Nitrous oxide emission during wastewater treatment. *Water Res.*, **43**, 4093–4103.

Kayee P., Sonthiphand P., Rongsayamanont C. and Limpiyakorn T. (2011). Archaeal amoa genes outnumber bacterial amoa genes in municipal wastewater treatment plants in Bangkok. *Microbial Ecol.*, **62**, 776–788.

Kim T.-S., Kim H.-S., Kwon S. and Park H.-D. (2011). Nitrifying bacterial community structure of a full-scale integrated fixed-film activated sludge process as investigated by pyrosequencing. *J. Microbiol. Biotechnol.*, **21**, 293–298.

Kim T.-S., Jeong J.-Y., Wells G. F. and Park H.-D. (2013). General and rare bacterial taxa demonstrating different temporal dynamic patterns in an activated sludge bioreactor. *Appl. Microbiol. Biotechnol.*, **97**, 1755–1765.

Klausen M. M., Thomsen T. R., Nielsen J. L., Mikkelsen L. H. and Nielsen P. H. (2004). Variations in micro-colony strength of probe-defined bacteria in activated sludge flocs. *FEMS Microbiol. Ecol.*, **50**, 123–132.

Knowles P. (1982). Denitrification. *Microbiol. Rev.*, **146**, 43–70.

Kong Y. H., Nielsen J. L. and Nielsen P. H. (2004). Microautoradiographic study of *Rhodocyclus*-related polyphosphate accumulating bacteria in full-scale enhanced biological phosphorus removal plants. *Appl. Environ. Microbiol.*, **70**, 5383–5390.

Kong Y. H., Nielsen J. L. and Nielsen P. H. (2005). Identity and ecophysiology of uncultured actinobacterial polyphosphate-accumulating organisms in full-scale enhanced biological phosphorus removal plants. *Appl. Environ. Microbiol.*, **71**, 4076–4085.

Kragelund C., Thomsen T. R., Mielczarek A. T. and Nielsen P. H. (2011). Eikelboom's morphotype 0803 in activated sludge belongs to the genus *Caldilinea* in the phylum Chloroflexi. *FEMS Microbiol.Ecol.*, **76**, 451–462.

Kristiansen R., Nguyen H. T. T., Saunders A. M., Nielsen J. L., Wimmer R., Le V. Q., McIlroy S. J., Petrovski S., Seviour R. J., Calteau A., Nielsen K. L. and Nielsen P. H. (2013). A metabolic model for members of the genus *Tetrasphaera* involved in enhanced biological phosphorus removal. *ISME J.*, **7**, 543–554.

Kwon S., Kim T.-S., Yu G. H., Jung J.-H. and Park H.-D. (2010). Bacterial community composition and diversity of a full-scale integrated fixed-film activated sludge system as investigated by pyrosequencing. *J. Microbiol. Biotechnol.*, **20**, 1717–1723.

Larsen P., Nielsen J. L., Otzen D. and Nielsen P. H. (2008a). Amyloid-like adhesins produced by floc-forming and filamentous bacteria in activated sludge. *Appl. Environ. Microbiol.*, **74**, 1517–1526.

Larsen P., Nielsen J. L., Svendsen T. C. and Nielsen, P. H. (2008b). Adhesion characteristics of nitrifying bacteria in activated sludge. *Water Res.*, **42**, 2814–2826.

Lee N., Nielsen P. H., Andreasen K. H., Juretschko S., Nielsen J. L., Schleifer K.-H. and Wagner M. (1999). Combination of fluorescent *in situ* hybridization and microautoradiography – a new tool for structure-function analyses in microbial ecology. *Appl. Environ. Microbiol.*, **65**, 1289–1297.

Levin G. V. and Shapiro J. (1965). Metabolic uptake of phosphorus by wastewater organisms. *J. Water Pollut. Control Fedn.*, **37**, 800–821.

Li D. H. and Ganczarczyk J. J. (1990). Structure of activated sludge flocs. *Biotechnol. Bioeng.*, **35**, 57–65.

Li D. H., Ganczarczyk J. J. and Jenkins D. (1986). Physical characteristics of activated-sludge flocs. *CRC Critical Revs Environ. Control*, **17**, 53–87.

Limpiyakorn T., Fuerhacker M., Haberl R., Chodanon T., Srithep P. and Sonthiphand P. (2013). amoA-encoding archaea in wastewater treatment plants: a review. *Appl. Microbiol. Biotechnol.*, **97**, 1425–1439.

Lopez-Vazquez C. M., Oehmen A., Hooijmans C. M., Brdjanovic D., Gijzen H. J., Yuan Z. and van Loosdrecht M. C. M. (2009). Modeling the PAO-GAO competition: effects of carbon source, pH and temperature. *Water Res.*, **43**, 450–462.

Loy A., Horn M. and Wagner M. (2003). probeBase: an online resource for rRNA-targeted oligonucleotide probes. *Nucleic Acids Res.*, 31, 514–516.

Loy A., Maixner F., Wagner M. and Horn M. (2007). probeBase—an online resource for rRNA-targeted oligonucleotide probes: new features. *Nucleic Acids Res.,* **35**, D800-D804.

Lucker S., Wagner M., Maixner F., Pelletier E., Koch H., Vacherie B., Rattei T., Damste J. S. S., Spieck E., Le Paslier D. and Daims H. (2010). A Nitrospira metagenome illuminates the physiology and evolution of globally important nitrite-oxidizing bacteria. *Proc. Natl Acad. Sci. USA,* **107**, 13479–13484.

Manz W., Wagner M., Amann R. and Schleifer K.-H. (1994). *In situ* characterization of the microbial consortia active in two wastewater treatment plants. *Water Res.,* **28**, 1715–1723.

Marsh T. L. (1999). Terminal restriction fragment length polymorphism (T-RFLP): an emerging method for characterising diversity among homologous populations of amplification products. *Current Opinion Microbiol.,* **2,** 323–327.

Maszenan, A. M., Seviour R. J., Patel B. K. C., Schumann P., Burghardt J., Tokiwa Y. and Stratton H. M. (2000). Three isolates of novel polyphosphate-accumulating Gram-positive cocci, obtained from activated sludge, belong to a new genus, Tetrasphaera gen. nov., and description of two new species, *Tetrasphaera japonica* sp, nov and *Tetrasphaera australiensis* sp nov. *Internl J. Syst. Evolut. Microbiol.,* **50**, 593–603.

McIlroy S. J., Albertsen M., Andresen E. K., Saunders A. M., Kristiansen R., Stokholm-Bjerregaard M., Nielsen K. L. and Nielsen P. H. (2013a). 'Candidatus Competibacter'-lineage genomes retrieved from metagenomes reveal functional metabolic diversity. *ISME J.* In press.

McIlroy S. J., Kristiansen R., Albertsen M., Karst S. M., Rossetti S., Nielsen J. L., Tandoi V., Seviour R. J. and Nielsen P. H. (2013b). Metabolic model for the filamentous 'Candidatus Microthrix parvicella' based on genomic and metagenomic analyses. *ISME J.,* **7**, 1161–1172.

McMahon K. D., Dojka M. A., Pace N. R., Jenkins D. and Keasling J. D. (2002). Polyphosphate kinase from activated sludge performing enhanced biological phosphorus removal. *Appl. Environ. Microbiol.,* **68**, 4971–4978.

McMahon K. D., Garcia Martin H. and Hugenholtz P. (2007). Integrating ecology into environmental biotechnology. *Current Opinion Biotechnol.,* **18**, 287–292.

McMahon K. D., He S. and Oehmen A. (2010). The microbiology of phosphorus removal. In: Microbial Ecology of Activated Sludge, R. J. Seviour and P. H. Nielsen (eds), IWA Publishing, London, 281–320.

Mielczarek A. T., Nguyen H. T. T., Nielsen J. L. and Nielsen P. H. (2013). Population dynamics of bacteria involved in enhanced biological phosphorus removal in Danish wastewater treatment plants. *Water Res.,* **47**, 1529–1544.

Mino T., Arun V., Yoshiaki T. and Matsuo T. (1987). Effect of phosphorus accumulation on acetate metabolism in the biological phosphorus removal process. In: Advances in Water Pollution Control – Biological Phosphate Removal from Wastewaters, R. Ramadori (ed.), Pergamon Press, Oxford, UK, p. 27.

Morgan-Sagastume F., Nielsen J. L. and Nielsen P. H. (2008). Substrate-dependent denitrification of abundant probe-defined denitrifying bacteria in activated sludge. *FEMS Microbiol. Ecol.,* **66**, 447–461.

Muyzer G. (1999). DGGE/TGGE a method for identifying genes from natural ecosystems. *Current Opinion Microbiol.,* **2**, 317–322.

National Research Council (2007). The New Science of Metagenomics: Revealing the Secrets of Our Microbial Planet. National Academies Press, Washington, DC.

Nielsen P. H. and Keiding K. (1998). Disintegration of activated sludge flocs in presence of sulfide. *Water Research,* **32**, 313–320.

Nielsen P. H., Lemmer H. and Daims H. (2009a). FISH Handbook of Wastewater Treatment. IWA Publishing, London, UK.

Nielsen P. H., Kragelund C., Seviour R. J. and Nielsen J. L. (2009b). Identity and ecophysiology of filamentous bacteria in activated sludge. *FEMS Microbiol. Revs.,* **33**, 969–998.

Nielsen P. H., Mielczarek A. T., Kragelund C., Nielsen J. L., Saunders A. M., Kong Y., Hansen A. A. and Vollertsen J. (2010). A conceptual ecosystem model of microbial communities in enhanced biological phosphorus removal plants. *Water Res.,* **44**, 5070–5088.

Nielsen P. H., Saunders A. M., Hansen A. A., Larsen P. and Nielsen J. L. (2012). Microbial communities involved in enhanced biological phosphorus removal from wastewater – a model system in environmental biotechnology. *Current Opinion Biotechnol.,* **23,** 452–459

Novak J. T., Love N. G., Smith M. L. and Wheeler E. R. (1998). The effect of cationic salt addition on the settling and dewatering properties of an industrial activated sludge. *Water Environ. Res.,* **70**, 984–996.

Oehmen A., Lopez-Vazquez C. M., Carvalho G., Reis M. A. M. and van Loosdrecht M. C. M. (2010). Modelling the population dynamics and metabolic diversity of organisms relevant in anaerobic/anoxic/aerobic enhanced biological phosphorus removal processes. *Water Res.,* **44**, 4473–4486.

Ofiteru I. D., Lunn M., Curtis T. P., Wells G. F., Criddle C. S., Francis C. A. and Sloan W. T. (2010). Combined niche and neutral effects in a microbial wastewater treatment community. *Proc. Natl Acad. Sci. USA,* **107**, 15345–15350.

Osaka T., Yoshie S., Tsuneda S., Hirata A., Iwami N. and Inamori Y. (2006). Identification of acetate- or methanol-assimilating bacteria under nitrate-reducing conditions by stable-isotope probing. *Microbial Ecol.,* **52**, 253–266.

Park H. D., Wells G. F., Bae H., Criddle C. S. and Francis C. A. (2006). Occurrence of ammonia-oxidizing archaea in wastewater treatment plant bioreactors. *Appl. Environ. Microbiol.,* **72**, 5643–5647.

Purkhold U., Pommerening-Röser A., Juretschko S., Schmid M. C., Koops H. P. and Wagner M. (2000). Phylogeny of all recognized species of ammonia oxidizers based on comparative 16S rRNA and amoA sequence analysis: implications for molecular diversity surveys. *Appl. Environ. Microbiol.,* **66**, 5368–6382.

Ranasinghe P. D., Satoh H., Oshiki M., Oshima K., Suda W., Hattori M. and Mino, T. (2012). Revealing microbial community structures in large- and small-scale activated sludge systems by barcoded pyrosequencing of 16S rRNA gene. *Water Sci. Technol.,* **66**, 2155–2161.

Sanapareddy N., Hamp T. J., Gonzalez L. C., Hilger H. A., Fodor A. A. and Clinton S. M. (2009) Molecular diversity of a North Carolina wastewater treatment plant as revealed by pyrosequencing. *Appl. Environ. Microbiol.*, **75**, 1688–1696.

Schuppler M., Mertens F., Schoen G. and Goebel U. B. (1995). Molecular characterization of nocardioform actinomycetes in activated sludge by 16S rRNA analysis. *Microbiol.,* **141**, 513–521.

Seviour R. J. and Nielsen P. H. (2010). Microbial Ecology of Activated Sludge. IWA Publishing, London, UK.

Skennerton C. T., Angly F. E., Breitbart M., Bragg J., He S., McMahon K., Hugenholtz P. and Tyson G. W. (2011). Phage encoded H-NS: a potential Achilles heel in the bacterial defence system. *PLoS One* **6**, e20095.

Snaidr J., Amann R., Huber I., Ludwig W. and Schleifer K.-H. (1997). Phylogenetic analysis and in situ identification of bacteria in activated sludge. *Appl. Environ. Microbiol.,* **63**, 2884–2896.

Strous M., Pelletier E., Mangenot S., Rattei T., Lehner A., Taylor M. W., *et al.* (2006). Deciphering the evolution and metabolism of an anammox bacterium from a community genome. *Nature*, **440**, 790–794.

Suwa Y., Norton J. M., Bollmann A., Klotz M. G., Stein L. Y., Laanbroek H. J., Arp D. J., Goodwin L. A., Chertkov O., Held B., Bruce D., Detter J. C., Tapia R. and Han C. S. (2011). Genome sequence of *Nitrosomonas* sp strain al212, an ammonia-oxidizing bacterium sensitive to high levels of ammonia. *J. Bacteriol.,* **193**, 5047–5048.

Tandoi V., Jenkins D. and Wanner J. (2005). Activated Sludge Separation Problems: Theory, Control Measures, Practical Experiences. IWA Publishing, London, UK.

Thomsen T. R., Nielsen J. L., Ramsing N. B. and Nielsen P. H. (2004). Micromanipulation and further identification of FISH-labelled microcolonies of a dominant denitrifying bacterium in activated sludge. *Environ. Microbiol.,* **6**, 470–479.

Thomsen T. R., Kong Y. and Nielsen P. H. (2007). Ecophysiology of abundant denitrifying bacteria in activated sludge. *FEMS Microbiol. Ecol.,* **60**, 370–382.

Verstraete W., Wittelbolle L., Heylen K., Vanparys B., de Vos, P. van de Wiele T. and Boon, N. (2007). Microbial resource management: the road to go for environmental biotechnology. *Eng. Life Sci.* **7**, 117–126.

Wagner M., Amann R., Lemmer H. and Schleifer K.-H. (1993). Probing activated sludge with oligonucleotides specific for Proteobacteria: inadequacy of culture-dependent methods for describing microbial community structure. *Appl. Environ. Microbiol.,* **59**, 1520–1525.

Wagner M., Erhart R., Manz W., Amann R., Lemmer H., Wedi D. and Schleifer K.-H. (1994). Development of an rRNA-targeted oligonucleotide probe specific for the genus Acinetobacter and its application for *in situ* monitoring in activated sludge. *Appl. Environ. Microbiol.,* **60**, 792–800.

Wagner M., Rath G., Koops H. P., Flood J. and Amann R. (1996). *In situ* analysis of nitrifying bacteria in sewage treatment plants. *Water Sci. Technol.,* **34**, 237–244.

Wagner M., Nielsen P. H., Loy A., Nielsen J. L. and Daims H. (2006). Linking microbial community structure with function: fluorescence *in situ* hybridization-microautoradiography and isotope arrays. *Current Opinion Biotechnol.,* **17**, 83–91.

Wang X., Hu M., Xia Y., Wen X. and Ding K. (2012). Pyrosequencing analysis of bacterial diversity in 14 wastewater treatment systems in China. *Appl. Environ. Microbiol.,* **78**, 7042–7047.

Wanner J. (1994). Activated Sludge: Bulking and Foaming Control. CRC Press.

Wells G. F., Park H. D., Yeung C. H., Eggleston B., Francis C. A. and Criddle C. S. (2009) Ammonia- oxidizing communities in a highly aerated full-scale activated sludge bioreactor: betaproteobacterial dynamics and low relative abundance of Crenarchaea. *Environ. Microbiol.,* **11**, 2310–2328.

Wentzel M. C., Lotter L. H., Loewenthal R. E. and Marais G. v. R. (1986). Metabolic behavior of *Acinetobacter* spp. in enhanced biological phosphorus removal – a biochemical model. *Water SA*, **12**, 209.

Wexler M., Richardson D. J. and Bond P. L. (2009). Radiolabelled proteomics to determine differential functioning of *Accumulibacter* during the anaerobic and aerobic phases of a bioreactor operating for enhanced biological phosphorus removal. *Environ. Microbiol.*, **11**, 3029–3044.

Wilmes P., Andersson A. F., Lefsrud M. G., Wexler M., Shah M., Zhang B., Hettich R. L., Bond P. L., Verberkmoes N. C. and Banfield J. F. (2008). Community proteogenomics highlights microbial strain-variant protein expression within activated sludge performing enhanced biological phosphorus removal. *ISME J.*, **2**, 853–864.

Wilmes P. and Bond P. L. (2004). The application of two-dimensional polyacrylamide gel electrophoresis and downstream analyses to a mixed community of prokaryotic microorganisms. *Environ. Microbiol.*, **6**, 911–920.

Winogradsky S. (1890). On the nitrifying organisms. *Annales de l'Insitut Pasteur*, **4**, 257–275.

Yang C., Zhang W., Liu R. H., Li Q., Li B. B., Wang S. F., Song C. J., Qiao C. L. and Mulchandani A. (2011). Phylogenetic diversity and metabolic potential of activated sludge microbial communities in full- scale wastewater treatment plants. *Environ. Sci. Technol.*, **45**, 7408–7415.

Ye L. and Zhang T. (2013). Bacterial communities in different sections of a municipal wastewater treatment plant revealed by 16S rDNA 454 pyrosequencing. *Appl. Microbiol. Biotechnol.*, **97**, 2681–2690.

Yu K. and Zhang T. (2012). Metagenomic and metatranscriptomic analysis of microbial community structure and gene expression of activated sludge. *PLoS One* **7**: doi: 10.1371/journal.pone.0038183.

Zhang T., Ye L., Tong A. H. Y., Shao M.-F. and Lok S. (2011). Ammonia-oxidizing archaea and ammonia-oxidizing bacteria in six full-scale wastewater treatment bioreactors. *Appl. Microbiol. Biotechnol.*, **91**, 1215–1225.

Zhang T., Shao M.-F. and Ye L. (2012). 454 pyrosequencing reveals bacterial diversity of activated sludge from 14 sewage treatment plants. *ISME J.*, **6**, 1137–47.

Zilles J. L., Peccia J., Kim M.-W., Hung C.-H. and Noguera D. R. (2002). Involvement of *Rhodocyclus*- related organisms in phosphorus removal in full-scale wastewater treatment plants. *Appl. Environ. Microbiol.*, **68**, 2763–2769.

Zumft W. G. (1992). The denitrifying Prokaryotes. In: The Prokaryotes. A Handbook on the Biology of Bacteria, Ecophysiology, Isolation, Identification, Applications, H. G. Trüper, M. Dworkin, W. Harder and K. H. Schleifer (eds), Springer-Verlag, Berlin, pp. 554–582.

Chapter 5

Nitrogen

Wendell O. Khunjar (USA), Paul A. Pitt (USA), Charles B. Bott (USA), Kartik Chandran (USA)

5.1 INTRODUCTION

Nitrogen is a key macronutrient whose speciation and availability is known to impact terrestrial, freshwater and marine environments (Vitousek *et al.*, 1997). Prior to the Industrial Revolution, conversion of N_2 gas (un-reactive N) to reactive forms (forms that directly or indirectly support biological growth) was primarily achieved by biological N_2 fixation mediated by bacteria and archaea (Galloway *et al.*, 2004). Since the Industrial Revolution, anthropogenic production and discharge of reactive N to the environment is estimated to have increased from 15 Tg/y in the 1860s to approximately 259 Tg/y in 2011. This increase was largely driven by food and energy production (Penuelas, 2012). While human metabolism processes only approximately 5% of the global reactive N pool, a large fraction of the metabolized N is associated with discharges to sensitive estuaries and coastal waters (Larsen *et al.*, 2007). Localized accumulation of reactive N species has been linked to water pollution issues including eutrophication and algal blooms which can lead to hypoxia and aquatic fauna die-off (Gruber & Galloway, 2008). Treatment of wastewater prior to discharge to these water bodies remains an effective strategy for directly removing the reactive N fraction associated with human metabolism. This chapter addresses N removal as related to AS treatment and discusses emerging concepts and innovative N removal and recovery technologies.

5.1.1 N in domestic wastewater

Of the nutrients excreted by humans, 85 to 90% of the reactive N can be found in urine with the remainder associated with feces (Larsen & Gujer, 1996). The excreted N is primarily in the form of ammonia/ammonium or organically-bound N (organic N). The N content of domestic wastewater is also influenced by food wastes that are directly discarded into the sewer and by commercial personal care products. Because diet and water usage vary from place to place, the N concentration in domestic wastewater can range from 20 to 150 mg N/L, with the lower concentrations being associated with regions that have more ready access to potable water supply.

The N in domestic wastewater is largely bioavailable (Figure 5.1) and can stimulate eutrophication in coastal waters and estuaries if untreated before discharge (Howarth & Marino, 2006). Consequently, wastewater treatment has evolved in many developed and developing nations to address both N and P removal to help limit primary production. While N can be removed from wastewater by physical–chemical and biological means, AS processes have become the most commonly applied approach for removing wastewater N.

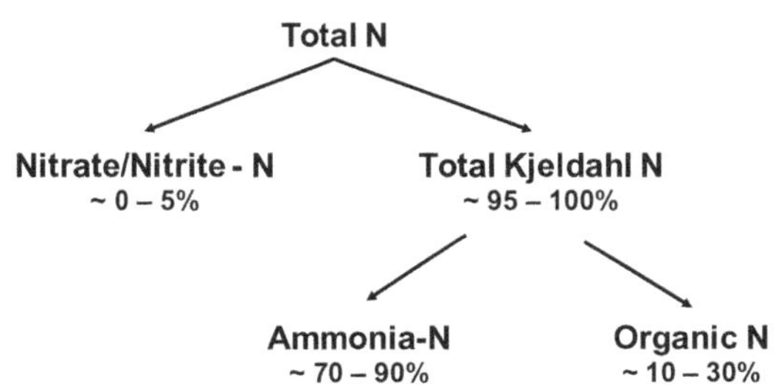

Figure 5.1 Domestic wastewater influent N speciation.

5.2 THE N CYCLE

N can be present in the environment at oxidation states ranging from –III to +V. N_2 gas (oxidation state = 0) constitutes approximately 79% of the atmosphere and is largely unavailable to support the growth of most organisms. N_2 gas can be converted to ammonia N(–III) by nitrogenase enzymes in N_2-fixing bacteria, by photochemical reactions and through the industrial Haber–Bosch process (Galloway *et al.*, 2004). In contrast to N_2, ammonia N is readily available for supporting growth since it does not need to be reduced prior to assimilation. Ammonia N is typically incorporated into biomass to form organic compounds containing N (organic N). Upon cell-death and lysis, organic N can be released and re-converted to ammonia (ammonification).

In addition to being incorporated into biomass, ammonia can be used as an electron donor for supporting energy synthesis. In the process of nitrification, ammonia N is oxidized by aerobic ammonia oxidizing bacteria (AerAOB) or archaea (AerAOA) primarily to nitrite N(+III). Nitrite oxidizing bacteria (NOB) can then convert nitrite N to nitrate N(+V), also typically under aerobic conditions. Both nitrite N and nitrate N can be used as electron acceptors during the process of denitrification. During denitrification, nitrite N and/or nitrate N are converted to N_2 gas, predominantly by heterotrophic bacteria, which use organic carbon compounds for biomass synthesis. Anaerobic ammonia N oxidation (anammox) is another major process in the N cycle, which facilitates the conversion of ammonia N and nitrite N to N_2 gas and nitrate N by autotrophic organisms (AnAOB). An overview of the major pathways of the N cycle is presented in Figure 5.2.

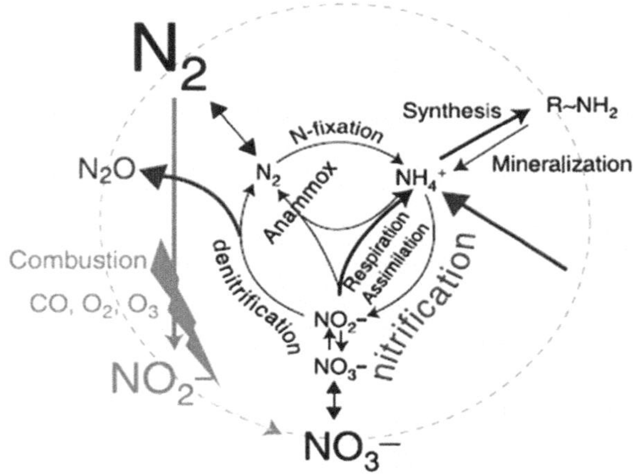

Figure 5.2 Overview of the N cycle. From Klotz *et al.*, 2008.

5.3 HISTORICAL ASPECTS OF BIOLOGICAL N REMOVAL

The AS process was originally developed as an alternative to malodorous anaerobic treatment processes for removing bulk organic carbon from wastewaters (Warrington, 1882). During the early years of process development (1870s to 1910s) researchers found that extended aeration of sewage transformed ammonium to nitrate (Ardern & Lockett, 1914); however, apart from providing an index of wastewater treatment, the process of nitrification was largely unexploited until the middle of the 20th century when nutrients from wastewater were linked with the proliferation of eutrophication (Henze *et al.*, 2008). At this point, research into biological N removal focused on the two-step process of nitrification–denitrification so that reactive N (primarily ammonia, nitrite and nitrate) could be converted to N_2 gas.

Research by Sawyer and Bradney (1945), Downing *et al* (1964), Johnson (1966) and Ludzak and Ettinger (1962) led to the development of a two-step N removal processes that combined denitrification and nitrification (Sawyer & Bradney, 1945; Ludzack *et al.*, 1962; Wuhrmann, 1962; Downing *et al.*, 1964; Johnson, 1966;

McCarty *et al.*, 1969; Balakrishnan & Eckenfelder, 1970; Barnard, 1973, 2006). These findings were further developed by Balakrishnan and Eckenfelder (1970) and Barnard (1973) to create multi-stage processes (e.g. modified Balakrishnan–Eckenfelder, modified Ludzak Ettinger, Bardenpho) that utilized alternating redox conditions to facilitate carbon, N and P removal from the mainstream flow (Figure 5.3) (Balakrishnan & Eckenfelder, 1970; Barnard, 1973). In these multi-stage AS systems, RAS and internal recycle streams are used to provide nitrifier retention and to manipulate the mass of N that is denitrified.

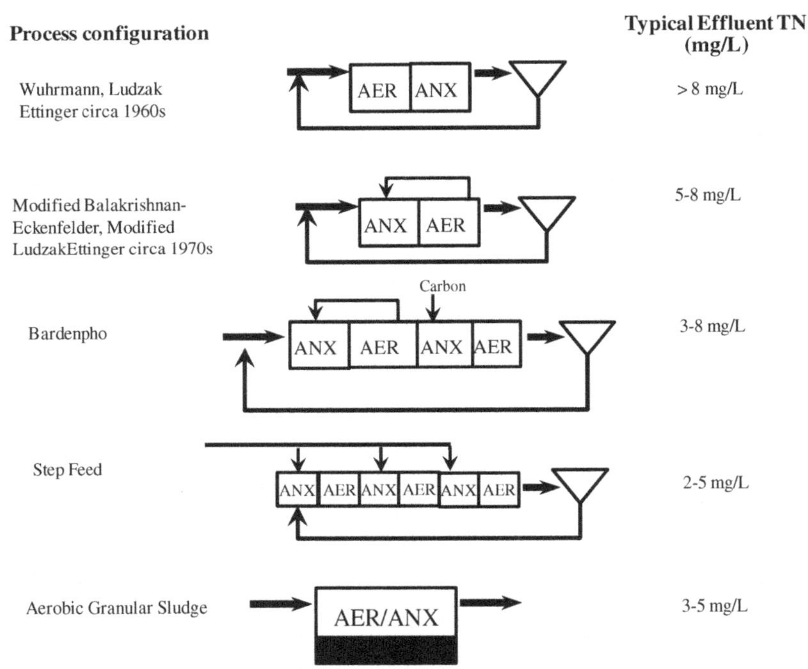

Figure 5.3 Biological N removal approaches. Adapted from Barnard, 2006. (AER = Aerobic; ANX = Anoxic.)

Utilizing similar principles, step-feed AS emerged as a robust strategy for maintaining N removal in facilities faced with high peak flows (e.g. combined sewer systems during wet weather events). This process helps to prevent biomass washout. Fixed film systems and other processes that decouple HRT and SRT (e.g. integrated fixed film AS (IFAS), aerobic granular sludge) have also recently emerged as effective methods for high rate biological N removal in a single reactor. These processes utilize the same fundamental principles as flocculent AS systems; however, biofilm growth on media or in granules allows for the development of dense populations that allow several reactions to occur in the same space (Figure 5.3).

N removal from sidestream flows is also being increasingly implemented. These flows can account for 5 to 30% of the total N load to the mainstream process. Removal of this sidestream N using biological processes can help facilities meet stringent effluent discharge limits. The sidestream treatment processes can utilize conventional nitrification/denitrification, shortcut N removal or deammonification.

5.4 CONVENTIONAL N REMOVAL

Mainstream N removal in WWTPs has traditionally relied on the processes of nitrification and denitrification (Figure 5.4). In this strategy, ammonia is oxidized to nitrate which is then denitrified to N_2 gas. Complete nitrification requires 4.57 g O_2/g NH_3-N and 7.14 g alkalinity/g NH_3-N, while denitrification requires approximately 4 to 6 g COD/g NO_3^--N. When coupled in a multi-stage process such as the MLE process, denitrification can recover 2.86 g O_2/g NO_3-N denitrified and 3.6 g alkalinity/g NO_3-N denitrified.

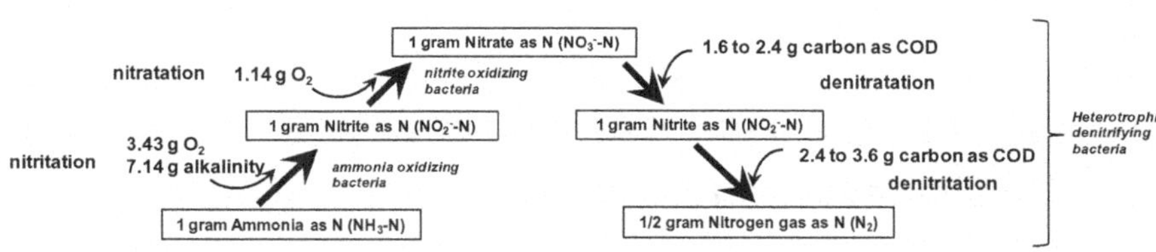

Figure 5.4 Overview of nitrification and denitrification (g = gram).

Engineered nitrification is a 2-step aerobic process. The 1st step, nitritation, is catalyzed by AerAOB and/or AerAOA. The 2nd step (nitratation) is catalyzed by NOB. These organisms primarily display chemolithoautotrophic metabolism, deriving their energy from inorganic chemical sources (lithotrophy) and utilize CO_2 as their primary carbon source (autotrophy). Because of the need to fix CO_2, the growth rates and biomass yields of nitrifiers are low compared to chemorganoheterotrophic organisms. Equation 5.1 shows the combined processes of ammonia and nitrite oxidation.

$$NH_4^+ + 1.9O_2 + 0.069CO_2 + 0.017HCO_3^- \rightarrow 0.017C_5H_7O_2N + 0.98NO_2^- + 0.97H_2O + 1.8H^+ \qquad (5.1)$$

In seminal work, Downing *et al.* (1964) demonstrated that the process of nitrification is limited by the maximum specific growth rate of the nitrifiers and that this rate is highly susceptible to environmental conditions such as changes in temperature, dissolved oxygen concentration and inhibition by toxicants. A consequence of the low growth rate and biomass yield is the need to provide sufficient aerobic SRT for nitrifiers to ensure that biomass washout does not occur.

Aerobic SRT requirements vary with temperature, but typical values range from 10 to 20 d at 10°C and 4 to 7 d at 20°C (Tchobanoglous *et al.*, 1991).

Expanding upon this work, Nowak *et al.* (1995) and Chandran (Chandran & Smets, 2000, 2005; Chandran *et al.*, 2008) developed two-step approaches for mathematically describing the process of nitrification that considered the growth dynamics of AerAOB and NOB separately and allowed for a more realistic representation of the nitrifier microbial consortia.

Denitrification is the reduction of N oxides to N_2 gas. These reactions are catalyzed by heterotrophic organisms in the absence of oxygen. In them, N oxides are utilized as electron acceptors and the reduced organic carbon compounds in wastewater are used as electron donors. If needed the wastewater reduced carbon content can be supplemented with compounds such as methanol, glycerol and fermentation products to achieve low effluent N concentrations (<5 mg N/L). Equation 5.2 describes denitrification assuming that the organic carbon present in wastewater (generalized as $C_{10}H_{19}O_3N$) is used as the electron donor.

$$C_{10}H_{19}O_3N + 10NO_3^- \rightarrow 5N_2 + 10CO_2 + 3H_2O + NH_3 + 10OH^- \tag{5.2}$$

5.5 INNOVATIVE N REMOVAL APPROACHES

5.5.1 Simultaneous nitrification and denitrification

Traditionally, AS systems have been designed and operated to achieve nitrification and denitrification in separate reactions zones. Recently, there has been significant effort to stimulate the simultaneous occurrence of nitrification and denitrification (SND). By achieving these two steps in the same basin, it is possible to reduce the required reactor volume and at the same time possibly also reduce operating costs by using the carbon generated from cell lysis to drive some of the denitrification.

To achieve stable SND, facilities must balance the oxygen demand for nitrification with the anoxic conditions required for denitrification. Operation at low DO set points (<1 mg/L) can help facilitate SND. Recently, much interest has arisen in the use of dynamic DO control strategies to facilitate SND (e.g. ammonia based DO control) (Rieger *et al.*, 2012). In these dynamic systems, effluent water quality parameters (e.g. NH_3 and/or NO_x) are used to help modulate basin DO levels.

5.5.2 Shortcut N removal

Shortcut N removal is another option that can be used for removing N from wastewaters. In this, ammonia is oxidized to nitrite (nitritation) which is then denitrified to N_2 gas (denitritation)/(Figure 5.5). Nitritation requires 3.43 g O_2/g NH_3-N and 7.14 g alkalinity/g NH_3-N, while denitritation requires approximately 3 to 4 g COD/g NO_2^--N.). By avoiding the conversion of nitrite to nitrate, this process can theoretically result in 25% oxygen savings and a 40% electron donor savings compared to conventional nitrification/denitrification. The promotion of nitrite accumulation to provide operational savings contrasts with historical approaches in which nitrite accumulation was viewed as an indicator of poor nitrification performance and a cause of operating problems such as excessive chlorine demand during effluent disinfection.

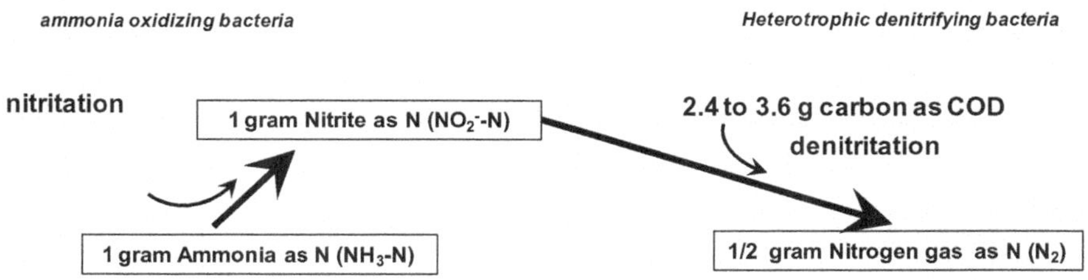

Figure 5.5 Overview of nitritation and denitritation (g = gram).

Recognizing the potential for savings by utilizing the shortcut approach, researchers and engineers at TU Delft and Water Board ZHEW partnered with the Cities of Utrecht and Rotterdam, The Netherlands, to pilot test and then install full-scale nitritation/denitritation processes for the treatment of sidestream flows that contain high N loads. This process, **S**ingle reactor system for **H**igh activity **A**mmonium **R**emoval **O**ver **N**itrite (SHARON) has since been implemented at least 11 full-scale facilities all over the world (Hellinga *et al.*, 1998). In these processes, NOB repression is achieved using a combination of free ammonia (FA) concentrations, DO and short aerobic SRT that wash out the NOBs. Following this, a supplemental carbon source such as methanol is typically added to drive denitrification (Hellinga *et al.*, 1998).

There is currently extensive research on the use of mainstream nitritation/denitritation processes for N removal. In these systems, low ammonia concentrations and low temperatures limit the ability to use free ammonia and low aerobic SRT for controlling NOB growth. Recent research has indicated that successful NOB repression in the mainstream flow will require a combination of factors that include the maintenance of residual ammonia concentrations, the use of intermittently high DO concentrations, a rapid transition from aerobic to anoxic conditions and a tight control of SRT control (Regmi *et al.*, 2013).

5.5.3 Deammonification

Another recent approach for removing N is the process of deammonification in which approximately 50% of the influent ammonia is converted into nitrite (Wett, 2007). Nitrite and ammonia are then consumed through anaerobic ammonia oxidation to produce N_2 gas (Equation 5.3) and a small amount of nitrate. Compared to conventional nitrification/denitrification, this process can theoretically save 62.5% of the oxygen required and all of the electron donor (Figure 5.5).

$$NH_4^+ + 1.32\,NO_2^- + 0.066\,HCO_3^- + 0.13\,H^+ \rightarrow 0.26\,NO_3^- + 1.02\,N_2 + 0.066\,CH_2O_{0.5}N_{0.15} + 2.03\,H_2O \quad (5.3)$$

Originally, the deammonification process was developed as a two-step sidestream process (on anaerobically digested sludge centrate) in which nitritation was performed in a separate reactor (e.g. SHARON) (van Loosdrecht & Salem, 2006). The partially nitrified wastewater was then combined with non-nitrified centrate (containing ammonia) to facilitate N removal. Since then, single reactor approaches have been developed in which, partial nitrification and anammox can be carried out in a compact footprint (Sliekers *et al.*, 2002; Wett, 2007; Christensson *et al.*, 2013).

Deammonification has been employed for treatment of sidestream flows that contain high nutrient loads (e.g. ANAMMOX/CANON (Sliekers *et al.*, 2002), DEMON (Wett, 2007), AnitaMOX (Christensson *et al.*, 2013) CLEARGREEN (Khunjar *et al.*, 2013) (Figure 5.6). In these approaches, repression of nitrite oxidation is achieved by controlling a combination of factors including free ammonia concentration, free nitrous acid concentration, temperature, pH, DO and aerobic SRT (Gabarro *et al.*, 2012). SBRs, granular up-flow reactors and MBBRs have been used for this purpose.

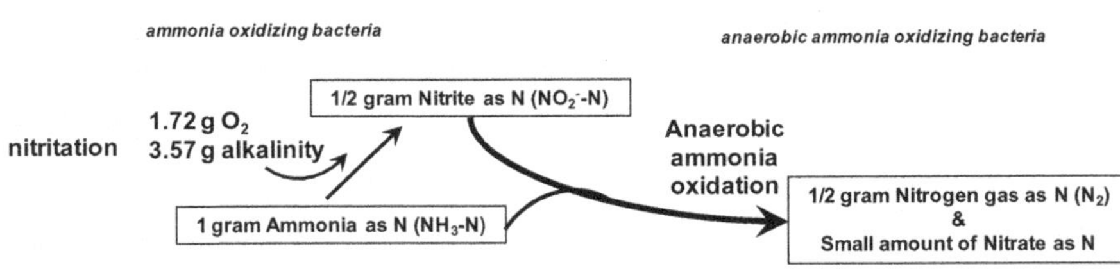

Figure 5.6 Overview of deammonification (g = gram).

Extensive research has been performed to determine the recipe for performing deammonification in the mainstream flow (Wett *et al.*, 2013). To achieve the nitrite shunt pathway in mainstream systems, NOB repression depends on maintaining residual ammonia concentrations, intermittently high DO concentrations, rapid transition to anoxic conditions and tight SRT control (Regmi *et al.*, 2013); however, in addition to suppressing NOB activity, mainstream deammonification requires retention of the anammox bacteria. This can be achieved by utilizing fixed film media, bioaugmentation and SRT decoupling devices such as settlers or hydrocyclones.

5.5.4 Nitrate-dependent anaerobic methane oxidation (N-DAMO)

Nitrate-dependent anaerobic methane oxidation is a recently discovered process in which specialized bacteria and/or archaea couple anaerobic methane oxidation with nitrite and nitrate reduction (Figure 5.7) (Ettwig *et al.*, 2008). This process is currently being investigated as a strategy for removing N in wastewater using the methane generated from anaerobic processes (Shen *et al.*, 2012). It is also being investigated as a complementary process to deammonification, in which nitrate is produced by anammox bacteria then reduced to N_2 gas using dissolved methane (Kampman *et al.*, 2012).

$$3CH_4 + 8NO_2^- + 8H^+ \rightarrow 3CO_2 + 4N_2 + 10H_2O \tag{5.4}$$

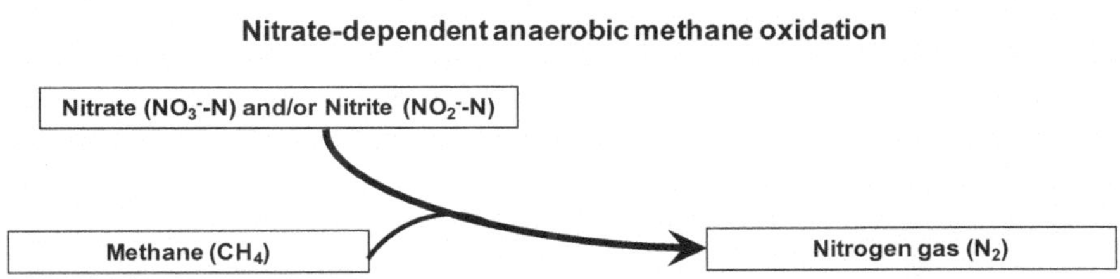

Figure 5.7 Overview of N-DAMO.

5.6 EMERGING TOPICS IN BIOLOGICAL N REMOVAL

5.6.1 Nitrogen oxide production and emission during nitrification and denitrification

In addition to generating N_2 gas, the processes of nitrification and denitrification can also result in the emission of gaseous N compounds such as nitric oxide (NO) and nitrous oxide (N_2O). N_2O is a greenhouse gas, which is about 300 times as potent as CO_2 based on a 100 years lifespan (IPCC, 2001). N_2O is also one of the most powerful ozone depleting substances generated this century. Until 2008, it was universally accepted that biological N removal strategies uniformly led to an increase in N_2O emissions, solely due to the process of denitrification. This viewpoint was further described using emission factors that were widely adopted by the EPA and IPCC (IPCC, 2001).

During the past 5 years, there have been significant developments in identifying several novel mechanisms for N_2O production in nitrifying AS systems. It has been shown that both AerAOB and AerAOA also can contribute to N_2O production (Klotz & Stein, 2008; Santoro *et al.*, 2011). N_2O production by AerAOB is a manifestation of the recovery response of AerAOB from system shocks, such as cycling between aerobic and anoxic conditions, possibly under non-limiting nitrite concentrations (Yu & Chandran, 2010). Therefore, BNR process designs and/or operating conditions that incorporate frequent changes in the physiological state of the AerAOB or denitrifying bacteria are associated with higher N_2O emissions (Kampschreur *et al.*, 2008; Ahn *et al.*, 2010). Furthermore, during denitrification, Lu and Chandran (2010) showed that organic electron donors that promote high rates of denitrification (such as ethanol) may also promote higher N_2O and NO emissions than low-rate electron donors (such as methanol) (Lu *et al.*, 2011).

5.6.2 Structure and function of chemoorganoheterotrophic denitrification

Methanol is the most widely used organic carbon and electron donor used for enhancing denitrification, especially in organic carbon limited wastewaters required to produce low effluent total N concentrations. The primary reason for the widespread use of methanol is that it is less expensive than alternative commercial chemicals such as glycerol, ethanol or acetate. However, the specific rates of denitrification obtained when using methanol are much lower than for ethanol or acetate (Baytshtok *et al.*, 2009). Furthermore, methanol is used by a very low 'active' fraction of AS microorganisms for denitrification (Baytshtok *et al.*, 2009). These factors, combined with the fact that methanol is flammable have stimulated the use of alternative carbon sources for denitrification in WWTPs that need to meet low N limits (Cherchi *et al.* 2009).

Using advanced molecular biological tools such as stable isotope probing, it has been shown that there are two principal communities of methanol assimilating denitrifying bacteria in AS. These include the more conventionally studied *Hyphomicrobium* spp. and the recently isolated *Methyloversatilis* spp. (Baytshtok *et al.*, 2009). While *Hyphomicrobium* spp. are obligate methylotrophs and cannot use other carbon sources, *Methyloversatilis* spp. are more versatile and can denitrify at much higher rates with ethanol (Baytshtok *et al.*, 2009). The molecular basis of the unique ability of *Methyloversatilis* spp. to use both methanol and ethanol is the presence of an alcohol dehydrogenase with a broad substrate base (Lu *et al.*, 2011).

These findings imply that the nature of the carbon source used for denitrification dictates the structural and functional ecology of activated sludge. Therefore, strategies for carbon addition can be better informed by directly diagnosing and quantifying the active fraction and kinetics of carbon-specific denitrification.

5.6.3 Refractory dissolved organic N

Refractory dissolved organic N (rDON) is a subset of soluble organic N present in wastewater effluents. rDON has emerged as a topic of interest for facilities that are required to meet strict effluent TN limits (<3 mg/L), since rDON can represent between 0.5 and 1 mg N/L of the effluent N (Sattayatewa *et al.*, 2010). rDON can be comprised of N in polysaccharides, proteins, lipids, nucleic acids, and natural organic matter (NOM) (Shon *et al.*, 2006). Removal of rDON from AS plant effluent can be achieved using enhanced coagulation and flocculation and activated carbon treatment (Chen *et al.*, 2011). A fraction of the rDON (up to 10%) can also be converted by advanced oxidation processes such as ozonation or H_2O_2–UV treatment to a form that is susceptible to further biological transformation (Simsek *et al.*, 2013).

5.7 N REMOVAL IN THE FUTURE

Historically, N management has been an energetically intensive process. The energy input to fix 1 g N_2 gas to ammonia and then reconvert that ammonia to N_2 gas is approximately 43 to 108 kJ (Finan *et al.*, 2002; Mulder, 2003). Recovery and reuse of reactive N is viewed as a strategy that can decrease our reliance on non-renewable resources (natural gas) for anthropogenic N fixation. Options for recovering N include gas stripping (Collivignarelli *et al.*, 1998), ion exchange (Sprynskyya *et al.*, 2005), chemical precipitation with P as struvite (Latimer *et al.*, 2012) and electrodialysis (Ippersiel *et al.*, 2011) (Table 5.1).

Table 5.1 Technologies available to recover N.

Technology	**Operating conditions**		**Pre-treatment required**	**Chemical input**	**N Recovery efficiency, %**
	Temp. (°C)	**pH**			
Electrodialysis	10–40	<8.0	Solid–liquid separation	Electricity	>90
Adsorption/ion exchange	10–40	<8.0	Solid–liquid separation	Adsorbent	>90
Liquid–gas stripping	>80°C	>9.5	pH and temperature adjustment	Sodium hydroxide	90
Struvite crystallization	25–40	8–9	Solid–liquid separation	Sodium, magnesium or calcium hydroxide	20–30

Source: Adapted from Latimer *et al.* (2012).

These recovery technologies would typically be applied for sidestream treatment and could provide some cost savings by reducing the aeration requirements during nitrification and reducing the amount of supplemental carbon required for denitrification. These N recovery technologies currently act as complementary technologies to AS processes for N management in treatment facilities and for helping to stabilize the AS process performance by reducing intermittent upsets from high nutrient loads in recycle streams.

Even with recovery and reuse, there will still be a need to convert reactive N to un-reactive forms. Conversion processes such as deammonification and N-DAMO are likely to be more fully integrated into mainstream AS wastewater treatment as they can result in significant operational savings. It is also anticipated that source separation and decentralized treatment options will become more prevalent because these technologies facilitate resource recovery.

5.8 REFERENCES

Ahn J., Kim S., Park H., Rahm B., Pagilla K. and Chandran K. (2010). N_2O emissions from activated sludge processes, 2008–2009: results of a national monitoring survey in the United States. *Environ. Sci. Technol.*, **44**(12), 4505–4511.

Ardern E. and Lockett W. T. (1914). Experiments on the oxidation of sewage without the aid of filters. *J. Soc. Chem. Industr.*, **33**(10), 523–539.

Balakrishnan S. and Eckenfelder W. W. Jr. (1970). Nitrogen removal by modified activated sludge processes. *J. Sanitary Enging Div., Amer. Soc. Civil Engng.*, **96**, 501–512.

Barnard J. L. (1973). Biological denitrification. *Water Pollut. Control*, **72**(6), 705–720.

Barnard J. (2006). Biological nutrient removal: where we have been, where we are going? Proc. WEFTEC, 2006, Washington DC, 25.

Baytshtok V., Lu H., Park H., Kim S., Yu R. and Chandran K. (2009). Impact of varying electron donors on the molecular microbial ecology and biokinetics of methylotrophic denitrifying bacteria. *Biotechnol. Bioeng.*, **102**(6), 1527–1536.

Chandran K. and Smets B. F. (2000). Applicability of two-step models in estimating nitrification kinetics from batch respirograms under different relative dynamics of ammonia and nitrite oxidation. *Biotechnol. Bioeng.*, **70**(1), 54–64.

Chandran K. and Smets B. F. (2005). Optimizing experimental design to estimate ammonia and nitrite oxidation biokinetic parameters from batch respirograms. *Water Res.*, **39**, 4969–4978.

Chandran K., Hu Z. and Smets B. F. (2008). A critical comparison of extant batch respirometric and substrate depletion assays for estimation of nitrification biokinetics. *Biotechnol. Bioeng.*, **101**(1), 62–72.

Chen B., Kim Y. and Westerhoff W. (2011). Occurrence and treatment of wastewater-derived organic nitrogen. *Water Res.*, **45**(15), 4641–4650.

Cherchi C., Onnis-Hayden A., Shawabkeh I. E. and Gu A. Z. (2009). Implication of utilizing different carbon sources for denitrification in wastewater treatments. *Water Environ. Res.*, **81**(8), 788–799.

Christensson M., Ekstrom S., Andersson Chan A., Le Vaillant E. and Lemaire. (2013). Experience from start-ups of the first ANITA Mox plants. *Water Sci. Technol.*, **67**(12), 2677–2684.

Collivignarelli C., Bertanza G., Baldi M. and Avezzu F. (1998). Ammonia stripping from MSW landfill leachate in bubble reactors: process modeling and optimization. *Water Management Res.*, **16**(5), 455–466.

Downing A. L., Painter H. A. and Knowles G. (1964). Nitrification in the activated sludge process. *J. Proc. Inst. Sewage Purif.,* **63**, 130–158.

Ettwig K. F., Shima S., van de Pas-Schoonen K. T., Kahnt J., Medema M. H., Op den Camp H. J., Jetten M. S. and Strous M. (2008). Denitrifying bacteria anaerobically oxidize methane in the absence of Archaea. *Environ. Microbiol.,* **10**(11), 3164–3173.

Finan T. M., O'Brian M. R., Layzell D. B., Vessey J. K., Newton W. (eds) (2002). Nitrogen Fixation: Global Perspectives. Proc.13th International Congress on Nitrogen Fixation. New York, NY, USA, CABI Publishing.

Gabarro J., Ganigue R., Gich F., Ruscalleda M., Balaguer M. D. and Colprim J. (2012). Effect of temperature on AOB activity of a partial nitritation SBR treating landfill leachate with extremely high nitrogen concentration. *Bioresource Technol.,* **126C**, 283–289.

Galloway J. N., Dentener F. J., Capone D. G., Boyer E. W., Howarth R. W., Seitzinger S. P., Asner G. P., Cleveland C. P., Green P. A., Holland E. A., Karl D. M., Michaels A. F., Porter J. H., Townsend A. R. and Vöosmarty C. J. (2004). Nitrogen cycles: past, present, and future. *Biogeochem.,* **70**, 153–226.

Gruber N. and Galloway J. N. (2008). An earth-system perspective of the global nitrogen cycle. *Nature,* **451**(7176), 293–296.

Hellinga C., Schellen J., Mulder J. W. and van Loosdrecht M. C. M. (1998). The SHARON process: an innovative method for nitrogen removal from ammonium-rich waste water. *Water Sci. Technol.,* **37**(9), 135–142.

Henze M., van Loosdrecht M. C. M. and Ekama G. A. (2008). Biological Wastewater Treatment: Principles, Modelling and Design. International Water Association, London, UK.

Howarth R. W. and Marino R. (2006). Nitrogen as the limiting nutrient for eutrophication in coastal marine ecosystems: evolving views over three decades. *Limnol. Oceanog.,* **51**(1), 364–376.

IPCC (2001*).* Climate Change 2001: The Scientific Basis. Cambridge, UK.

Ippersiel D., Mondor M., Lamarche F., Tremblay F., Dubreuil J. and Masse L. (2011). Nitrogen potential recovery and concentration of ammonia from swine manure using electrodialysis coupled with air stripping. *J. Environ. Management,* **95**, S165–S169.

Johnson W. K. (1966). Removal of nitrogen by biological treatment. Adv. Water Qual. Improvement, Water Resources Symposium No. 1, Univ. Texas Press, 178–189.

Kampman C., Hendrickx T. L. G., Zhang T., Luecken F. A., van Alen T. A., de Camp H. J. M. O., Jetten M. S. M., Zeeman G., Buisman C. J. N. and Temmink T. (2012). Denitrification with dissolved methane from anaerobic digestion: a novel opportunity for sewage treatment. *IWA LET,* 2012, Brisbane, Australia.

Kampschreur M. J., Tan N. C., Kleerebezem R., Picioreanu C., Jetten M. S. M. and van Loosdrecht M. C. M. (2008). Effect of dynamic process conditions on nitrogen oxides emission from a nitrifying culture. *Environ. Sci. Technol.,* **42**(2), 429–435.

Khunjar W. O., Bilyk K., Bruton T., Yi P., Latimer R., Pitt P. A., Mehta S., Desmottes C., Graveleau L. and Grandstaff J. (2013). Use of free ammonia and dissolved oxygen for process control of a sidestream deammonification sequential batch reactor. Proc. 86th WEFTEC, 2013, Chicago, IL, USA.

Klotz M. G. and Stein L. Y. (2008). Nitrifier genomics and evolution of the nitrogen cycle. *FEMS Microbiol. Lett.,* **278**(2), 146–156.

Larsen, T. A. and Gujer W. (1996). Separate management of anthropogenic nutrient solutions (human urine). *Water Sci. Technol.,* **34**(3–4), 87–94.

Larsen T. A., Maurer M., Udert K. M. and Lienert J. (2007). Nutrient cycles and resource management: implications for the choice of wastewater treatment technology. *Water Sci. Technol.,* **56**(5), 229–237.

Latimer R., Nguyen V., Smeby M., Vadiveloo E., Pitt P. A., Harris R., Porter R., Elmendorf H. and Richards T. (2012). Pilot testing nutrient recovery from WAS streams for struvite control and recycle load reduction. Proc. 85th WEFTEC, 2012, New Orleans LA, 489–511.

Lu H. and Chandran K (2010). Diagnosis and quantification of glycerol assimilating denitrifying bacteria in an integrated fixed-film activated sludge reactor via 13C DNA stable-isotope probing. *Environ. Sci. Technol.*, **44**(23), 8943–8949.

Lu H., Nuruzzama F., Ravindhar J. and Chandran K. (2011). Alcohol dehydrogenase expression as a biomarker of denitrification activity in activated sludge using methanol and glycerol as electron donors. *Environ. Microbiol.,* **13**(11), 2930–2938.

Ludzack F. J. and Ettinger M. B. (1962). Controlled operation to minimize activated sludge effluent nitrogen. *J. Water Pollut.Control Fedn.,* **34**(9), 920–931.

McCarty P. L., Beck L. and Amant P. S. (1969). Biological denitrification of wastewaters by addition of organic materials. Proc. 24th Purdue Industr. Waste Conf., Lafayette, IN, 1271–1285.

Mulder A. (2003). The quest for sustainable nitrogen removal technologies. *Water Sci. Technol.,* **48**(1), 67–75.

Nowak O., Svardal K. and Schweighofer P. (1995). The dynamic behaviour of nitrifying activated sludge systems influenced by inhibiting wastewater compounds. *Water Sci. Technol.,* **31**(2), 115–124.

Penuelas J., Sardans J., Rivas-Ubach A. and Janssens I. A. (2012). The human-induced imbalance between C, N and P in Earth's life system. *Global Change Biol.,* **18**, 3–6.

Regmi P., Holgate B., Miller M., Bunce R., Wett B., Murthy S. and Bott C. (2013). NOB repression in mainstream makes two-stage deammonification and nitrite-shunt possible. Proc. WEF/IWA Nutrient Removal and Recovery 2013: Trends in Resource Recovery and Use, Vancouver, Canada.

Rieger L., Bott C., Balzer W. and Jones R. M. (2012). Model-based aeration systems design – case study Nansemond WWTP. Proc. 85th WEFTEC, 2012, New Orleans, 2445–2471.

Santoro A. E., Buchwald C., McIlvin M. R. and Casciotti K. L. (2011). Isotopic signature of N_2O produced by marine ammonia-oxidizing archaea. *Science,* **333**, 1282–1285.

Sattayatewa C., Pagilla K., Sharp R. and Pitt P. A. (2010). Fate of organic nitrogen in four biological nutrient removal wastewater treatment plants. *Water Environ. Res.,* **82**(12), 2306–2315.

Sawyer C. N. and Bradney L. (1945). Rising of activated sludge in final settling tanks. *Sewage Works J.,* **17**, 1191–1209.

Shen L. D., He Z. F., Zhu Q., Chen D. Q., Lou L. P., Xu, P. Zheng X. Y. and Hu B. L. (2012). Microbiology, ecology, and application of the nitrite-dependent anaerobic methane oxidation process. *Front. Microbiol.,* **3**, 269.

Shon K. H., Vigneswaran S. and Synder S. (2006). Effluent organic matter (EfOM) in wastewater: constituents, effects and treatment. *Environ. Sci. Technol.,* **36**(4), 327–374.

Simsek H., Wadhawan T. and Khan E. (2013). Overlapping photodegradable and biodegradable organic nitrogen in wastewater effluents. *Environ. Sci. Technol.,* **47**(13), 7163–7170.

Sliekers A. O., Derwort N., Gomez J. L., Strous M., Kuenen J. G. and Jetten M. S. M. (2002). Completely autotrophic nitrogen removal over nitrite in one single reactor. *Water Res.,* **36**(10), 2475–2482.

Sprynskyya M., Lebedynets M., Zbytniewski R., Namieśnik J. and Buszewski B. (2005). Ammonium removal from aqueous solution by natural zeolite, Transcarpathian mordenite, kinetics, equilibrium and column tests. *Sep. Purif. Technol.,* **46**(3), 155–160.

Tchobanoglous G., Burton F. L. and Eddy M. (1991). Wastewater Engineering: Treatment, Disposal, and Reuse. New York, McGraw-Hill.

van Loosdrecht M. C. M. and Salem S. (2006). Biological treatment of sludge digester liquids. *Water Sci. Technol.,* **53**(12), 11–20.

Vitousek P. M., Aber J. D., Howarth R. W., Likens G. E., Matson P. A., Schindler D. W., Schlesinger W. H. and Tilman D. G. (1997). Human alteration of the global nitrogen cycle: sources and consequences. *Ecol. Applic.,* **7**(3), 737–750.

Warrington R. (1882). Some aspects of recent investigations on nitrification. *J. Royal Soc. Arts,* **30**, 532–544.

Wett B. (2007). Development and implementation of a robust deammonification process. *Water Sci. Technol.,* **56**(7), 81–88.

Wett B., Omari A., Podmirseg S. M., Han M., Akintayo O., Gomez Brandon M., Murthy S., Bott C., Hell M., Takacs I., Nyhuis G. and O'Shaughnessy M. (2013). Going for mainstream deammonification from bench to full scale for maximized resource efficiency. *Water Sci. Technol.,* **68**(2), 283–289.

Wuhrmann K. (1962). Nitrogen removal in sewage treatment process. Proc.15th Intl Cong. Limnol., Madison, WI, USA, 580–596.

Yu R. and Chandran K. (2010). Strategies of *Nitrosomonas europaea* 19718 to counter low dissolved oxygen and high nitrite concentrations. *BioMed. Central Microbiol.,* **10**(70), 1–11.

Chapter 6

Phosphorus removal in activated sludge

James Barnard (USA), Yves Comeau (Canada)

6.1 INTRODUCTION

The interest in P removal in AS systems was greatly stimulated by the observation of severe eutrophication in some water bodies, resulting in limnological studies that determined that, in most inland water bodies, P was the limiting nutrient (Vollenweider, 1968). This was especially true in the Johannesburg, South Africa, area where the position of the conurbation on the continental divide resulted in the discharge of P-containing effluent to receiving reservoirs to the north and south with consequential excessive growth of cyanobacteria. The speedy recovery of Lake Washington, Seattle, Washington, USA when wastewater was diverted convinced researchers that improvement was possible if the P discharged to the water bodies could be reduced. In North America, the restoration of Lake Erie depicted in 'The Lorax' by Dr. Seuss, is an excellent example of this recovery:

> You're glumping the pond where the Humming-Fish hummed!
> No more can they hum, for their gills are all gummed.
> So I'm sending them off. Oh, their future is dreary.
> They'll walk on their fins and get woefully weary
> in search of some water that isn't so smeary.
> *I hear things are just as bad up in Lake Erie.*

The work of Vollenweider (1968) led to an agreement between Canada and the USA to remove P from discharges to the Great Lakes (Great Lakes Water Quality Agreement, 1972). Widespread addition of chemicals such as ferric salts and alum followed to precipitate P where it could be removed with the WAS. The remainder of this chapter is dedicated to the development of EBPR processes in AS.

6.2 EARLY HISTORY

The first observations of EBPR in full-scale plants were made independently by two research groups, Srinath *et al.* (1959) in India and Alarcon (1961) in the USA. Srinath *et al.* noted that P was released in the under-aerated upstream zone of a plug-flow reactor but thought of this as a problem that needed more oxygen, which the plant could not supply. Levin and Shapiro (1965) experimented in the laboratory with P uptake under aerobic conditions and release under anaerobic conditions and devised the first practical EBPR process called PhoStrip in which the P was taken up by organisms in a plug-flow high rate plant that did not nitrify (Figure 6.1; 6.4a). From the final clarifier, the RAS was passed through a thickener with more than 30 h HRT, where the RAS thickened while depleting the oxygen through fermentation, upon which the P was released to the supernatant. Lime was used to precipitate the higher concentration of P from the supernatant while the underflow, 'stripped' of P, was returned to the aeration basin. It was observed that after the P was stripped, the bacteria would again take up P during aeration. More incidents were reported by Vacker *et al.* (1967), Witherow (1970), and Milbury *et al.* (1971). All of these plants were non-nitrifying plants with plug-flow configurations. Milbury *et al.* also reported problems of P release at the inlet zone because of the inability of the aeration system to supply enough air and noted that this 'problem' was experienced at all the plug-flow plants that reported 'luxury uptake' of P.

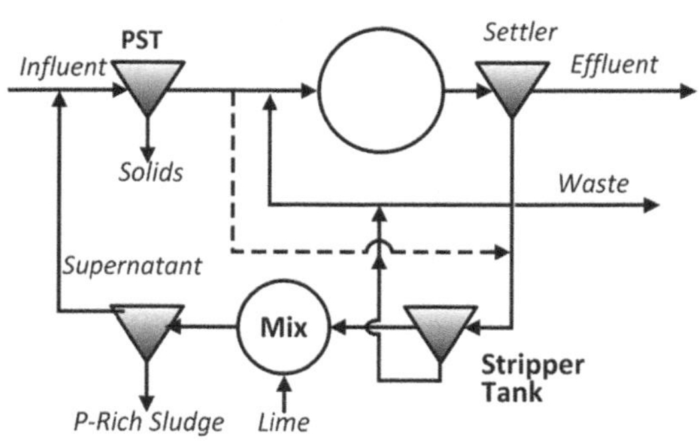

Figure 6.1 The PhoStrip process.

6.3 DEVELOPMENT OF BIOLOGICAL NUTRIENT REMOVAL (BNR)

The water reclamation project for the city of Windhoek, Namibia, and severe eutrophication around Johannesburg, South Africa emphasized the need for nutrient removal from wastewater and led to Barnard (1974a, b) developing the 4-stage N removal process now known as the Bardenpho process at the laboratories of the National Water Research Institute (NWRI) in Pretoria, South Africa. A 100 m^3/d pilot plant consisting of the four stages was created by partitioning an existing structure using movable gates (Figure 6.2). The dead zone resulted from the need to allow changes to the active zone sizes by moving the partitions. Two holes of 25 mm in the partition between the 2nd anoxic and the dead zone were provided to even the hydraulic pressure on both sides of the partition. N removal in excess of 90% was observed in the pilot plant. P profiles through the plant showed that the effluent orthophosphate was consistently removed from about 9 mg P/L in the influent to less than 0.2 mg P/L in the filtered effluent, while the pilot plant was operated in this configuration. P was released to over 30 mg P/L in the 2nd anoxic zone, followed by uptake in the last aeration zone. Laboratory tests in plug-flow reactors showed that P could be reduced to low levels only when this release was observed. When operating a four-stage bench-scale unit in the laboratory, a close relationship was observed between the concentration of nitrate in the effluent and the P (Figure 6.3).

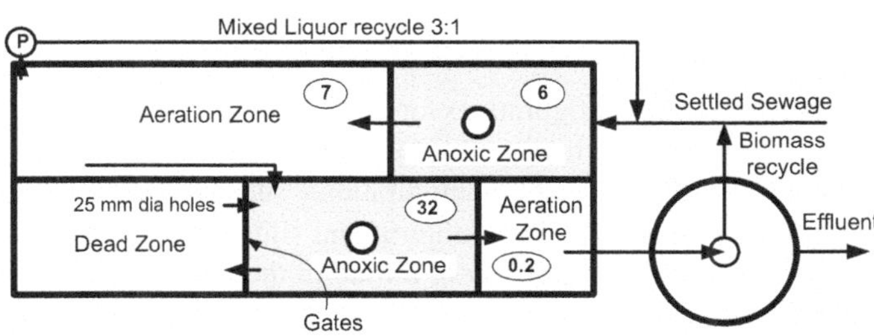

Figure 6.2 Physical layout of pilot plant.

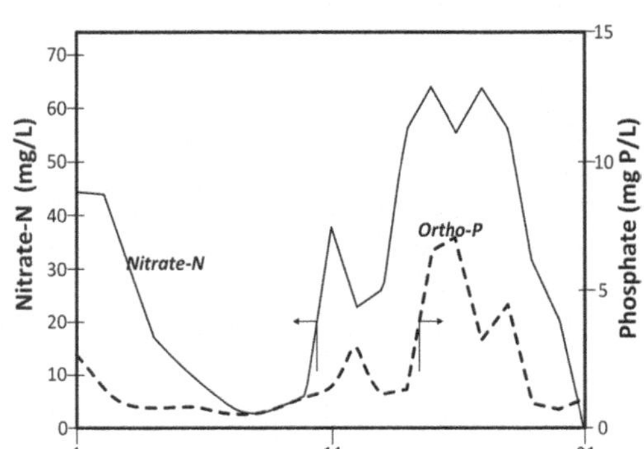

Figure 6.3 Correlation between nitrate and orthophosphate.

Noting that all plants that observed biological P uptake reported a release of P in a zone ahead of the main aeration zone, Barnard (1974a, b) postulated that for biological P removal to occur, it is necessary that the mixed liquor pass through an anaerobic phase, free of nitrate and DO where P is released, followed by an aerated zone where it will be taken up. Unaware that the inadvertent connection between the 2nd anoxic zone and the dead zone of the pilot plant provided this condition, Barnard (1974a, b) proposed that anaerobic conditions could best be obtained ahead of the 1st anoxic zone, on the assumption that the four-stage N removal plant would leave little nitrate in the RAS. The process could not be explained. However, the role of nitrate was convincing evidence that it was a biological process. Nicholls (1975) experimented successfully with this concept at the temporary Alexandra Wastewater Treatment Plant (WWTP) in Johannesburg when switching off aerators near the inlet to create a low DO condition similar to that observed in other full-scale plants, so it was decided to add formal anaerobic stages to the 750,000 PE 4-stage Johannesburg, Goudkoppies WWTP, then under construction. This rapid application of the BNR technology encouraged other authorities in the area to use the process because the need was evident, even though there was yet no regulation that required P removal.

6.4 PROCESS CONFIGURATIONS FOR BNR

The development of the first configurations for BNR was influenced by the regulations that required ammonia reduction at all plants in South Africa. Barnard (1976) proposed a number of BNR process configurations, such as the Phoredox (Figure 6.4b) for a non-nitrifying plant. The configurations illustrated of other flow sheets enabled BNR by allowing the reduction of nitrate in the return sludge to minimize interference in the anaerobic zone. The Modified Bardenpho plant (Figure 6.4c) was used for the original plant designs for the Goudkoppies WWTP in Johannesburg and the Cape Flats WWTP in Cape Town, South Africa, both of them treating 750,000 PE. The three-stage Bardenpho process (Figure 6.4d) was used with plants having surface aerators, which allowed a high degree of simultaneous nitrification and denitrification (SND), which reduced nitrate in the RAS (van Huyssteen *et al.*, 1990).

The University of Cape Town (UCT) process (Figure 6.4e) was developed in the laboratory by Marais *et al.* (1983) in response to their inability to reduce the effluent nitrate to a level that would be compatible with EBPR, especially when using the three-stage Phoredox process. Since Marais *et al.* used fine bubble aeration, there was little SND and, therefore, excessive nitrate in the RAS. In the UCT process, the RAS was returned to the anoxic zone for denitrification, and mixed liquor was recycled from the end of the anoxic zone, where the nitrate concentration could be controlled prior to being returned to the anaerobic zone. This was followed by the Modified UCT (MUCT) process shown in Figure 6.4f, which used the effluent from the anaerobic zone for denitrification of the RAS before recycling the RAS to the anaerobic zone. The main anoxic zone was used for denitrification of the mixed liquor recycled from the aeration basin. The UCT and MUCT flow diagrams resulted in the mixed liquor concentration in the anaerobic zone being about 50% of that in the main plant. Due to the lack of volatile fatty acids (VFAs) in the influent to the Westbank (now Westside Regional Wastewater Treatment Plant [WRWTP], BC, Canada), primary sludge was fermented (Section 6.5) and supernatant passed to the anaerobic zone. Since there was little readily biodegradable chemical oxygen demand (rbCOD) in the feed – most of it was passed to the anoxic zone – while 5–10% of it was used to assist in the denitrification of the RAS in a pre-anoxic zone (Figure 6.4g). The city of Johannesburg (Nicholls, 1975) proposed the Johannesburg process (Figure 6.4h) to reduce nitrate in the RAS by using a pre-anoxic zone for the RAS ahead of the anaerobic zone for denitrification through endogenous respiration in the higher MLSS in this zone. This was developed to reduce the excess nitrate resulting from RAS recycle rates of more than double the inflow rate to the plant instead of the more normal 60–90% of the inflow rate.

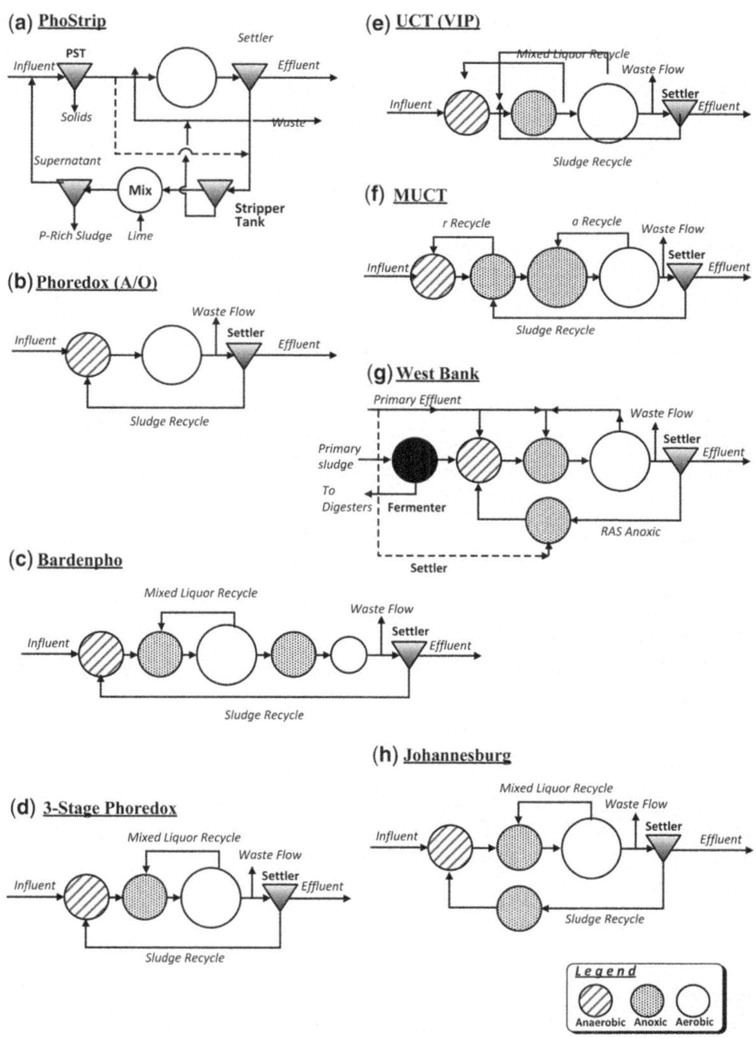

Figure 6.4 Various flow sheets for EBPR.

6.5 ACID FERMENTATION FOR PRODUCTION OF VFAS

6.5.1 Fermentation of primary sludge

While studying the PhoStrip process, Fuhs and Chen (1975) suggested that certain bacteria – later collectively referred to as polyphosphate accumulating organisms (PAOs) – can take up excessive amounts of P when supplied with acetate as feed. This was best done in an anaerobic zone free of nitrate and DO to favor the PAOs over other faster growing heterotrophs that could use DO or nitrate as electron acceptors. This hypothesis explained why nitrate in the RAS could pose a problem and was justification for the processes that reduced the nitrate in the RAS. It also showed that plants with an adequate supply of VFAs in the influent would be more amenable to EBPR. During the design of the Kelowna Wastewater Treatment Facility, BC, Canada, in 1978, it was determined that there would not be enough rbCOD in the influent that could be fermented to VFAs in the anaerobic zone, and thus, primary sludge was passed to an existing sludge holding tank in the form of a large primary sludge thickener with an SRT of approximately 6 d. VFAs that formed in the sludge blanket through acid fermentation were released to the supernatant, which was discharged to the anaerobic zone of the main plant, where it successfully augmented the supply of VFAs. This was later referred to as a Static Fermenter (Figure 6.5a). Oldham and Stevens (1984) clearly showed that without a fermenter, there would be little P removal in this plant. Rabinowitz and Oldham (1986) examined the design needs concerning the ideal SRT for maximizing the fermentation of organic solids to VFAs without conversion to methane. Barnard (1984) suggested that primary sludge could be accumulated in primary tanks and allowed to ferment and release VFAs when some of the underflow was recycled to the influent, a concept referred to as Activated Primary tanks (Figure 6.5b).

Primary sludge could also be fermented in a dedicated fermenter, and the sludge then passed to a thickener where the sludge could be elutriated (Figure 6.5c). In these fermenters, the object is to retain primary sludge for a sufficient period (4–8 d) at ambient temperatures to undergo acid fermentation but wash out the slower growing methanogenic organisms. After separation, the supernatant containing the VFAs could then be passed to the anaerobic zone. In the past thirty years, primary sludge fermenters have been used in BNR plants in Canada, the USA, Europe, South Africa, Australia, and New Zealand. There seems to be more success with primary sludge fermenters in colder regions where less fermentation takes place in the collection system. At the Westside plant, BC, Canada, most of the primary effluent is passed to the anoxic zone, and only fermenter supernatant is passed to the anaerobic zone (Figure 6.4g). In the latest development at the Douglas L. Smith Middle Basin Wastewater Treatment Plant,

Overland Park, Kansas, USA, an upflow sludge blanket acid fermenter is being used with great success.

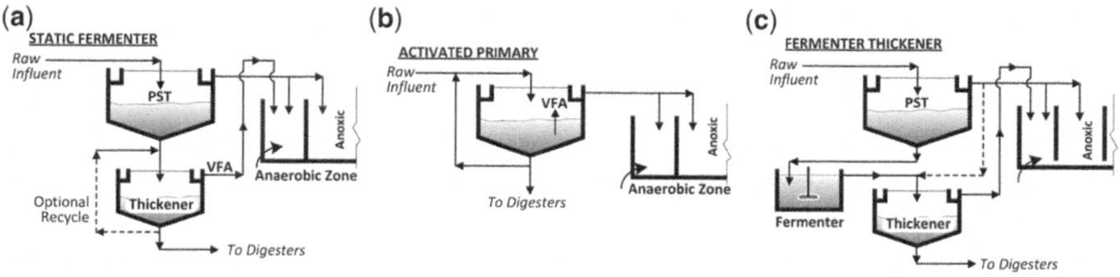

Figure 6.5 Types of primary sludge fermenters.

6.5.2 Fermentation of MLSS or RAS

This concept appears counterintuitive to the purpose of producing VFAs, since fermenting of mixed liquor containing the P that was taken up in the aerobic zone will release all the P and could be considered as secondary release. The pilot plant shown in Figure 6.2, in which simultaneous N and P removal was first observed, had no anaerobic zone and insufficient VFAs in the feed to encourage P removal, yet the P in the 2nd anoxic zone was released to more than 30 mg P/L, which was followed by uptake in the reaeration zone to less than 0.2 mg P/L. Barnard (1985) later raised the possibility that mixed liquor from the 2nd anoxic zone passed through two 25 mm holes in the partition between this zone and the perceived dead zone in which mixed liquor was fermented (Figure 6.2). The pulsing action of the mixer in the anoxic zone was likely responsible for this transfer of mixed liquor to the fermenter and of fermented sludge back to the anoxic zone, leading to the release of orthophosphate in the 2nd anoxic zone. The excellent performance of the plant ceased when the pilot was reconfigured and there was no longer a connection between the two zones. Stroud and Martin (2001) converted a plant to present the configuration shown in Figure 6.6 which resulted in excellent P removal. In the basic Bardenpho plant, they passed the RAS through an anoxic zone and then through an anaerobic stage with a portion of the RAS being fermented and returned to the anaerobic zone. Note that the anaerobic zone does not receive any of the VFAs from the influent. Vollertsen *et al.* (2005) fermented 10–15% of the RAS before returning it to the anaerobic zone and this assisted in P removal in more than 40 Danish treatment plants. Clark improved P removal in a five-stage plant without an adequate supply of influent VFAs by switching off a mixer in one of the anaerobic zones and permitting mixed liquor to settle to the floor and ferment (Clark & Neethling, 2009). The mixer was switched on once a day for 15 min. With the mixer off, mixed liquor floated over the heavier fermenting sludge and exchanged some fresh substrate and VFAs. P was reduced from 9 mg/L in the influent to less than 0.5 mg/L in the final clarifier effluent. Barnard *et al.* (2010) used this technique to improve P removal in the Henderson, Nevada, USA plant, which had no primary settling tanks. The VFAs in the influent were consumed by the addition of nitrate in the collection system for odor control. Effluent orthophosphate averaged approximately 0.1 mg P/L. This technique is now applied at a number of wastewater treatment plants, some of which were not designed for P removal. The only rational explanation for these phenomena is that when the PAOs enter the fermenter zone, they take up as much VFAs as would be allowed by their storage of polyphosphates (poly-Ps) and then maintain themselves during the period under fermentation, while the other heterotrophic bacteria decay and ferment to supply the necessary VFAs. Studies are under way to determine the mechanisms and simulate these phenomena in mathematical models.

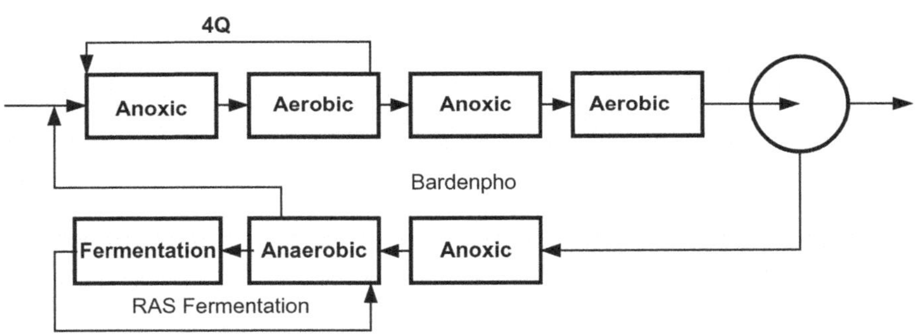

Figure 6.6 Stroud RAS fermentation.

6.6 SECONDARY RELEASE OF P

Barnard (1984) observed that P may be released in EBPR plants in the absence of an electron acceptor, such as nitrate or DO, and of VFAs. It was not possible to take up P thus released by further aeration because no prior uptake of VFAs took place. This was referred to as secondary P release. Experience showed that additional VFAs would be required to take up the P again. Secondary release may happen in the anoxic zones when there is insufficient nitrate or when the zone is oversized relative to the required denitrification capacity, or in final clarifier sludge blankets. It has been suggested that under such conditions, the PAOs will use some of the stored energy in the poly-Ps for cell maintenance, releasing P to the liquid phase. Since this was not associated with an uptake of VFAs, P uptake in the aeration zone cannot take place. Secondary release of P should be avoided in both design and operation of plants. P may also be released through endogenous respiration (Oldham & Stevens, 1984) when the uptake is completed in a plug-flow aeration zone some distance before the end of the aerobic zone of the basin. This may indicate that the aerobic SRT is too long. Ideally, the uptake of P should be completed near the end of the aeration zone.

6.7 HISTORICAL AND SCIENTIFIC PERSPECTIVE

6.7.1 Intensive research

The high degree of indirect reuse of water in what is now the Gauteng Province in South Africa resulted in salinity of the drinking water reservoirs of up to 800 mg/L, which eliminated chemical precipitation as an alternative for P removal. The initial biological mechanisms proposed by Fuhs and Chen (1975) were followed by intensive research at the National Institute for Water Research (NIWR), UCT, the city of Johannesburg, and shortly afterwards, the University of British Columbia and other laboratories to uncover more information on the behavior of the PAOs. Gerber *et al.* (1986) demonstrated that the PAOs can only take up the VFAs acetate and propionate, and that all other forms of biodegradable COD must first be converted to these acid salts before uptake is possible. Siebritz *et al.* (1983), Wentzel *et al.* (1985), Nicholls (1987), Wentzel *et al.* (1990), and Randall *et al.* (1994) showed that since most wastewater contained insufficient VFAs, the influent rbCOD could be fermented in the anaerobic zones to produce VFAs. The undesirable presence of nitrate or DO in the anaerobic zone would serve as an electron acceptor for ordinary heterotrophic organisms (OHOs) to use the rbCOD preferentially but, in addition, would prohibit acid fermentation in the anaerobic zone. With sufficient VFAs in the influent, strictly anaerobic conditions are not necessary for P removal (Barnard, 1984). There is not sufficient retention time in the anaerobic zone to ferment slowly degradable organic matter, but it could take place in side-stream fermenters of part of the mixed liquor from an EBPR plant, as was previously discussed.

Many of the concepts presented previously for the optimization of EBPR processes are summarized in Figure 6.7. These concepts can be grouped around the following principles: (1) maximizing the availability of VFAs in the anaerobic zone by having a favorable influent composition that can be improved by primary sludge fermentation, AS fermentation, or addition of acetate from favorable industrial wastes; (2) minimizing the addition of nitrate and oxygen to the anaerobic zone by denitrifying the AS and the RAS and by limiting vortices, cascades and excessive mixing; (3) precipitating orthophosphate by coagulant addition (aluminum (Al), iron (Fe), or lime) at various points of the process (mainstream anaerobic zone, return sludge non aerated zone, final effluent); and (4) capturing secondary settling tank SS by granular-, micro-, or ultra-filtration.

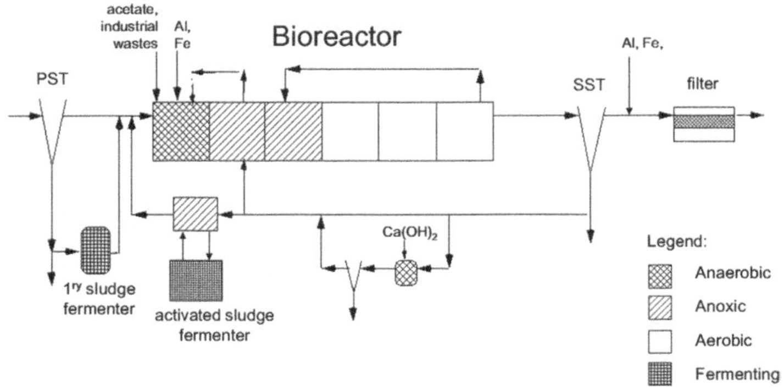

Figure 6.7 Overview of EBPR process optimization concepts (PST: primary settling tank; SST: secondary settling tank).

6.7.2 Microbiology

A debate took place between researchers in the 1970s and early 1980s (Arvin, 1983) on whether the enhanced EBPR was a chemical (biologically induced) precipitation (Menar & Jenkins, 1969) or a biological phenomenon (Yall *et al.*, 1970; Fuhs & Chen, 1975; Barnard, 1976; Marais *et al.*, 1983). The chemical theory could not explain the role of nitrate. Buchan (1983) published electron-microscope pictures of the accumulated intracellular poly-P granules in the organisms, which appeared to settle the debate in favor of the biological mechanism. In trying to explain observed EBPR in high-rate plug-flow plants, Harold (1966) studied poly-P accumulation by pure cultures by the 'luxury uptake' and the 'overplus accumulation' phenomena, neither of which take place in EBPR systems. Indeed, the luxury uptake of P is observed when microorganisms are exposed to a lack of a nutrient such as N or sulfur in the presence of sufficient P, while overplus accumulation is observed following the sudden exposure to P after a period of deprivation.

The essential requirements for EBPR were gradually recognized as being the exposure of the sludge to a zone with the absence of DO and nitrate (anaerobic conditions) in the presence of sufficient VFAs that could be stored intracellularly by PAOs, alternating with zones with the presence of DO or nitrate (aerobic or anoxic conditions). Concurrently with the development of the BNR processes around these concepts, there was much interest in explaining the underlying mechanisms involved.

PAOs are responsible for EBPR. These organisms are able to store 3 types of intracellular reserves: poly-P (both a P and energy reserve), resulting in an efficient removal of P from the wastewater to be treated, and two forms of carbon reserves, poly-β-hydroxyalkanoates (PHAs; a bacterial lipid), and glycogen. Under anaerobic conditions other fast-growing microorganisms cannot utilize these substrates while PAOs store VFAs as PHAs obtaining the energy required from breaking up stored poly-P compounds and glycogen. While the dissolved VFA concentration decreases under anaerobic conditions, those of P (Figure 6.8) Mg and K increase. When PAOs are exposed to either anoxic (absence of DO but presence of nitrate) or aerobic conditions, PHA reserves are then consumed for growth, supplying the necessary energy for taking up and storing P as high energy poly-P bonds and glycogen (Figure 6.9). Poly-P enriched biomass is then wasted from the process for efficient P removal by the treatment system.

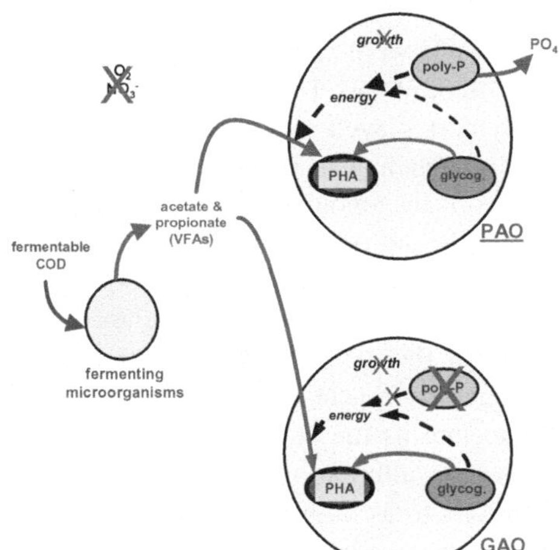

Figure 6.8 Simplified biochemical model for PAO, GAO, and fermenting organism metabolism under anaerobic conditions.

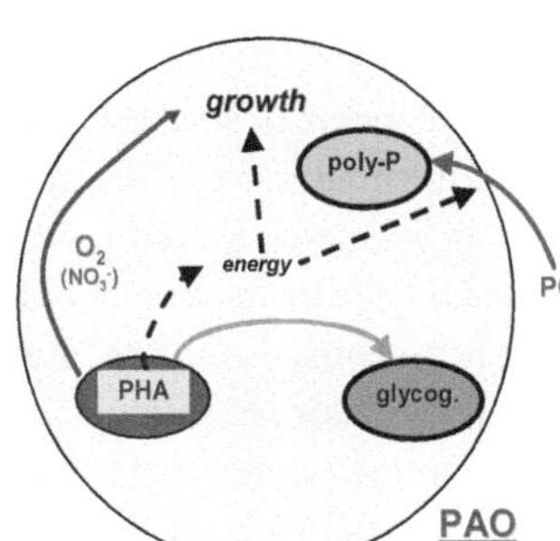

Figure 6.9 Simplified biochemical model for PAO metabolism under aerobic (or anoxic) conditions.

The mass of P taken up from the influent is related to the mass of VFA available to the PAOs. Gerber *et al.* (1986) demonstrated that the PAOs can take up only acetic and propionic acid, which are the end products of acid fermentation. Acid fermentation takes place naturally in the slimes in sewer systems and force mains and is more common in warm climates than in cold climates. Most wastewater influents do not contain sufficient VFAs for sustaining P removal; however, Marais *et al.* (1983) demonstrated that most of the rbCOD in the influent can be fermented to VFAs in the anaerobic zone by the facultative heterotrophic organisms. This is another reason why the absence of DO and nitrate is essential.

6.7.3 Biochemical models

Biochemical models for PAOs were proposed according to the Comeau–Wentzel model (Comeau *et al.*, 1986; Wentzel *et al.*, 1986), in which poly-P and PHA intracellular storage polymers played a central role, and the Mino model (Mino *et al.*, 1995), in which glycogen was identified as a third central intracellular polymer. Experimental work by Smolders *et al.* (1994a, b) and others confirmed the role of glycogen in PAOs and opened the way to integrate the competition of GAOs that stored PHAs and glycogen but not poly-P (Filipe *et al.*, 2001; Oehmen *et al.*, 2007) (Figures 6.8, 6.9). Microbiological studies on EBPR conducted with culture-based techniques in the 1970s and 1980s indicated that *Acinetobacter* species could easily be grown on acetate (Fuhs & Chen, 1975). Mixed enriched cultures developed on acetate as the sole carbon source (Wentzel 1988, 1989a) ('enhanced EBPR cultures') or alternating acetate and propionate (Lu *et al.*, 2006) allowed enrichment of PAOs to as high as 90% of the population but have yet to succeed in isolating them in pure cultures. Fluorescent *in situ* hybridization (FISH) techniques with 16S rRNA-targeted probes showed that in laboratory and full-scale biomass, *Acinetobacter* played a minor role in EBPR processes and were not PAOs (Mino *et al.*, 1998; Oehmen *et al.*, 2007). PAO species identified using molecular techniques were named *Accumulibacter* bacteria and *Tetrasphaera* (Seviour *et al.*, 2003; Oehmen *et al.*, 2010; Nielsen *et al.*, 2012).

Four main types of PAOs have been described according to their metabolic abilities (electron acceptor and fermentation) (Oehmen *et al.*, 2010). PAO1 are denitrifiers that can use DO or, in its absence, reduce nitrate to N_2 gas. PAO2 are partial denitrifiers that can use DO or, in its absence, denitrify nitrite to N_2 gas. PAO3 are strict aerobes that can use only DO. PAO4 are aerobic and fermenting organisms that can use DO or, in its absence, ferment organic matter into VFAs.

A complete genome of *Accumulibacter* was obtained from two distinct enriched cultures of PAOs from the USA and Australia (Martin *et al.*, 2006). This information was used to determine the presence or absence of genes coding for enzymes of various metabolic processes such as the transport of P and the synthesis or degradation of poly-P, PHA, and glycogen. Such information was useful for indicating the potential biochemical pathways of the cultures analyzed.

Elemental sulfur storage by PAOs was recently characterized in an EBPR SBR treating saline wastewater containing sulfate (Wu *et al.*, 2013). Sulfur PAOs

(SPAOs) thus appear to be able to store as many as four types of storage polymers: poly-P, PHAs, glycogen, and elemental sulfur.

6.7.4 GAO/PAO competition

In the AS process, GAOs can accumulate PHAs and glycogen under anaerobic conditions, but not poly-P (Figure 6.8). GAOs thus compete with PAOs for VFA storage into PHAs and can negatively affect the efficiency of P removal in WWTPs under conditions that favor them. A strong correlation has been reported between the P release/VFA uptake ratio (P mol/C mol) and the abundance of *Accumulibacter* in a number of AS biomasses (Oehmen *et al.*, 2007). For example, 0.5 P mol released per C mol of VFA taken up was associated with an 80% abundance of *Accumulibacter* in AS while a ratio of 0.08 Pmol/Cmol was associated with an absence of this organism, this latter condition being a strong indication of an abundance of the competing GAOs.

These organisms were first named G bacteria because they could grow on glucose (Cech & Hartman, 1990), then tetrad-forming organisms (TFOs; Tsai & Liu, 2002). The name GAOs, initially proposed by Mino *et al.* (1995), is now favored. Various groups of GAOs differ in their ability to use the electron acceptors DO, nitrate and nitrite. Four types of GAOs were described for mathematical modeling by Oehmen *et al.* (2010). Two of these types are *Competibacter* that either can or cannot denitrify (respectively GB, standing for G bacteria and DGB), and two are *Defluviicoccus*, which can only denitrify nitrate to nitrite or cannot denitrify (DDEF and DEF, respectively). This can have a significant effect on a BNR system if the remaining VFAs are no longer available for denitrification and can lead to limited denitrification and consequently more recycle of nitrate to the anaerobic zone.

Conditions that favor PAO predominance over GAOs are mainly: low carbon substrate concentration, absence of chemical precipitants, high pH, low temperature, low DO concentration in the aeration basin and low nitrite concentration in the mixed liquor.

While it has been reported that a COD/P ratio of greater than 25 g COD/g P favors EBPR (Metcalf and Eddy Inc., 2003), from a PAO–GAO competition point of view, an influent rbCOD/TP ratio between 10 to 20 g COD/g TP favors PAOs while a ratio greater than 50 g/g favors GAOs, especially when a sufficient proportion of COD is in the form of VFAs (Oehmen *et al.*, 2007). The availability of propionate or the simultaneous presence of acetate and propionate was shown to favor PAOs despite some GAOs being able to use both of these substrates (Oehmen *et al.*, 2007). At a low residual acetate concentration, such as in a full-scale anaerobic zone, as opposed to a high concentration in a rapidly fed SBR under unaerated conditions, PAOs were shown to have a competitive advantage over GAOs due to their better active transport capacity (Schuler & Jenkins, 2003; Burow *et al.*, 2007; Tu & Schuler, 2013). This could explain why PAOs have an advantage in conventional AS systems leading to their uptake of all the VFAs needed to sustain them while the GAOs will take up mostly what remains. However, PAOs should have enough polyphosphates for this purpose which would result from their uptake of P in the aerobic zone, in turn minimizing the amount of remaining VFAs available to GAOs. This may be the reason why in some full scale plants it is found more efficient to add an external source of VFAs, such as from a primary sludge fermenter to the anaerobic zone, and to add only a fraction of the primary effluent flow to the anaerobic zone while by-passing the remainder directly downstream to the first anoxic zone, improving denitrification (notably by DPAOs) and limiting GAO abundance (Figure 6.5). Another situation that would play in favor of such a process configuration when the primary effluent is added to the anoxic zone would be when the influent wastewater does not undergo much fermentation in the sewer system and contains significant amounts of fermentable substrate but little VFAs, as is commonly reported in colder climates. The requirement for preferential anaerobic VFA removal by PAOs could also explain why the simultaneous addition of coagulants (iron or aluminum) to an EBPR sludge could reduce the availability of orthophosphate and limit the storage of polyphosphates, thereby limiting the uptake of VFAs by PAOs and favoring the development of GAOs (de Haas *et al.*, 2001).

A pH > 7.25 was shown to favor PAOs over GAOs, due to the higher energy requirement for the transport of VFAs at a higher pH, a condition that is more easily met with the availability of poly-P reserves (Filipe *et al.*, 2001; Schuler & Jenkins, 2003; Serafim *et al.*, 2002; Oehmen *et al.*, 2005, 2007). Above pH 8.0, however, overall rates of reaction are reduced and the competition of P precipitation with calcium may become significant (Maurer & Boller, 1999), a condition that may take place in the micro-environment of aerobic granular sludge (Mañas *et al.*, 2011).

Mixed liquor temperature <20°C was shown to favor PAOs over GAOs while temperatures >28°C tend to favor GAOs (Brdjanovic *et al.*, 1998). Some treatment plants in the Las Vegas, Nevada, USA area display excellent P removal capacity during the year except in August when the temperature is >28°C (Drury, 2013). PAOs appear to have a greater tolerance to low DO concentrations than GAOs, as indicated by their predominance in EBPR processes operated at DO <2.5 mg/L (Carvalheira *et al.*, 2013). This characteristic can be exploited to decrease aeration in EBPR processes, potentially resulting in operational savings.

A high nitrite concentration in EBPR processes was shown to negatively affect PAOs more than GAOs (Saito *et al.*, 2004; Jiang *et al.*, 2006).

In summary, an EBPR plant can be operated to favor PAOs by operating at the lowest SRT that will allow nitrification at all times of the year, with good DO control and with pH values higher than 6.8. In warm climates, the SRT may have to be lower than 4 d during the summer.

6.8 DEVELOPMENT OF MATHEMATICAL MODELS

Initial guidelines to design EBPR processes in the late 1970s were empirical and based on nominal hydraulic retention times for the anaerobic, anoxic, and aerobic zones. With the recognition of the detrimental effect of nitrate addition to the anaerobic zone and the need for rbCOD, a parametric model was developed that could be used to predict reasonably well the N and P removal by an EBPR process (Siebritz *et al.*, 1983; Ekama, 1983). In trying to understand the role of not only an 'EBPR biomass' but also that of organisms specifically responsible for EBPR, Wentzel *et al.* (1988) cultivated an enriched culture of PAOs fed with acetate as the sole carbon source. An enhanced culture kinetic model was then developed (Wentzel *et al.*, 1989b), which formed the basis for developing a simplified enhanced culture steady-state model to account for the fact that many PAO processes were virtually complete and to include the processes of aerobic and denitrifying OHOs, of nitrifiers, and chemical P precipitation (Wentzel *et al.*, 1990). Under the conditions used for their enhanced cultures, the PAOs had not exhibited much denitrification activity, and the model was largely used as the basis for the Activated Sludge Model No. 2 (ASM2) (Gujer *et al.*, 1995). Other researchers had demonstrated the denitrification capacity of PAOs (DPAOs; e.g. Vlekke *et al.*, 1988), and this behavior was considered in the updated model ASM2d (Henze *et al.*, 1999). Models similarly built around PAO and DPAO, OHO, and nitrifier behavior, as included in ASM2d were proposed by various groups (University of Cape Town Activated Sludge Model [UCTPHO], by Hu *et al.*, 2007a, b; General model by Barker and Dold, 1997; ASM3-P by Rieger *et al.*, 2001).

A metabolic model approach was used by researchers of the Technological University of Delft (TUD) to account not only for one storage polymer (PHA) as in ASM2d but also for glycogen to better represent the metabolism of PAOs (Murnleitner *et al.*, 1997; Meijer, 2004). This approach was used to explain and model the microbial ecology of PAO and GAO subgroups under various environmental and operational conditions (Oehmen *et al.*, 2010). Various commercial wastewater treatment simulators have incorporated ASM-type models and are overviewed by Olsson and Newell (1999). They include AQUASIM, BioWin, EFOR, GPS-X, SIMBA, STOAT, and WEST. Information for each can be found on their specific websites.

6.9 P REMOVAL IN AEROBIC GRANULAR SLUDGE

Aerobic granular sludge (AGS) is a variation of the activated sludge process with extremely good settling characteristics was developed in SBRs. The AGS process exhibited EBPR with faster settling and denser granules populated by PAOs (aerobic and denitrifying PAOs) and lighter granules with competing GAOs (Bassin *et al.*, 2012). Selective top and bottom wastage of granules allowed very high P removal efficiency and also out-competed GAOs. The Nereda® AGS process has continued to be developed in the Netherlands since 2003 to reduce energy, chemical consumption and space requirements (van der Roest *et al.*, 2011) and is now applied at full scale in a few locations for EBPR (Winkler *et al.*, 2012). There is still some uncertainty as to the mechanism by which the P is removed. Mañas *et al.* (2011) found biologically induced precipitation of P as hydroxyapatite (HAP; $Ca_5(PO_4)_3OH)_{(s)}$) in the core of granules. They concluded that biologically induced precipitation was responsible for 45% of the overall P removal in the operating conditions tested, with pH varying from 7.8 to 8.8. Li *et al.* (2012) compared an upflow SBR after producing granular sludge with a conventional EBPR plant using the same feed and found that the former achieved 80% P removal, while the latter removed 90%. The P release and uptake rates for the latter were considerably higher than that of the former, while the diversity and the level of PAOs in flocculent sludge were much higher than those in the granular sludge. This appeared to support the findings of others that a substantial portion of the P in the granules consists of HAP. Barr (2010) fed two different substrates to similar granular sludge SBRs in the laboratory and found that while one of the units produced granules that were white and well-defined, the other produced a yellowish granule with lots of protrusions. The white granules appeared as compact, smooth, dense aggregates dominated by 97.5% *Accumulibacter* (PAOs), and the yellow granules appeared as loose, rough, irregular aggregates with a mixed microbial population of 12% *Accumulibacter* and 58% *Competibacter* (GAOs), among other bacteria. The composition of the wastewater appears to play an important role in the types of granules formed.

6.10 RELIABILITY OF EBPR

The reliability of the EBPR process has often been questioned. This question may be hypothetical since EBPR is usually accompanied by standby chemical addition, especially when a stringent effluent standard must be met so as to compensate for unavoidable events such as a power failure or the failure of mechanical or control equipment. While EBPR can reliably reduce the orthophosphate concentration to well below 0.1 mg/L, some form of filtration with or without polymer would be required to coagulate and remove particulate and colloidal P. Furthermore, the degree of P removal will depend on the reliable availability of VFAs. Since their availability may be seasonable, some operators may prefer to add inorganic chemicals to the AS plant for short periods, rather than to augment the VFAs for this period. Some plants have readily available industrial sources of rbCOD in the influent wastewater or that may be added. At the Bushkoppies plant in Johannesburg, RSA (Osborn *et al.*, 1986), a yeast factory discharges a waste that is beneficial for EBPR. Because of past patent issues, many plants were constructed in the USA without fermenters to augment VFAs and this resulted in unreliable P removal. The Kalispell, Montana, USA plant has fermentation of primary sludge and, although it is in a cold climate, it can achieve annual average TP values of between 0.11 and 0.15 mg/L with filtration but no chemical addition (Neethling *et al.*, 2005).

Some conclusions on reliability can be drawn from the Executive Summary of a WERF report by Neethling *et al.*, (2005):

- EPBR is capable of achieving effluent P concentrations less than 0.1 mg P/L for months, less than 0.03 mg/L for a week, and even below 0.02 mg P/L for several days.
- Optimizing EBPR by adding a carbon source such as rbCOD or acetate is a viable method for reliably achieving effluent orthophosphate concentrations of as low as 0.02 mg P/L
- P release in the anaerobic zone can vary within and between plants from 20 to 70 mg P/L, with no apparent correlation to the effluent P concentration.
- Secondary release of P due to poor design or operation could result in an increase in effluent P and may require more influent VFA addition to achieve good P removal.
- The presence of GAOs in full-scale facilities does not necessarily preclude the plant from achieving low effluent P concentrations.
- For greater reliability, backup chemicals are required to ensure levels below 0.5 mg P/L of total P.
- Recycle streams containing P may affect the reliability without additional sources of VFAs.

The fermentation of a portion of the mixed liquor for EBPR is increasingly applied to improve reliability. This can be achieved through pumping approximately 10% of the mixed liquor from the anaerobic zone to a fermenter with an SRT of approximately 2 d and returning the supernatant to the anaerobic zone. It can also be attempted by switching off a mixer in one of the anaerobic cells most of the time. Typically, it would only be required to turn them on for no more than 15 min/d. If there are three cells, using the 2nd one would produce the best results. With only two anaerobic cells, switching off the first one may yield better results.

6.11 RESOURCE RECOVERY

P is a limited resource that is essential to all life forms and irreplaceable. The P in wastewater effluents is primarily discharged to water bodies and becomes dispersed in a form that cannot be recovered in the future. High quality P ore is being used up at an unsustainable rate and may last no more than 50 years. In addition, the supply of P is limited to only a few countries of which Morocco and China have more than 80% of the world's total. Lower grade ore resources may last for another 200 years, but the cost of production will increase at the same time that the population of the world will top the 10 to 12 billion mark. Conservation and recovery of P should be given a high priority. In European countries such as Sweden and the Netherlands, targets are being set for achieving approximately 70% recovery of P from wastewater. During the uptake of P by the PAOs, Mg and K are co-transported across the cell walls. During anaerobic digestion, the cells decay and release P, Mg and K. Since there is an abundance of ammonia during digestion of sludge, there is a great propensity for struvite formation, depending on the pH of the sludge. With EBPR processes, more than 95% of P is transferred to the sludge, and about 20% of that amount appears in the return stream after dewatering of the digested sludge. The remainder is either in the organic form, precipitated in the sludge as metal phosphates, or tied up with struvite in the digester. Of the P in the return streams, it is possible to recover up to 90% as struvite ($MgNH_4PO_4 \cdot 6H_2O$). This also recovers about 20% of the N. The recovery could be doubled by stripping the P and Mg from the WAS under anaerobic conditions before digestion and diverting it to the recovery process. Any of a number of proprietary processes can be used to form pure crystalline struvite crystals which are an excellent slow release fertilizer much prized by some users. Alternatives to P recovery as struvite are composting, sludge drying and pelletizing, or land application of sludge based on the P requirements. When EBPR is used, between 40 and 50% of the P in the product will be available to plants. When such sludge is incinerated, the P will remain in the ash so the same percentage would be recoverable.

6.12 REFERENCES

Alarcon G. O. (1961). Removal of Phosphorus from Sewage. MS Thesis, John Hopkins Univ., Baltimore, M D.

Arvin E. (1983) Observations supporting the removal by biologically mediated chemical precipitation – a review. *Water Sci. Technol.,* **15**(3/4), 43–63.

Barker P. S. and Dold P. L. (1997). General model for biological nutrient removal activated-sludge systems: model presentation. *Water Environ. Res.*, **69**(5), 969–984.

Barnard J. L. (1974a). Cut P and N without chemicals. *Water Wastes Eng.*, **11,** 33–36.

Barnard J. L. (1974b). Cut P and N without chemicals. *Water Wastes Eng.*, **11**, 41–44.

Barnard J. L. (1976). A review of biological phosphorous removal in activated sludge process, *Water SA*, **2**(3), 136–144.

Barnard J. L. (1984). Activated primary tanks for phosphate removal. *Water SA*, **10**(3), 121.

Barnard J. L. (1985). The Role of Full Scale Research in Biological Phosphate Removal. Proc. Univ. British Columbia Conf. on New Directions and Research in Waste Treatment and Residuals Management, Vancouver, BC, Canada, June 23–28.

Barnard J., Houweling D., Analla H. and Steichen M. (2010). Fermentation of Mixed Liquor for Phosphorus Removal. Proc. 83rd Ann. Tech. Exhibition and Conf. Water Environ. Fedn., New Orleans, LA.

Barr J. (2010). The Microbial Ecology of Granular Sludge in Enhanced Biological Phosphorus Removal. PhD. Thesis, School of Chemical Engineering, Univ. Queensland. Australia.

Bassin J. P., Winkler M. K. H., Kleerebezem R., Dezotti M. and van Loosdrecht M. C. M. (2012). Improved phosphate removal by selective sludge discharge in aerobic granular sludge reactors. *Biotechnol. Bioeng.*, **109**(8), 1919–1928.

Brdjanovic D., Logemann S., van Loosdrecht M. C. M., Hooijmans C. M., Alaerts G. J. and Heijnen J. J. (1998). Influence of temperature on biological phosphorus removal: process and molecular ecological studies. *Water Res.*, **32**(4), 1035–1048.

Buchan L. (1983). Possible biological mechanisms of phosphorus removal. *Water Sci. Technol.*, **15**, 87.

Burow L. C., Kong Y. H., Nielsen J. L., Blackall L. L. and Nielsen P. H. (2007). Abundance and ecophysiology of *Defluviicoccus* spp., glycogen-accumulating organisms in full-scale wastewater treatment processes. *Microbiol.*, **153**, 178–185.

Carvalheira M., Oehmen M., Carvalho G. and Reis M. A. M. (2013). The Effect of Dissolved Oxygen Concentration on the Competition between Polyphosphate Accumulating Organisms and Glycogen Accumulating Organisms. Proc. IWA-WEF Nutrient Removal and Recovery Conf., Vancouver, Canada, July 28–31.

Cech J. S. and Hartman P. (1990). Glucose induced break down of enhanced biological phosphate removal. *Environ. Technol.*, **11**, 651–656.

Clark K. and Neethling J. B. (2009). Operation of 5-Stage Bardenpho Process with Clarifier/filters to Achieve Low Effluent Phosphorus. Presented at *WEFTEC 2009 Workshop W216, WEF/WERF Real-World LOT Nutrient Removal: Available Limits and Statistical Reliability*, October 11, 2009.

Comeau Y., Hall K. J., Hancock R. E. W. and Oldham W. K. (1986). Biochemical model for enhanced biological phosphorus removal. *Water Res.*, **20**, 1511–1521.

de Haas D. W., Wentzel M. C. and Ekama G. A. (2001). The Use of simultaneous chemical precipitation in modified activated sludge systems exhibiting biological excess phosphate removal – Part 6: Modelling of simultaneous chemical-biological P removal – review of existing models. *Water SA*, **27**, 135–150.

Drury D. (2013).Personal communication.

Ekama G. A., Siebritz I. P. and Marais G. v. R. (1983). Considerations in the process design of nutrient removal activated sludge processes. *Water Sci. Technol.*, **15**(3/4), 283–318.

Filipe C. D. M., Daigger G. T. and Grady C. P. L. Jr. (2001). A metabolic model for acetate uptake under anaerobic conditions by glycogen accumulating organisms: stoichiometry, kinetics, and the effects of pH. *Biotechnol. Bioeng.*, **76**, 17–31.

Fuhs G. W. and Chen M. (1975). Microbiological basis of phosphate removal in the activated sludge process for the treatment of wastewater. *Microbial Ecol.*, **2**(2), 119–138.

Gerber A. and Mostert E. H., Winter C. T. and de Villiers R. H. (1986). The effect of acetate and other short-chain compounds on the kinetics of biological nutrient removal processes. *Water SA*, **12**, 7.

Gujer W., Henze M., Mino T., Matsuo T., Wentzel M. C. and Marais G. v. R. (1995). The activated-sludge model No. 2 – biological phosphorus removal. *Water Sci. Technol.*, **31**(2), 1–11.

Harold F. M. (1966). Inorganic polyphosphates in biology: structure, metabolism and function. *Bacteriol. Revs*, **30**, 772–784.

Henze M., Gujer W., Mino T., Matsuo T., Wentzel M. C., Marais G. v. R. and van Loosdrecht M. C. M. (1999). Activated sludge model No. 2d ASM2d. *Water Sci. Technol.*, **39**(1), 165–182.

Hu Z.-R., Wentzel M. C. and Ekama G. A. (2007a). A general kinetic model for biological nutrient removal activated sludge systems: model development. *Biotechnol. Bioeng.*, **98**(6), 1242–1258.

Hu Z.-R., Wentzel M. C. and Ekama G. A. (2007b). A general kinetic model for biological nutrient removal activated sludge systems: model evaluation. *Biotechnol. Bioeng.*, **98**(6), 1259–1275.

Jiang Y. F., Wang B. Z., Wang L., Chen J. M. and He S. B. (2006). Dynamic response of denitrifying poly-P accumulating organisms batch culture to increased nitrite concentration as electron acceptor. *J. Environ. Sci. Hlth, Part a – Toxic/Hazardous Substances & Environ. Eng.*, **41**(11), 2557–2570.

Levin G. V. and Shapiro J. (1965). Metabolic uptake of phosphorus by wastewater organics. *J. Water Pollut. Control Fedn.*, **37**(6), 800–821.

Li X., Gao D., Liang H., Liu L. and Fu Y. (2012). Phosphorus removal characteristics of granular and flocculent sludge in SBR. *Appl. Microbial Technol.,* **94**(1), 231–236.

Lu H., Oehmen A., Virdis B., Keller J. and Yuan Z. (2006). Obtaining highly enriched cultures of candidatus Accumulibacter phosphatis through alternating carbon sources. *Water Res.*, **40**(20), 3838–3848.

Mañas A., Biscans. B. and Spérandio M. (2011). Biologically induced phosphorus precipitation in aerobic granular sludge process. *Water Res.*, **45**(12), 3776–3786.

Marais G. v. R., Loewenthal R. E. and Siebritz I. P. (1983). Review: observations supporting phosphate removal by biological excess uptake. *Water Sci.Technol.*, **15**(3/4), 15–41.

Martin H. G., Ivanova N., Kunin V., Warnecke F., Barry K. W., McHardy A. C., Yeates C., He S. M., Salamov A. A., Szeto E., Dalin E., Putnam N. H., Shapiro H. J., Pangilinan J. L., Rigoutsos I., Kyrpides N. C., Blackall L. L., McMahon K. D. and Hugenholtz P. (2006). Metagenomic analysis of two enhanced biological phosphorus removal (EBPR) sludge communities. *Nature Biotechnol.*, **24**(10), 1263–1269.

Maurer M. and Boller M. (1999). Modelling of phosphorus precipitation in wastewater treatment plants with enhanced biological phosphorus removal. *Water Sci. Technol.*, **39**(1), 147–163.

Meijer S. C. F (2004). Theoretical and Practical Aspects of Modeling Activated Sludge Processes. PhD. Thesis, Delft Univ.of Technol., ISBN 90-9018027-3.

Menar A. B. and Jenkins D. (1969). The Fate of Phosphorus in Wastewater Treatment Processes. The Enhanced Removal of Phosphate by Activated Sludge. Proc. 24th Industr. Waste Treat. Conf., Purdue Univ. West Lafayette, IN, USA.

Metcalf and Eddy Inc. (2003). *Wastewater Engineering: Treatment and Reuse*, McGraw-Hill, New York.

Milbury W. F., McCaluley D. and Hawthorne C. H. (1971). Operation of conventional activated sludge for maximum phosphorus removal. *J. Water Pollut. Control Fedn*, **43**(9), 1890–1901.

Mino T., Arun V., Tsuzuki Y. and Matsuo T. (1987). Effect of Phosphorus Accumulation on Acetate Metabolism in the Biological Phosphorus Removal Process, In: Adv. Water Pollution Control, R. Ramadori (ed.), Pergamon Press, Oxford, UK, pp. 27–38.

Mino T., Liu W. T., Kurisu F. and Matsuo T. (1995). Modeling glycogen-storage and denitrification capability of microorganisms in enhanced biological phosphate removal processes. *Water Sci. Technol.*, **31**(2), 25–34.

Mino T., van Loosdrecht M. C. M. and Heijnen J. J. (1998). Microbiology and biochemistry of the enhanced biological phosphate removal process. *Water Res.*, **32**(11), 3193–3207.

Murnleitner E., Kuba T., van Loosdrecht M. C. M. and Heijnen J. J. (1997). Integrated metabolic model for the aerobic and denitrifying biological phosphorus removal. *Biotechnol. Bioeng.*, **54,** 434–450.

Neethling J. B., Bakke B., Benisch M., Gu A., Stephens H., Stensel H. D. and Moore R. (2005). Factors Influencing the Reliability of Enhanced Biological Phosphorus Removal. WERF Report 01-CTS-3.

Nicholls H. A. (1975). Full scale experimentation on the new Johannesburg extended aeration plants. *Water SA*, **1**(3), 121.

Nicholls H. A. (1987). Improvement of the Stability of the Biological Phosphorus Removal Process at the Johannesburg Northern Works. Proc. IAWPRC International Specialized Conf. on Biological Phosphate Removal from Wastewater, Rome, Italy, pp. 261–272.

Nielsen P. H., Saunders A. M., Hansen A. A., Larsen P. and Nielsen J. L. (2012). Microbial communities involved in enhanced biological phosphorus removal from wastewater – a model system in environmental biotechnology. *Current Opinion Biotechnol.*, **23**(3), 452–459.

Oehmen A., Vives M. T., Lu H., Yuan Z. and Keller J. (2005). The effect of pH on the competition between polyphosphate accumulating organisms and glycogen-accumulating organisms. *Water Res.*, **39**(15), 3727–3737.

Oehmen A., Lemos P. C., Carvalho G., Yuan Z., Keller J., Blackall L. L. and Reis M. A. M. (2007). Advances in enhanced biological phosphorus removal: from micro to macro scale. *Water Res.*, **41**(11), 2271–2300.

Oehmen A., Lopez-Vazquez C. M., Carvalho G., Reis M. A. M. and van Loosdrecht M. C. M. (2010). Modelling the population dynamics and metabolic diversity of organisms relevant in anaerobic/anoxic/aerobic enhanced biological phosphorus removal processes. *Water Res.*, **44**(15), 4473–4486.

Oldham W. K. and Stevens G. M. (1984). Initial operating experiences of nutrient removal process (modified bardenpho) at Kelowna, British Columbia. *Can. J. Civil Eng.*, **11**, 474.

Olsson G. and Newell B. (1999). Wastewater Treatment Systems. Modelling, Diagnosis and Control. IWA Publishing, London, UK.

Osborn D. W., Lotter L. H., Pitman L. H. and Nicholls H. A. (1986). Enhancement of Biological Phosphate Removal by Altering Process Feed Composition. Water Research Commission, WRC Report. Pretoria, South Africa.

Rabinowitz B. and Oldham W. K. (1986). Excess biological phosphorus removal in the activated sludge process using primary sludge fermentation. *Can. J. Civil Eng.*, **13**, 345–351.

Randall A. A., Benefield L. D. and Hill W. E. (1994). The effect of fermentation products on enhanced biological phosphorus removal, poly-phosphate storage and microbial population dynamics. *Water Sci. Technol.*, **30**(6), 213–219.

Rieger L., Koch G., Kuhni M., Gujer W. and Siegrist H. (2001). The EAWAG bio-P module for activated sludge model No. 3. *Water Res.*, **35**(16), 3887–3903.

Saito T., Brdjanovic D. and van Loosdrecht M. C. M. (2004). Effect of nitrite on phosphate uptake by phosphate accumulating organisms. *Water Res.*, **38**(17), 3760–3768.

Schuler A. J. and Jenkins D. (2003). Enhanced biological phosphorus removal from wastewater by biomass with different phosphorus contents, Part 1: Experimental results and comparison with metabolic models. *Water Environ. Res.*, **75**(6), 485–498.

Serafim L. S., Lemos P. C. and Reis M. A. M. (2002). Effect of pH control on EBPR stability and efficiency. *Water Sci. Technol.*, **46**(4–5), 179–184.

Seviour R. J. Mino T. and Onuki M. (2003) The microbiology of biological phosphorus removal in the activated sludge systems. *FEMS Microbiol. Revs.*, **27**(1), 99–127.

Siebritz I., Ekama G. A. and Marais G. v. R. (1983). Biological Excess Phosphorus Removal in the Activated Sludge Process. Research Rept W47, Dept. Civil Eng., Univ. Cape Town, South Africa.

Smolders G. J., van der Meij J., van Loosdrecht M. C. M. and Heijnen J. J. (1994a). Model of the anaerobic metabolism of the biological phosphorus removal process: stoichiometry and pH influence. *Biotechnol. Bioeng.*, **43**(6), 461–470.

Smolders G. J., van der Meij M. J., van Loosdrecht M. C. M. and Heijnen J. J. (1994b). Stoichiometric model of the aerobic metabolism of biological phosphorus removal process. *Biotechnol. Bioeng.*, **44**(7), 837–848.

Srinath E. G., Sastry C. A. and Pillai S. C. (1959). Rapid removal of phosphorus from sewage by activated sludge. *Water Waste Treat.*, **11**, 410.

Stroud R. and Martin C. (2001). South Carey Water Reclamation Facility's Nutrient Removal Modifications and Reduction Success, Proc. WEFTEC, 2001. Town of Carey, NC, USA.

Tsai C. S. and Liu W. T. (2002). Phylogenetic and physiological diversity of tetrad-forming organisms in deteriorated biological phosphorus removal systems. *Water Sci. Technol.*, **46**(1–2), 179–184.

Tu Y. and Schuler A. (2013). Metabolism of Acetate Transport in EBPR and Biokinetics for PAOs and GAOs. Proc. IWA-WEF Nutrient Removal and Recovery Conf., Vancouver, Canada, July 28–31.

Vacker P., Connell C. H. and Wells W. N. (1967). Phosphate removal through municipal wastewater treatment at San Antonio, Texas. *J. Water Pollut. Control Fedn*, **39**(5), 750–771.

van der Roest H. F., de Bruin L. M. M., Gademan G. and Coelho F. (2011). Towards sustainable wastewater treatment with Dutch Nereda® technology. *Water Practice Technol.*, doi:10.2166/wpt.2011.059.

van Huyssteen J. A., Barnard J. L. and Hendriksz J. (1990). The Olifantsfontein nutrient removal plant. *Water Sci. Technol.*, **22**(7/8), 1.

Vlekke G. J. F. M., Comeau Y. and Oldham W. K. (1988). Biological phosphate removal from wastewater with oxygen or nitrate in sequencing batch reactors. *Environ. Technol. Lett.*, **9**, 791–796.

Vollenweider R. A. (1968). Scientific Fundamentals of the Eutrophication of Lakes and Flowing Waters, with Particular Reference to Nitrogen and Phosphorus as Factors in Eutrophication. Paris: Organization for Economic Co-Operation and Development. 1968. Technical Report DAS/CS 1/68.27, p. 250.

Vollertsen J., Petersen G. and Borregaard V. R. (2005). Hydrolysis and Fermentation of Activated Sludge to Enhance Biological Phosphorus Removal, Proc. IWA Specialized Conf. on Nutrient Management in Wastewater Treatment Processes and Recycle Streams, Krakow, Poland, 18–21, September 2005.

Wentzel M. C., Dold P. L., Ekama G. A. and Marais G. v. R. (1985). Kinetics of biological phosphorus release. *Water Sci. Technol.*, **17**, 57.

Wentzel M. C., Lötter L. H., Loewenthal R. E., and Marais G. v. R. (1986). Metabolic behaviour of *Acinetobacter* spp. in enhanced biological phosphorus removal – A biochemical model. *Water SA*, **12**, 209–224.

Wentzel M. C., Loewenthal R. E., Ekama G. A., and Marais G. v. R. (1988). Enhanced Polyphosphate Organism Cultures in Activated Sludge Systems: Part I. Enhanced Culture Development. *Water SA*, **14**(2), 81–92.

Wentzel M. C., Loewenthal R. E., Ekama G. A. and Marais G. v. R. (1989a). Enhanced polyphosphate organism cultures in activated sludge systems. Part II. Experimental behavior. *Water SA*, **15**(2), 71–88.

Wentzel M. C., Dold P. L., Ekama G. A., and Marais G. v. R. (1989b). Enhanced polyphosphate organism cultures in activated-sludge systems. 3. Kinetic-model. *Water SA*, **15**(2), 89–102.

Wentzel M. C., Ekama G. A., Dold P. L. and Marais G. v. R. (1990). Biological excess phosphorus removal – Steady-state process design. *Water SA*, **16**(1), 29–48.

Winkler M. K. H., Kleerebezem R., and van Loosdrecht M. C. M. (2012). Integration of anammox into the aerobic granular sludge process for main stream wastewater treatment at ambient temperatures. *Water Res.*, **46**(1), 136–144.

Witherow J. L. (1970). Phosphorus Removal in Activated Sludge. Proc. 24th Industr. Waste Conf., Purdue Univ., West Lafayette, IN, USA. 1169.

Wu D., Ekama G. A., Lu H., Chui H.-K., Liu W.-T., Brdjanovic D., van Loosdrecht M. C. M. and Chen G. H. (2013). A new biological phosphorus removal process in association with sulfur cycle. *Water Res.*, **47**(9), 3057–3069.

Yall I., Boughton W. H. and Sinclair N. A. (1970). Biological uptake of phosphorus by activated sludge. *Appl. Microbiol.*, **20**, 145–150.

Chapter 7

Micro-pollutant removal

Hansruedi Siegrist (Switzerland), Adriano Joss (Switzerland), Thomas A. Ternes (Germany)

7.1 INTRODUCTION

Today more than 100,000 different chemicals are registered in the European Union (EU), of which some 30,000 are distributed in the market in quantities in excess of one ton per year (Giger, 2002). During production and disposal, as well as after regular use, a significant portion of these substances will unavoidably enter the aquatic environment. Through constantly improving methods of chemical analysis, organic compounds are increasingly being detected in water bodies and in sewage sludge in low concentration ranges (μg/L and ng/L and lower); these are designated organic micropollutants.

Only in the last two decades has the focus of environmental chemistry research been extended from the more 'classic' environmental micropollutants such as PCBs, DDT, dioxins and pesticides to the so called 'emerging pollutants', although these have been released for much longer. Prominent among emerging pollutants are pharmaceuticals, hormones, and cosmetic ingredients (PPCP) as well as biocides that enter the environment mainly through regular domestic use and in municipal wastewater. Additional compounds leaching out from electrical products (flame retardants) or used to create inert surfaces (perfluorinated compounds) also give reason for concern. Hospital wastewaters contribute significantly but not to a major extent, to the total pharmaceutical and biocide loads. Compounds not readily degraded in municipal wastewater treatment plants (WWTP) are discharged to the receiving waters either in the dissolved states or sorbed onto suspended solids (TSS). Therefore, the removal efficiency of WWTPs is crucial in determining the contamination of the receiving water bodies. In surface water and during underground passage, further elimination is observed by photochemical, sorption and biological processes but micropollutants can still enter groundwater where they can act as tracers for wastewater pollution. Some polar pollutants (e.g. lipid regulators, carbamazepine and iodinated X-ray contrast media) might even pass through common drinking water treatment processes.

A German study (Ternes, 1998) investigated the occurrence of 55 pharmaceutically active ingredients and 9 metabolites in the discharge of 49 sewage treatment plants and in the receiving water bodies. In total, 36 active agents and 5 metabolites were detected in the WWTP effluent in concentrations of up to several μg/L. Even in surface waters, some pollutant (e.g. beta blockers and anti-epileptic drugs) peak concentrations were measured in excess of 1 μg/L.

There has been much discussion in recent years of new environmental effects, such as fish feminization (Jobling & Tyler, 2003). This is partly a result of the chronic introduction of endocrine disrupting chemicals. These include, along with the body's natural hormones, pharmaceuticals used for their hormonal effect, such as contraceptives and anti-diabetic drugs. A hormonal side effect is also attributed to certain pharmaceuticals such as β-sitosterol (control of cholesterol level, also called phytoestrogen as it is widely distributed in plants) and clenbuterol (asthma relief agent). Endocrine effects are reported also for nonylphenol (surfactant metabolite) and industrial chemicals such as bisphenol A and phthalates.

7.2 FATE OF MICROPOLLUTANTS IN AS TREATMENT

The removal of trace substances in WWTPs depends on the level of the biological treatment and the efficiency of the TSS separation methods. In the last forty years, biological wastewater purification has been adapted step by step to the tightening effluent conditions for COD, TSS and nutrient removal. The most important micropollutant elimination processes are:

- Sorption to TSS in mechanical and biological treatment, and removal by sedimentation and filtration;
- Stripping by aeration, which is negligible for the considered micropollutants since they are mostly large, lipophilic or polar compounds with low volatility; and
- Biological degradation and transformation by the microorganisms in AS; this may achieve biological mineralization but is mostly limited to compound transformation (Wick *et al.*, 2011a).

For *sorption* of organic trace substances, a distinction is made between:

- *Absorption*: hydrophobic interactions of the aliphatic and aromatic groups of a compound with the lipophilic cell membrane of the microorganisms and the lipid fractions of the sludge (compounds with high octanol–water partition coefficients (K_{OW}), e.g. tonalide in Figure 7.1).

- *Adsorption*: electrostatic interactions of positively charged groups of chemicals with the negatively charged surfaces of the microorganisms (e.g. norfloxacin in Figure 7.1).

The sorbed concentration of a substance C_{sorb} (µg/L), can be expressed by the Freundlich model or for wastewater TSS by a simplified linear model (Joss *et al.*, 2006b; Wick *et al.*, 2011b). It is dependent upon the sorption constant K_{d} (L/g_{TSS}), the TSS concentration or the amount of sludge produced (SP; g_{TSS}/L) to which the substance can adhere, and the dissolved substance concentration C_{diss} (µg/L):

$$C_{\text{sorb}} = K_{\text{d}} \cdot SP \cdot C_{\text{diss}} \tag{7.1}$$

Simplified, for a fully mixed system the removal efficiency η_{sorb} (–) is:

$$\eta_{\text{sorb}} = \frac{C_{\text{sorb}}}{C_{\text{diss}} + C_{\text{sorb}}} = \frac{K_{\text{d}} \cdot \text{SP}}{1 + K_{\text{d}} \cdot \text{SP}} \tag{7.2}$$

K_{d} can be roughly estimated from the octanol–water distribution coefficient for non-polar compounds. For polar and charged compounds with electrostatic interactions, it must be determined by means of sorption tests. A substance with a low K_{OW} that sorbs relatively well to TSS is the antibiotic norfloxacin. The sorption is based to a large extent on electrostatic interactions between the positively charged amino group of norfloxacin and the negatively charged surfaces of the microorganisms. In a study carried out in the Zurich WWTP, Golet *et al.* (2003) confirmed that more than 80% of the norfloxacin in the primary effluent was sorbed to the secondary sludge (Figure 7.1), while removal in primary treatment was significantly lower.

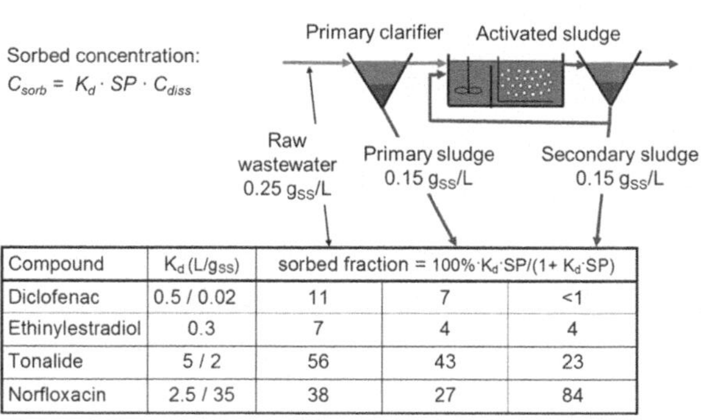

Compound	K_d (L/g_{SS})	sorbed fraction = 100%·K_d·SP/(1+ K_d·SP)		
Diclofenac	0.5 / 0.02	11	7	<1
Ethinylestradiol	0.3	7	4	4
Tonalide	5 / 2	56	43	23
Norfloxacin	2.5 / 35	38	27	84

Figure 7.1 Sorption of some compounds in raw wastewater, primary and secondary sludge based on the K_d values given in Joss *et al.* (2006a) and Golet *et al.* (2003) assuming no biological degradation. The sorbed fraction is related to the inlet of the primary clarifier and the inlet to biology, respectively. K_d values are given for primary and secondary sludge TSS.

The reason for this is that micro-organisms in the waste secondary sludge represent the greater proportion of the TSS compared to primary sludge, resulting in a relatively high sorption constant $K_d \approx 35\ L/g_{TSS}$. For the primary sludge the sorption constant of norfloxacin is much lower ($K_d \approx 2.5\ L/g_{TSS}$), because in spite of having similar TSS concentrations, the primary sludge contains fewer microorganisms but has instead a large lipid fraction. Thus, only 27% norfloxacin is sorbed to the primary sludge. The hydrophobic musk fragrance tonalide sorbs more strongly to the lipophilic primary sludge than to the secondary sludge (Figure 7.1).

With other substances, such as the anti-inflammatory diclofenac and estrogens the proportion sorbed is significantly smaller (Figure 7.1). K_d values for other pharmaceuticals, personal care products and biocides are given in Joss *et al.* (2006a) and Wick *et al.* (2011b).

For membrane bioreactors (MBRs), the effluent is virtually free of TSS and higher airflows per volume of treated wastewater are required for membrane fouling control than in conventional WWTP. Therefore the removals by sorption and stripping (see below) are higher.

Stripping due to AS aeration is mostly rather low because of the low volatility of the large and rather hydrophilic compounds under consideration. For the dimensionless Henry's constant $H < 1$, describing the equilibrium between the gas ($C_{air,off}$) and water phase (C_{diss}), the rising gas bubble is in equilibrium with the dissolved concentration and:

$$C_{air,off} = H \cdot C_{diss} \tag{7.3}$$

The stripping efficiency is therefore only dependent on the Henry constant and the airflow. For the relatively volatile musk fragrance tonalide, having a dimensionless Henry's coefficient of $H_{Tonalide} \sim 0.005$ and an air to water flow ratio $q_{air} = Q_{air}/Q_{water} = 8–10$ for conventional plants and 20–30 for MBRs, the stripping efficiency is:

$$\begin{aligned}\eta_{stripp} &= \frac{H \cdot C_{diss} \cdot q_{air}}{C_{diss} + H \cdot C_{diss} \cdot q_{air}} = \frac{H \cdot q_{air}}{1 + H \cdot q_{air}} \\ &\cong H_{Tonalide} \cdot q_{air} = 0.04 - 0.15\end{aligned} \tag{7.4}$$

Biological degradation and transformation of a substance can take place under aerobic, anoxic and/or anaerobic conditions. Due to their low concentrations (ng/L to μg/L), single substrate growth on micropollutants is not assumed possible and we have to distinguish between (Siegrist *et al.*, 2003):

- Co-metabolism, when the bacteria only transform but do not mineralize the trace substance and do not use it as a carbon and or energy source; and
- Mixed substrate growth, in which the bacteria use the trace substance as a carbon and energy source, and hence can totally mineralize it.

Both removals are thought to arise through the chance affinity of a trace substance with the bacterial enzymes in the AS. To some extent, the transformation or mineralization increases with the sludge age because slow growing bacteria can become established, thereby increasing the microbial diversity (Figure 7.2).

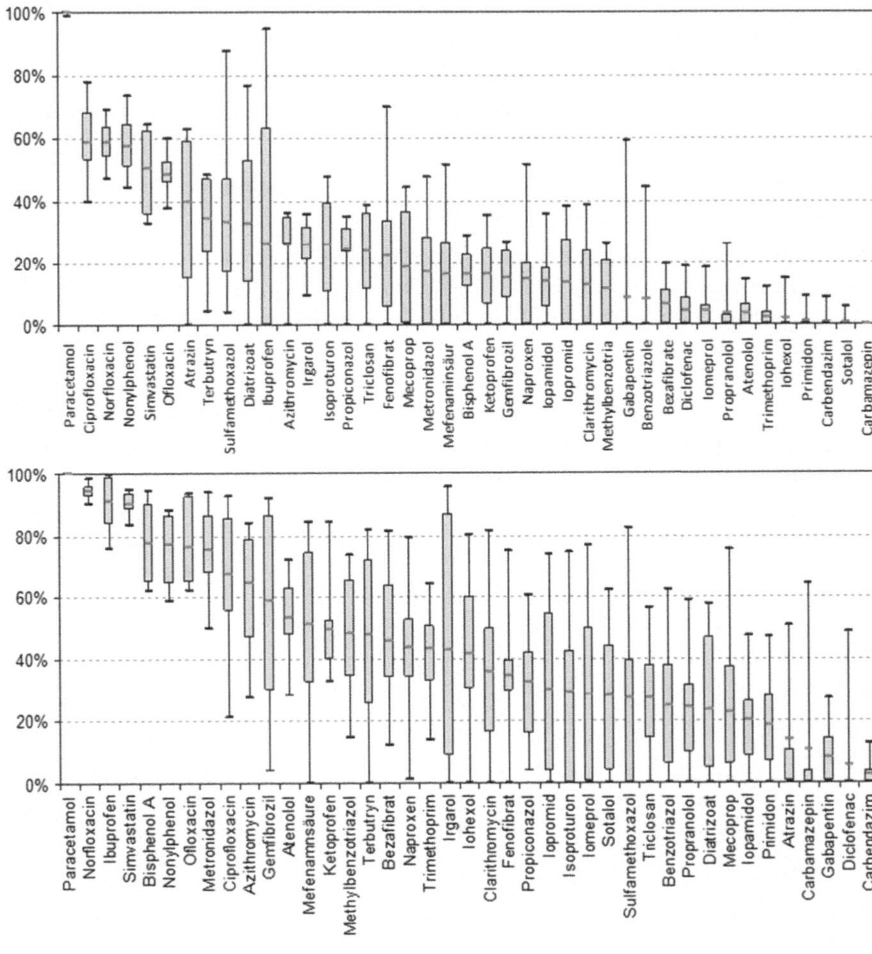

Figure 7.2 Biological degradation of a compound depends on the sludge age. Comparison of the elimination of different pharmaceuticals and day care products in a BOD removal plant (above) and nitrifying plant below (Margot *et al.*, 2011).

Degradation also depends on the redox state of the process. This is demonstrated by the contraceptive 17α-ethinylestradiol (Figure 7.3), which requires nitrifying conditions for degradation and is hardly degraded under anoxic and anaerobic conditions (Andersen *et al.*, 2003).

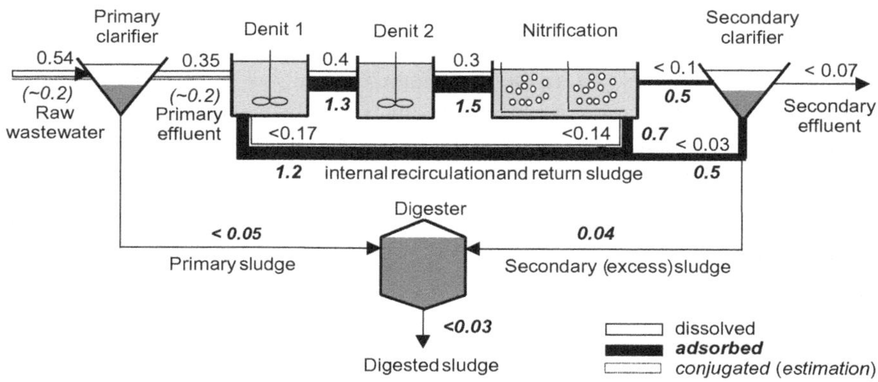

Figure 7.3 Substance flow and breakdown of the contraceptive 17α-ethinylestradiol in the Wiesbaden, Germany WWTP (Andersen *et al.*, 2003). The data are in g/d. The value in the inflow includes both dissolved 17α-ethinylestradiol and its conjugated form.

While membrane bioreactors (MBRs) sometimes have significantly higher sludge ages than conventional plants their micropollutant removal is only slightly increased above sludge ages of 15 d (Figure 7.4).

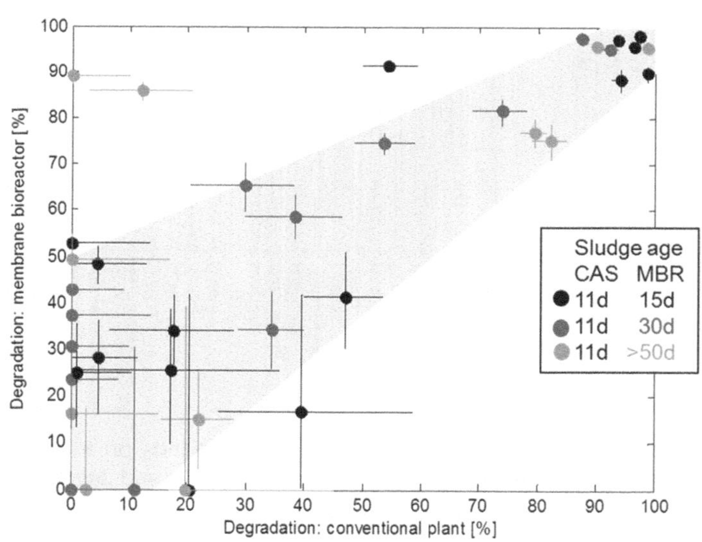

Figure 7.4 Degradation of PPCPs in an MBR (sludge age 15, 30, >50 d, black, dark grey, grey circles) compared to those in a conventional AS system (sludge age 11 d) fed with the same wastewater (bars: 95% confidence interval) (Joss *et al.*, 2006a).

Due to the low concentrations of trace substances, the elimination occurs mostly at a reaction rate r_{biol} (µg/L, d), which can be described as the product of the first order reaction constant k_{biol} (L/g_{TSS}, d), the AS concentration, X_{SS} (g_{TSS}/L) and the dissolved concentration (µg/L) (Wick *et al.*, 2009; Joss *et al.*, 2006b):

$$r_{\text{biol}} = k_{\text{biol}} \cdot X_{\text{TSS}} \cdot C_{\text{diss}} \tag{7.5}$$

Because of this, a compartmentalized aeration basin results in lower discharge concentrations compared to a single completely mixed basin.

Joss *et al.* (2006b) present a classification of degradable pharmaceuticals and personal care products based on first order removal rate coefficients in typical nutrient removal WWTPs (Figure 7.5).

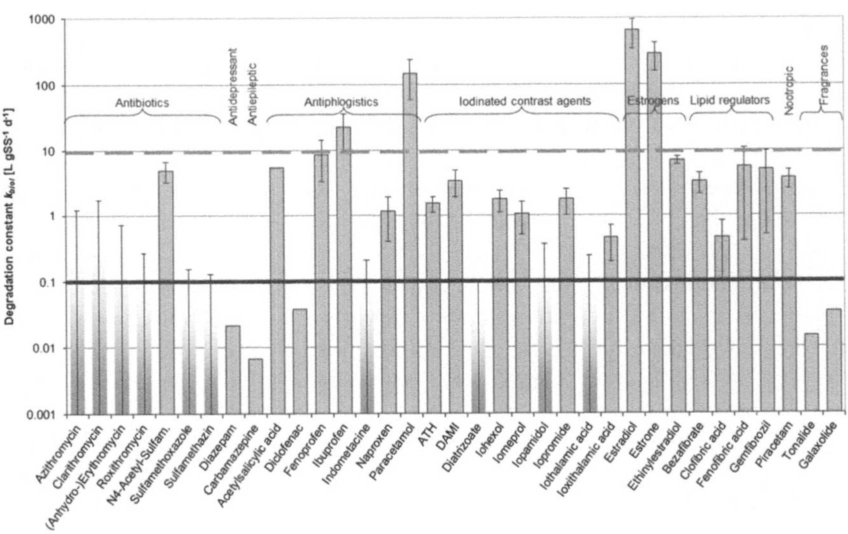

Figure 7.5 Kinetic degradation constants of 35 pharmaceuticals, hormones and PCPPs observed in sludge from municipal nutrient-removal WWTPs. The error bars indicate the 95% confidence interval. The lines at k_{biol} 0.1 and 10 indicate the limits of <20% and >90% removal for a nitrifying AS system with pre-denitrification (sludge age 11–15 d). The fading grey bars indicate high uncertainty because the measured concentrations are close to the detection limit (Joss *et al.*, 2006b).

7.3 BIOLOGICAL TRANSFORMATION PRODUCTS

Even though transformation products (TPs) have gained increasing interest as water contaminants, only a few studies have investigated the formation and fate of biological transformation products of pharmaceuticals in contact with biologically active matrices such as AS or sediments (DWA, 2014). One reason for this is the challenge of the structural elucidation of TPs present at low concentrations in natural matrices. Sophisticated analytical techniques are needed such as hybrid high-resolution mass spectrometry and nuclear magnetic resonance (NMR) (Kormos *et al.*, 2009). With a few exceptions of very simple transformation reactions (e.g. hydrolysis of amides and esters), quadrupole mass spectrometry and even high resolution mass spectrometry MS (e.g. LC-LTQ-Orbitrap MS) have not been sufficient to obtain or confirm the chemical structures of TPs, even though the parent compound is known. The TP structure suggestions based on exact masses and mass fragments have to be confirmed by alternative analytical methods or by specific chemical reactions that take place with the new functional moieties formed. Possibilities of analytical methods include a range of nuclear magnetic resonance spectroscopic methods (NMR) and to a much lesser extent infrared spectroscopy. A drawback of both types of technique is the elevated quantity and the high purity needed for isolated standards. In cases where no authentic standard is available and only the MS spectra of the TPs have been obtained, we might better define the suggestions of the TP chemical structures as 'tentative identification' unless other confirming criteria are available for the proposed chemical structure. A comprehensive overview of the literature up to 2008 on the detection and identification of pharmaceutical TPs has been provided by Celiz *et al.* (2009).

Several recent studies have indicated that the majority of pharmaceutical TPs formed under aerobic conditions have a slightly modified molecular structure with increased polarity due to the introduction of hydroxyl-, carboxyl- or keto-moieties (Neilson & Allard, 2007; Helbling *et al.*, 2010). Based on the similarity of their molecular structure to the parent compound, a significant number of TPs are expected to possess comparable biological activities to their chemical precursors (van Zelm *et al.*, 2010). However, the enhanced polarity should improve the permeability of these compounds during several water treatment processes such as adsorptive filtration (e.g. activated carbon), underground soil passage or bank filtration. As a consequence, the likelihood increases that TPs are contaminating groundwater and drinking water (Kormos *et al.*, 2010).

Several enzyme-catalysed reactions seem to be quite common in the transformation of pharmaceuticals. These include: mono- and dihydroxylation, alcohol and aldehyde oxidation, ester and amide hydrolysis, *N*-dealkylation, *N*-deacetylation and decarboxylation.

As an example the results of the interesting and complex codeine transformation study by Wick *et al.* (2011a) are described in Figure 7.6. This example indicates that during biological treatment a combination of biotic, enzymatic controlled reactions, and abiotic reaction lead to the formation of a wide range of transformation products.

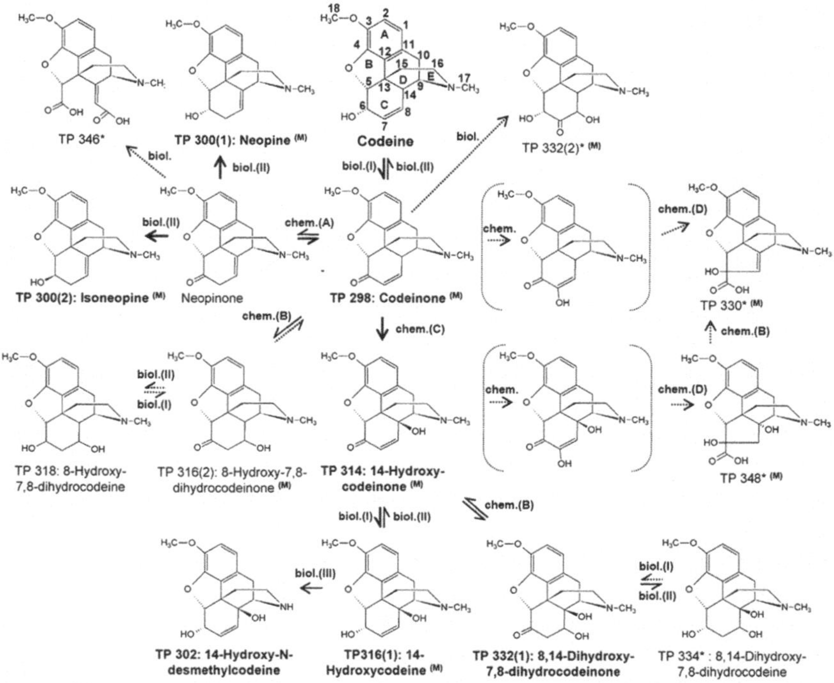

Figure 7.6 Proposed transformation pathways of codeine determined from laboratory batch experiments with codeine and codeine TPs in diluted AS as well as in autoclaved controls under aerobic conditions. Major TPs are printed in bold. The dashed arrows indicate predicted reactions which could not be directly confirmed in batch experiments. Tentative TP chemical structures are marked with an asterisk. The index (M) indicates that the corresponding TPs were also detected for morphine (O–CH_3 at C3 replaced by OH). Proposed intermediates in the formation of TPs are shown in parenthesis. The roman numerals I–III and the letters A–D indicate biotic and abiotic reactions, respectively: I) oxidation, II) reduction, III) N-demethylation; (A) double bond shift, (B) nucleophilic addition, (C) hydroxylation, (D) benzylic acid rearrangement.

7.4 MEASURES TO BE TAKEN TO IMPROVE MICRO-POLLUTANT REMOVAL AND THEIR EFFECT ON AS TREATMENT

Many micropollutants and their intermediates are polar and are only biologically degraded or transformed to a small degree or not at all; furthermore, their sorption is also limited (Wick *et al.*, 2011a). They are only partly removed in the WWTPs and are discharged to the receiving water in the plant effluent. Using a mass flux model, which includes chemical consumption, transformation and excretion in the human body, removal in the WWTP, as well as the hydraulic conditions in the receiving waters during dry weather conditions, the concentration ranges in Swiss river systems were modeled and validated with measurements (Ort *et al.*, 2009). Downstream of more than 100 WWTPs the concentration of eco-toxicologically critical compounds exceeded the proposed environmental quality standards because of the low dilution of the treated wastewater. This situation is likely to become even more critical with climate change, since drier and warmer summers are expected.

Parallel to *measures at the source*, for example eco-labeling (Agerstrand *et al.*, 2009; Klaschka *et al.*, 2004), ban of critical compounds and separate treatment of concentrated waste streams (e.g. hospital wastewater), *technical measures at the WWTPs* using advanced treatment methods are required to protect sensitive receiving waters and water resources. Two main additional treatment steps have been investigated and tested in full scale: ozonation with biological post treatment and powdered activated carbon (PAC) addition as a post-treatment step with a contact tank and a clarifier followed by a polishing filter (Hollender *et al.*, 2009; Metzger, 2010; Boehler *et al.*, 2012; Abegglen *et al.*, 2012).

Ozonation of the biologically treated wastewater with 3–10 gO_3/m^3 wastewater for a background dissolved organic carbon (DOC) of 5–15 mg/L is a cost-effective method for >90% oxidation of pharmaceuticals and pesticides; however reactive oxidation by-products are formed mainly by reactions with background DOC (Krauss *et al.*, 2010; Abegglen *et al.*, 2009; Benner & Ternes, 2009) and these have to be degraded in a biofilm system, for example a sand filter following the ozonation step (Huber *et al.*, 2005; Hollender *et al.*, 2009; Zimmermann *et al.*, 2011). PPCPs are normally not detectable, and their eco-toxicological effect is removed, whereas the iodinated contrast media were only partly oxidized (Ternes *et al.*, 2003; DWA K8.1, 2013). The effectiveness of ozone depends on the background load of the wastewater with dissolved organic carbon compounds and the chemical properties of the residual substances (Huber *et al.*, 2002). During ozonation, hydroxyl radicals are produced from interaction with the background DOC, which also degrade chemicals that are not attacked by the ozone (Hollender *et al.*, 2009; Zimmermann *et al.*, 2011).

Operation and investment costs (including sand filtration) are in the range of about 10 US-cents/m^3 of wastewater for a plant size of 100,000 PE. The energy expenditure for ozonation, including the pure oxygen production (16–18 kWh/kgO_3) is about 0.1 kWh/m^3 and is therefore significant in comparison with the total energy consumption of a municipal WWTP.

Ozonation has a strong disinfecting effect and achieves the EU bathing water standards for *E. coli* (<500 counts per 100 mL) (DWA KA 8.1, 2008; Abegglen *et al.*, 2009). However, ozonation requires *well-functioning nitrification* because nitrite reacts rapidly with ozone, consuming 3.4 gO_3/g NO_2^--N thus significantly increasing the ozone dosage.

Powdered activated carbon (PAC) addition: with a pilot plant consisting of flocculation in a 0.5 h retention time contact reactor, a 2 h detention time sedimentation tank, a PAC detention time of a few days and post-filtration (required to prevent PAC loss), it was demonstrated that the addition of 10 g PAC/m^3 (5–10 g DOC/m^{-3}) to biologically treated wastewater removes up to 70% of most of the investigated pharmaceuticals and X-ray contrast agents (Metzger, 2010). Recycling the used PAC to biological treatment, to achieve a two step countercurrent use of the PAC, increases removals up to 90% and provides a 40–50% reduction of the background DOC (Boehler *et al.*, 2012). The PAC process has already been in full scale operation for several years for the decolorization of municipal wastewater with a substantial contribution from the textile industry. Today, several WWTPs in Southern Germany have an additional PAC step in their operation. For efficient coagulation and sedimentation of the PAC, about 0.4 g Fe/g PAC of coagulant has to be added yielding a sludge production of 15–20 g TSS/m^3. Capital and operating costs are in the range of 15 US-cents/m^3 for a plant size of 100,000 PE.

Boehler *et al.* (2012) successfully tested a direct dosage of 15 gPAC/m^3 (0.1 gFe/gPAC as flocculant) on a two-layer sand filter with a backflush interval of 1 d and, after recycling of the filter backwash to the AS for countercurrent use of PAC, similar removal efficiencies were achieved as with a new post treatment step.

The recycling of PAC increases the secondary sludge production by about 10%, leading to a lower AS SRT, but this is partly compensated for by a decreased sludge volume index. Corrosion problems due to PAC application can be reduced by using high density polyethylene or stainless steel pipes and pumps. The primary energy of PAC application (30–35 kWh/kgPAC) results in 0.3 to 0.7 kWh/m^3 wastewater for 10–20 g PAC/m^3.

Recently, *granulated activated carbon filters* (GAC) are increasingly being investigated due to the low regeneration energy required. Initial results are promising, but are not yet sufficient to discuss process design. For longer filter runs (8000–10,000 bed volumes, corresponding to 50 g GAC/m^3) a biologically well treated wastewater with low TSS concentrations (sand-filter or clarifiers with low effluent TSS) and enough oxygen to maintain aerobic conditions is required for partial biological regeneration of the GAC filter.

7.5 CONCLUSIONS AND OUTLOOK

The removal of micropollutants by conventional municipal wastewater treatment due to sorption, stripping and biological degradation is incomplete and a multitude of compounds are only slightly transformed. To significantly reduce eco-toxicological effects of micropollutants and their transformation products in receiving waters and water resources, additional treatment steps (e.g. ozonation or activated carbon adsorption) are required in combination with measures taken at the source (chemical production and application as well as separation and treatment of concentrated wastewater streams). Ozonation requires complete nitrification with low effluent nitrite concentrations. Activated carbon may reduce the AS SRT if the spent PAC is recycled to the biological treatment and the waste sludge, with its increased micropollutant content is not suitable for agricultural use. Additional measures in wastewater treatment should be based on careful micro-pollutants mass flux studies and their environmental effects to identify critical catchment areas so that relevant hot spots are addressed.

7.6 REFERENCES

Abegglen C. and Siegrist H. (2012). *Mikroverunreinigungen aus kommunalem Abwasser, Verfahren zur weitergehenden Elimination auf Kläranlagen* (Micro-pollutants from municipal wastewater, processes for advanced treatment), Bundesamt für Umwelt (BAFU), Bern, Umweltwissen No. 1214. http://www.bafu.admin.ch/publikationen/publikation/01661/index.html

Abegglen C., Escher B., Hollender J., Koepke S., Ort C., Peter A., Siegrist H., von Gunten U. and Zimmermann S. (2009). *Ozonung von gereinigtem Abwasser, Schlussbericht Pilotversuch Regensdorf* (Ozonation of biologically treated wastewater, final report of full-scale experiments at the WWTP Regensdorf), Bundesamt für Umwelt, Bern. http://www.bafu.admin.ch/gewaesserschutz/03716/11218/11223/index.html

Agerstrand M., Wester M. and Ruden C. (2009). The Swedish Environmental Classification and Information System for Pharmaceuticals – An empirical investigation of the motivations, intentions and expectations underlying its development and implementation. *Environ. Internl.*, **35**(5), 778–786.

Andersen H., Siegrist H., Halling-Sørensen B. and Ternes T. A. (2003). Fate of estrogens in a municipal sewage treatment plant. *Environ. Sci. Technol.*, **37**, 4021–4026.

Benner J. and Ternes T. A. (2009). Ozonation of propranolol: formation of oxidation products. *Environ. Sci. Technol.*, **43**(13), 5086–5093.

Boehler M., Zwickenpflug B., Hollender J., Ternes T., Joss A. and Siegrist H. (2012). Removal of micropollutants in municipal wastewater treatment plants by powder-activated carbon, *Water Sci.Technol.*, **66**(10), 2115–2121.

Celiz M. D., Tso J. and Aga D. S. (2009). Pharmaceutical metabolites in the environment: analytical challenges and ecological risks. *Environ. Toxicol. Chem.*, **28**(12), 2473–2484.

DWA Arbeitsgruppe KA-8.1 (2008). Anthropogene Spurenstoffe im Wasserkreislauf, DWA, Hennef 2008, ISBN 978-3-940173-74-4.

DWA Arbeitsgruppe KA 8.1 (2014). *Leitfaden zur Bewertung der Relevanz von Transformations producten im Wasserkreislauf* (Compendium to evaluate the relevance of transformation products in the urban water cycle), in preparation.

Giger W. (2002). Dealing with risk factors. *EAWAG news*, **53**, 3–5.

Golet E., Xifra I., Siegrist H., Alder A. and Giger W. (2003).Environmental exposure assessment of fluoroquinolone antibacterial agents from sewage to soil. *Environ. Sci. Technol.*, **37**, 3243–3249.

Helbling D. E., Hollender J., Kohler H. P. and Fenner K. (2010). Structure-based interpretation of biotransformation pathways of amide-containing compounds in sludge-seeded bioreactors. *Environ. Sci. Technol.*, **44**, 6628–6635.

Hollender J., Zimmermann S., Koepke S., Kraus M., McArdell C., Ort C., Singer H., von Gunten U. and Siegrist, H. (2009). Elimination of organic micropollutants in a municipal WWTP upgraded with a full-scale post-ozonation followed by sand filtration. *Environ. Sci. Technol.*, **43**, 7862–7869

Huber M., Canonica S., Park G. and von Gunten U. (2002) Oxidation of pharmaceuticals during ozonation and advanced oxidation processes. *Environ. Sci. Technol.*, **37**, 1016–1024.

Huber M., Göbel A., Joss A., Hermann N., Löffler D., McArdell C., Ried A., Siegrist H., Ternes T. and von Gunten U. (2005). Oxidation of pharmaceuticals during ozonation of municipal wastewater effluents: a pilot study. *Environ. Sci. Technol.*, **39**, 4290–4299.

Jobling S. M. and Tyler C. (2003). Endocrine disruption in wild freshwater fish. *Pure Appl. Chem.*, **75**, 2219.

Joss A., Carballa M., Kreuzinger N., Siegrist H. and Zabczynski S. (2006a). Chapter 6: Wastewater treatment. In: Human Pharmaceuticals, Hormones and Fragrances – The Challenge of Micropollutants in Urban water management, T. Ternes and A. Joss (eds), IWA Publishing, London, ISBN10 1843390930.

Joss A., Zabczynski S., Göbel A., Hoffmann B., Löffler D., McArdell C., Ternes T., Thompsen A. and Siegrist H. (2006b). Biological degradation of pharmaceuticals in municipal wastewater treatment: proposing a classification scheme. *Water Res.*, **40**, 1686–1696.

Klaschka U., Liebig M., Moltmann J. F. and Knacker T. (2004). Potential environmental risks by cleaning hair and skin. Eco-label – a possibility to reduce exposure to personal care products. In: Pharmaceuticals in the Environment. Sources, Fate, Effects and Risks, K. Kümmerer (ed.), 2nd edn., Springer Verlag, Berlin, pp. 411–430.

Kormos J. L., Schulz M., Wagner M., Ternes T. (2009). Multistep approach for the structural identification of biotransformation products of iodinated X-ray contrast media by liquid chromatography/hybrid triple quadrupole linear ion trap mass spectrometry and (1)H and (13)C nuclear magnetic resonance. *Anal. Chem.*, **81**(22), 9216–9224.

Kormos J. L., Schulz M., Kohler H. P. and Ternes T. (2010). Biotransformation of selected iodinated X-ray contrast media and characterization of microbial transformation pathways. *Environ. Sci. Technol.*, **44**(13), 4998–5007.

Krauss M., Longree P., Van Houtte E., Cauwenberghs J. and Hollender J. (2010). Assessing the fate of nitrosamine precursors in wastewater treatment by physicochemical fractionation. *Environ. Sci. Technol.*, **44**(20), 7871–7877.

Margot J., Magnet A., Thonney D., Chevre N., de Alencastro F. and Rossi L. (2011). Traitement de micropolluant dans les eaux usées, rapport final sur les essais pilotes a la step de Vidy, Lausanne, Bundesamt für Umwelt, Bern. http://www.bafu.admin.ch/gewaesserschutz/03716/11218/11223/index.html

Metzger S. (2010). Einsatz von Pulveraktivkohle zur weitergehenden Reinigung von kommunalem Abwasser. Dissertation at TU Berlin, Oldenbourg Industrieverlag München, ISBN 978-3-8356-3231-8.

Neilson A. H. and Allard A. S. (2007). Environmental Degradation and Transformation of Organic Chemicals. CRC Press, Boca Raton, FL, USA.

Ort C., Hollender J., Schaerer M. and Siegrist H. (2009). Model-based evaluation of reduction strategies for micropollutants from wastewater treatment plants in complex river networks. *Environ. Sci. Technol.*, **43**, 3214–3220.

Siegrist H., Joss A., Alder A., McArdell C., Göbel A. and Ternes T. (2003). Micropollutants – New challenge in wastewater disposal, *EAWAG news*, **57**, 7–10.

Ternes T. (1998). Occurrence of drugs in German sewage treatment plants and rivers. *Water Res.*, **32**, 3245–3260.

Ternes T., Stüber J., Herrmann N., McDowell D., Ried A., Kampmann M., and Van Zelm R., Huijbregts M. A. J. and van de Meent D. (2010). Transformation products in the life cycle impact assessment of chemicals. *Environ. Sci. Technol.,* **44**, 1004–1009.

Wick A., Fink G., Joss A., Siegrist H. and Ternes T. (2009). Fate of betablockers and psycho-active drugs in conventional wastewater treatment. *Water Res.*, **43**, 1060–1074.

Wick A., Wagner M. and Ternes T. (2011a). Elucidation of the transformation pathway of the opium alkaloid codeine in biological wastewater treatment. *Environ. Sci. Technol.*, **45**, 3374–3385.

Wick A., Marincas O., Moldovan Z. and Ternes T (2011b). Sorption of biocides, triazine and phenylurea herbicides, and UV-filters onto secondary sludge. *Water Res.*, **45**, 3638–3652.

Zimmermann S., Wittenwiler M., Hollender J., Krauss K., Ort C., Siegrist H. and von Gunten U. (2011). Kinetic assessment and modeling of an ozonation step for full-scale municipal wastewater treatment: micropollutant oxidation, by-product formation and disinfection. *Water Res.*, **45**, 605–617.

Chapter 8
Aeration and mixing

Martin Wagner (Germany), Michael K. Stenstrom (USA)

8.1 INTRODUCTION

The AS process is the most common biological treatment process in the world for both municipal and industrial wastewater. It exists in a variety of process modifications, from short retention time processes that remove only carbonaceous pollutants through high purity oxygen processes to long retention time processes for nutrient removal. For wastewater treatment plants with advanced nutrient removal in particular, adequately dimensioned, reliable aeration systems and mixing devices are an essential requirement for achieving high efficiency and, at the same time, economical operation. Aeration is usually the most energy intensive part of the AS process with 50 to 80% of the power consumption being used for aeration and mixing (Hwang & Stenstrom, 1985; DWA, 2013). Power consumption provides great motivation for plant designers and operators to optimize aeration systems in the AS process.

8.2 DEVELOPMENT OF MODERN AERATION AND MIXING SYSTEMS

The history of aeration in wastewater technology is understandably closely linked to the history of the AS process, since aeration of the AS is essential. Ardern and Lockett (1914, 1915), the inventors of the AS process, used perforated pipes (1/8 in. orifices placed 8 in. apart) during their first investigations on aeration (Arden & Lockett, 1914, Part II, p. 1125). Soon after, ceramic filter plates for fine-bubble aeration were tested in slightly larger tanks. In parallel investigations, Jones and Attwood (cited in Kinnicutt *et al.*, 1919, p. 384) carried out comprehensive industrial-scale experiments at the wastewater treatment plant in Worcester, UK. This research provided an important contribution to the development of fine-bubble aeration.

In 1925, eleven years after the invention of the AS process, Imhoff compiled all of the aeration systems common at the time in his publication *Fortschritte der Abwasserreinigung* (*Advances in Wastewater Treatment*). It is interesting to note that the aeration systems described then were based on the same principles applied today. Besides fine-bubble aeration with ceramic plates, many systems using surface aerators are described, such as the Haworth paddle, the Bolton cone and the Kessener brush. Another example is the Hurd reactor, a diffused aeration process in which the ceramic plate diffusers were arranged parallel to the flow direction. Between the 1920s and the 1940s, additional investigations advanced the knowledge of aerobic degradation processes in the AS tank. It was found that the oxygen demand did not depend on the oxygen concentration in the tank, above a minimum concentration, but rather on the biosolids concentration. It was concluded that it would be sensible to keep the DO concentration precisely at the level sufficient for microbiological degradation. Prior to this conclusion, oxygen was added in excess at most plants; limiting the DO concentration to a minimum saved energy. In the following years there were calls for the application of control systems in aeration processes (von der Emde, 1998).

Between 1950 and 1970, there were significant advances in the further development of aeration technology. Ceramic filter plates, which had been most commonly used, were increasingly replaced by tubes and domes. In spite of advances made in diffused aeration technology, there were still large numbers of plants in Europe with surface aerators, especially cone (vertical turbine) aerators. At the time, construction simplicity and robustness were the key benefits of using these systems. There were also calls for research on new, economical processes, which simulated the development of the Wuhrmann high-load AS process, reduced aeration and the Pasveer oxidation ditch (von der Emde, 1998).

The increases in energy prices of the 1970s stimulated greater use of diffused aeration systems. Since then, AS plants have been operated with increasingly reduced process loading, thus improving the energy efficiency of diffused aeration systems compared to surface aeration systems. In addition, new materials started to be used, in particular plastic foils and membranes to generate very fine bubbles. Fine bubbles, in combination with the full-floor distribution of the diffusers, increased oxygen transfer efficiency by increasing bubble contact time. The application of surface aeration systems has declined significantly during recent decades (von der Emde, 1998) both in Europe and North America.

The development of oxygen transfer measurements was another decisive contribution to the further development of aeration systems. Quantitative, robust techniques were needed to evaluate the performance of different aeration systems. The first oxygen transfer measurements were carried out by Kessener and Ribbius (1934), who noted differences in efficiency and were the first to note that the presence of organic materials reduced the transfer rates for diffused aeration more than for surface aeration. Eckenfelder (1959) later created a parameter to define the reduction in oxygen transfer rate due to the presence of contaminants, and called it the α-factor; this factor is widely used today in designing and specifying aeration systems. During the following decades, testing methodology has been further developed, so today there are several standard methods for measuring the oxygen transfer rates of aeration systems in Europe (prEN 12255-15, 2003; DWA, 2007) and a single standard in the USA (ASCE, 2007). Standard guidelines for translating these rates to process conditions and process water testing also exist (ASCE, 1996).

8.3 AERATION SYSTEMS

8.3.1 General information

Diffused and surface aeration systems are classified according to their layouts (Table 8.1). The use of coarse bubble systems, especially those with diffusers mounted on tank walls or in wide bands has rapidly declined over the past 25 years, due to their low efficiency and high power cost. These have been replaced with higher efficiency systems that spread the air as uniformly as possible across the tank bottom, and are called 'full floor coverage'. These systems usually employ fine pore diffusers because they create fine bubbles through small pores or orifices in tube, disc, plate, panel or strip geometries. Efficiency is more a function of diffuser surface area and how well the air is distributed rather than diffuser material or geometry. Diffuser development has been rapid for fine pore diffusers, with a range of materials of various shapes and sizes being employed. Ceramic diffusers have been largely replaced with organic 'membrane' materials such as EPDM, polyurethane and silicone. Emphasis has been placed on resistance to fouling, scaling and membrane deterioration by interaction with contaminants in the wastewater. Less development of surface aerators has occurred but the pitch-bladed turbine with four rectangular, flat blades at a 30 to 45° angle to the liquid surface (commonly used in North America) has been replaced with a propeller of more complex geometry which is more efficient and sensitive to liquid level.

Conventional AS tanks in municipal wastewater treatment plants usually limit water depths to 5–6 m. The High Purity Oxygen AS Process (HPOAS) is often used at liquid depths to 9 m when equipped with low speed surface aerators with bottom impellers. The need to design compact plants is making deeper aeration tanks more popular and there are AS tanks more than 12 m deep in Bonn (Germany), Helsinki (Finland) and Stockholm (Sweden). Deeper aeration tanks are more often found in industrial AS processes. The longer ascent path of the air bubbles in deep tanks theoretically allows lower specific air flow rates to dissolve the same amount of oxygen in the water. The higher hydrostatic pressure requires greater blower pressure. In a theoretical examination Günkel-Lange (2013) found that these two effects approximately cancelled out. Yunt and Hancuff (1980) in extensive clean water testing of more than 10 aeration systems, observed an equal trade-off between depth and blower power, except for jet aerators. Therefore,

Table 8.1 Layouts of diffused aeration and surface aeration systems.

Aeration system type	Equipment arrangement	Comments
Diffused or subsurface	Coarse bubble diffusers or spargers mounted to tank walls or in wide bands, spiral roll, cross roll, ridge and furrow.	Excellent mixing but inefficient because high liquid velocities reduce bubble retention time. Largely replaced by more efficient systems but sometimes still used since they require little or no in-tank maintenance. Medium cooling and VOC stripping rate.
	Full-floor coverage, using either a coarse bubble system or more often, fine pore tubes, discs, plates, membrane panels or strips.	Higher efficiency especially for dense diffuser coverage. The method of choice for most new installations or retrofits. Plant designs must anticipate redundancy and tank dewatering capability to perform periodic diffuser cleaning. Lowest cooling and VOC stripping rate. Recent use in oxidation ditch systems with mechanical mixers.
Mechanical aeration	High speed	Floating aerators ranging in power from 2 to 60 kW, with propeller RPM generally 900 to 1200, without reduction gears. Medium efficiency with high rates of cooling and VOC stripping. Limited to depths of 2–3 m unless used with draft tubes. Can be quickly installed without extensive construction and engineering.

	Low speed	Floating or pier mounted aerators ranging in power from 7–150 kW, with propeller RPM generally 40–80. Higher efficiency than high speed aerators with the highest rates of cooling and VOC stripping. Limited to depths of 4–5 m unless used with draft tubes or lower impellers. Require gear reduction and design engineering.
Combined systems	Turbines	Peer mounted aerators with power up to 150 kW or more. Uses a blower to provide air which is released below a mixing propeller that shears coarse bubbles into fine bubbles. Requires gear reduction and design engineering.
	Jets and aspirators	Usually mounted on a piping system near the tank bottom. Uses a pump to force liquid through a venturi-like sparger with air from a blower introduced at the throat of the venturi. Turbulence produces fine bubbles.
Separation of mixing and aeration	Oxidation ditch systems, including some proprietary designs.	Tank geometry in the form of a race track with a circulating flow pattern, which facilitates the use of anaerobic, anoxic and oxic zones. Horizontal low speed brush or rotor mechanical aerators; some recent designs use a horizontal, low speed blade pump with fine pore diffusers.

shallow and deep tanks may be considered approximately equal in terms of power requirements. Mixed results have been observed with α-factors. Pöpel *et al.* (1999) found that water depth did not influence the α-value, while others (Hwang & Stenstrom, 1985) have observed a decrease in α-value with increasing depth.

8.3.2 Table of standard values for aeration systems

For designers and treatment plant owners, it would be very desirable to develop a database of the standard efficiencies of aeration systems. Such a standard would make aeration system selection much more easy and straightforward. The difficulty with such a database is that it could not include subsequent improvements in diffusers and unless routinely updated would not capture all site-specific conditions. Manufacturers might contest some of the results in the database.

Despite the concerns regarding a standard database, Pöpel and Wagner (1998) created one by conducting 80 oxygen transfer measurements in 30 WWTPs, and proposed a table of standard values for oxygen transfer rate and aeration efficiency in clean water for diffused and surface aeration systems and for different designs. The results are differentiated for full-floor arrangements of tubes, disks and plates as well as systems involving separating of mixing and aeration. With surface aeration systems, there are values for aerators in the mixing tank as well as aerators and brush aerators in the circulation tank. For diffused aeration under favorable conditions, the table includes the calculated average oxygen transfer rate at an air flow rate of 2 $m^3/(m^3 \cdot h)$. For systems with separated mixing and aeration, the oxygen transfer rate is calculated at an air flow rate of 1.5 $m^3/(m^3 \cdot h)$, due to specific layout conditions. The oxygen transfer efficiency of diffused aeration systems is calculated using the oxygen transfer and an assumed specific energy demand of the blowers of 5.5 $Wh/(m^3 \cdot m)$. To determine transfer rate and efficiency under process conditions, the α-value is assessed according to the respective process variant and the respective load case.

With surface aeration systems under favorable conditions, the average values for oxygen transfer rate and aeration efficiency are specified at a power density of 35 W/m^3. The α-value for converting the clean-water values into wastewater values was 0.9. The table defines the average conditions – for surface aeration as well as diffused aeration systems – to be 75% of the values under favorable conditions.

The results from the oxygen transfer measurements that were used to compile the original table for standard values date from before 1989. Due to the further development of membrane aeration diffusers and blowers and the enormous improvement in the efficiency of diffused aeration in particular, there are now values for specific oxygen transfer efficiency (%/m) and standard aeration efficiency (kg/kWh) about 40% higher than before 1989. However, these high values can only be achieved, if the diffusers degas optimally and, at the same time, the most efficient blowers are used. In case aeration systems with separate mixing and aeration are applied, the mixers have to ensure the penetration of the air bubble wall formed by aeration, since only in this way can the circular current in the bypass channels be generated. With surface aeration systems, there were no improvements in the efficiency compared to 1989.

Based on the approximately 40% higher values for oxygen transfer and aeration efficiency described above, a modified table for standard values for diffused aeration systems was compiled (Table 8.2). When using these modified values for selection of aeration systems, one should be aware that in individual cases the actual values of oxygen transfer rate and standard aeration efficiency might be higher or lower than the 'table values'. With diffused aeration systems, both factors can depend on the amount of air applied to the diffusers, the diffuser density, water depth, degassing behavior, type of compressed air generation, installation of throttles, etc. Tables 8.2 and 8.3 respectively show Pöpel's and Wagner's proposed standard values based on European experience.

Table 8.2 Modified table of standard values for diffused aeration systems.

System	Favorable		Average	
	SOTE (%/m)	SAE (kg/kWh)	SOTE (%/m)	SAE (kg/kWh)
	Clean-water conditions			
Full-floor coverage	8.0–8.7	4.2–4.5	6.0–7.0	3.3–3.4
Circulation and aeration	6.7–8.0	3.7–4.2	5.0–7.0	3.2–3.3

Table 8.3 Table of standard values for surface aeration systems.

System	Efficiency SAE (kg O_2/kWh)	
	Favorable	Average
Surface aerators in the mixing tank	1.7	1.3
Surface aerators in the circulation tank	2.1	1.6
Brush aerators the circulation tank	1.7	1.3

To achieve high oxygen transfer and thus economical oxygen transfer efficiencies, it is necessary to optimize the adjustment of diffusers, piping system and blower. Fine-bubble diffusers must diffuse the air evenly at all air flow rates to guarantee the best possible oxygen transfer.

Stenstrom and co-workers (Stenstrom, 2001; Rosso *et al.*, 2008) have created 'average' transfer efficiencies but these are based on observations at different SRTs. 'Low SRT' refers to systems with 1.5–4 d SRTs and 'high SRT' refers to systems with greater than >8 d SRT. Observations between 4 and 8 d are not common because treatment plant operators attempt to avoid the partial nitrification that can occur in this range. Table 8.4 is based in part on α – factors and additional information provided by Rosso *et al.* (2005). Table 8.4 supports Kessener and Ribbius's observation in 1934 that diffused aeration systems are more strongly impacted by organic contaminants than mechanical systems.

Table 8.4 Observed performance of aeration equipment in the USA.

Aerator type	SAE in clean water (kgO_2/kWh)	SAE in process (kgO_2/kWh)	
		at 'Low SRT' and 2 mg DO/L	at 'High SRT' and 2 mg DO/L
High speed surface	0.9–1.3	0.4–0.8	
Low speed surface	1.5–2.1	0.7–1.5	
Sparged turbine	1.2–1.8	0.4–0.6	0.6–0.8
Coarse bubble	0.6–1.5	0.3–0.7	0.4–0.9
Fine pore	3.6–4.8	0.7–1.0	2.0–2.6

Tables 8.2 to 8.4 are intended to show approximate ranges and expected values and should not be used for designing specific treatment plants. For this, site specific observations and sound engineering judgment should be used.

Blowers play a decisive role at high aeration efficiencies. Common blower types in wastewater treatment plants include rotary blowers, centrifugal or turbo compressors and, occasionally screw compressors. Rotary blowers are used for low to medium air flow rates and up to a maximum water depth of about 8 m. Large wastewater treatment plants with high air flow rates are the main field of application for turbo compressors) and less frequently, screw compressors.

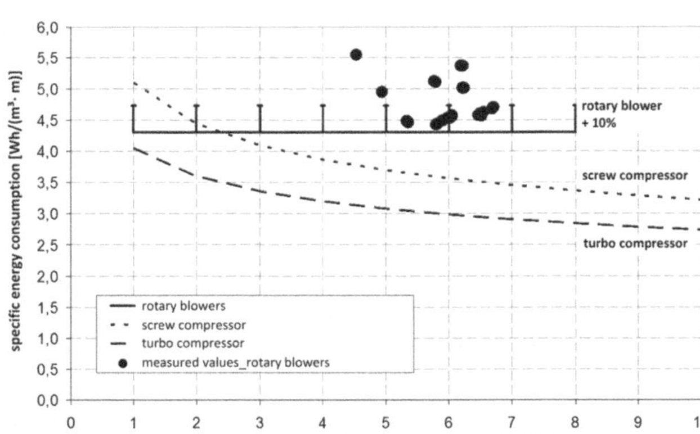

Figure 8.1 Specific energy demand of compressed air generators at 5000 m^3/h.

Figure 8.1 shows an example of the variation of specific energy demand (Wh/($m^3 \cdot m$) with discharge pressure (mWC) for various blower types at an air flow rate of 5000 m^3/h (Wagner, 2001). Besides the idealized curves, unpublished data for rotary blowers are shown; these were collected by the Institute IWAR, Technische Universität, Darmstadt, Germany in various compliance tests over the past several years. Some of the measured values are within 10% of the ideal value of 4.3 Wh/($m^3 \cdot$ m) but others are higher, leading to lower than predicted efficiencies.

8.4 APPROACHES FOR THE OPTIMIZATION OF AERATION SYSTEMS

The components of a diffused aeration system are compressed air generation, piping system (including fittings), diffusers and process measurement and control technology. Optimizing such a system requires the optimization of a large number of components and is usually more complex than for surface aeration systems. Possible approaches are as follows:

- Dimensioning for different oxygen demand loads
- Adjustments for seasonal changes in the MLSS concentration
- Adjustment for the DO concentration
- Control of compressed air generation
- Countermeasures for efficiency-reducing effects on diffusers

8.4.1 Dimensioning of different oxygen demand loads

Operation experience over the last twenty years has shown that many aeration systems are oversized because (Jardin, 2001):

- Optimistic assumptions were made of future loads
- Aeration systems were sized using the same forecasting as the treatment plant structures
- Aeration systems were sized for unfavorable conditions and for too infrequent loads
- Aeration systems were sized with too large, and overlapping safety factors

The cumulative effects of the factors described above have led to considerable over-sizing of aeration systems, so the most frequent operating mode is often in an energetically unfavorable range. This produces an inefficient aeration system with the following problems (Jardin, 2001):

- At the lower end of the operating range, it is hard to regulate the aeration system and the aeration tank DO concentration is far too high, particularly at night. Energy is wasted and high DO concentrations can negatively impact denitrification.
- Low aeration rates lead to low aeration basin mixing intensities, which increase the risk of biosolids settling. Intermittent aeration may be helpful here.
- Over sizing unnecessarily increases the capital and operating costs.

Specifying equipment with the greatest 'turn up' or 'turn down' capabilities is desirable.

To avoid these problems, it is recommended that the aeration system be sized for a range of load cases, starting with the determination of oxygen demand (OUR). The DWA Merkblatt 229-1 (2013) states that the OUR should be determined for the following four load cases:

- *Load case 1: average oxygen demand for the actual situation OUR, aM:* this load case is used to determine the annual energy demand. The oxygen demand is determined using the annual average of the expected loads and the mean annual water temperature.
- *Load case 2: maximum oxygen demand in the actual situation OUR, max:* this load case is used for dimensioning the aeration and mixing facilities and represents either the oxygen demand at times with the highest wastewater temperature or in the winter months with the lowest wastewater temperature and directly after startup.
- *Load case 3: minimum oxygen demand in the actual situation OUR, min:* this load case is used for dimensioning the aeration and mixing facilities and is necessary for sizing the blowers and for verification of the minimum air flow rate to the diffusers directly after start-up. In addition, this load case is required to determine the required working range of surface aerators, if necessary in combination with additional mixers. If the minimum oxygen demand is accounted for, there is a risk that the aerated zone oxygen input will lead to impairment of denitrification performance.
- *Load case 4: oxygen demand for the predicted situation and, if need be, for expansion:* the oxygen demands mentioned above have to be determined for both the design situation and for planned expansions. These parameters are to allow for future flexibility within the planning horizon.

The α-value should be carefully selected for each load case; the lowest α-value is applied to the maximum load case and higher α-factors are applied to average and minimum load cases; α-factors that are too high can result in undersizing the aeration system and α-factors that are too low can result in oversizing the aeration system (Günkel-Lange, 2013). In the case of diffused aeration, α-factors can be estimated quantitatively using efficiency measurements by off-gas testing (Redmon *et al.,* 1983)

The α-factors increase when the loading decreases and vice-versa, but quantitative data are scarce. Leu *et al.* (2009) measured α-factors and loads at an 8-d SRT nutrient removal plant near Los Angeles, CA and found that, over a 24-h period, α-factors changed from 0.3 at highest loading to 0.55 at lowest loading.

Table 8.5 shows the recommended German values for minimum, average and maximum α-factors in fine-bubble diffused aeration systems for various load cases and process variants (Günkel-Lange, 2013). The α-factors can be selected on the basis of treatment goals, process loading, SRT and process variant. Under favorable conditions and with surface aerators, an α-value of 0.9–0.95 or even 1.0 have been observed in Germany (DWA, 2013).

Table 8.5 Recommended α-factors for maximum, average and minimum load cases process variants for fine-bubble diffused aeration systems.

Process	α_{min} **(Maximum load case)**	α_{ave} **(Average load case)**	α_{max} **(Minimum load case)**
Continuously operated denitrification (simultaneous, intermittent, alternating, upstream)	0.60	0.75	0.85
SBR process for N removal	0.50	0.65	0.80
MBR process (MLSS ~ 12 g/L, SRT = 25 d)	0.50	0.60	0.70
Simultaneous aerobic stabilization	0.70	0.80	0.90
Carbon removal	0.35	0.50	0.60

Source: Günkel-Lange (2013).

In American experience the maximum values in Table 8.5 are rarely observed. The value of 0.85 for continuous flow denitrification is much higher than the 0.55 found by Leu *et al.* (2009) and generally reported by Rosso *et al.* (2005). In Stenstrom's experience α-factors for surface aerators are rarely greater than 0.8 except for high power density applications (80 W/m^3) such as in the HPO-AS process (Tzeng *et al.*, 2003).

An alternative to designing at fixed points is to use a commercially available process simulator. A range of site specific conditions with various assumptions for α-factors can then be evaluated.

After determining the OURs and the corresponding α-factors for the various load cases, the standard clean water oxygen transfer rate (SOTR) (at a water temperature of 20°C and a pressure of 1013.25 hPa) can be calculated. In the DWA-M 229-1 (2013) approach, the effects of dissolved salts, water temperature, tank depth and plant elevation are considered for design. A similar set of calculations and recommendations are provided in the ASCE Standard (2007).

$$SOTR = \frac{f_d \cdot \beta \cdot C_{S,20}}{\alpha \cdot \left(f_d \cdot \beta \cdot C_{S,T} \cdot \dfrac{p_{atm}}{1.013} - C_X \right) \cdot \theta^{(T_W - 20)}} \cdot OV_h \left(\frac{kg}{h} \right)$$

where: f_d: depth factor ($f_d = 1 + (d_e/20.7)$) for fine-bubble aeration (dimensionless), d_e: effective depth of submergence (m), β: salt factor (dimensionless, in domestic wastewater ≈ 1), $C_{S,20}$: oxygen saturation concentration at 20°C (mg/L), $C_{S,T}$: oxygen saturation concentration at T (mg/L), p_{atm}: atmospheric pressure (bar), C_X: oxygen concentration during operation mg/L), θ: temperature correction factor with $\theta =$ 1.024 (dimensionless), α-value (dimensionless). The elevation correction parameter should only be used for plants higher than 600 m. Methodology is not standardized between North America and Europe, with equilibrium oxygen concentration being represented as $\beta C^*_{\infty,20}$ or $\beta C^*_{\infty,T}$ and calculated from the non-linear estimation of the DO versus time data measured in a clean water reaeration test (ASCE, 2007). Using a depth factor to estimate C^*_∞ is inconsistent with the ASCE standard (2007).

Using the required oxygen transfer rates, it is possible to calculate the required air flow rates and the number of diffusers. It is customary to evaluate different numbers of diffusers to determine the trade-offs between capital and operating costs.

8.4.2 Adjustment to seasonal changes in MLSS concentration

It is possible to reduce the energy demand of an aeration system by adjusting to seasonal changes in MLSS concentration. Since the required SRT can decrease with increasing water temperature, it is possible in the summer to operate only part of the aeration basin tank volume or to operate the entire aeration basin volume at a reduced MLSS concentration. In both cases the oxygen demand will decrease significantly. Figure 8.2 is an example of the savings possible for a plant treating the wastewater from 100,000 PE, (according to the ATV-Arbeitsblatt A 131 standard (ATV-DVWK, 2000). In this example at a wastewater water temperature of 20°C, the MLSS concentration can be reduced from 3.15 kg/m³ to 1.5 kg/m³ and result in 10% savings of the energy cost of aeration (Jardin, 2001).

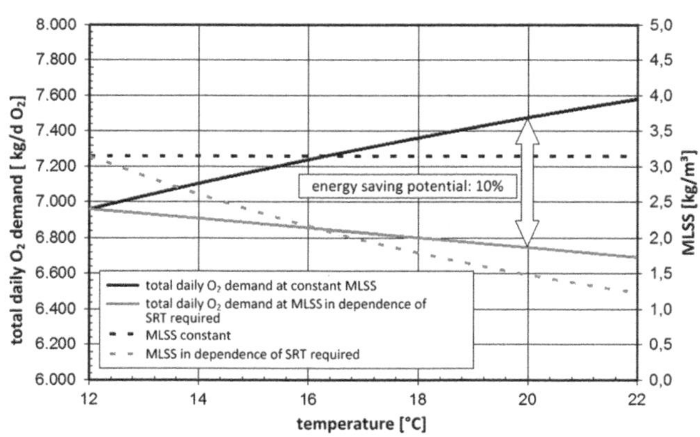

Figure 8.2 Oxygen demand and required MLSS concentration as a function of water temperature (Jardin, 2001).

Another energy and cost savings potential arises from equalization or feed flow management with storage tanks. On the one hand, peak loads can be buffered, thus making it possible to reduce the impact factor in the dimensioning of the aeration system. On the other hand, energy costs can be reduced by storing wastewater during times of high power rates then working off the stored flow during times of less expensive, off-peak power rates. Similarly, storage and management of process water from sludge treatment can contribute to energy savings in aeration. Leu *et al.* (2009) analyzed a treatment plant with equalization and evaluated several strategies for cost reduction, including the timing of return flows from dewatering digested sludge.

8.4.3 Adjustment AS tank oxygen concentration according to the treatment goal

The rate of nitrification directly depends on the aeration basin DO (Stenstrom & Poduska, 1980). To provide stable and efficient nitrification, a DO concentration of 2 mg/L is often recommended. However, experience, has shown that during warm summer months, stable nitrification can occur at DO concentrations considerably lower than 2 mg/L. For such cases, the oxygen transfer rate can be regulated according to the target discharge ammonia concentration. As an example, a wastewater treatment plant with 800,000 pe can save up to 20% in energy costs (mean annual cost savings = 12%) by using this method. However at these low DO concentrations nitrification rate is sensitive to even small changes in DO concentration and a sensitive, reliable DO control system is required.

Caution is also required to avoid the formation of bulking and floating sludge due to excessive growth of *Microthrix parvicella* and other filamentous organisms that can grow at low DO concentrations. Each plant has to find a suitable compromise between reliable operation and energy costs.

8.4.4 Control of compressed air generation

The goal of controlling the generation of compressed air is to produce the required air flow rate at the lowest power. Various control strategies are available. With direct DO control, blower power directly depends on the aeration basin DO concentration. When the DO concentration falls below the set point, the air flow rate is increased by increasing the motor frequency/compressor speed. Adjusting blower speed is preferable to using a control valve because the minimum blower discharge pressure can be obtained.

From the energetic point of view, direct DO control would seem to be the best possible solution. However, this control system becomes problematic when (as is common in medium- to large-sized plants) several tanks are supplied with different air flow rates, through a common pipe. Thus, direct DO control is mainly used in small plants.

The most common system for constant pressure control attempts to provide a constant, fixed pressure in the main compressed-air line. If individual aeration zones or tanks require more air, the appropriate control valve on the main pipe is opened to increase the air flow rate. This reduces the pressure in the main pipe, which the blower control system detects and responds to by increasing blower output until the set point pressure is restored. The set point pressure is typically adjusted to provide the maximum needed air flow rate over a wide range of loading conditions. This means that the set point pressure is often significantly higher than required and consequently, the control gate valves usually are not fully opened, thus wasting energy.

Variable pressure control systems dynamically adjust the pressure in the main air pipe according to air requirements. In this way the blowers operate at the minimum required pressure. This type of control system can save up to 4% of the cost for aeration compared to the constant pressure control method (Holzenthal, 2003).

DO control systems are conceptually simple but often work poorly. Constant hunting around the DO set point is often observed and the need to maintain DO probes is sometimes difficult for plant operators. As a result, many DO control systems are set to manual operation with power wastage.

8.4.5 Measures to avoid efficiency reduction

Fine-bubble membrane diffusers deteriorate in performance over time and lead to reduced aeration system efficiency. Rosso and Stenstrom (2006) showed that the oxygen transfer efficiency of fine-pore diffusers decreases significantly during the first 24 months of operation before reaching a stable level. Loock (2009) found that the drop in efficiency was primarily caused by changes in the membrane material and the formation of inorganic coatings in the diffuser membrane slits or orifices. These changes are almost always manifested in increasing diffuser back pressure. The oxygen transfer efficiency may also decrease, and the blower energy demand increases leading to increased operating cost. Diffuser cleaning will require additional costs.

Several methods are available for cleaning aeration membranes (Loock, 2009):

- High-pressure cleaning with conventional high-pressure nozzle (fan nozzle).
- High-pressure cleaning with rotating point nozzle.
- Manual cleaning with hand brush.
- Injection of formic acid into the air distribution system.

The cleaning method efficiency depends on the diffuser material. High-pressure cleaning with a rotating point nozzle is particularly effective and is generally preferred over conventional high-pressure (fan) nozzles which can damage the diffuser membrane. When only biofouling has occurred, membrane diffusers can usually be cleaned manually with a hand brush. Formic or acetic acid injection into the piping system to dissolve membrane deposits can improve efficiency, but can also produce negative results, such as the formation of reaction products that increase the deposit thickness and further narrow the membrane diffuser slots. This effect was observed at both EPDM and silicon membranes (Loock, 2009). Injecting HCl gas has been used with success but has largely been abandoned in the US due to safety concerns over transporting HCl gas cylinders. Ceramic diffusers can be cleaned to remove precipitates by applying 15% hydrochloric acid to the diffuser surface and then brushing and rinsing. Pressure washing after simple mopping has recently been shown to be very successful for polyurethane strip diffusers (Stenstrom *et al.*, 2013)

Some cleaning methods risk damaging the diffuser systems. Hydrochloric acid cannot be used on membranes and can cause corrosion on some types of stainless steel (type 304 in particular). If the wand of a pressure washer is held too close to a membrane diffuser, it can cut the membrane, requiring diffuser replacement. Submerging the diffuser system with 0.3 m of water and then holding the pressure washer wand above the water level is a safe and effective method of pressure washing (Stenstrom *et al.*, 2013). When tanks are dewatered, UV from sunlight can damage the diffusers. Membrane diffusers, especially polyurethane membranes, should be protected from sunlight by keeping tap water or secondary effluent above the diffusers to at least 0.3 m depth. In some cases, new diffusers show a fast pressure rise within the first 4–6 weeks after start of operation, which cannot be restored by cleaning (Masutani & Stenstrom, 1990). After this initial period, pressure increases can be monitored to determine cleaning needs. In some cases, diffusers may show a rapid increase in pressure to pre-cleaning levels. The effort for mechanical cleaning (shut-down of the tanks, emptying, cleaning, etc.) can be significant and designers should anticipate the need to rapidly dewater and refill aeration tanks to facilitate diffuser cleaning. Also a greater range of turn up and turn down capability can be used to extend periods between cleaning.

8.5 AERATION SYSTEMS IN COLD AND WARM CLIMATE REGIONS

When planning and operating wastewater treatment plants in different climate regions, one should pay special attention to a variety factors that may have small impacts in temperate zones. In Germany, wastewater temperatures range from 10 to 20°C while in California, wastewater temperatures are often above 20°C. World-wide, wastewater temperatures can in the range 3–27°C (Metcalf & Eddy, 2003). Because wastewater temperature affects all biological processes, including oxygen demand and oxygen transfer efficiency, it is of critical importance in the sizing of wastewater treatment plants.

The type of aeration system strongly affects the equilibrium wastewater temperature because of evaporation. Surface aerators have the greatest cooling potential and fine pore diffusers usually provide the least cooling because they have the lowest air flow rate. Talati and Stenstrom (1990) showed that the type of aeration made a 4°C difference between fine pore and surface aerators treating domestic wastewater in a cold climate. The cooling rate is influenced by ambient humidity and wind speed. The location of the plant and the type of wastewater will determine the need for cooling. Municipal wastewaters seldom benefit from cooling, whereas petroleum refinery wastewaters and some other industries benefit from cooling during the warmer times of the year. Good practice is to avoid surface aerators in cold climates, and several manufactures provide diffused aeration systems that can function in lagoons with uneven depths.

The following planning and operation recommendations are valid for aeration systems in both cold and warm climates:

- Several references exist which can be consulted to predict equilibrium aeration basin temperature from climate conditions as well as process configurations (Argamann & Adams, 1977; Talati & Stenstrom, 1990; Sedory & Stenstrom, 1995; Scherfig *et al.,* 1996 and Gillot & Vanrolleghem, 2003). They use a variety of techniques to predict the impact of various environmental factors and process conditions on aeration basin temperature.
- The α-factor does not vary with water temperature (Mueller *et al.,* 2002).
- In cold climates, condensate tends to collect in air piping and suitable drains and freeze protection must be provided. Icing on the surfaces of the motor and floats of surface aerators often occurs. Suitable strip heaters are needed in extreme cases, for floating surface aerators; otherwise ice formation may destabilize the aerator, causing it to turn over and fail. Generally, blowers and associated equipment should be installed in a building. Spray from surface aerators should not be allowed to strike or impinge on handrails, walkways or other equipment (Wagner & Günkel, 2010).

The following recommendations should be followed for climates that produce water temperatures >25°C (Wagner & Günkel, 2010):

- In the planning phase for carbon-only removal systems, it is often difficult to avoid nitrification so the aeration system should be sized to provide sufficient oxygen for nitrification.
- If blowers are installed outside they should at least be covered by a roof.
- When there is a risk of sandstorms, sand traps should be installed on the blower houses' air intakes.
- Compressed air temperature should not exceed diffuser manufacturers' recommendations or 60°C to prevent diffuser membrane damage.

8.6 MIXING SYSTEMS

8.6.1 Types of mixing systems

In AS systems required to achieve low N and P concentrations, portions of aeration tanks may be unaerated and mixing is required to prevent biosolids settling. The main task of the mixers is to keep the AS flocs in suspension at a constant temperature and concentration throughout the tank, and to do so without unnecessarily shearing the flocs or creating backmixing from aerated zones. These functions can be carried out by the mixing devices listed in Table 8.6. The items in the column 'function' highlighted in bold type are the main tasks of the respective mixing device.

Table 8.6 Mixing devices commonly used in AS treatment (major function in bold).

Type of mixing device	Layout	Function
Propeller mixer/ low-speed), installed horizontally or vertically	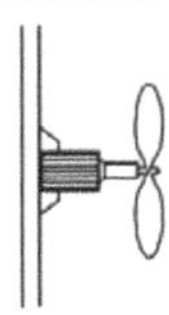	Homogenization **Suspension**
Propeller mixer (high-speed)	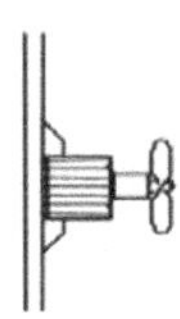	Homogenization **Suspension** (higher viscosity fluids)
Blade agitator	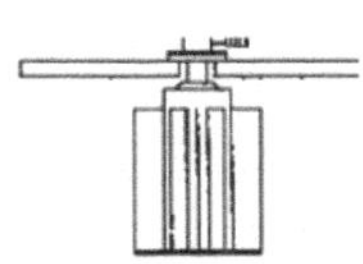	Homogenization **Suspension Dispersion**
Hyperboloid mixer	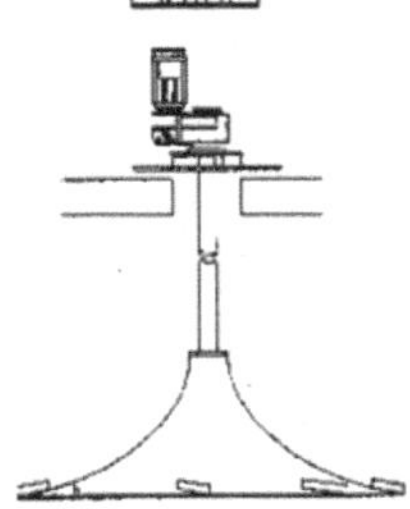	**Homogenization Suspension Dispersion**

Source: Höfken (1993).

Mixing devices are used in anaerobic tanks (BioP removal), anoxic stages (denitrification) and aerobic stages (nitrification). Anaerobic and anoxic tanks are typically mixed with either horizontally or vertically-mounted low-speed or high-speed propeller mixers. Mixing devices and aeration equipment are sometimes both installed in deep aerobic tanks; oxidation ditches with nitrification and denitrification can also have both aerators and mixers. There is an increasing trend to use a horizontal mixer and fine pore diffusers in oxidation ditches.

Rectangular tanks: in rectangular or square anaerobic tanks, with vertically mounted propeller mixers, the water flow direction can either be from the surface to the bottom or vice versa. The main advantage of this design is that the necessary changes in flow direction from vertical to horizontal occur at the water surface and can be observed to detect unwanted turbulence. If the propeller is close to the water surface, spray may be formed and floating solids may accumulate in the middle of the tank. Horizontal submersible mixers mounted on the wall of a rectangular tank will mix up the AS flocs with a 'corkscrew effect'. When the mixing devices are installed inappropriately, dead zones may occur and allow AS accumulation on the tank floor, especially in corners. Hyperboloid mixers provide energy input close to the tank bottom and can prevent such accumulation. They can be installed on bridges or peers or with submersible motors.

Rectangular and circular anoxic tanks are mixed similarly. Mixing devices are generally not needed for aerobic tanks with full-floor aeration systems or wall-mounted coarse bubble diffusers. Fine pore diffusers cannot break up a jet formed by an influent pipe. Such pipes should be terminated with tees or baffles or turned toward the tank bottom.

Circular tanks: horizontal high-speed propeller mixers and hyperboloid mixers are used for mixing in circular anaerobic tanks. Vortexes should be avoided through baffles at the tank walls and/or distribution piping at the tank center. Circular anoxic and aerobic tank mixer options include either fixed or bridge-mounted, vertical or horizontal propeller devices. The vertical mixers can have a draft tube. In aerobic basins, diffusers can be installed for additional oxygen transfer.

Table 8.7 Selection of mixing devices in circulation tanks.

Type of mixing device	Layout	Function
Blade agitator	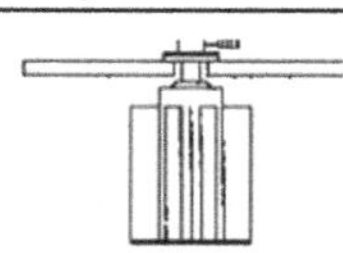	Suspension, (aeration with additional diffusers)
Propeller mixers (conveying horizontally)	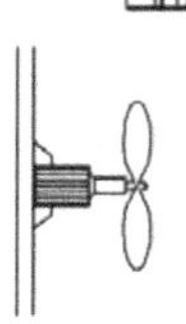 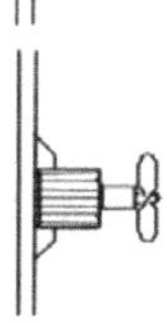	Suspension, (aeration with additional diffusers)

Source: Höfken (1993).

Circulation tank: circulation tanks or 'race track' tanks are usually with anaerobic, anoxic and aerobic zones. Table 8.7 shows mixing devices commonly used in circulation tanks, indicating that generally the tanks are equipped with blade agitators and horizontally conveying propeller mixers. For aeration, diffused aeration systems with tubes, disks and plates as well as membrane panels or strips are used. Horizontal brush or rotor aerators are still used for aeration and mixing, but are becoming less popular.

8.6.2 Dimensioning of mixing facilities

Design of the actual mixing devices is generally carried out by the manufacture using information on the mixing requirements and their placement provided by the design engineer as follows:

- To avoid sludge accumulation on the tank floor, the near tank bottom velocities must be 0.30 m/s or greater for AS aeration tanks. Greater velocities will be required for digesters or holding tanks, and need to be based on site-specific conditions. 'thick' sludge.
- Specification of minimum power density (W/m^3) for the purpose of avoiding bottom sludge accumulation is usually not adequate because different types of mixing devices and varying tank depths with equal power densities can produce different near-bottom velocities.
- Mixing device manufacturers must be advised of the air flows from any air diffusers in the tank.
- Bottom sludge accumulation at tank walls and corners must be prevented by construction details.
- Mixing devices must be reliable and able to deal with the impact of solids such as fibrous materials.
- Horizontal mixers, especially low-speed mixers with large diameter, often called 'banana blade' mixers are sensitive to the direction of approaching flow. Such mixers should only be mounted in straight sections of the tank, where the approaching liquid does not exert a torque at an angle to the mixer shaft. The existence of such a torque will reduce bearing life.
- Mixing devices should be designed and installed so that they can be readily removed and, if necessary, modified.

8.7 PERSPECTIVES AND OUTLOOK

Aeration systems for AS wastewater treatment plants are technically mature. If an aeration system is designed with the flexibility to satisfy all of the relevant loading cases discussed in this paper, its operation should be reliable and energy-efficient. However, further optimization measures should be pursued to develop, dimension and operate aeration systems that are even more energy-efficient. With surface aeration systems, optimization strongly depends on the manufacturer. For diffused aeration systems, further development of individual components such as compressed air generation, pipe fittings, diffusers and process measuring and control technology are needed. Properties of membranes used for diffusers are still evolving.

Blower development is essential for the energetic optimization of diffused aeration systems. New rotary and turbo blower designs can reduce power demand by 15% compared to older equipment. Also, the use of variable speed drives, even for centrifugal blowers, is gaining acceptance and shows promise for saving energy. Variable speed drives may be especially useful in cases where older blowers are used with replacement aeration systems that require less air.

The design of air piping systems must be optimized to achieve low pressure losses. Increasing electricity cost can justify piping systems with lower pressure drop. Diffuser technology should be further developed to provide even air distribution across their entire perforated/slotted area for their entire operating life time (approximately 5–10 years). These goals can best be met through the development of new membrane materials which resist degradation (hardening, softening) and improved perforation (orifice or slit) designs which do not clog, shrink or elongate over time. The development of new techniques for *in-situ* diffuser cleaning should be a high priority. Even though aeration system process measurement and control technology is technically mature, there is still room for further optimization by real-time adjustment of air flow supply to the current oxygen demand.

Additional knowledge of oxygen transfer mechanisms in wastewater and on the impacts of contaminants on membrane materials will be necessary to further improve the efficiency of diffused and surface aeration systems. Several universities, including the Institute IWAR at the Technische Universität, Darmstadt, and the University of California Los Angeles and Irvine campuses are pursuing these research issues. The impact of MLSS concentration on oxygen transfer rate, the solids retention time (Henkel, 2010) and the process loading rate (Günkel-Lange, 2013) are of interest.

8.8 REFERENCES

Ardern E. A. and Lockett W. T. (1914, 1915). The oxidation of sewage without the aid of filters, parts I, II, III. *J. Soc. Chem. Industr.*, **33**(10), 523–539; **33**(23), 1122–1125; **34**(18), 937–943.

Argamann Y. and Adams C. F. (1977). Comprehensive temperature model for aerated biological systems, *Prog. Water Technol.*, **9**(1/2), 397–409.

ASCE (1996). Standard Guidelines for In-process Oxygen Transfer Testing. Amer. Soc. Civil Eng., New York, NY, USA.

ASCE (2007). Measurement of Oxygen Transfer in Clean Water. Amer. Soc. Civil Eng., New York, NY, USA.

ATV-DVWK (2000). *ATV-DVWK-Arbeitsblatt A 131: Bemessung von einstufigen Belebungsanlagen* (ATV-DVWK standard A 131: Dimensioning of single-stage activated sludge plants). Gesellschaft zur Förderung der Abwassertechnik e.V.-GFA, Hennef, Germany.

DWA (2007). *DWA-Merkblatt M 209: Messung der Sauerstoffzufuhr von Belüftungseinrichtungen in Belebungsanlagen in Reinwasser und in belebtem Schlamm* (DWA advisory leaflet M 209: measurement of the oxygen transfer in activated sludge aeration tanks with clean water and in mixed liquor). DWA, Hennef, Germany.

DWA (2013). *DWA-Merkblatt M 229-1: Systeme zur Belüftung und Durchmischung von Belebungsanlagen, Teil 1: Planung, Ausschreibung und Ausführung* (DWA advisory leaflet M 229-1: Aeration and mixing. systems for activated sludge plants, Part 1: Planning, invitation of tenders and accomplishment). DWA, Hennef, Germany.

Eckenfelder W. W. Jr. (1959). Factors affecting the aeration efficiency of sewage and industrial wastes. *Sewage Industr. Wastes*, **31**(1), 60–70.

Gillot S. and Vanrolleghem P. A. (2003) Equilibrium temperature in aerated basin—comparison of two prediction models. *Water Res.*, **37**(3), 3742–3748.

Günkel-Lange T. (2013). *Sauerstoffzufuhr und α-Werte feinblasiger Belüftungssysteme beim Belebungsverfahren - Abhängigkeiten und Bemessungsempfehlungen* (Oxygen transfer and α-values of fine bubble aeration systems at the activated sludge process–dependencies and recommendations). PhD Thesis, IWAR Schriftenreihe, No. 221, Technische Universität Darmstadt, Germany.

Henkel J. (2010). Oxygen Transfer Phenomena in Activated Sludge. PhD Thesis, IWAR Schriftenreihe, No. 210, Technische Universität Darmstadt, Germany.

Höfken M. (1993). Die Bedeutung der Rührtechnik in der Abwasserreinigung. In: *Abwasserreinigung; Alte Probleme – Neue Lösungen* (The meaning of mixing systems in the field of waste water treatment. In: Waste water treatment: Old problems – new solutions). *Conf. Proc. LSTM-Seminar*, June 23–24, Sulzbach-Rosenberg, Germany.

Holzenthal K. (2003). *Gleitdruckregelung von Klärwerksverdichtern* (Sliding pressure control of compressors in wastewater treatment plants). *Korrespondenz Abwasser*, **50**(10), 1157–1161.

Hwang H. J. and Stenstrom M. K. (1985). Evaluation of fine-bubble alpha factors in near-full scale equipment, *J. Water Pollut. Control Fedn.*, **57**(12), 1142–1151.

Imhoff K. (1925). *Fortschritte der Abwasserreinigung* (Advances in wastewater treatment), 1st edn, Carl Heymann, Berlin, Germany.

Jardin N. (2001). *Auswirklungen der neuen ATV-DVWK Arbeitsblätter A 131 und A 198 (Entwurf) auf die Berechnung des Sauerstoffbedarfs* (Effects of ATV-DVWK standards A 131 and A 198 (draft) on the calculation of oxygen demand). In: *WAR Schriftenreihe*, 134 edn, 64. Darmstädter Seminar – Abwassertechnik: Neues zur Belüftungstechnik, Technische Universität Darmstadt, Darmstadt, Germany.

Kessener H. J. N. H. and Ribbius F. J. (1934). Comparison of aeration systems for the activated sludge process. *Sewage Works J.*, **6**(3), 423–443.

Kinnicutt L. P., Winslow C-E. A. and Pratt R. W. (1919). Sewage Disposal, 2nd edn, Wiley and Sons, New York.

Leu S.-Y., Rosso D., Larson L. E. and Stenstrom M. K. (2009). Real-time monitoring aeration efficiency in the activated sludge process and methods to reduce energy consumption, *Water Environ. Res.*, **81**(12), 2471–2481.

Loock P. (2009). *Veränderung der Leistungsfähigkeit feinblasiger Membranbelüftungselementeunter abwassertechnischen Betriebsbedingungen* (Alteration in performance quality of fine bubble membrane aeration systems under operating conditions). PhD thesis, IWAR Schriftenreihe, No. 202, Technische Universität Darmstadt, Germany.

Masutani G. and Stenstrom, M. K. (1990). Fine Pore Diffuser Fouling: The Los Angeles Studies, University of California Los Angeles, ENG 90-02, 1–145.

Metcalf and Eddy (2003). Wastewater Engineering – Treatment and Reuse. 4th edn, McGraw-Hill, New York, NY, USA.

Mueller J. A., Boyle W. C. and Pöpel H. J. (2002). Aeration: Principles and Practice. CRC Press LLC, Boca Raton, FL, USA.

Pöpel H. J. and Wagner M. (1998). *Sauerstoffeintrag moderner Belüftungssysteme, Teil I Druckbelüftung und Teil II Oberflächenbelüftung* (Oxygen transfer of modern aeration systems, part I diffused aeration systems and part II surface aeration systems). *Korrespondenz Abwasser*, **45**(4), 453–457 and 582–590.

Pöpel H. J., Wagner M. and Weidmann F. (1999). *Einfluss der Einblastiefe auf den Sauerstoffeintrag–Teil 2–Abwasser* (Influence of depth of submergence on oxygen transfer–part 2–wastewater). Final report, unpublished.

Redmon D. T., Boyle W. C. and Ewing L. (1983). Oxygen transfer efficiency measurements in mixed liquor using off-gas techniques, *J. Water Pollut. Control Fedn*, **55**, 1338–1347.

prEN 12255-15 (2003). Wastewater Treatment Plants - Part 15: Measurement of the oxygen transfer in clean water in activated sludge aeration tanks. European standard.

Rosso D. and Stenstrom M. K. (2006). Economic implications of fine-pore diffuser aging. *Water Environ. Res.*, **78**(8) 810–815.

Rosso D., Iranpour R. and Stenstrom M. K. (2005). Fifteen years of off-gas transfer efficiency measurements on fine pore aerators: key role of sludge age and normalized air flux. *Water Environ. Res.*, **77**(3), 266–273.

Rosso D., Larson, L. and Stenstrom M. K. (2008). Aeration of large-scale municipal wastewater treatment plants: state of the art, *Water Sci. Technol.*, **57**(7), 973–978.

Scherfig J., Schleisner L., Brønd S. and Kilde N. (1996). Dynamic temperature changes in WWTPs, *Water Environ. Res.*, **68**(2), 143–151.

Sedory P. E. and Stenstrom M. K. (1995). Dynamic model for the prediction of wastewater aeration basin temperature, *J. Environ. Engr., Amer. Soc. Civil Eng.*, **121**(9), 609–618.

Stenstrom M. K. (2001). Aeration Systems, 25 Years of Experience, www.seas.ucla.edu/stenstro/ (accessed 28 February 2014)

Stenstrom M. K. and Poduska R. A. (1980). The effect of oxygen concentration on nitrification, *Water Res.*, **14**(6), 643–649.

Stenstrom M. K., Willman L., Migsich N. and Leland T. (2013). My diffuser goes up to eleven (actually twelve). Proc. 86th WEFTEC Conf., Chicago, IL, October 7.

Talati S. N. and Stenstrom M. K. (1990) Aeration basin heat loss, *J. Environ. Engr., Amer. Soc. Civil Eng.*, **116**(1), 70–86.

Tzeng C. J., Iranpour R. and Stenstrom M. K. (2003). Modeling and control of oxygen transfer in the HPO activated sludge process. *J. Environ. Eng., Amer. Soc. Civil Eng.*, **129**(5), 402–411.

von der Emde W. (1998). *Geschichte des Belebungsverfahrens* (History of the activated sludge process). *Korrespondenz Abwasser*, **45**(6), 1086–1101.

Wagner M. (2001). *Neue Tendenzen bei der Belüftungstechnik* (New trends in aeration technology). In: WAR Schriftenreihe, 134 edn, 64. Darmstädter Seminar–Abwassertechnik: Neues zur Belüftungstechnik, Technische Universität Darmstadt, Darmstadt, Germany, pp. 1–30.

Wagner M. and Günkel T. (2010). *Belüftungssysteme in kalten und warmen Klimaten* (Aeration systems in cold and warm climates). In: Leitfaden zur Abwassertechnologie in anderen Ländern, 2nd edn, Exportorientierte Forschung und Entwicklung auf dem Gebiet der Wasserver- und–entsorgung, Teil II: Abwasserbehandlung und Wasserwiederverwendung, Ruhr-Universitaet Bochum, Lehrstuhl für Siedlungswasserwirtschaft und Umwelttechnik, Bochum, Germany, pp. 22–31.

Yunt F. W. and Hancuff T. O. (1980). Aeration Equipment Evaluation – Phase I, Clean Water Results, USEPA Grant No. 14-12-150, Los Angeles County Sanitation Districts, Los Angeles, CA.

Chapter 9

Air emissions

Jay R. Witherspoon (USA), Michael D. Short (Australia), Kate Simmonds (New Zealand), Ben van den Akker (Australia), Ewa Madon (Australia), Richard M.Stuetz (Australia)

9.1 INTRODUCTION

Interest in air emissions from AS has evolved since the early 1950s and today there are regulations for a range of compounds that drive air quality assessments, strategies, control options and management programs. Typical air emission compounds include small or trace amounts of odorous compounds, air toxics (including hazardous air pollutants – HAPs), greenhouse gasses (GHGs), and volatile organic compounds (VOCs). Air emission compounds vary with individual wastewater treatment plant (WWTP) influent characteristics, the type of WWTP including the AS treatment configuration and plant discharges. Studies have shown that approximately 4% of total VOCs in a WWTP influent stream can be emitted during biological treatment processes.

This chapter will explore air quality regulatory drivers and how WWTPs have responded to these, will detail the currently regulated air emissions and will provide an overview of how air emissions are assessed through sampling, source testing, modeling and emission factors. We will present a discussion of the innovations used today to measure and inventory air emissions and recommendations of those needed in the future to address new and emerging air quality concerns and compounds.

9.2 REGULATIONS AND LEGISLATION

While regulation and legislation has driven innovations in measurement, monitoring and treatment of air emissions from AS plants, these regulations vary both between and within countries. It is often the case that a country or region will have an overarching air quality framework, which feeds localized legislation. An example of this localized regulation is California, USA, which developed and passed its own air toxics inventory and control regulations in 1991 based on State-driven risk management plans for acutely hazardous compounds and tightened the Federal acceptable ambient air quality standards.

Odorous emissions are typically regulated at the local level based on a balance between public nuisance thresholds and acceptable ambient air quality standards. GHG regulations are currently being reviewed and emissions inventories are being prepared in response to looming regulatory programs aimed at reducing net environmental impacts. California is again leading the USA in this area with its State-wide GHG emissions inventory and is now starting to consider potential regulatory programs and control requirements.

Table 9.1 summarizes the key air emissions legislation across four global regions (UK, Europe, USA and Australia). To allow ease of comparison, legislation within this Table is separated according to the four key emissions types most relevant to AS (i.e. odor, VOCs, air toxics and GHGs).

9.3 AS EMISSIONS MECHANISMS

Proper characterization of emissions is important, because emission levels trigger regulatory requirements such as the need for controls and permits. WWTPs are many and varied; however, groups of processes typically have similar functions. Figure 9.1 presents a generalized WWTP process schematic. Upstream and downstream processes create a complex problem of interrelated cause and effect within systems that impact the emissions potential from each unit process. For example, covering a preliminary treatment process may retain pollutants such as VOCs in solution at that location, but may enhance VOC emissions from downstream processes. For the purpose of estimating air emissions, wastewater treatment processes are typically grouped into categories based on their emission mechanisms and operational characteristics:

Collection systems (liquid–gas phase mass transfer and ventilation rates);
Physical unit operations (preliminary treatment, sedimentation tanks and tertiary filtration);
Biological treatment (AS, suspended and attached growth);
Solids processing (stabilization, conditioning, dewatering, thickening, combustion sources, flares and digester gas burners, boilers, biosolids incineration, internal combustion engines and turbines, and fuel storage tanks);
Disinfection (risk management plan and process safety management, formation and emissions of trihalomethanes, and other emission mechanisms).

For this chapter, the focus is on AS emissions with particular reference to: (1) emission mechanisms; and (2) the key factors affecting emissions.

Table 9.1 Air emissions regulation and legislation summary

Country/ Region	Regulation overview	GHGs	Air toxics	VOCs	Odors
USA	Air quality pollution legislated federally since 1955 – *Air Pollution Control Act 1955*. Various amendments led to current *Federal Clean Air Act Amendments 1990*. Act sets National Ambient Air Quality Standards for 6 common air pollutants (carbon monoxide, ground-level ozone, lead, N oxides, particulate matter and sulfur dioxide). Regulated by National EPA. Each state can monitor according to its own standards if more stringent than national standards.	Regulated by USEPA under the *Federal Clean Air Act Amendments 1990.* EPA facilitated GHG Reporting Program developed under the Greenhouse Gas Reporting Rule.	Regulated by EPA, the National Emission Standards for Hazardous Air Pollutants (NESHAPS) established under *Federal Clean Air Act Amendments 1990*. California Air Toxics 'Hot Spots' Information and Assessment Act, 1987.	Regulated by EPA under the *Federal Clean Air Act Amendments 1990.* Standards only set for industrial settings. Occupational Safety and Health Administration regulates formaldehyde.	Regulated at state and local levels.

(*Continued*)

Table 9.1 Air emissions regulation and legislation summary (*Continued*).

Country/ Region	Regulation overview	GHGs	Air toxics	VOCs	Odors
Australia	First consolidated air pollution legislation formed part of NSW *Environmental Offences and Penalties Act 1989.* The National Environmental Protection Council (NEPC) consisting of Commonwealth, State and Territory Ministers produced *National Environment Protection (Ambient Air Quality) Measure* – established air quality standards to apply in states and territories for 6 pollutants (particles, ozone, sulfur dioxide, nitrogen dioxide, carbon monoxide and lead). The National Pollutant Inventory (NPI) helps to track emissions nationally.	*National Greenhouse and Energy Reporting* Act provides a national framework for energy and emissions reporting. Administered by the Clean Energy Regulator.	*National Environment Protection Measure (Air Toxics)* established by NEPC and implemented/ regulated by each state.	Emissions regulated by individual states but must comply with requirements of Air NEPM.	Regulated at a state level.

Europe	European Union has 13 Council Directives and 3 Council Decisions concerned with regulating air quality. The EU Commission introduces ambient air quality standards and thresholds but national governments can set more rigorous standards, or set standards for additional pollutants. The European Environment Agency (EEA) provides environmental information to member countries.	Covered by Montreal Protocol or *The Decision on Monitoring Carbon Dioxide and Other Greenhouse Gas Emissions.*	Regulated at national and local levels.	Member states required to control emissions from 'installations'. *The Directive of Volatile Organic Compounds.*	Regulated at national and local levels.
UK	*The Environment Act 1995* initiated *Air Quality Strategy for England, Scotland, Wales and Northern Ireland 2000* and Addendum 2003. Each country responsible for domestic policies that are in line with strategy. *Water Industry Act* establishes emission-reporting requirements under the Pollution Inventory. Also member of EU.	*Carbon Reduction Commitment* – cap and trade scheme. EA regulates Pollution Inventory reporting protocol.	Regulated at country and local levels.	VOCs are regulated through the UK *Pollution Prevention and Control Regulations* and *Air Quality Strategy for England, Scotland, Wales and Northern Ireland, 2000.*	Regulated at country and local levels.

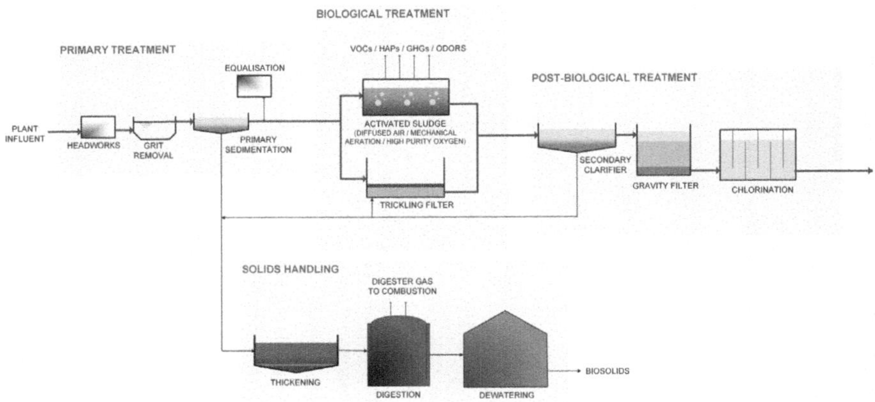

Figure 9.1 Typical WWTP process schematic.

9.3.1 AS aeration basins overview

Wastewater organics serve as carbon sources for biological treatment that removes substances that exert an oxygen demand, with the oxidized end products of such treatment reactions primarily being CO_2. The systems used to supply oxygen to the AS reactor (e.g. diffused air, high purity oxygen (HPO) and mechanical aeration systems) can influence significantly the level of emissions.

9.3.1.1 Emission mechanisms

Emission mechanisms for the AS treatment process include surface volatilization and mass transfer from the wastewater to the rising air bubbles during aeration followed by ventilation. In addition, VOC emissions can be emitted from aerosol particles primarily during mechanical aeration/agitation and from emissions associated with weir drops. Other mechanisms, such as biodegradation and solids adsorption, can compete for VOCs and may significantly reduce emission loads. Highly degradable VOCs can have lower air emission rates than those that are not readily degradable.

9.3.1.2 Key factors affecting emissions

The key variables that affect emissions for AS systems are:

Gas-to-liquid ratio: for diffused air systems, a higher ratio of airflow to wastewater flow will significantly increase emissions. Fine bubbles also provide larger water–air surface area relative to coarse bubbles, yielding higher VOC mass transfer rates for a given air flow rate (Desing, 1991).

The lower gas-to-liquid ratio required by fine bubble aeration outweighs the increased VOC mass transfer rate which results in lower overall VOC stripping rates. For HPO systems, which operate at low gas-to-liquid ratios, emissions are typically much lower.

Biodegradation rate: a higher VOC biodegradation rate equates to lower emissions. Highly degradable compounds such as acetone and methanol will be quickly degraded upon entering the aeration basin, resulting in relatively low emissions towards the front of the reactor and a negligible amount downstream.

Aerator characteristics: for mechanical aeration systems, the rotational characteristics, size and power of the aerators will affect emissions.

Adsorption: compounds that sorb to particulates can be removed in secondary clarification. However, subsequent biodegradation or desorption (emissions) may occur during sludge processing, making sorption a difficult mechanism to characterize by direct measurement.

Other factors: other factors that affect emissions are detention time, basin surface area, diffuser type and other system design parameters.

The following two subsections provide examples of source testing program results for odors, VOCs and air toxics under California Local and State Regulations. The program used direct air emission measurements to determine industry-based air emission factors and represent typical values for processes under normal operating conditions (i.e. not optimized for emissions reduction).

9.3.2 Air emissions inventory programs

The California South Coast Air Quality Management District enacted a rule in 1991 that required Publicly Owned Treatment Works (POTWs) with capacities of greater than 10 mgd (≈40 MLD) to complete inventories of VOC and odorous emissions (known as the Joint Emissions Inventory Program (JEIP)) to assess their contribution relative to other known VOC emission categories. More than 1100 air-phase and 800 liquid-phase samples were collected and analyzed for VOCs at 22 POTWs to estimate emissions from liquid and solids processes. Table 9.2 summarizes the average emissions from each unit process and shows that diffused air aeration was a significant contributor to the total fraction of influent VOCs emitted (average of 28.2%), whereas mechanically aerated AS had a much lower emission factor (6.5%).

Southern California WWTP operators established a Pooled Emission Estimating Program (PEEP) to develop a statewide inventory of site-specific air toxic emissions for over 400 substances to enable the assessment, management and communication of the health risks to the public of emissions exposure. The purpose of the PEEP is to provide JPA members with a standardized estimation method for air emission from liquid, solid and gas handling processes (Leong *et al.*, 1992).

Table 9.2 JEIP average emissions summary for AS processes.

Unit process	VOC emissions, lb/year/influent mgd ADWF – (kg/year/influent MLD)[1]
AS – Diffused air	190 (326)
AS – Mechanical aeration	27 (46)
AS – High purity oxygen	5 (9)

[1]VOC emissions from wastewater treatment processes vary significantly depending on factors such as influent composition, operating factors, design of processes and so on.

9.4 IMPACTS AND TREATMENT OF EMISSIONS

9.4.1 Odorous emissions

Composition: odorous emissions from WWTPs often contain sulfur compounds (e.g. hydrogen sulfide (H_2S) at <1–5 ppm and mercaptans at ppb levels), VOCs (e.g. fatty acids, aromatic, aliphatic and chlorinated hydrocarbons, terpenes, aldehydes and ketones), ammonia and occasionally N derivatives (Vincent, 2001). Trimethylamine and dimethylamine are usually considered as priority odorants in WWTPs because of the high N content of municipal wastewaters (Gostelow *et al.*, 2001).

Treatment: a variety of measures exist for reducing odors associated with AS processes, beginning with influent concentration controls, covering and venting to an odor abatement device (Lebrero *et al.*, 2011) such as carbon and chemical scrubbers and biotechnological devices (e.g. biofilters and bioscrubbers); these odor abatement devices are often also suitable for controlling VOC emissions.

9.4.2 Air toxics and VOCs

Composition: air toxics are pollutants that exist as particles or as a gas, and include volatile and semi-volatile organic compounds and heavy metals. The primary gaseous air toxics emitted include xylenes, methylene chloride, toluene, ethyl benzene, chloroform, tetrachloroethylene, benzene, perchloroethylene and naphthalene. Examples of air toxics associated with particles include heavy metals such as cadmium, mercury, chromium and lead compounds, as well as the semi-volatile organic compounds such as polycyclic aromatic hydrocarbons. Among wastewater treatment processes, AS is one of the most significant VOC emission sources. However, since the total mass emissions from AS plants are typically a small component of an air basin emissions inventory, WWTPs are generally not required to control these emissions unless the AS process is causing a public odour nuisance.

Treatment: VOCs from AS removed through stripping, adsorption and biodegradation. Namkung and Rittmann (1987) concluded that biodegradation was the major mechanism for the removal of non-chlorinated VOCs such as toluene, benzene, ethylbenzene and methylene chloride, and stripping was the main removal mechanism for chlorinated compounds such as chloroform, tetrachloroethylene and trichloroethylene. The same control and capture schemes described above for odor control above can be used for VOCs and air toxics.

9.4.3 GHG emissions

GHG emissions from wastewater processing have received increased international attention in recent years. While wastewater GHG emissions are either 'direct' (scope 1) or 'indirect' (scopes 2 and 3), (McGuckin *et al.*, 2013), the focus here is on direct emissions. The two GHGs of primary interest for wastewater operations are methane (CH_4) and nitrous oxide (N_2O). Direct CO_2 emissions from wastewater are considered biogenic and are, therefore, not included in GHG emission inventories (Doorn *et al.*, 2006); although recent research has challenged this approach (Law *et al.*, 2013).

9.4.3.1 CH_4

Wastewater CH_4 emissions are generally reported as the fraction of CH_4 produced per unit COD load and occur at all stages of the wastewater management chain from collection and treatment, to sludge handling and disposal. The IPCC *Guidelines* take a first principles approach to estimating wastewater CH_4 emissions based on the maximum CH_4 yield from a given quantity of organics and a correction factor to account for the likely realization of this maximum CH_4 production along a given treatment and discharge pathway (Doorn *et al.*, 2006). Key variables for CH_4 generation in wastewater systems appear to be temperature, oxidation–reduction potential, O_2 and the organic carbon/COD load (Czepiel *et al.*, 1993; Daelman *et al.*, 2012); temperature may have a dual effect since higher temperatures increase the rate of methanogenesis while lowering CH_4 solubility.

Compared to N_2O, CH_4 emissions from wastewater systems have been much less well researched and there are relatively few published studies (Wang *et al.*, 2011a; Ren *et al.*, 2013; Willis *et al.*, 2012; Daelman et al., 2013). For WWTPs without onsite anaerobic digestion, the majority of onsite CH_4 emissions appear to originate in the upstream sewer networks (Guisasola *et al.*, 2008; Foley *et al.*, 2009, 2011; Willis *et al.*, 2012) and are simply ventilated during turbulent wastewater processing and/or aeration (e.g. influent works, aerated grit tanks, secondary aeration basins).

Data on CH_4 emissions from AS are scarce; the limited data currently available suggest that there is some temporal variability in emissions (approximately one order of magnitude) at both daily and sub-daily timescales (Daelman *et al.*, 2012, 2013). The intensity of CH_4 emissions from AS systems is likely to be a function of wastewater strength, sewer network configuration and HRT, combined with the potential for upstream CH_4 flux from sewers, inlet works and preliminary treatment processes. Recent research suggests that where the potential for such upstream CH_4 flux is high, CH_4 emissions from AS basins are relatively low (Wang *et al.*, 2011a). At the same time, where turbulent and/or aerated upstream processes are not present at WWTPs, CH_4 emissions from AS can be high; although influent CH_4 may also be oxidized during AS aeration rather than being emitted (Daelman *et al.*, 2013; Ho *et al.*, 2013). Recent research investigating several engineered treatment interventions (air filters) to mitigate CH_4 emissions from wastewater facilities has proven largely unsuccessful (Daelman *et al.*, 2013).

9.4.3.2 N_2O

N_2O generation during wastewater treatment is primarily biological, with ammonia oxidizers and denitrifying microorganisms the primary facilitators (Kampschreuer *et al.*, 2009; Wunderlin *et al.*, 2012). Recent work (Stein *et al.*, 2012) suggests a possible role for 'methanotrophic nitrification' in wastewater N_2O production, although this remains unexplored. Past thinking that heterotrophic denitrification was the dominant source of wastewater N_2O has recently been replaced by an emerging consensus view that ammonia-oxidizing microorganisms are largely responsible for generated N_2O (e.g. Ahn *et al.*, 2010a; Law *et al.*, 2012). The key variables for N_2O production in nitrifying AS appear to be transient or highly dynamic process operation, aeration regime and/or dissolved O_2 concentration, and ammonia and nitrite concentrations (Chandran, 2012). For denitrifying AS, the key variables are most likely oxygen inhibition, low COD/N ratio (Law *et al.*, 2012) and possibly copper availability (Zhu *et al.*, 2013).

N_2O emissions are most commonly reported as the fraction of N_2O-N produced per unit total N load, or less commonly per unit N removed during treatment. Several comprehensive reviews of wastewater N_2O emissions – both from laboratory- and full-scale work – have been published in recent years (Dotro *et al.*, 2011; Foley *et al.*, 2011; Rassamee *et al.*, 2011). The range of full-scale N_2O emission factors reported in these studies varies widely (0–25% of the N load) in line with the highly variable nature of wastewater operations surveyed (e.g. process configuration and operation, season and climate, wastewater type, etc.); however, N_2O emissions are generally in the order of 0–5% of N load. Adding another layer of complexity to this already high degree of variability is the fact that N_2O production is highly dynamic in both space and time (Ahn *et al.*, 2010a, b). Importantly, this spatio-temporal variability exists not only between treatment plants, but also extends to individual reactors (Figure 9.2). For the relatively few fully enclosed aeration basins (Daelman *et al.*, 2013), the variability in N_2O emissions is less critical, but for the majority of unenclosed plants, it presents an important monitoring and control challenge.

At best, the tremendous variability in N_2O emissions from AS processes makes it difficult to determine single universal emission factors; at worst, it makes these values irrelevant. Similar sentiments have been put forward in recent times, with several authors calling into question the usefulness of the current "universal" emission factor approach and suggesting instead a more dynamic, mechanistic or "bottom-up" approach to emissions estimation and mitigation based on a better understanding of the key drivers for N_2O production (Ahn *et al.*, 2010a; Chandran, 2012; Daelman *et al.*, 2013). A future shift away from single estimate universal N_2O emission factors will make it increasingly important to validate emissions on a plant-by-plant basis, particularly if suggested future N_2O emissions crediting policy is implemented (Wang *et al.*, 2011b). No research to date has investigated engineered solutions/retro-fits to mitigate N_2O emissions (e.g. chemical/biological denitrifying filters/scrubbers). The most promising emissions control measures are likely to be tighter AS process control and/or performance optimization, combined with future emissions policy incentives.

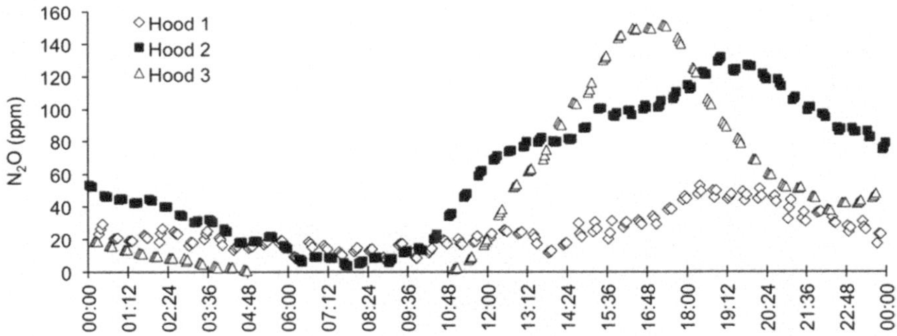

Figure 9.2 N_2O emission data over a 24-h period from 3 sampling locations across a single South Australian AS sludge reactor (gas flow rates ≈20–60 L/min).

9.5 TECHNIQUES USED TO ASSESS EMISSIONS

A number of different methodologies are currently used to quantify odorous, gaseous and volatile air emissions from AS.

Sensorial Odor Measurement: sensorial odor perception, together with analytical measurements, constitutes the most common approach for odor characterization. When combined, sensory and analytical measurements characterize odors in terms of their perceived effects and their chemical composition (Gostelow *et al.*, 2001).

Gas and liquid phase grab sampling: gas phase grab samples can be collected from chambers and wind tunnels floating on the reactor surface. Collected samples can either be analyzed immediately, or transported to laboratories for later analysis. Knowledge of the gas flow rate is required to calculate mass emission rates. Liquid phase grab sampling can also be performed by sub-sampling the reactor volume followed by headspace equilibration and measurement. Liquid phase sampling requires knowledge of gas flow rate and analyte mass transfer coefficient to derive emission data. Due to the inability of both grab sampling approaches to adequately capture temporal dynamics, continuous online monitoring is the method of choice.

Online monitoring: emissions from individual WWTPs are highly dynamic in both space and time, owing to diurnal changes in sewage strength, non-ideal reactor hydraulics, changes in operational parameters and seasonal factors. Characterizing spatio-temporal variability is essential when quantifying emissions and is best achieved by taking online measurements over the operational range of the WWTP. Various portable online instruments capable of monitoring compounds of interest are currently available and include infra-red analyzers, field gas chromatographs, and so on.

Sample Capture: most WWTPs are not enclosed and, therefore, floating hoods are required to capture representative gas samples. Figure 9.3 shows an online monitoring system capable of monitoring the off-gas sampled from multiple surface hoods anchored along an aerated zone of a plug-flow AS reactor. For non-aerated zones, the injection of a weak flowing sweep-air across the headspace of the hood is required to measure gaseous wastewater emissions.

Figure 9.3 Continuous emissions monitoring at a South Australian AS plant (developed by The University of Queensland and SA Water Corporation). (1) hood; (2) captured gas inlet; (3) gas temperature sensor; (4) pressure sensor; (5) flow meter; (6) gas exhaust; (7) capillary wick to remove condensation from flow meters; (8) solenoid gas valve for multiple hood sampling; (9) programmable logic controller to control sampling and data logger; (10) gas sample conditioning system; (11) infrared N_2O, CH_4 and O_2 gas analyzer.

9.6 CONCLUSIONS

Knowledge of AS process air emissions including their composition, measurements, inventorying, controlling needs, and human health and property impacts has evolved greatly over the past fifteen years. Globally, we are seeing AS process emissions being regulated on the basis of the following four main component categories: odors, VOCs, air toxics, and GHGs.

Even though AS air emissions may constitute the major air emissions source within a WWTP, they are typically small contributors to the overall quality of an air basin. Studies have shown that typical, very large WWTPs with AS processes (greater than 378 MLD) are minor air emissions contributors to an air basin VOC, smog, air toxics, or GHG emissions inventories. Consequently, WWTPs are not required to control their air emissions, unless the AS process (or other odor emitting processes) is causing a nuisance odor impact to the surrounding community.

Several countries, including the USA, Australia and Singapore, are leading the way in the development of GHG emission measurement approaches and in the inventorying of GHG emissions. Globally, the Intergovernmental Panel on Climate Change (IPCC) has issued emissions estimation *Guidelines* for national GHG inventories which include AS processes. The IPCC default GHG emission factors for (CH_4 and N_2O) from AS processes are considered conservative and research to date indicates that they do not adequately represent actual emissions in many instances. For example, the IPCC default emission factor for CH_4 is based on daily per-capita BOD generation rates and for N_2O is based on a single limited field study from 1993; both approaches are inadequate to establish robust estimates of AS process GHG emissions as required today.

Recent studies on AS process GHG emissions have shown that the default IPCC methodology can greatly overestimate CH_4 emissions and can either under- or overestimate N_2O emissions depending on the AS process configuration. More research is required to better understand the key mechanisms involved in N_2O production during AS treatment so that it is possible to allow GHG emissions mitigation through better process design, operation and control. Options for retrofitting and/or augmenting poor performing plants to minimize process GHG emissions also warrant further investigation.

In the near future, it is possible that additional emissions components including new compounds and AS aerosols will be identified. For these, it is likely that the traditional regulatory format of measurement, inventory, impact and control will be followed. However, this traditional format will have to be modified for addressing trace-level emissions; in addition a new suite of research will be required to assess impacts (long-term and acute) and to define acceptable levels. New measurement approaches will be needed to allow for the detection and resolution of these emerging trace-level emissions. Rigorous assessments of the human health and environmental impacts of these new compounds will also be required to avoid unnecessary regulatory burden on wastewater operations.

9.7 REFERENCES

Ahn J. H., Kim S., Park H., Rahm B., Pagilla K. and Chandran K. (2010a). N_2O emissions from activated sludge processes, 2008–2009: results of a national monitoring survey in the United States. *Environ. Sci. Technol.*, **44**(12), 4505–4511.

Ahn J. H., Kim S., Park H., Katehis D., Pagilla K. and Chandran K. (2010b). Spatial and temporal variability in atmospheric nitrous oxide generation and emission from full-scale biological nitrogen removal and non-BNR processes. *Water Environ. Res.*, **82**(12), 2362–2372.

Chandran K. (2012). Greenhouse Nitrogen Emissions from Wastewater Treatment Operation: Phase I: Molecular Level Through Whole Reactor Level Characterization, Water Environ. Res. Foundn (WERF), USA and IWA Publishing, UK. ISBN: 9781780404813.

Czepiel P. M., Crill P. M. and Harriss R. C. (1993). Methane emissions from municipal wastewater treatment processes. *Environ. Sci. Technol.*, **27**(12), 2472–2477.

Daelman M. R. J., van Voorthuizen E. M., van Dongen U. G. J. M., Volcke E. I. P. and van Loosdrecht M. C. M. (2012). Methane emission during municipal wastewater treatment. *Water Res.*, **46**(11), 3657–3670.

Daelman M. R. J., van Voorthuizen E. M., van Dongen L. G. J. M., Volcke E. I. P. and van Loosdrecht M. C. M. (2013). Methane and nitrous oxide emissions from municipal wastewater treatment – results from a long-term study. *Water Sci. Technol.*, **67**(10), 2350–2355.

Desing W. E. (1991). Aeration Systems Selection and Specification to Minimize Toxic Air Emissions. 65th Ann. Meeting Central States Water Pollut. Control Assoc., Fontana, WI, USA.

Doorn M. R. J., Towprayoon S., Vieira S. M. M., Irving W., Palmer C., Pipatti R. and Wang C. (2006). Wastewater Treatment and Discharge, In: 2006 IPCC Guidelines for National Greenhouse Gas Inventories: Vol. 5 – Waste, The National Greenhouse Gas Inventories Programme, The Intergovernmental Panel on Climate Change, H. S. Eggleston, L. Buendia K. Miwa, T. Ngara, and K. Tanabe (eds), Hayama, Kanagawa, Japan, pp. 6.1–6.28. ISBN 4-88788-032-4.

Dotro G., Jefferson B., Jones M., Vale P., Cartmell E. and Stephenson T. (2011). A review of the impact and potential of intermittent aeration on continuous flow nitrifying activated sludge. *Environ. Technol.*, **32**(15), 1685–1697.

Foley J., Yuan Z. and Lant P. (2009). Dissolved methane in rising main sewer systems: field measurements and simple model development for estimating greenhouse gas emissions. *Water Sci. Technol.*, **60**(11), 2963–2971.

Foley J., Yuan Z., Keller, J., Senante E., Chandran K., Willis J., Shah A., van Loosdrecht M. C. M. and van Voorthuizen E. (2011). N_2O and CH_4 Emission from Wastewater Collection and Treatment Systems: Tech. Rept, Global Water Res. Coalition (GWRC), London, UK. ISBN: 9789077622247.

Gostelow P., Parsons S. A., and Stuetz R. M. (2001). Odour measurement in sewage treatment: a review. *Water Res.*, **35**, 579–597.

Guisasola A., de Haas D., Keller J. and Yuan Z. (2008). Methane formation in sewer systems. *Water Res.*, **42**(6–7), 1421–1430.

Ho A., Vlaeminck S. E., Ettwig K. F., Schneider B., Frenzel P. and Boon N. (2013). Revisiting methanotrophic communities in sewage treatment plants. *Appl. Environ. Microbiol.*, **79**(8), 2841–2846.

Kampschreur M. J., Temmink H., Kleerebezem R., Jetten M. S. M. and van Loosdrecht M. C. M. (2009). Nitrous oxide emission during wastewater treatment. *Water Res.*, **43**(17), 4093–4103.

Law Y., Ye, L., Pan Y. and Yuan Z. (2012). Nitrous oxide emissions from wastewater treatment processes. *Philos. Trans. Royal Soc. B.*, **367**, 1265–1277.

Law Y., Jacobsen G. E., Smith A. M., Yuan Z. and Lant P. (2013). Fossil organic carbon in wastewater and its fate in treatment plants. *Water Res.*, **47**(14), 5270–5281.

Lebrero R., Bouchy L., Stuetz R. M. and Muñoz R. (2011). Odor assessment and management in wastewater treatment plants: a review. *Critical Rev. Environ. Sci. Technol.*, **41**(10), 915–950.

Leong L. Y. C., Regan M. M., Kuo J.-F. and Wong E. (1992). An overview of the pooled emission estimation program (PEEP) for POTWs. *Environ. Prog.*, **11**(4), 278–287.

McGuckin R., Oppenheimer J., Badruzzaman M. and Jacangelo J. G. (2013) Toolbox for Water Utility Energy and Greenhouse Gas Emission Management, Water Res. Foundn. ISBN: 978-1-60573-185-8.

Nakasone H. (1986). Study of aeration at weirs and cascades. *J. Environ. Eng.*, **113**(1), 64–81.

Namkung, E. Rittman, B. E. (1987). Evaluation of bisubstrate secondary utilization kinetics by biofilms, *Biotechnol. Bioeng.*, **29**(2), 355–342.

Rassamee V., Sattayatewa C., Pagilla K. and Chandran K. (2011). Effect of oxic and anoxic conditions on nitrous oxide emissions from nitrification and denitrification processes. *Biotechnol. Bioeng.*, **108**(9), 2036–2045.

Ren Y. G., Wang J. H., Li H. F., Zhang J., Qi P. Y. and Hu Z. (2013). Nitrous oxide and methane emissions from different treatment processes in full-scale municipal wastewater treatment plants. *Environ. Technol.*, **33**(21), 2917–2927, doi: 10.1080/09593330.2012.696717.

Stein L. Y., Roy R. and Dunfield P. F. (2012). Aerobic methanotrophy and nitrification: processes and connections. In: Encyclopedia of Life Sciences (eLS), J. Battista et al. (ed.), John Wiley & Sons, Ltd, Chichester, doi: 10.1002/9780470015902.a0022213.

Vincent A. J. (2001). Sources of odors in wastewater treatment. In: Odours in Wastewater Treatment: Measurement, Modeling and Control (Chapter 4). R. Stuetz and F. B. Frechen (eds), IWA, London, UK.

Wang J., Zhang J., Xie H., Qi P., Ren Y. and Hu Z. (2011a). Methane emissions from a full-scale A/A/O wastewater treatment plant. *Bioresource Technol.*, **102**(9), 5479–5485.

Wang J. S., Hamburg S. P., Pryor D. E., Chandran K. and Daigger G. T. (2011b). Emissions credits: opportunity to promote integrated nitrogen management in the wastewater sector. *Environ. Sci. Technol.*, **45**(15), 6239–6246.

Willis J., Shah A., Yuan Z., Sharma K., Joiner M. and Bocarro R. (2012). Methane Evolution from Wastewater Conveyance, Water Environ. Res. Foundn and IWA Publ. ISBN: 9781780404691.

Wunderlin P., Mohn J., Joss A., Emmenegger L. and Siegrist H. (2012). Mechanisms of N_2O production in biological wastewater treatment under nitrifying and denitrifying conditions. *Water Res.*, **46**(4), 1027–1037.

Zhu X., Chen Y., Chen H., Li X., Peng Y. and Wang S. (2013). Minimizing nitrous oxide in biological nutrient removal from municipal wastewater by controlling copper ion concentrations. *Appl. Microbiol. Biotechnol.*, **97**(3), 1325–1334.

Chapter 10

Activated sludge solids separation

Jiří Wanner (Czech Republic), Andrea Jobbágy (Hungary)

10.1 REQUIREMENTS AND MEASUREMENT OF SEPARATION

10.1.1 Requirements for good AS separation

Separation of biomass from treated effluent is crucial in AS wastewater treatment. Although processes such as membrane filtration have been used in full-scale systems in the last three decades, the conventional technology of gravity sedimentation in secondary clarifiers is still the most common and economically feasible option for municipalities and industrial enterprises worldwide. This process requires that the biomass can (i) flocculate (ii) settle and thicken by gravity. Poor solids separation may lead to major operational problems:

- Biomass escaping with the treated effluent deteriorates its quality with respect to TSS, BOD_5, COD, N and P.
- A dilute RAS stream may not allow maintenance of the required biomass concentration and SRT control in the reaction basin(s); severe biomass loss may even empty the bioreactor.
- A dilute WAS stream may result in hydraulic overloading of the sludge handling facilities.

To ensure safe operation, well settling AS should: (i) settle fast, with zone settling velocities (ZSVs) of ≥3 m/h; (ii) not occupy an excessive volume after settling and thickening; (iii) leave a clear supernatant after settling (TSS) ≤15 mg/L; (iv) not rise to the surface for at least a 2–3 h period after settling (Wanner, 1994a; Jenkins *et al.*, 2004).

10.1.2 Basic measurements

Settling in secondary clarifiers can be best simulated by ZSV measurement in transparent cylinders so that the velocity of the sludge–supernatant interface can be observed (Wanner, 1994a). Tall, wide cylinders give more realistic ZSV values. A settling curve (a plot of sludge height *vs.* time) usually has three distinct regions: I – reflocculation, II – zone settling, and III – transition and compaction. ZSV is calculated from the slope of the curve in region II. In wastewater treatment practice the most common way to characterize settleability is the sludge volume index, SVI, (mL/g) defined as:

$$\mathrm{SVI} = \frac{V_{30}}{X}$$

where V_{30} (mL/L] is the volume of settled sludge after 30 min sedimentation in a 1 L cylinder, and X is the mixed liquor SS (MLSS) concentration in g/L. The SVI is affected by MLSS concentration, by settled sludge volume after 30 min and by wall effects in the cylinder. In some countries, settling test conditions are standardized. Common standardization methods include:

- Stirred Sludge Volume Index, SSVI; the settling cylinder is equipped with a slowly rotating stirring device (White, 1975).
- SVI at a standard MLSS concentration, for example $\mathrm{SVI}_{3.5}$ = SVI at 3.5 g/L.
- Diluted Sludge Volume Index, DSVI; dilute the sludge so that after 30 min the settled sludge volume is ≤200 mL/L.

On the basis of SVI, AS can be classified as (i) well settling (SVI <100 mL/g), (ii) 'light' sludge (SVI = 100–200 mL/g), and (iii) bulking sludge (SVI >200 mL/g) (Wanner, 1998).

There are no generally accepted methods for quantifying foaming problems. Possibly the best measure of the amount of foam formed in aeration basins and the extent of resulting problems is the scum index SI (%) (Pretorius & Läubscher, 1987) where

$$\mathrm{SI} = \frac{\text{mass of biomass in foam}}{\text{total mass of biomass}} \times 100$$

The portion of biomass in foam can also be estimated by floatation in a batch floatation cell (80 mm diam., 500 mm high) at an aeration rate of 10 $m^3/m^3 \cdot h$. Floatation is repeated on the sample and floated material is collected repeatedly until the sample no longer foams. SI values range from 0–0.5% (negligible) to >15% (severe).

10.1.3 Microscopic examination of floc structure

Microscopic examination is useful for evaluating the causes of solids separation problems. Good settling AS is formed by largely compact, firm, roughly spherical flocs that have an average size of ≥100 µm. Lower density flocs accompanied by increased numbers of filaments growing inside the flocs can result in open-structured flocs that can cause poor AS sedimentation. Flocs can also influence separation properties. The dominance of very small (usually 20–80 µm) often quite compact flocs, may indicate a pinpoint floc problem while the presence of individual cells or just small aggregates indicates dispersed growth (Section 10.2). The presence of filamentous microorganisms is evaluated by the microscopic examination of AS from the point of view of their interference with floc sedimentation (open floc structure or bridging by filaments). The manuals for microscopic examination of AS usually provide protocols for the examination of flocs and the influence of filaments (e.g. Eikelboom & van Buijsen, 1981; Jenkins *et al.*, 1993, 2004).

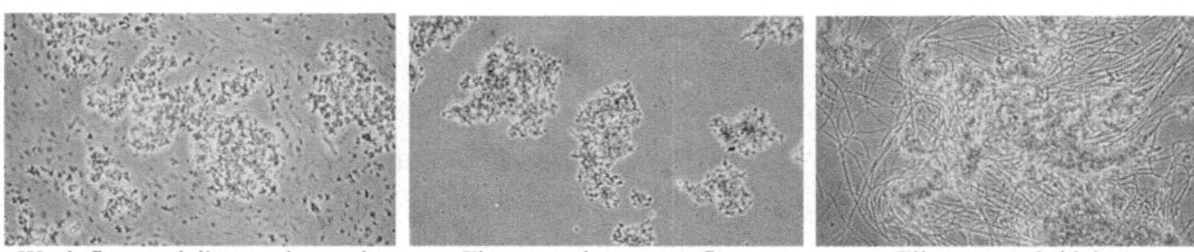

Weak flocs and dispersed growth | Firm, round, compact flocs | Filamentous bridging

Figure 10.1 Characteristic AS floc structures (photographs by M. G. Richard).

Palm *et al.* (1980) showed that the SVI of AS was a function of the length of filaments extending from the flocs or freely floating in the bulk liquid (TEFL). Because TEFL measurement is time consuming, a rapid, simple subjective scoring method for filament abundance was developed (Jenkins *et al.* 1993, 2004). Schuler and Jassby (2007) showed that there appears to be a generally continuous (and often nearly linear) relationship between filament content and SVI.

Microscopic examination is useful for diagnosing the cause of floating or foaming solids because biological foam, especially on a secondary clarifier surface can look like floating sludge. However, biological foams contain much greater amounts of foam-associated filaments while the floating sludge filamentous organism population is similar to that in the mixed liquor.

10.2 AS SEPARATION PROBLEMS

The following six major microbial biomass quality problems can be distinguished:

Microfloc-dispersed growth. The AS contains many microorganisms present as individual cells or small aggregates (≤10 µm) dispersed in the bulk liquid. Their sedimentation rate is too low for them to be removed by gravity sedimentation and zone settling does not occur in either an SVI test or in the secondary clarifiers. Dispersed growth causes turbid effluents and dilutes RAS.

Microfloc-pinpoint floc. The AS contains many poorly settling 50–100 µm aggregates together with flocs that zone settle rapidly. The secondary clarifier effluent TSS is high and the SVI is very low (<75 mL/g). AS is more prone to pinpoint floc formation when there is an insufficient amount of the exocellular materials that flocculate them and in the absence of filamentous organisms.

Viscous (or non-filamentous) bulking sludge contains excessive amounts of extracellular biopolymers which impart a jelly-like consistency to the sludge. Because these polymers are hydrophilic, the AS becomes highly water retentive (even viscous) and has low settling and compaction velocities. Because the biopolymers can be surface active, the AS may foam on aeration (Wanner, 2002).

Filamentous bulking is caused by an excessive growth of filamentous organisms that interfere with AS settling and compaction. High SVIs and major operational problems result (Wanner, 1994a). These include reduced RAS SS and MLSS concentrations, secondary clarifier overloading, high effluent TSS levels and poor WAS thickening and dewatering characteristics.

Biological foaming is really a floatation process. It is caused by the excessive growth of dispersed filamentous microorganisms that have hydrophobic cell walls and may also produce biosurfactants. These properties allow them to attach to air bubbles and rise with them to the water surface to form stable foam. These 'foams' always contain higher levels of the foam-causing filamentous microorganisms than the mixed liquor from which it was generated.

Rising sludge is a form of floatation, usually in secondary clarifiers, caused by the formation of N_2 gas in the flocs through denitrification. This phenomenon can also be observed during the conduct of the SVI test.

10.3 FILAMENTOUS BULKING CONTROL METHODS

10.3.1 Theory and causes of filamentous bulking

The microbial composition of AS results from strong competition for energy, carbon sources, nutrients and electron acceptors. It is influenced by intrinsic factors such as SRT, substrate type, oxygen and nutrient concentrations in the reactor and cultivation conditions (oxic, anoxic, anaerobic); and by external factors such as wastewater composition (substrate type), temperature and pH.

Filaments can interfere with the sedimentation and compaction of AS flocs by forming *diffuse flocs* or by *bridging between the flocs.* Filaments that grow inside flocs can produce flocs with a diffuse open structure that allows for water retention inside them. Other filaments protrude from flocs into the bulk liquid and when present in large numbers can mechanically interfere with floc compaction.

AS settling properties are influenced both by the nature and abundance of filamentous microorganisms and by the weighting effect of various floc-forming microorganisms. Some floc-forming organisms form large dense clusters that can enhance settling rate. These include the phosphate accumulating organisms (PAOs) in enhanced biological phosphate removal (EBPR) systems and nitrifying bacteria (Larsen *et al.,* 2006, 2008) In addition, the settling properties of AS in EBPR systems can be improved by increased density of flocs resulting from inorganic poly-P precipitates (Schuler *et al.,* 2001).

The extent and control of filamentous bulking depends on a knowledge of the types of the dominant filamentous microorganisms and of their basic properties and growth strategies. More than 30 morphotypes have been observed microscopically in municipal wastewater AS (Eikelboom & van Buijsen, 1981, 2000; Jenkins *et al.,* 1993, 2004). Many more are encountered in industrial wastewater treatment plants (Eikelboom & Geurkink, 2002). Some filament types can be reliably characterized using morphology (e.g. *Candidatus* Microthrix parvicella, *Thiothrix* spp., some Mycolata (Nocardia or nocardioform actinomycetes). Others (e.g. Nostocoida limicola which affiliates with at least five different bacterial phyla) require the use of culture-independent molecular methods such as Fluorescent *In Situ* Hybridization (FISH). Knowledge of the treatment plant design, operation and wastewater characteristics are important in selecting the appropriate control measure(s).

10.3.1.1 The most important filamentous microorganisms

Surveys based on morphological identification.

The most abundant filamentous species in municipal AS plants in Europe, South Africa and Australia are *Candidatus* Microthrix parvicella, type 0041, type 0092 and Mycolata (Blackbeard *et al.* 1986; Seviour *et al.,* 1994; Wanner *et al.,* 1998; Krhutkova *et al.,* 2002). In the USA the top four morphotypes observed are Mycolata, type 1701, type 021N and type 0041 (Jenkins *et al.,* 1993, 2004). The difference between these lists is primarily caused by the difference in selective pressure on the microbial population exerted by differences in plant design, operating conditions and wastewater characteristics (Wanner *et al.,* 1987a, b).

Surveys based on FISH probing.

The few surveys performed using both conventional identification and FISH analysis have all been mostly on industrial wastewater treatment plant samples (Nielsen *et al.* (2010) and by Mielczarek *et al.* (2012)). The probe-defined filamentous bacteria had low diversity with *Candidatus* M. parvicella and species in the Chloroflexi phylum (e.g. type 0803, type 0092) being most abundant. Filamentous population types were stable in each plant over 3 years and their relative abundance was unique to each plant giving a characteristic ‘fingerprint’. No significant correlation between the presence and abundance of individual types and plant design or process parameters was found, supporting the conclusion from previous surveys that the filamentous organism distribution and abundance is related more to their nutritional requirements than to other factors. The most abundant filamentous species or types found by FISH analysis surveys are summarized in Table 10.1. (Wanner *et al.*, 2010; Eikelboom & Geurkink, 2002).

Table 10.1 Ranking of most abundant filamentous bacteria causing bulking in municipal and industrial AS.

Filamentous organism	Municipal plants	Industrial plants
Alphaproteobacteria (N. limicola-like)		1
Betaproteobacteria (type 0803-like)	7	
Thiothrix eikelboomii (type 021N)		2
Thiothrix type I, II		2
Bacteroidetes (type 0092-like)	6	
Haliscomenobacter hydrossis	4	
Chloroflexi (type 1851, type 0041/0675 and type 1701)	2	4
TM7-related (type 0041/0675), *Curvibacter*-related bacteria/type 1701)	3	
Candidatus M. parvicella		
Mycolata (Nocardioforms*)	5	3

(*) Mycolata are mostly foaming microorganisms but, often are present in significant amounts in bulking AS.

10.3.2 Principles of selection

The current knowledge on filamentous and floc-forming microorganisms allows bulking control measures to be taken on a sound scientific basis. These 'bioengineering' or 'biotechnological' methods are specifically targeted at the biological factors that govern the competition of floc-forming and filamentous microorganisms. Another group of so-called non-specific methods are aimed at dealing with the effects of excessive filament growth rather than the cause(s) of their growth.

10.3.2.1 Bioengineering bulking control methods

(1) External factors

Wastewater composition. Wastewater is a mixture of substrates, nutrients and micronutrients which supports filamentous and floc-forming organism growth as well as continuously inoculating this consortium with microorganisms.

Readily biodegradable substrate (rbCOD). Low molecular weight organic compounds (e.g. monosaccharides, alcohols, volatile fatty acids and amino acids) can be directly utilized by bacteria. rbCOD makes up approx. 10–20 % of the COD in municipal wastewater and favors the growth of filamentous microorganisms such as type 021N, *Sphaerotilus natans*, N. limicola and possibly *H. hydrossis*. Special attention has to be paid to AS design for wastewaters containing food industry wastewaters. Reduced sulfur compounds can support the growth of S-metabolizing filaments such as type 021N/*Thiothrix* or *Beggiatoa* spp.

Bioengineering controls for bulking caused by rbCOD are:

(a) Maintain high concentration gradient of rbCOD in AS tank (plug flow, compartmentalization, contact zone *kinetic selection*, see below)
(b) Provide high DO at the head-end of the AS tank
(c) If it is not possible to provide enough oxygen in the head-end, it is better to operate it under anoxic or anaerobic conditions (*metabolic selection*, see below)

Particulate, slowly biodegradable substrates. A considerable fraction of municipal wastewater organic matter is high molecular weight organic material present as colloids and SS. These materials must be hydrolyzed by extracellular enzymes to compounds that are chemically similar to the rbCOD in wastewater before becoming available to microorganisms.

Bioengineering controls for bulking related to high particulate concentrations are:

(a) Maintain a low particulate substrate content in the AS flocs by operating at high a SRT (high MLSS).
(b) Compartmentalize the aerobic zone to prevent release of hydrolysis products throughout this zone.

Some particulate substrates can support the growth of specific filamentous microorganisms. Fats and grease selectively concentrate in foams and may support the growth of some Mycolata. Long-chain fatty acids are specific substrates for *Candidatus* Microthrix parvicella (Andreasen & Nielsen, 1997, 2000; McIlroy *et al.*, 2013). When the wastewater contains high fat/oil/grease (FOG) levels the primary treatment units should be equipped with efficient grease trap skimming equipment. FOG separators directly at the source can also reduce the severity of foaming. Foaming caused by *Candidatus* Microthrix parvicella can be controlled by dosing Al^{3+} salts to the AS (Section 10.6).

Inoculation of AS systems from wastewaters. The sewer system and the wastewater treatment plant are a single system. Processes in sewers can affect AS community composition, especially in extended sewer systems and in systems with a large force-main component. Here rbCOD can be used by bacteria growing in sewer slimes which slough off and inoculate the AS. A typical example of this problem is the inoculation of AS with *Beggiatoa* or *Thiothrix* from septic sewer wall growth. The control of these problems consists of improved sewer design and operation to avoid septic conditions (e.g. proper sewer slopes, better ventilation, nitrate addition). If control measures cannot be taken in the collection system, the use of anaerobic contact zones in the aeration tank should be avoided.

Temperature, pH, toxicity. When AS is preceded by preliminary and primary treatment it is necessary to assess these parameters on the wastewater entering the AS rather than on the raw wastewater. The values of these parameters can be affected by processes such as (i) stripping in step screens and aerated grit chambers; (ii) pH changes because of fermentation in primary clarifiers; (iii) neutralization and precipitation reactions; (iv) coagulation of larger molecules in primary clarifiers.

(2) Intrinsic factors

SRT. SRT impacts the individual microbial species distribution in AS. Low SRT may wash-out species whose net growth rate is lower than the dilution rate *D* (=1/SRT) while high SRT values favors slow-growers. Because the limiting SRT value for many filamentous microorganisms is close to that of nitrifying bacteria, a filament 'wash-out' strategy can also wash-out the nitrifiers (unless their population is stabilized by some type of side-stream bioaugmentation process (Krhutková *et al.*, 2006; Parker & Wanner, 2007) which maintains a higher nitrifier population than that associated with the mainstream SRT.

Reactor substrate concentration. AS cultured in reactors with *substrate concentration gradients* can acquire the following 'selector' effect features (i) high substrate consumption rate; (ii) high oxygen (electron acceptor) uptake rate; (iii) enhanced floc-former growth. Substrate concentration gradients can be achieved by compartmentalizing the entire reactor volume or just its head-end. The compartmentalized part is called a *selector*. The substrate concentration gradient in a selector may induce both *kinetic selection* and *metabolic selection together with the storage selection*. Most modern bulking control in all types of reactor (oxic, anoxic and anaerobic) uses a substrate concentration gradient. *Kinetic selection* in AS systems with a substrate concentration gradient can be explained by the different growth strategies of filamentous and floc-forming microorganisms. While many filaments are K_S-strategists, most floc-formes are μ-max strategists (Figure 10.5). Examples of filamentous K_S-strategists are type 021N, *Thiothrix*, Nostocoida limicola, type 1851 (Jenkins, 1992).

Metabolic selection requires aerated and non-aerated compartments and *storage selection* requires a concentration gradient to induce unbalanced growth (Figure 10.2) and time for stored substrate metabolism (storage capacity regeneration). It is economical to provide the regeneration in a regeneration zone on the RAS stream (Figure 10.3).

Selector design is largely based on practical experience verified by long-term operation. Some basic rules can be found in the following (Jenkins *et al.*, 1993; Wanner, 1994a, b; 1998; Martins *et al.*, 2004).

Dissolved oxygen (DO). Some filamentous microorganisms (e.g. *Sphaerotilus natans*, type 1701, *H. hydrossis*) exhibit a high affinity for DO at low concentrations because of their low half-saturation coefficient values (K_{DO}). For completely mixed systems or for the first aerobic compartment of compartmentalized aeration basins the boundary between 'bulking' and 'non-bulking' DO concentrations depends on organic loading (Palm *et al.,* 1980); the boundary DO concentration increases as organic loading increases. In municipal AS system DO concentrations <0.5 mg/L are prone to low DO bulking, especially when there is rbCOD in the mixed liquor (Jenkins *et al.,* 1993; Wanner, 1994a).

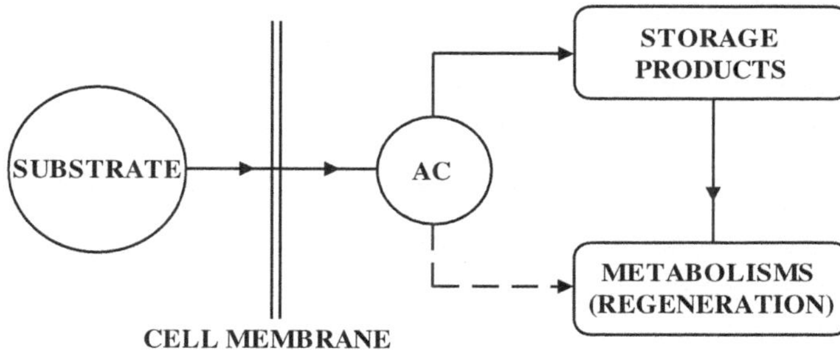

Figure 10.2 Substrate accumulation/storage and metabolism with unbalanced growth (AC-temporary substrate accumulation).

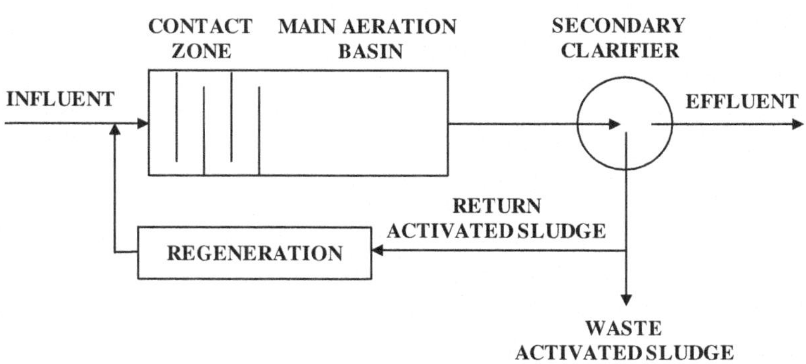

Figure 10.3 AS with compartmentalized contact zone and regeneration of RAS.

Nutrients. Because some filaments have higher affinities for nutrients (N, P, micronutrients), their addition for bulking control is sometimes necessary. Like DO bulking, the extent of problems caused by 'low nutrient filaments' depends on the organic substrate flux into the flocs. N deficiency can be masked by its form (particulate organic N *vs.* ammonia N). Guidelines for nutrient dosing to industrial plants include achieving final effluent levels of: ≥1 mg inorganic N/L and ≥0.2 mg soluble orthoP/L (Wanner, 1994a).

pH. None of the most common filamentous microorganisms are favored by extreme pH values. Some fungi prefer low pH but the pH decrease caused by nitrification in municipal wastewater treatment plants does not produce fungal bulking.

Temperature. Temperature affects all biochemical process rates and oxygen solubility so that elevated temperatures can result in low DO bulking. Significant seasonal shifts in dominant filamentous types have been observed. Thus *Candidatus* Microthrix parvicella, which is dominant in winter, can be replaced by Mycolata, type 0041 or Nostocoida limicola in summer.

10.3.2.2 Control measures using knowledge of filament ecophysiology (metabolic selection)

Filamentous organisms differ in their ability to use electron acceptors other than oxygen and in the chemical and physical nature (soluble or particulate) of their primary substrates (Wanner *et al.,* 2010). The following filamentous bacteria groups grow on soluble substrates and are common in systems treating soluble industrial wastewaters: *Alphaproteobacteria* (e.g. type Nostocoida limicola), *Gammaproteobacteria* (e.g. *Thiothrix,* type 021N).

Other filaments grow in systems degrading particulate substrates (e.g. low organic load municipal wastewater). They use few substrates, have low substrate uptake rates, low or no storage capacity, and produce many exoenzymes. They include *Chloroflexi*, *Aquaspirillum*-related species and TM7-related species (such as types 0041, 1701 and 1851), *H. hydrossis* and related species and *Candidatus* Microthrix parvicella.

A third group consists of the foam-forming Mycolata (primarily Gordonia and Skermania), which are primarily associated with plants receiving high (particulate) lipid levels. They are hydrophobic and can form nuisance foams.

Oxygen is the terminal electron acceptor for most filamentous bacteria, although some (e.g. *Alphaproteobacteria)* can use nitrate and nitrite usually with lower growth rates. *Candidatus* Microthrix parvicella can *take up* long chain fatty acids under anaerobic conditions, but can only *grow* in the presence of nitrate or oxygen (Wanner *et al.,* 2010; McIlroy *et al.,* 2013).

10.3.2.3 Non-specific, abiotic bulking control methods

All of the above methods selectively control excessive filamentous organism growth. Non-specific bulking control methods are aimed at suppressing the problems caused by filaments or at their complete elimination.

Elimination of the filaments is appropriate when 'floc bridging' causes bulking. The filamentous organisms that protrude from the flocs can be selectively killed by adding toxic compounds (most commonly chlorine or NaOCl) to the RAS or the aeration basin. Most of the floc-forming organisms will survive inside the flocs (Jenkins *et al.,* 1993). In many European countries with very strict environmental limits for chlorinated organic compounds, chlorination of AS is viewed as a remedial method rather than a permanent solution to bulking problems (Wanner, 1994a).

Increasing sedimentation velocity. Mixed liquor sedimentation velocity can be increased by increasing the specific weight of the flocs by:

- Adding mineral suspensions to the mixed liquor (e.g. soil suspension, primary effluent particles by reducing primary settling tank capacity)
- Adding digested sludge to the mixed liquor (Kraus, 1945; Hatfield, 1959)
- Forming dense floc 'cores' by mineral salt (e.g. ferric salts) addition to mixed liquor (Wanner *et al.*, 2000).
- Dosing polymeric flocculants, especially when the secondary clarifier is equipped with flocculation zone.

Secondary clarifier improvements. Features of secondary clarifiers that make them more amenable to handling bulking sludge include: (i) an inlet zone for efficient sludge flocculation; (ii) deep tanks to provide enough 'buffering' capacity; (iii) reduction of hydraulic and mechanical disturbances in the settled sludge layer to prevent sludge re-suspension (Ekama *et al.,* 1997; ATV Design Standard A131, 2000; Wanner, 2006).

10.3.3 Practical measures for controlling filamentous bulking

10.3.3.1 Bioreactor configuration for filamentous bulking control

Many bulking problems result from inappropriate bioreactor configuration. An understanding of the influence of the growth conditions on full-scale filament control had its genesis in the early 1970s when it was found that the AS cultivation method could affect the culture. AS grown in SBRs had much lower SVIs and filamentous organism populations than AS grown in continuous flow completely stirred tank reactors (CSTRs). In lab studies Chudoba *et al.* (1973a) compared the settling characteristics of AS from a CSTR with those of three differently staged (4, 8, 16) aerated reactors and found that the SVI and the filamentous organism abundance both decreased significantly with an increasing number of compartments. It was concluded, that the CSTR encouraged excessive filament growth whereas 'higher concentration gradients of substrate' caused non-filamentous organism growth and suppressed filament growth.

On the strength of these results Chudoba *et al.* (1973b) first introduced the concept of a selector into the engineering vocabulary defining it as 'that part of the aeration system along which a substantial concentration gradient of the substrate exists'. They stated clearly that the task of a selector was to suppress filament growth. It is important to understand that even though the word selector is singular it was meant to be a series of tanks (Figure 10.4) constructed as either a number of small CSTRs in series or by the effective compartmentalization provided by a plug flow reactor. Chudoba *et al.* (1973a) also noted that the substrate gradient leveled off earlier than the specific respiration rate, possibly due to substrate storage.

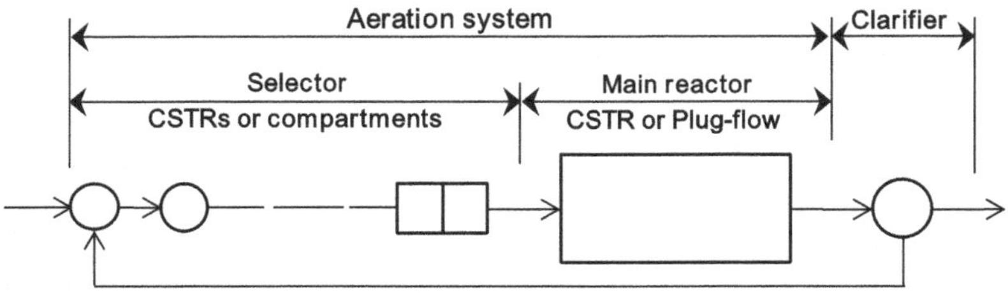

Figure 10.4 Selector AS system configuration (after Chudoba *et al.* 1973b).

Selectors can fail if they are too large or too small. When the compartment volumes are too large they resemble a CSTR and do not induce the selector effect; when they are too small there is danger that rbCOD can bleed through into the main aeration basin (Marten & Daigger, 1997).

The advantage that floc-forming bacteria have over filaments at high substrate concentration is related to their relative maximum specific growth rates (μ_{max}) and half saturation coefficients (K_s), as predicted by Chudoba *et al.* (1973b) and illustrated in Figure 10.5. Experiments have consistently shown that floc-formers have higher μ_{max} and K_s values than filamentous bacteria (Chudoba *et al.*, 1985; van Niekerk *et al.*, 1987, 1988; Kappeler & Gujer, 1994; Smets *et al.*, 1994). Biomass from a CMAS with an aerobic selector (SCMAS) proved to have significantly higher μ_{max} and K_s values than biomass from a single CMAS system for both biogenic (van Niekerk *et al.*, 1988), and xenobiotic substrates (Smets *et al.*, 1994). These results reveal the necessity of using the proper reactor configuration when conducting lab- or pilot-scale design studies.

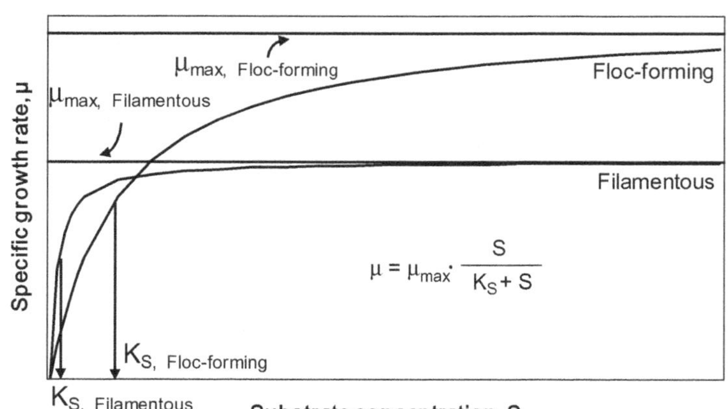

Figure 10.5 Illustration of growth kinetics (K_s and μ_{max}) of floc-forming and filamentous bacteria.

The first full-scale selectors were installed in the 1980s. Daigger and Nicholson (1990) compared the performance of four full-scale nitrifying plants that had selectors with the following configurations and selection mechanisms: two systems with aerated channels, one system with a six-stage anaerobic selector, and one system with a CSTR-like anoxic selector that received recirculated mixed liquor. Despite the considerable differences in layout and operation, the SVI values for three of the four plants during an 11–18 month test period were similar (70–80% <100 mL/g and 100% <200 mL/g) and much lower than those experienced prior to selector installation (as high as 500–600 mL/g). In the selector system that did not perform well it was later found that the high SVI was caused by excessive growth of *Candidatus* Microthix parvicella. The plant had mechanical surface aerators and there were anoxic zones in between the aerators that encouraged its growth.

Jobbágy *et al.* (1999) conducted lab-scale onsite selector experiments at the North-Budapest WWTP. On the basis of SVI data shown in Figure 10.6a: (i) the full-scale AS reactor was suitably modeled by a CSTR; (ii) all selector systems tested had lower SVIs than a single CSTR; (iii) the aerated selector system performed best, and (iv) the anoxic selector system with a staged aerated reactor gave slightly lower and more stable SVI values than the anoxic selector with the CSTR-like aerated basin. In a full-scale experiment, Train I of the plant was equipped with an anoxic selector that had the same 1:7 volume ratio to the aerated basin as in the lab-scale systems. Figure 10.6b shows that the SVIs in the selector system were much lower and more stable than those in the unmodified aeration trains. Nitrate concentrations in all non-aerated selectors were close to zero and simultaneous P-release was detected.

Marten and Daigger (1997) pointed out that anoxic selector loading should be decreased with decreasing temperature because the slower COD uptake rates at lower temperatures could allow substrate 'bleed through' to the main reactor.

By the early 2000s selectors had become common both for upgrading plants and in new construction. Because of their advantages, anoxic and anaerobic selectors were more common than aerobic selectors. Parker *et al.* (2004) surveyed 21 North American plants incorporating anoxic or anaerobic selectors, and compared their performance with Czech (Krhutkova *et al.,* 2002), Hungarian (Jobbágy *et al.,* 1999), Danish (Andreasen & Sigvardsen, 1996) and Italian (Davoli *et al.,* 2002) results. The SVI values of anaerobic selector systems were significantly lower and more stable (50th and 90th percentiles of 88 mL/g and 116 mL/g) than those from anoxic selector systems (50th and 90th percentiles of 121 mL/g and 184 mL/g). Similar conclusions from the Danish survey (Andreasen & Sigvardsen, 1996) were attributed to the presence of denser flocs containing 'clusters of Bio-P bacteria' (later confirmed by Schuler *et al.,* 2001). The SVIs of Czech and Danish plants with anaerobic zones were much higher than those in US plants with anaerobic selectors and similar in value to those from US plants with anoxic selectors. This could be because the US anaerobic selector plants were not designed for N removal, so they did not have large anoxic zones with high rates of mixed liquor recycle.

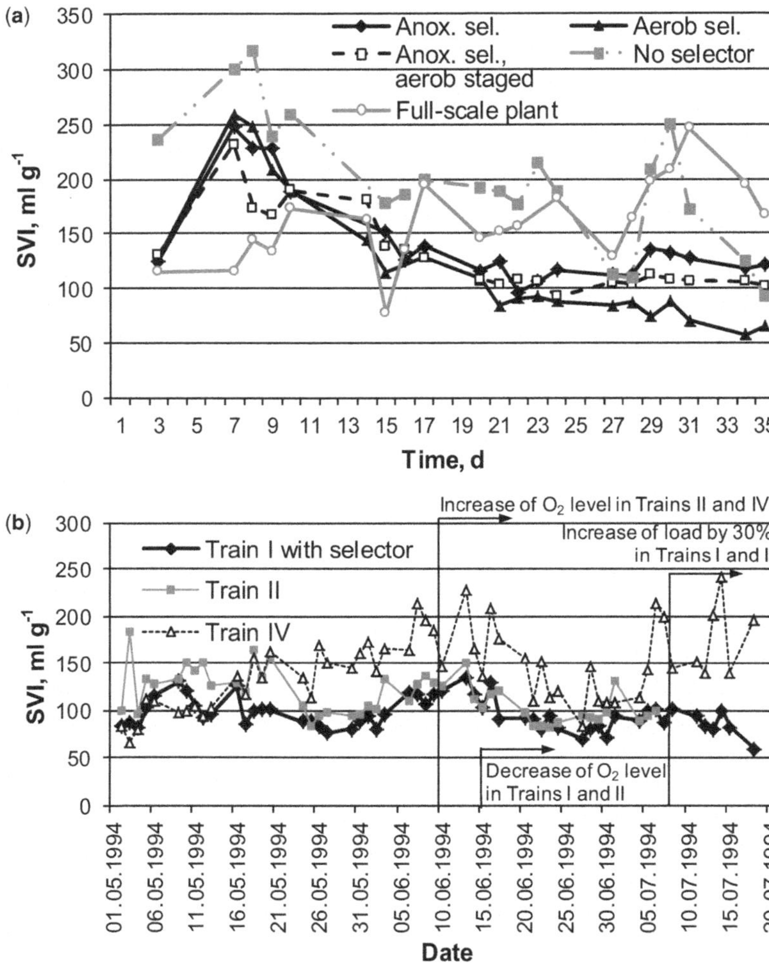

Figure 10.6 SVI values from (a) lab-scale and (b) full-scale experiments (redrawn from Jobbágy *et al.*, 1999).

10.3.3.2 Adjustment of DO level

While filamentous organisms seem to grow better than non-filamentous bacteria at 'low DO' concentrations, the dividing line between 'high' and 'low' DO is not as clear as the differences between selector and non-selector systems. Sezgin *et al.* (1978) showed that the bulk DO concentration was not the DO to which the majority of microorganisms in a floc were exposed (Figure 10.7a). The DO gradient through a floc and therefore the outcome of competition between the zoogloeal and filamentous bacteria is influenced by factors such as substrate flux and utilization through the floc, the floc diameter and size distribution, and the relative amounts of the competing microorganisms present. Therefore, although floc-formers have an advantage above a 'critical point' (Figure 10.7b), it is hardly possible to define a bulk DO level that ensures the repression of filaments under all conditions.

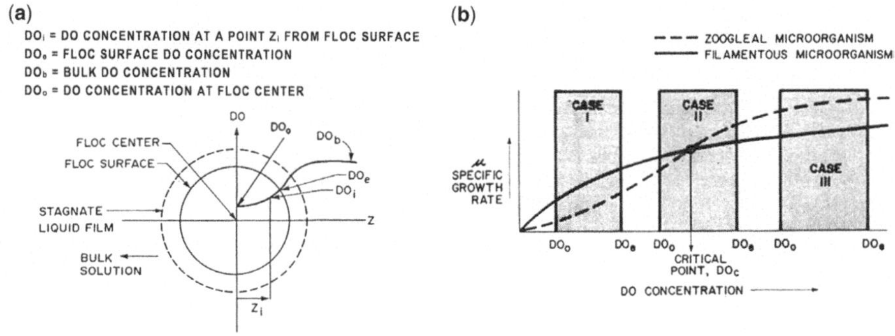

Figure 10.7 (a) DO concentration in and around an AS floc; (b) postulated effect of DO concentration on filamentous and zoogloeal organism growth rates (Sezgin *et al.*, 1978).

Efforts have been made to find a 'universally' safe bulk DO level that prevents low DO filamentous bulking. The generally accepted value is >2 mg/L (Sezgin *et al.*, 1978; Daigger & Nicholson, 1990; Parker *et al.*, 2004). However, higher or lower levels may be appropriate depending on the nature of the wastewater (Section 10.3.2.1) and actual growth rate (Section 10.3.3.1). When low DO bulking is experienced under intentionally anoxic conditions the cause can usually be traced to either an insufficient substrate gradient (Grau & Wanner, 1992) or to the presence of DO in the selector either from the mixed liquor being recycled for denitrification or by unintentional back mixing from the aeration basin through an improperly designed baffle wall.

Oxygen transfer from the atmosphere through the surface of open, full-scale, gently mixed anoxic selectors is typically insignificant but can be important when the concentration of influent rbCOD gets low (Plósz *et al.*, 2003) due to, for example dilution with stormwater or snow melt, especially at its generally low availability (Kappeler & Gujer, 1994; Tardy *et al.*, 2012). Jobbágy *et al.* (2012) detected DO levels up to 0.2 mg/L in the anoxic basins of the North-Budapest WWTP in wintertime. Since this was accompanied by increasing SVI values a floating seal was installed on the basin surface (Figure 10.8).

Excessive or inappropriate mixing (e.g. with mixers that entrain air) may enhance oxygen penetration significantly (Plósz *et al.*, 2003); laboratory and pilot-scale reactors are especially prone to this problem, because of their greater surface area to volume ratio and this has to be taken into consideration in their design and operation (Jobbágy *et al.*, 2000b; Martins *et al.*, 2004).

Figure 10.8 Anoxic reactor surface seal cover; North-Budapest WWTP (photograph by A. Jobbágy).

10.4 CONTROL OF MICROFLOC FORMATION

Wahlberg and Keinath (1995) developed sampling and analysis methods for differentiating between physical floc disruption (pin-point floc) and bioflocculation interference (dispersed growth).

Dispersed growth can be caused by factors such as toxicity, including low pH and overchlorination during bulking control (Jenkins *et al.,* 1993), high monovalent to divalent cation ratio (Higgins & Novak, 1997), low SRT (Bisogni & Lawrence, 1971), poorly biodegradable surfactants and high temperature (Parks *et al.,* 2000). Resolution of dispersed growth problems requires the cause of dispersion to be eliminated.

Pin-point flocs can be aggregated into larger well-settling flocs by eliminating the source(s) of turbulence in the process (e.g. violent aeration, mixed liquor pumping, tortuous mixed liquor piping, and hydraulic drops) that break them up. Parker *et al.* (2004) showed that incorporating a flocculation zone between the aeration tank and the secondary clarifier or within the clarifier itself can improve its efficiency of TSS capture.

Flocs can be strengthened and pin-point floc controlled by altering conditions to allow a small number of filamentous organisms to grow. Marten and Daigger (1997) added a small amount of air into an anoxic selector, which 'performed too well' and produced pin-point flocs. This technique promoted a moderate filament growth (Daigger & Nicholson, 1990; Grady *et al.*, 2011) that reduced elevated effluent TSS levels.

In a six-stage AS system treating chemical industry wastewater, high effluent TSS concentrations were attributed to weak pin-point flocs with no filaments (Jobbágy *et al.*, 2000a). When the influent was split between the first two basins, instead feeding only to the first stage, the concentration of toxic organics was reduced and the growth of some filaments was encouraged, and the effluent TSS concentrations decreased significantly (Figure 10.9).

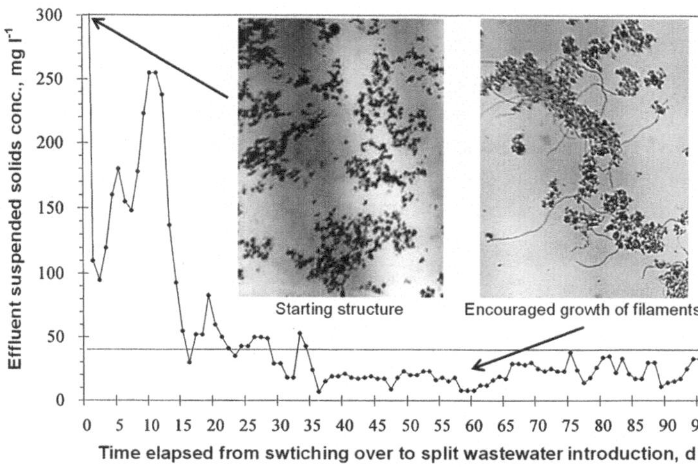

Figure 10.9 Effect of splitting influent feed on filament growth and effluent TSS level (Jobbágy *et al.* 2000a).

10.5 CONTROL OF VISCOUS BULKING

Viscous bulking can be attributed to the overproduction of exocellular polysaccharides (Jenkins *et al.*, 2004) that can produce a water-retentive slime that makes the flocs viscous and produces sticky, high solids foam. This problem can be detected by staining a sample with Indian ink (Jenkins *et al.*, 2004), as illustrated in Figure 10.10. Since the main cause of viscous bulking is severe nutrient (N and/or P) deficiency, which does not allow the formation of normal cells, municipal wastewater treatment plants rarely experience this problem. The problem, however, is common in industrial wastewater treatment plants that treat N and/or P deficient wastes.

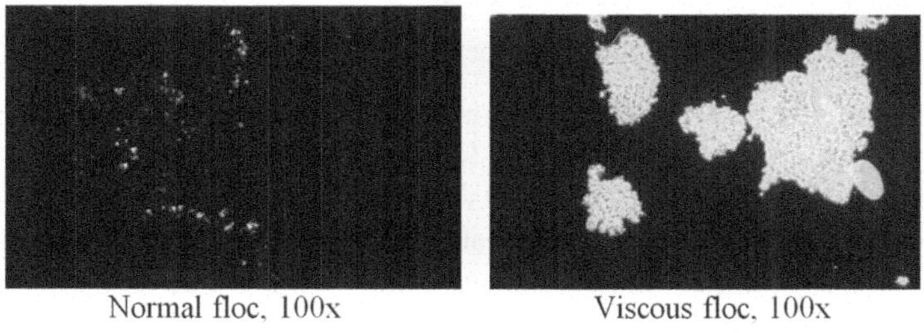

Figure 10.10 Identification of overproduction of exocellular polysaccharides by India ink staining (microphotographs provided by David Jenkins).

The rbCOD that forms exocellular polysaccharides under nutrient deficient conditions can also be taken up and internally stored as glycogen when the AS contains glycogen accumulating organisms (GAOs). Jobbágy *et al.* (2002) used this mechanism to devise a process for treating nutrient deficient wastewater (containing acetic acid, sugar and wine) in a laboratory anaerobic/aerobic AS system without nutrient addition. SVI values in a fully aerobic system increased up to 600–800 mL/g, while in the anaerobic/aerobic system the SVI remained below 250 mL/g. Figure 10.11 shows the effect of reactor configuration and wastewater type on the distribution of intracellular and extracellular polysaccharides.

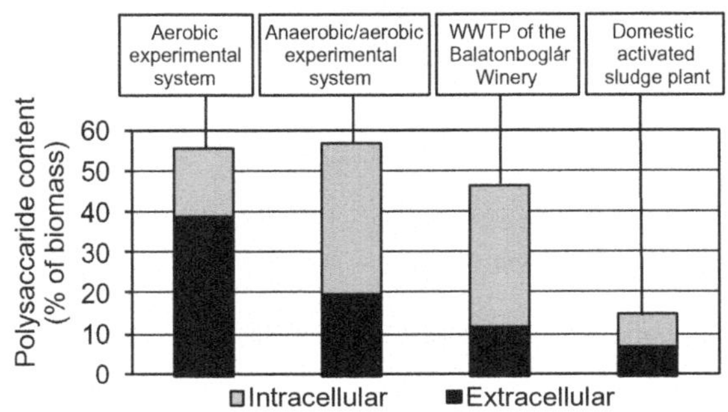

Figure 10.11 Polysaccharide distribution as a function of reactor configuration and wastewater type. (data from Jobbágy *et al.*, 2002 and Kiss *et al.*, 2011).

When the anaerobic/aerobic AS configuration had been implemented in full-scale to treat a nutrient deficient winery wastewater at the Balatonboglár Winery, Hungary (Kiss *et al.*, 2011), SVIs in the range 70–200 mL/g were achieved without nutrient addition while treating an influent COD, which varied between 620 and 10,900 mg/L, and producing an effluent COD <320 mg/L (which was well below the 1000 mg/L effluent limit). The AS biomass contained abundant amounts of GAOs.

10.6 CONTROL OF AS FOAMING

Control measures for AS foaming have been summarized by Wanner *et al.*, (1994a).

Manipulation of SRT. Most foam-forming filamentous microorganisms have lower net growth rates than floc-forming bacteria so that by lowering the SRT the slower growing filaments can be washed out of the AS microbial community. The practical applicability of this method is limited because the SRTs, at which washout occurs, are very low (1.5–3 d depending on wastewater temperature). This means that the washout of these filaments will be accompanied by the washout of nitrifying bacteria.

Use of chemicals and antifoam agents. Chlorination of RAS is not effective for foaming control because the foam-forming microorganisms are selectively retained in floating biomass at concentrations higher than those in the mixed liquor or RAS. More promising results have been achieved by spraying a fine mist of concentrated chlorine solution, or sprinkling powdered calcium hypochlorite directly onto the foam surface. The application of commercial antifoam agents can be costly and is unreliable. Once thick biological foam has become established, it cannot be collapsed by antifoams. One of the most successful recent methods of *Candidatus* Microthrix parvicella-caused biological foam control is to add polyaluminum chloride (PAX) directly into mixed liquor or RAS. PAX doses in the range of 2–3 g Al/kg MLSS, d in the RAS (Seviour & Nielsen, 2010) have been effective.

Foam skimming. The properties of the foam forming bacteria offer the possibility of skimming the foam from the water surface. The selective removal of floating biomass that contains a higher concentration of foam-forming filaments than the mixed liquor effectively reduces the SRT of the filaments which helps to eliminate the foam-formers. Mechanical skimming is usually performed in the secondary clarifiers with effluent weirs protected by baffles against escaping floating biomass. The skimmers should act over the entire water surface and if possible, the skimmed foam should be handled separately from the WAS, especially when it is anaerobically digested (to avoid digester foaming). Experience from Czech wastewater treatment plants has shown that the floating biomass can be stabilized by lime and then mechanically dewatered before final disposal.

Selective floatation. This method uses the same principles as foam skimming but the floatation occurs in an aeration chamber placed between the main aeration basin and the secondary settling tank (Pretorius & Läubscher, 1987), or directly in the RAS stream or the aeration basin itself (Parker *et al.*, 2003; 2014).

Water sprays. Biological foams can be destabilized by substantial dilution. Dilution opens the dams that filamentous microorganisms form in the liquid films surrounding the gas bubbles, so that the bubbles can collapse more quickly. The amount of water sprayed over the foam surface in aeration basins or in secondary settling tanks needs to be much higher than in conventional water sprays used for detergent foam control and the amount of water needs to be adjusted empirically for each particular plant. Because the collapsed foam re-enters the mixed liquor or the RAS, this method does not eliminate them from the AS system. Thus water sprays should only be applied in emergency cases.

10.7 FUTURE OUTLOOK

Although biotechnological tools used for enhancing solids separation from the effluent of AS wastewater treatment have undergone enormous development during the past three decades, it can be assumed that neither all the requirements nor all the possibilities have been exhausted. Besides ensuring possible metabolic selection conditions, future research should certainly go towards better control of kinetic parameters such as substrate profile and DO level under changing influent quality and quantity, including shock or seasonal events such as rain and snow melt. Due to the tendencies of reduced water consumption and consequently increased retention time in the sewer systems, it is foreseen that selectors for overall or periodical shortage of readily biodegradable carbon source should be developed together with the appropriate operation control strategy.

It can be predicted that free 'sculpturing' of AS flocs, in order to establish and maintain the desired ratio of filamentous and floc-forming bacteria in the long term, will be possible. Careful bioreactor staging throughout the whole system and the possibility of split introduction of both the influent and recycled streams, fitted to the changing conditions, may play an important role in this respect. Regarding utilization of DO concentration for floc shaping, it can be presumed that development will follow in adjusting the bulk DO level in aerated reactors according to influent readily biodegradable substrate concentration and elimination rates for keeping the aeration just as low as possible while avoiding the undesired growth of filaments.

On the basis of current knowledge, it can be predicted that more care will be taken to avoid oxygen recirculation into anoxic selectors as well as to possibly minimize the oxygen penetration through the surface of the unaerated basins, especially when the influent readily biodegradable substrate level is low. Application of novel processes or compartments for deoxygenation of the recycled mixed liquor as well as development of surface-sealing can be expected. Should effluent nitrate levels not be kept low during cold seasons, aeration of the anoxic selector may give better performance; therefore application of alternatively aerated anoxic selectors may gain ground in these cases.

It can be foreseen that further efforts will be taken to find biotechnological tools such as anaerobic/aerobic reactor configuration to decrease the requirement for external nutrient sources or chemicals dosed to avoid deterioration of floc structure by filamentous and viscous bulking. Applications of conditions for enhanced biological carbon removal (EBCR) processes at stringent effluent requirements are expected to be developed.

Biological foam will not be as common an operational problem as it is today if the principles of biological selection are combined with proper foam removal. The main goal will be to avoid foam recycling in the AS system. Foam-forming filaments can be also removed by selective floatation. In addition, the dosage of aluminum salts seems to be a reliable and consistent solution against foams caused by *Candidatus* Microthrix parvicella.

Since wastewater quality, especially when industrial wastewaters are present, may significantly influence the kinetic behavior, appropriately designed, comparative pilot-scale and full-scale experiments may be highly encouraged to find the optimum solutions. There has been a common belief that these experiments could also lead to gain the full potential of biotechnological possibilities both in construction and in cost savings, as well as in finding additional factors that may affect the efficiency. Since progress may lay in discovering and identifying the problems, publicity of poor performance is one of the keys of future development, just like the fruitful cooperation of microbiologists and engineers of both hydraulics and biotechnology.

10.8 REFERENCES

Andreasen K. and Sigvardsen L. (1996). Experiences with sludge settleability in different process alternatives for nutrient removal. *Water Sci. Technol.*, **33**(12), 137–146.

Andreasen K. and Nielsen P. H. (1997). Application of microautoradiography to the study of substrate uptake by filamentous microorganisms in activated sludge. *Appl. Environ. Microbiol.*, **63**, 3662–3668.

Andreasen K. and Nielsen P. H. (2000). Growth of *Microthrix parvicella* in nutrient removal activated sludge plants: studies of *in situ* physiology. *Water Res.*, **34**, 1559–1569.

ATV Design Standard A131 (2000). Dimensioning of Single-Stage Activated Sludge Plants. ATV-DWA, Germany, Hennef.

Bisogni J. J.Jr. and Lawrence A. W. (1971). Relationships between biological solids retention time and settling characteristics of activated sludge. *Water Res.*, **5**(9), 753–763.

Blackbeard J. R., Ekama G. A., and Marais G. v. R. (1986). A survey of filamentous bulking and foaming in activated sludge plants in South-Africa. *Water Pollut. Control*, **85**, 90–100.

Chudoba J., Ottova V. and Madera V. (1973a). Control of activated sludge filamentous bulking-I. Effect of the hydraulic regime or degree of mixing in an aeration tank. *Water Res.*, **7**(8), 1163–1182.

Chudoba J., Grau P. and Ottova V. (1973b). Control of activated-sludge filamentous bulking-II. selection of microorganisms by means of a selector. *Water Res.*, **7**(10), 1389–1406.

Chudoba J., Blaha J. and Madera V. (1974). Control of activated sludge filamentous bulking-III. Effect of sludge loading. *Water Res.*, **8**(4), 231–237.

Chudoba J., Cech J. S., Farkac J. and Grau P. (1985). Control of activated sludge filamentous bulking: experimental verification of a kinetic selection theory. *Water Res.*, **19**(2), 191–196.

Daigger G. T. and Nicholson G. A. (1990). Performance of four full-scale nitrifying wastewater treatment plants incorporating selectors. *Res. J. Water Pollut. Control Fedn.*, **62**(5), 676–683.

Davoli D., Madoni P., Guglielmi L., Pergetti M. and Barilli S. (2002). Testing the effect of selectors in the control of bulking and foaming in full scale activated-sludge plants. *Water Sci. Technol.*, **46**(1–2), 495–498.

Eikelboom D. H. and van Buijsen H. J. J. (1981). Microscopic Sludge Investigation Manual. Report A94a, IMG-TNO, Delft.

Eikelboom D. H. (2000). Process Control of Activated Sludge Plants By Microscopic Investigation. IWA Publishing, London.

Eikelboom D. H. and Geurkink B. (2002). Filamentous micro-organisms observed in industrial activated sludge plants. *Water Sci. Technol.*, **46**, 535–542.

Ekama G. A., Barnard J. L., Gunthert F. W., Krebs P., McCorquodale J. A., Parker D. S. and Wahlberg E. J. (1997). Secondary Settling Tanks: Theory, Modeling, Design and Operation. Scientific and Technical Report No. 6, IAWQ London.

Grady C. P. L., Daigger G. T., Love N. G. and Filipe C. D. M. (2011). Biological Wastewater Treatment. 3rd edn, CRC Press LLC, Boca Raton, FL, USA.

Grau P. and Wanner J. (1992). Process theory: biochemistry, microbiology, kinetics, and activated sludge quality. In: Activated Sludge Process Design and Control: Theory and Practice, W. W. Eckenfelder and P. Grau (eds), Technomic Publishing, Lancaster, PA, USA, pp. 1–35.

Hatfield W. D. (1959). Modifying plant processes to handle shock loads. *Wastes Eng.*, **30**, 612.

Higgins M. J. and Novak J. T. (1997). The effect of cations on the settling and dewatering of activated sludges: laboratory results. *Water Environ. Res.*, **69**, 215.

Jenkins D. (1992). Towards a comprehensive model of activated sludge bulking and foaming. *Water Sci. Technol.*, **25**, 215–230.

Jenkins D., Richard M. and Daigger G. T. (1993). Manual on Causes and Control of Activated Sludge Bulking and Foaming. Lewis Publishers, Chelsea, MI, USA, 224 pages.

Jenkins D., Richard M. G. and Daigger G. T. (2004). Manual on the Causes and Control of Activated Sludge Bulking, Foaming, and Other Solids Separation Problems, 3rd edn, CRC Press LLC, Boca Raton, FL, USA, 216 pages.

Jobbágy A., Garai G., Farkas F., Sevella B. and Oszoly T. (1999). Enhanced nitrogen removal at the Northpest wastewater treatment plant. Proc. 8th IAWQ Conference on Design, Operation and Economics of Large Wastewater Treatment Plants, Budapest, Hungary, September 6–9, 255–262.

Jobbágy A., Németh N., Altermatt R. H. and Samhaber W. M. (2000a). Encouraging filament growth in an activated sludge treatment plant of the chemical industry. *Water Res.*, **34**(2), 699–703.

Jobbágy A., Simon J. and Plósz B. (2000b). The impact of oxygen penetration on the estimation of denitrification rates in anoxic processes. *Water Res.*, **34**(9), 2606–2609.

Jobbágy A., Literáthy B. and Tardy G. (2002). Implementation of glycogen accumulating bacteria in treating nutrient-deficient wastewater. *Water Sci. Technol.*, **46**(1–2), 185–190.

Jobbágy A., Palkó Gy., Weinpel T. and Makó M. (2012). Comparative studies on the differently operated trains of the North-Budapest wastewater treatment plant. *Water Sci. Technol.*, **65**(10), 1801–1808.

Kappeler J. and Gujer W. (1994). Development of a mathematical model for 'aerobic bulking'. *Water Res.*, **28**(2), 303–310.

Kiss B., Bakos V., Liu W.-T. and Jobbágy A. (2011). Full-scale use of glycogen-accumulating organisms for excess biological carbon removal. *Water Environ. Res.*, **83**(9), 855–864.

Kraus L. S. (1945). The use of digested sludge and digester overflow to control activated sludge bulking. *Sewage Works J.*, **7**(6), 1177.

Krhutkova O., Ruzickova I. and Wanner J. (2002). Microbial evaluation of activated sludge and filamentous population at eight nutrient removal plants during year 2000. *Water Sci. Technol.*, **46**(1–2), 471–478.

Krhutková O., Novák L., Pachmanová L., Benáková A., Wanner J. and Kos M. (2006). *In situ* bioaugmentation of nitrification in the regeneration zone: practical application and experiences at full-scale plants. *Water Sci. Technol.*, **53**, 39–46.

Larsen P., Eriksen P. S., Lou M. A., Thomsen T. R., Kong Y. H., Nielsen J. L. and Nielsen P. H. (2006). Floc-forming properties of polyphosphate accumulating organisms in activated sludge. *Water Sci. Technol.*, **54**, 257–265.

Larsen P., Nielsen J. L., Svendsen T. C. and Nielsen P. H. (2008). Adhesion characteristics of nitrifying bacteria in activated sludge. *Water Res.*, **42**, 2814–2826.

Marten W. L. and Daigger G. T. (1997). Full-scale evaluation of factors affecting the performance of anoxic selectors. *Water Environ. Res.*, **69**(7), 1272–1281.

Martins A. M. P., Heinen J. J. and van Loosdrecht M. C. M. (2004). Bulking sludge in biological nutrient removal system. *Biotechnol. Bioeng.*, **86**(2), 125–135.

McIlroy S. J., Kristiansen R., Albertsen M., Karst S. M., Rossetti S., Nielsen J. L., Tandoi V., Seviour R. J. and Nielsen P. H. (2013). Metabolic model for the filamentous 'Candidatus Microthrix parvicella' based on genomic and metagenomic analyses. *ISME J.*, **7**, 1161–1172.

Mielczarek A. T., Kragelund C., Eriksen P. S. and Nielsen P. H. (2012). Population dynamics of filamentous bacteria in Danish wastewater treatment plants with nutrient removal. *Water Res.*, **46**(12), 3781–3795.

Nielsen P. H., Mielczarek A. T., Kragelund C., Nielsen J. L., Saunders A. M., Kong Y., Hansen A. A. and Vollertsen J. (2010). A conceptual ecosystem model of microbial communities in enhanced biological phosphorus removal plants. *Water Res.*, **44**(17), 5070–5088.

Palm J. C., Jenkins D. and Parker D. S. (1980). Relationship between organic loading, dissolved oxygen concentration and sludge settleability in the completely-mixed activated sludge process, *J. Water Pollut. Control Fedn.*, **52**(10), 2484–2506.

Parker D. S. and Wanner J. (2007). Review of methods for improving nitrification through bioaugmentation. *Water Practice*, **1**(5), doi: 10.2175/193317707X256964.

Parker D. S., Geary S., Jones G., McIntyre L., Oppenheim S., Pedregon V., Pope R., Richards T., Voigt C., Volpe G., Willis J. and Witzgall R. (2003). Making classifying selectors work for foam elimination in the activated sludge process. *Water Environ. Res.*, **75**(1), 83–91.

Parker D. S., Appleton R., Bratby J. and Melcer H. (2004). North American performance experience with anoxic and anaerobic selectors for activated sludge bulking control. *Water Sci. Technol.*, **50**(7), 221–228.

Parker D., Bratby J., Esping E., Hull T., Kelly R., Melcer H., Merlo R., Pope R., Shafer T., Wahlberg E. and Witzgall R. (2014). A critical review of nuisance foam formation and biological methods for foam management or elimination in nutrient removal facilities. *Water Environ. Res.*, **86**(3), in press.

Parks J., Bordacs K. and Jenkins D. (2000). High temperature operation of an activated sludge plant treating pharmaceutical wastewater. Proc. Millenium Conf., Chartered Inst. Water Env. Management, Leeds, UK.

Plósz B. Gy., Jobbágy A. and Grady C. P. L. Jr. (2003). Factors influencing deterioration of denitrification by oxygen entering an anoxic reactor through the surface. *Water Res.*, **37**(4), 853–863.

Pretorius W. A. and Läubscher C. J. P. (1987). Control of biological scum in activated sludge plant by means of selective flotation. *Water Sci. Technol.*, **19**(5/6), 1003–1011.

Schuler A. J., Jenkins D. and Ronan P. (2001). Microbial storage products, biomass density, and settling properties of enhanced biological phosphorus removal activated sludge. *Water Sci. Technol.*, **43**(1), 173–180.

Schuler A. J. and Jassby D. (2007). The filament threshold effect on activated sludge bulking: artifact or reality? *Water Res.*, **41**(19), 4349–4356.

Seviour E. M., Williams C., Degrey B., Soddell J. A., Seviour R. J. and Lindrea K. C. (1994). Studies on filamentous bacteria from Australian activated sludge plants. *Water Res.*, **28**, 2335–2342.

Seviour R. and Nielsen P. H. (eds) (2010). Microbial Ecology of Activated Sludge. IWA Publishing, London.

Sezgin M., Jenkins D. and Parker D. S. (1978). A unified theory of filamentous activated sludge bulking. *J. Water Pollut. Control Fedn.*, **50**(2), 362–381.

Smets B. F., Ellis T. G., Brau S., Sanders R. W. and Grady C. P. L. (1994). Quantification of the kinetic differences between communities isolated from completely mixed activated sludge systems operated with or without selector using a novel respirometric method. *Water Sci. Technol.*, **30**(11), 255–261.

Tardy G. M., Bakos V. and Jobbágy A. (2012). Conditions and technologies of biological wastewater treatment in Hungary. *Water Sci. Technol.*, **65**(9), 1676–1683.

van Niekerk A. M., Jenkins D. and Richard M. G. (1987). The competitive growth of *Zoogloea ramigera* and Type 021N in activated sludge and pure culture – a model for low F:M bulking. *J. Water Pollut. Control Fedn.*, **59**(5), 262–272.

van Niekerk A. M., Jenkins D. and Richard M. G. (1988). A mathematical model of the carbon-limited growth of filamentous and floc-forming organisms in low F/M sludge. *J. Water Pollut. Control Fedn.*, **60**(1), 100–106.

Wahlberg E. J and Keinath T. M. (1995). Development of settling curves using SVI: an addendum. *Water Environ. Res.*, **67**, 872.

Wanner J., Chudoba J., Kucman K. and Proske L. (1987a). Control of activated sludge filamentous bulking. 7. Effect of anoxic conditions. *Water Res.*, **21**, 1447–1451.

Wanner J., Kucman K., Ottova V. and Grau P. (1987b). Effect of anaerobic conditions on activated sludge filamentous bulking in laboratory systems. *Water Res.*, **21**, 1541–1546.

Wanner J. (1994a). Activated Sludge Bulking and Foaming Control. Technomic Publishing Co., Inc., Lancaster, PA, USA.

Wanner J. (1994b). The implementation of bulking control in the design of activated sludge plants. *Water Sci. Technol.*, **29**, 193–202.

Wanner J. (1998). Process theory: biochemistry, microbiology, kinetics, and activated sludge quality control. Chapter 1. In: Activated Sludge Process Design and Control: Theory and Practice, W. W. Eckenfelder, Jr. (ed.), Technomic Publishing Co., Inc., Lancaster, PA, USA, pp. 1–57.

Wanner J. (2002). Control of filamentous bulking in activated sludge. In: Encyclopedia of Environmental Microbiology, G. Bitton (ed.), John Wiley & Sons, Inc., New York, NY, USA, pp. 1306–1315.

Wanner J. (2006). Aeration tank and secondary clarifier as one system. In: Activated Sludge Separation Problems: Theory, Control Measures, Practical Experiences. V. Tandoi, D. Jenkins and J. Wanner (eds), Scientific and Technical Report No. 16, IWA Publishing Co., London, pp. 47–61.

Wanner J., Ruzicková I., Jetmarová P., Krhutková O. and Paraniaková J. (1998). A national survey of activated sludge separation problems in The Czech Republic: filaments, floc characteristics and activated sludge metabolic properties. *Water Sci. Technol.*, **37**(4–5), 271–279.

Wanner J., Ruzickova I., Krhutkova O. and Pribyl M. (2000). Activated sludge population dynamics and wastewater treatment plant design and operation. *Water Sci. Technol.*, **41**, 217–225.

Wanner J., Kragelund C. and Nielsen P. H. (2010). Microbiology of Bulking. Chapter 7. In: Microbial Ecology of Activated Sludge, R. Seviour and P. H. Nielsen (eds), IWA Publishing, London, 191–214.

White M. J. D. (1975). Settling of Activated Sludge, Technical Report TR11, Water Research Centre, England.

Chapter 11

Secondary clarifiers

Denny S. Parker (USA), Wolfgang Günthert (Germany), Britt-Marie Wilén (Sweden)

11.1 INTRODUCTION

A challenge for this chapter was that, while the AS process is globally applied, the seminal developments in secondary clarification have not been widely dispersed around the globe. This occurs for secondary clarification in particular because it is impossible to make meaningful contributions at lab or pilot scale, since the work is not scalable. Most of the significant developments in secondary clarification have been found at full-scale, where the full complexity of the problem becomes evident. Due to this, secondary clarification development has primarily occurred in regions of the world where the AS process has been in use longest: Europe and North America. That is not to say significant contributions have not been made in research laboratories in many locations world-wide, but examination of our review will show those contributions primarily assist the understanding of full-scale findings, including the development of test and analysis methods.

Our approach was to debate the significant developments first, rather than start writing first. Between the papers already reviewed in Ekama *et al.* (1997) and ones we had accumulated for an eventual rewrite of that STR, there were over 1000 that had to be winnowed down using the editor's criterion for significance. Due to page limitations, even some from the final list could only be included in tabular form rather than detailed in the text.

11.2 SIZING AND RATING

11.2.1 Overview

Table 11.1 shows the main parameters impacting the two main tasks of settling tanks: clarification of the effluent and thickening of sludge. Site conditions such as water and air temperature and sludge characteristics, as well as density flows have also to be considered.

Table 11.1 Factors for the dimensioning of secondary clarifiers.

Flow (average, peak day, and peak wet weather)
Sludge characteristics
Hydraulic features and external tank dimensions
Surface area and overflow rate
Solids loading rate
Depth and retention time
RAS ratio

Since flow is dependent on the size and type of the sewer system, the next most influential factors are the sludge characteristics and the concentration of sludge in the AS tank. These are crucial for sizing and operation of secondary clarifiers. Many developments have been made to find methods to assess the sludge characteristics to determine their relation to the operation of the secondary clarifiers. From today's perspective we now know secondary clarifiers should produce a clear effluent with average effluent suspended solids (ESS) concentrations less than 10 mg/L, but they should also be able to compact the sludge and to store it at peak wet weather flows. Once size, in terms of surface area and depth, as well as influent and effluent position and sludge collection are determined, the operation has to be optimized.

There is a vast amount of literature in this field that has led to today's approach to the sizing and rating of secondary clarifiers, and Table 11.2 lists chronologically what we believe are the key publications during the last 100 years.

11.2.2 The first 50 years (1913–1963)

Most of the earliest studies of performance and operation of secondary clarifiers were carried out in North America. Already by the 1920s and 30s, sanitary engineers had tried to establish parameters to characterize the AS settling properties. Mohlman (1934) assessed the various methodologies applied at 16 Wastewater Treatment Plants (WWTPs), and introduced the SVI as a means to have a parameter that could be used to compare the sludge properties at different treatment plants. Later, standardized tests for settling performance were developed, to enable a more accurate determination of required clarifier area and depth. Camp (1945) described in a very theoretical way the different types of settling, from individual particle settling to hindered settling and

Table 11.2 Development of sizing and rating of secondary clarifiers.

Year	Parameters	Important findings	Reference
1934	Sludge Volume Index (SVI) introduced for the first time	Main parameter for dimensioning secondary clarifiers; recalculation of earlier types of indices for better comparison	Mohlman (1934)
1945	Measurements of velocities and suspended solids concentrations	Established flow pattern; density current, importance of fast sludge removal and outlet weir construction for good effluent quality	Anderson (1945)
1945	Behavior of the AS during settling	Introduction of settling and compaction and recommendation on settling test performance	Camp (1945)
1951	Secondary clarifier design criteria	First standards for 10 American states introduced	Wilson and Keinath (1990)
1957	Settling and compression	Analysis of critical concentrations at which compression (thickening) starts	Eckenfelder and Melbinger (1957)
1967	Settling assessment	Solids-flux theory introduced for the first time	Dick and Ewing (1967)
1968	Settling assessment	Broadly used relationship between sludge concentration and hindered settling velocity	Vesilind (1968)
1969	Hydraulic and sludge loading of secondary clarifiers	Importance to consider both sludge and hydraulic loading when designing secondary clarifiers	Pflanz (1969)

(Continued)

Table 11.2 Development of sizing and rating of secondary clarifiers (*Continued*).

Year	Parameters	Important findings	Reference
1970	Thickening of sludge	Assessment of different methods to solve solids flux equations – applicability to AS	Dick (1970)
1970	RAS control	Discontinuous RAS removal gives stable Mixed Liquor Suspended Solids (MLSS) concentration	Kalbskopf and Londong (1970)
1970	Flocculation zone at inlet	Floc aggregation in mildly stirred flocculation zone	Parker (1970)
1971	DSV	Introduction of Diluted Sludge Volume (DSV) giving new insights into settleability	Merkel (1971)
1977	Analysis of solids-flux theory	Description of the state-point analysis for the first time	Keinath *et al.* (1977)
1978	Assessment of secondary clarifier effluent quality	Relationship between sludge volume load and effluent suspended solids concentration	Billmeier (1978)
1984	Performance of scrapers	Density flow	Günthert (1984a)
1986	Assessment of solids loading rate for rectangular secondary clarifiers	Verification of the solids-flux procedure with full-scale data	Ekama and Marais (1986)
1990	Design criteria for secondary clarifiers	By 1990 approximately 28 out of 50 states had standards in place	Wilson and Keinath (1990)

1991	Standards for WWTP > 5000 pe	Including sludge storage zone and buffer zone for wet weather peak flows	ATV (1991)
1992	Coupling dynamic AS model with Computational Fluid Dynamics (CFD) clarifier model	First time coupling an AS model with a CFD model	Merrill *et al.* (1992)
1996	Design criteria for secondary clarifiers	Good performance over a large range of Surface Overflow Rates (SORs) and Solids Loading Rates (SLRs)	Parker *et al.* (1996)
1999	Clarification mechanisms in secondary clarifiers	Filtration of small particles through the sludge blanket; two mechanisms crosslinked separation and dispersed particle separation	Fuchs and Staudinger (1999)
2000	Standards for WWTP	Standard for secondary clarifiers, sludge removal system	ATV-DVWK (2000)
2001	Protocol for rating of secondary clarifiers	Standard protocol for evaluating and stress testing secondary clarifiers, widely applied internationally	Wahlberg (2001)
2008	SVI measurements at different temperatures	SVI is very sensitive to temperature differences between ambient and MLSS	Rössle and Pretorius (2008)

compaction, and recommended methods for performing batch-tests. The scaling-up problem was identified and the need recognized that for reliable settling tests, the settling columns must be at least as deep as the clarifier depth and the temperature must be similar to field conditions. Camp also introduced recommended design concepts based upon his theoretical concepts, but many engineers debated this work in formal discussions based on their practical experience. There were disagreements regarding the importance of tank depth, effluent withdrawal position and whether the sludge was exposed to shear forces leading to floc break-up in the inlet to the clarifiers. This tension between the expectations of theory and the practical results seen by practitioners with their field scale results persists to the present day.

The seminal work of Anderson (1945), then chief engineer of the Sanitary District of Chicago, on effluent launder placement influenced many to adopt launders placed within the tank (often cantilevered) rather than at the peripheral wall. In what must be the largest and most expensive clarifier pilot experiment ever performed, three 38 m diameter tanks were built, one with the older peripheral launder design and two with in-board launder designs. Even though the in-board design with two launders was superior in performance as shown in Table 11.3, and the engineer's agency went on to build hundreds of single in board launders in its AS plants. Anderson favored deeper tanks (4.2 m in circular tanks) with rapid sludge removal, so as to avoid anaerobic conditions.

Table 11.3 Anderson's results for impact of effluent weir placement.

Number of weir troughs	Weir position, % of tank radius	ESS, mg/L
2	60/76 (two locations)	10.6
1	76	13.8
1	100 (at tank wall)	18.2

Notes: SOR = 1.7 m/h, scraper sludge removal with RAS rate 20% of influent flow. Tank depth = 4.2 m at wall.

Eckenfelder and Melbinger (1957) further improved the settling tests by introducing stirring to better simulate the sludge removal rake action and to assess the critical concentration at which compression starts. They also discovered that hydraulic movement (e.g. stirring with a rake) breaks the stratification of the settled sludge and allows for a better thickening. Additionally, it was shown that the thickening time must be limited because otherwise uncontrolled denitrification in anaerobic zones of the sludge blanket lead to floatation of the sludge.

11.2.3 The second 50 years (1964–2013)

In the late 1960s the solids flux theory for thickening was introduced for AS for the first time, where the area was estimated from an analysis of thickening

(Dick & Ewing, 1967; Dick, 1970). It was demonstrated that different methods could be applied to solve the equations and to estimate the limiting solids flux. Vesilind (1968) derived the broadly used relationship between initial settling velocity and sludge concentration which is still universally adopted in CFD models. A wider application of the solids flux theory came with the state point analysis (Keinath *et al.*, 1977) which can also be applied for process operation once the system has been constructed. Ekama and Marais (1986) checked the solids flux theory by applying it to full-scale data from several plants, but found the need to apply a safety factor to de-rate the predicted solids loading to conform to field measurements.

Rather than the solids flux theories advanced in English speaking countries, a simpler approach was adopted in the Netherlands and Germany, relating the batch SVI measurement to the process of settling and thickening within the tank. In 1971, Merkel was the first to introduce the Diluted Sludge Volume (DSV). Based on a large number of measurements he could identify the relationship between SVI and Settled Sludge Volume (SV):

$$DSV = SVI * (300/SV)^{0.6}$$

The advantage of this approach was that the influence of sludge properties could be described in a simple formula. Based on works of different authors (Mohlmann, 1934; Camp, 1936; Schmidt-Bregas, 1958; Pflanz, 1969; Kalbskopf & Londong, 1970), the surface overflow rate q_A was related to the sludge volume rate q_{SV} and DSV established for secondary settling tanks (ATV-DVWK, 2000):

$$q_A = q_{SV}/DSV = q_{SV}/(MLSS * SV) \tag{11.1}$$

The ATV and solids flux secondary clarifier dimensioning approaches are compared in Ekama *et al.* (1997).

The earlier mentioned studies on secondary clarifier performance focused mainly on the thickening process, which was considered important from an operational point of view, as it sustained process stability; clarification was regarded as less important. With stricter effluent limits in the late 1960s and 70s, more emphasis was given to the effluent quality. Pflanz (1969) showed that both hydraulic loading rate and solids loading rates are important when designing secondary clarifiers, as high solids loading rates in combination with high hydraulic loading rates tend to give high effluent suspended solids concentrations.

The translation of water treatment flocculation concepts and kinetics to AS treatment was first realized in the work by Parker and co-workers (Parker, 1970; Parker *et al.*, 1971). Typically, AS floc size distributions were found to be bimodal, consisting of primary particles and flocs. The concept of a filament backbone as a primary mechanism establishing floc strength was introduced and proven for the first time. Floc breakup was shown to be either due to breakage of the filament backbone within the floc or due to surface shearing of small particles, with the latter being the most influential. Hence, turbulence in the AS tank may lead to an increased number of small particles that cannot settle. Water treatment kinetic equations for flocculation were adapted and new breakup equations developed based on bench scale work to allow sizing of a flocculation zone. It was predicted that, with placement of a flocculation zone in the clarifier inlet with correct energy input, clarification could be enhanced. In an Austrian study, the clarification mechanisms were studied by means of an automated settling and clarification device, which demonstrated that the filtration of small particles in the sludge blanket is also important (Fuchs & Staudinger, 1999).

In the 1990s, a design standard had been set for secondary clarifiers in more than half of the American states (Wilson & Keinath, 1990). In Germany, the ATV standard (1991) was revised to include a sludge storage zone and buffer zone for wet weather peak flow. During this time, the application of CFD models to better describe flow in clarifiers and to connect it to AS models was described for the first time (Merrill *et al.*, 1992). Indeed, these more refined analyses showed that deeper tanks provide better effluent quality. In North America, standardized protocols for stress testing of secondary clarifiers were introduced and became widely applied internationally (Wahlberg, 2001).

11.3 OPERATIONAL ASPECTS OF SECONDARY CLARIFIERS

Although the capacity of a WWTP is mainly determined by the size and design of the secondary clarifier, optimization of the operation has a great influence on the performance of the whole plant.

11.3.1 Managing mixed liquor with different sludge settling properties

One of the main tasks of secondary settling tanks is the separation and thickening of AS, which depends upon its settling properties. The surface overflow rate is the critical hydraulic parameter, which should be kept as low as possible in order to prevent suspended solids in the effluent. Deeper tanks can either accommodate higher surface overflow rates or will produce lower ESS at the same overflow rate for a given SVI (Merrill *et al.*, 1992). Higher SVI values in turn reduce the allowable surface overflow rate. The same trends can be shown with the ATV relationships shown in Section 11.2. Good settleability of the MLSS can be gained by low SVI and flocculation, with the result of highly concentrated returned AS. High SVI may be reduced by optimized conditions in the aeration tanks (Chapter 10), or by adding precipitants or coagulants to the inlet of the secondary clarifier. Provision for addition of a coagulant to the final clarifiers during wet weather operation eliminated the need for a battery of secondary clarifiers during the design of a large plant serving the Boston metropolitan area in North America (Landon *et al.*, 1996). In Germany new advice for clarifier operation was included in the DWA (2007a, b).

11.3.2 Operational strategies for dynamic flow rates

Typical practice for secondary treatment plants serving combined sewers is to limit the flow through the treatment plant to some multiple of the dry weather flow, with the multiple determined by governmental authorities. In combined systems, off line storage is typically provided to prevent extreme flows from reaching the plant, with reintroduction of the flow for treatment after the storm subsides. For treatment systems serving separated sewer systems, the governmental authorities usually proscribe treatment of all of the flow entering the plant. During the increase in flow in wet weather events, particularly at the beginning of the event, there is an inventory transfer of MLSS to the secondary clarifiers. Different strategies have been developed for managing this sludge inventory transfer to prevent plant upsets.

Observations of Born *et al.* (1999) on the influence of wet-weather inflow on secondary clarifiers can be seen in Figure 11.1. The plant is operated with a constant return sludge flow for low hydraulic loadings, and constant return sludge ratio at high loadings. A drop in the RAS SS concentration can be observed at the beginning of the wet weather flow (WWF), followed by higher ESS concentrations after the event due to a higher sludge blanket depth (SBD) and better thickening. To avoid high turbulence in the clarifier by the combined peak of hydraulic flow and RAS flow, it was found that the RAS flow should be adapted to inflow conditions at Peak WWF (PWWF), with a delay of 0.5–1 h in order to prevent immediate hydraulic overloading of the plant and potential loss of biomass. It is for this reason that ATV-DVWK (2000) recommends the inclusion of a sludge storage zone in the secondary clarifier as a buffer for the shifting of mixed liquor from the aeration tanks.

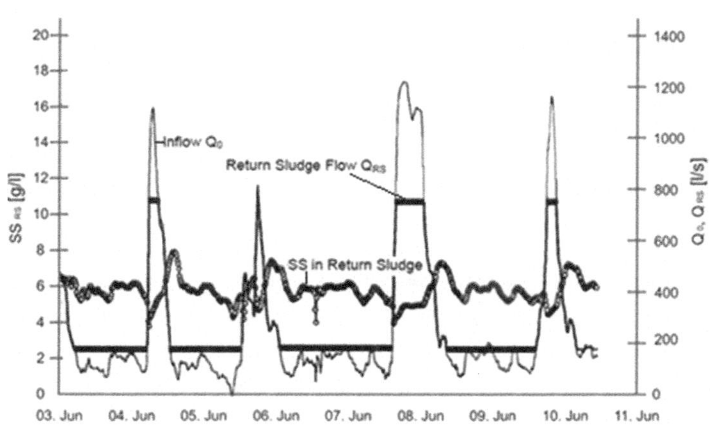

Figure 11.1 Wet weather sludge operation (Born *et al.*, 1999).

In North America, many plants are now being regulated to require more flow to pass through secondary treatment beyond that for which they were originally designed. For example, some state agencies originally had guidelines that limited flows passed through secondary treatment to a multiple of 1.5–3.0 times dry weather flows (Water Environment Federation, 2006). These flow values often did not accommodate practically experienced peak flows that plants have since become expected to treat, so operators developed different ways to manage peak flows, including: (1) maximizing in-service clarifiers, (2) minimizing sludge blanket depths (SBDs) before storm events to accommodate inventory transfers, (3) switching to step feed to minimize MLSS concentrations sent to the secondary clarifiers, (4) using selectors to maintain good settling and compaction conditions, (5) provision for coagulants to be added ahead of the secondary clarifiers to enhance settling and compaction, and (6) turning off aeration in the last portions of the aeration basins to allow in-tank settling (Water Environment Federation, 2006).

It is the management of wet weather flows and the adopted schemes for sludge inventory management during those extreme events that ultimately sets the allowable design MLSS concentration in the aeration basins as well as the solids loading rates (SLRs) impressed on the secondary clarifier during storm events. These schemes, plus the local cost of construction for aeration basins and secondary clarifiers, dominate decision making about the optimum combination of aeration volume and clarifier area (and depth). For instance, differences in practices and cost of construction have resulted in higher MLSS concentrations in countries employing the DWA design standards (typically 3–4 g/L), than in North America (typically 2–3 g/L). This difference also impacts the optimum design of the internals of secondary clarifiers as higher SLRs can impact internal clarifier hydraulics.

11.3.3 Influences of nitrification and biological nutrient removal

Floating solids in a secondary clarifier can be the result of a multitude of processes, among which are processes caused by biological nutrient removal, which were developed in the 1970s and 1980s, as they demand relatively high SRTs. Especially after the introduction of N removal in WWTPs, secondary clarifiers had to be operated with the awareness of potential uncontrolled denitrification. The formation of N_2 gas within the sludge blanket leads to decreased density of the sludge mass and decreased settling and compaction rates. In extreme situations sludge flotation can occur. In 1974, Clayfield investigated the sludge of Minworth and Coleshill treatment plants. It was found that the amount of N_2 necessary for floatation depends on the ambient pressure and biomass concentration.

The introduction of EBPR in AS plants in the 1980s led to different investigations on the behavior of P stored in the sludge. In particular, anaerobic conditions during stabilization and thickening were found to be sources of secondary P-release (Jardin & Pöpel, 1996). Studies of Mikola *et al.* (2009) at the Pihlajaniemi plant, Finland, agree with these results. Increased P-values in the effluent could only be observed when nitrate was missing completely. According to Born (1997), P-release is possible in EBPR below a minimum nitrate concentration of 1.5–2 mg/L. In order to avoid problems with secondary P-release, sufficient performance of the sludge removal system and limited sludge thickening times in the secondary clarifier are necessary and disadvantageous flow conditions (e.g. turbulence, density currents) have to be avoided.

11.4 RECTANGULAR SECONDARY CLARIFIERS

11.4.1 Overview

While circular clarifiers are very common in North America and Germany, rectangular clarifiers are mainly found in other parts of Europe. This kind of secondary clarifier can either be operated with a longitudinal or transversal flow. Longitudinal flow tanks are usually between 30 and 60 m in length (ATV-DVWK, 2000) and use either a suction system to remove the sludge from where it has settled, or a scraper mechanism which transports the sludge into a sludge hopper for further thickening. The sludge removal is therefore always parallel to or against flow direction. Suction sludge removal systems in North America have proven unpopular because of the ease with which their drive systems become misaligned with ensuing mechanical failures.

In a transverse flow system, the inflow is distributed over the long side of the rectangle, which results in a lower loading per unit width compared to longitudinal flow. Since those tanks are typically constructed with a width of 10 m and a range of up to more than 100 m, the vertical/horizontal flow rate is often higher than 1:2. Those systems are considered to be vertical flow systems (ATV-DVWK, 2000). The sludge removal in these systems is perpendicular to the flow direction. Transverse flow systems have mainly been constructed in the last twenty years, mostly because they can be constructed as deep tanks with smaller surface areas. Transverse flow systems have not been built in North America.

11.4.2 Overflow rate and depth

Camp's (1936) theoretical groundwork 'clarification theory for ideal continuous flow rectangular basin' divided an ideal rectangular settling tank into four zones: inlet, outlet, sludge, and sedimentation zone, and sedimentation would only happen in the last zone. He then derived that the removal rate r is determined by the ratio of the settling velocity v_0 which is needed to reach the sedimentation zone to the critical settling velocity v of all discrete particles and the hydraulic retention time, as follows:

$$r = v/v_0 = A * v/Q \tag{11.2}$$

In Camp's view, the critical design parameter is thus the surface overflow rate Q/A, which is independent of the tank's depth. Influences like density flows and overall energy input, however, lead to significantly different ESS. Lumley and Balmér (1991) suggested a division into different transport zones for better modeling of the processes, as different flow and sedimentation behavior within different layers of a rectangular tank were observed, thus reinforcing the role of tank depth in design and operation. In order to evaluate the effect of hydraulic and solids overload on ESS, Wahlberg *et al.* (1994) examined a rectangular secondary clarifier under high SLR conditions. Overall, it could be proven that high SLRs resulted in increasingly non-ideal flow, which is proportional to the amount of MLSS, resulting in increased ESS concentrations.

11.4.3 Sludge removal

Early clarifier developments for primary clarifier applications preceding the invention of the AS can be seen in Figure 11.2, showing the steps from manual to mechanized sludge collection, which later enabled the operation of aeration tanks and secondary clarifiers as a unit. The first rectangular tanks in AS applications had been constructed like primary tanks, with the sludge collection at the inlet end, based on the assumption that most of the sludge is settled shortly after the inlet. Gould (1943) presented a new type of settling tank at Bowery Bay, New York City, USA, with sludge collection at the far end, expecting better results as bottom currents and the sludge removal system are then directed in the same way. He found better performance (higher overflow rates) compared to sludge collection at the inlet end, and attributed this in part to more rapid removal of the sludge. The County Sanitation Districts of Los Angeles County built many of these tanks, and they have demonstrated acceptable effluent quality at very high surface overflow rates during performance testing (Wahlberg *et al.*, 1994). For long rectangular tanks (>40 m), Gould positioned the sludge hopper in the middle (Gould, 1950). Kalbskopf and Herter (1984) reported on secondary clarifiers with a blade scraper removal system and two sludge withdrawal locations, the first at the inlet baffle and the second after one-third of the tank length. In Europe, placing sludge hoppers at the inlet end has continued to be the prevailing practice.

Figure 11.2 Development of sludge collection systems at the Bolton (MA) WWTP showing manual collection on left and mechanized collection on the right (Dunbar, 1907).

In German rectangular clarifiers, three types of sludge collection are used today: blade scrapers, flight scrapers and suction collectors – whereas for transverse tanks, only suction collectors are installed. Flight scraper systems are most common (Figure 11.3 top), as they provide continuous return sludge for the aeration tank. However, the drive chain requires continuous maintenance. On the other hand, blade scrapers (Figure 11.3 bottom) are easier to maintain, with the drawback of discontinuous return sludge. In North America, flight scrapers are the predominant form, while mechanical misalignment issues have caused suction collectors to fall out of favor; blade scrapers are no longer used.

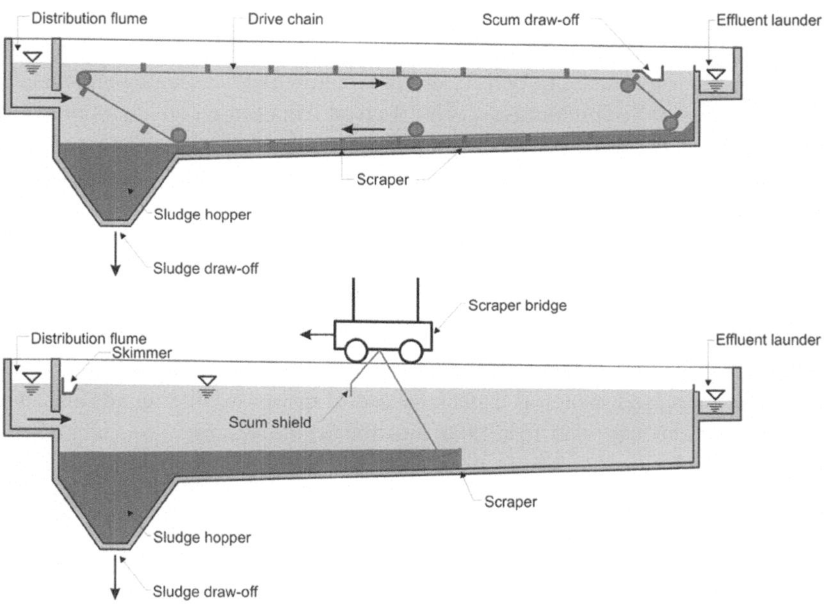

Figure 11.3 Blade scraper systems (bottom) and flight scraper system (top) (Günthert, 1985).

Investigations by Renner (1978) at the Emscher Mouth WWTP showed that MLSS in the aeration tanks were practically not affected by discontinuous sludge removal of a blade scraper system. Optimum velocity was found with 4–5 cm/s. Furthermore, a comparison with a parallel operated flight scraper proved that both systems lead to similar results in ESS, when optimized individually. The sludge removal rates can be calculated by formulas provided by the ATV-DVWK (2000).

11.4.4 Inlet structure

As Camp pointed out in 1936, inlet structures have a strong influence on effluent quality. Weidner (1967) presented the Stuttgarter-inlet, which was among the first to include baffles for energy dissipation. Kalbskopf and Herter (1984) presented the idea of a flocculation chamber for inclusion in rectangular tanks. By stirring at adjustable rotation speeds of 1.4–2.8/min a G value of 30–60/s is induced, promoting the flocculation of dispersed suspended solids. The mixed suspension then flows below an inlet baffle into the secondary clarifier. It was proved that the flocculation chamber had a considerable effect on the effluent quality. Krebs (1991) managed to demonstrate and explain the presence of a forward flow near the tank bottom and a backward flow at the tank surface using computational modeling. The major point of influence in secondary clarifiers was found to be the total energy input and consequently the position of the inlet structure.

A more detailed description of inlet design is given by Krebs *et al.* (1995) who pointed out the importance of density effects on flow fields as well as flocculation properties on the AS. Three criteria were derived from those considerations. First, the inlet should expand over the whole width of the tank, be positioned at the bottom and have a rather low aperture height for a minimum potential energy flux. Second, angle bars beyond the inlet should lead to the dissipation of energy and thereby decrease velocity fluctuations. Finally, the inlet volume should be dimensioned as a flocculation chamber.

Further improvements of existing secondary clarifiers can be achieved by the installation of vertical perforated walls according to Baumer *et al.* (1996). By slowing down the bottom current, the total sludge mass stored in the final settling tank can be increased, which is especially useful in wet-weather conditions. Very low inlet positions can lead to a failure of the whole system if a critical SVI is exceeded, which is why Armbruster and Pahl (2003) patented the idea of a vertically adjustable inlet structure. Variability of the inlet position guarantees optimized flow conditions for different hydraulic loadings. The idea of inlets closer to the bottom of the tank is still being discussed in DWA (2013). The DWA specialist committee suggests a distance of 1–1.5 m from the bottom. During wet-weather conditions the suspension directly flows into the thickening and sludge removal zone according to ATV-DVWK (2000), which leads to the lowest ESS concentrations.

11.4.5 Outlet structure

Outlet design of circular and rectangular final settling tanks is very similar, due to similar flow conditions at the end walls. An alternative to surface effluent launders was presented by Günthert and Deininger (1995) who suggested the use of submerged outlet tubes. Their advantages can be summarized as follows:

- A more uniform withdrawal from the effluent surface area
    - leading to a more uniform effluent distribution under windy conditions
    - leading to less uplift effect, resulting in sludge clouds.

- The removal of floatables is easier, because the surface is not interrupted by effluent launders.
- No operation problems with algae were observed.

11.5 CIRCULAR SECONDARY CLARIFIERS

11.5.1 Overview

The earliest applications in practice are not well documented, but some of the earlier AS applications, like the City of Indianapolis, IN, USA used circular units in the 1920s. More recent developments were recorded in the literature or in patents and were developed by attempting a more scientific approach to the concepts. These developments are recorded chronologically in Table 11.4. As can be seen, most of the innovations were demonstrated on a full-scale basis.

11.5.2 The first 50 years (1913–1963)

Novel developments occurred in sludge removal methods in the first fifty years, which replaced the conventional scraper designs that had steep slopes with central hoppers. These innovations provide the lineage to the favored sludge removal mechanisms of today. Prior to the development of the AS process, the spiral scraper was developed in Germany in about 1900, as shown in Figure 11.4 (left). It was later refined in Germany for application in the AS process and finally applied in North America in the 1980s.

The most common circular secondary clarifier initially adopted in North America was a simple modification of the designs used for primary clarification, using scrapers on floors sloped 1:12 to the center, where a sludge hopper would be placed (Figure 11.4, right). Similar designs are offered today, but are no longer the most widely used type.

Townsend and Brower (1930 and 1934), respectively chief engineer and plant manager of the Milwaukee Sewerage Commission, were concerned that conventional sludge removal with scrapers disturbed the sludge, so they devised a method for continuous removal of sludge across the tank diameter. Slowly moving pipes with multiple withdrawal points (Figure 11.5 top) allowed quick removal of sludge where it was deposited. This was described as a 'sucking' action and that term is the origin of today's 'suction sludge removal'. They termed the device, Tow-Bro, celebrating their partnership, a name in common use today. While the Tow-Bro design used a flat bottom to ensure their rotating pipes would withdraw sludge evenly, this apparently was not important to the engineers at the Dorr-Over company (Rankin, 1958), whose concept of suction removal is shown in Figure 11.5 (bottom). They provided riser pipes for sludge collection, while at the same time scrapers with a sloped floor moved sludge to a centrally located hopper. The combination of the two concepts for sludge removal in one mechanism was thought to encourage the 'relatively active' solids to be returned to the process, while the

Table 11.4 Circular clarifier innovations.

Year	Type	Innovation	Reference
1900 approx	Sludge Removal	Development of the spiral scraper	Dunbar (1907)
1930	Sludge removal	Rotating pipe with orifices for rapid sludge removal	Townsend and Brower (1930)
1953	Development of ports with gated deflectors	'ClariFlow' inlet induced flow exiting inlet to feed well in tangential direction	Walker (1953)
1958	Sludge removal	Riser pipes for rapid sludge removal	Rankin (1958)
1961	Feedwell inlet	First clear statement of energy dissipation inlet's purpose	Fitch (1961)
1980	Very deep tanks with inlets at the bottom	Very high SORs with floc blanket clarification	Resch (1980)
1980	Flocculation zone at inlet	Enlarged center wells with mixing for flocculation (full-scale)	Norris *et al.* (1982)
1983	Increase in sidewater depth	Practical demonstration that deeper tanks perform better	Parker (1983)
1984	Influence of sludge return rate on sludge collection	With hopper below inlet, RAS rate influences short circuiting	Günthert (1984b)
1984	Baffled peripheral launder reduces ESS	Baffle redirects end wall currents away from peripheral weirs	Crosby (1984)
1992	Influence of sidewater depth	CFD model confirms influence of sidewater depth on ESS	Merrill *et al.* (1992)
1996	Density current variations with hydraulic loading	Use of Doppler current meters for high accuracy field study	Deininger *et al.* (1996)
2007	Field comparison of spiral scraper to suction	Suction sludge removal operates with lower SBD and ESS	Moreno and Reed (2007)

'heavier and inert' solids reaching the hopper would be wasted. The combination of a thickening section and a sludge removal section persisted as a Dorr-Oliver offering into the 2000s.

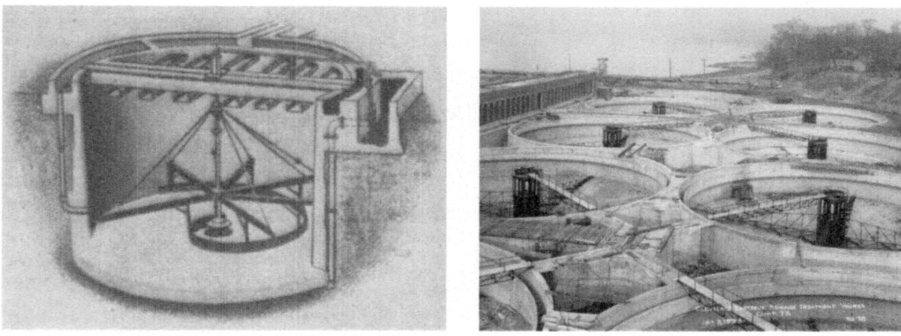

Figure 11.4 Left: Spiral scraper development in Germany (Dunbar, 1907) and right: conventional scrapers in clarifiers under construction in Cleveland in 1933.

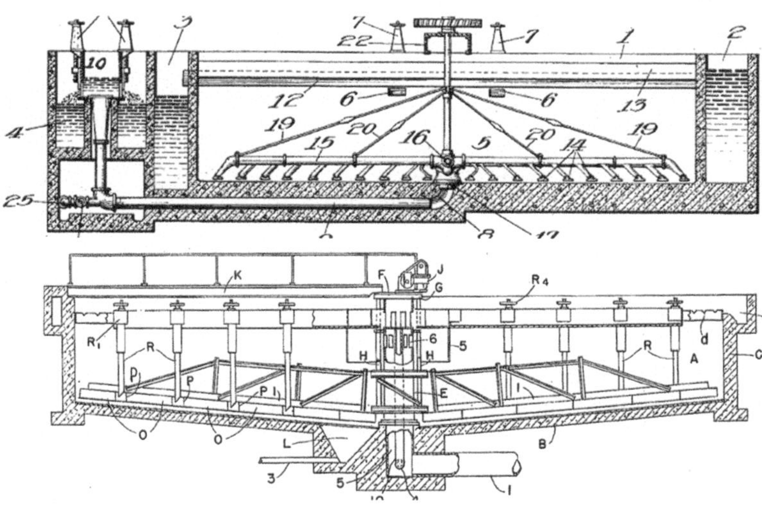

Figure 11.5 Early suction sludge removal designs for the replacement of conventional scrapers (Top: Townsend and Brower Patent, 1934, Bottom: Rankin Patent, 1958).

11.5.3 The second 50 years (1964–2013)

Parker and coworkers' work in a consulting firm environment allowed translation of the earlier lab scale version of a Flocculator Clarifier to be expanded to full-scale, followed by continuous contributions towards its improvement (Norris *et al.*, 1982; Parker, 1983; Parker & Stenquist, 1986; Merrill *et al.*, 1992; Parker *et al.*, 1996).

A design from the 1990s shown in Figure 11.6 depicts the large flocculator center well optimized for flocculation. The use of earlier innovations from the first fifty years of AS development are apparent, such as Anderson's inboard launders, Tow-Bro suction pickup with its required flat bottom, and an energy dissipation inlet (EDI) that emulated both Walker's flow directing port invention and Fitch's EDI features. Deeper clarifiers (up to 6 m SWD) were used based upon Parker and co-workers' full-scale experience and CFD model results. Field studies showed that these specific designs were able to support much higher SORs than had previously been used in North American practice. Resulting from this favorable experience, enlarged flocculator centerwells are now standard practice for new circular clarifiers in North America and further, they are often retrofitted into existing clarifiers to improve their performance. An independent study showed that in parallel testing of parallel Flocculator Clarifiers, theTow-Bro type suction pickup resulted in more efficient performance and lower SBDs than when using spiral sludge scrapers of North American design (Moreno & Reed, 2007).

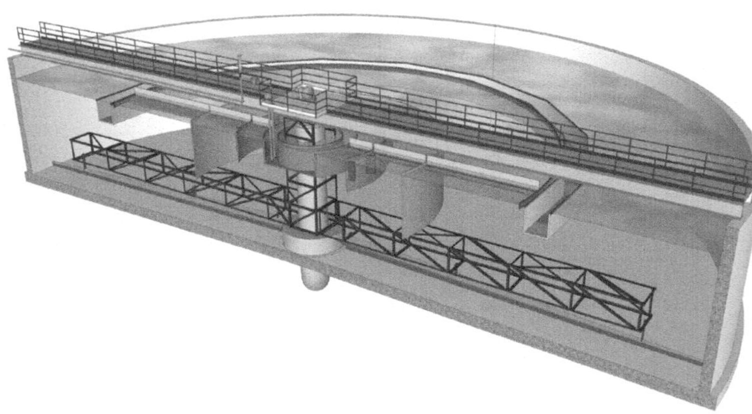

Figure 11.6 Features of the flocculator clarifier in the 1990s (Copyright Brown and Caldwell).

Another engineer, working in Germany, also found deep tanks (Dortmund) to have a great advantage (Resch, 1980) in that, when influent was introduced near the clarifier bottom in a funnel section, a blanket was introduced through which the effluent had to rise and was clarified through floc blanket clarification. These allowed higher SORs than typically employed in shallower tank designs in Germany. Also in Germany, Günthert (1984a) found that when the sludge collection hopper was placed below the inlet, the RAS rate determined the extent of short circuiting from the inlet to the outlet. This spawned later research on baffle development that included fixed and rotating types oriented to preventing this short-circuiting, as well as improving the RAS consistency of the intermittently produced sludge in spiral scraper type clarifiers.

11.6 FUTURE TRENDS

11.6.1 Overview

The development in secondary clarifier design is today, and will be more so in the future, driven by the improved CFD models, where increased computer capacity makes it possible to refine the models substantially, and to include the sludge movement. The first models were developed in the 1970s, and many have followed, with increasing complexity and correspondingly increased value to design engineers as opposed to being strictly of interest to researchers. CFD models are used both for design and for operational improvements of settling tanks already in place. Recently, the main challenge has been to incorporate flocculation mechanisms and discrete particle settling.

Process improvements will generally be related to increasing the loadings possible on secondary clarifiers and including the use of ballasting agents to increase the settling and compaction characteristics, as well as improvements in sludge conditioning just before entry into the secondary clarifier, to enhance its performance. A remaining research challenge is to have an objective comparison of the ratings of rectangular and circular tanks.

11.6.2 CFD models for design

Allowable SLRs, determined by one dimensional solids flux theory, have been found to overestimate the maximum flux by 20–25% (Ekama & Marais, 2004). This is because 1-D models do not take into account the complex hydrodynamic conditions in the secondary clarifier. It has long been recognized that CFD models can be a powerful tool to design inlet constructions such as baffles to improve the flow in the tank and hence get nearer to the secondary clarifier's maximum calculated capacity than can be determined from any one dimensional model (Ekama *et al.*, 1997). McCorquodale *et al.* (2005) introduced the first widely available 2-D CFD design tool for circular secondary clarifier design including flocculation and floc breakup reactions, discrete particle and zone settling, and compaction. When applying 2-D CFD models, batch settling tests and stress tests with the response to increased hydraulic and solids loadings are generally applied to calibrate and verify the model. With the advent of CFD models, Ekama and Marais (2004) concluded that such models automatically produce the correct rating for secondary clarifiers, whereas the solids flux (or 1-D) approach had to be de-rated to get similar results. Confirming this, Kelly *et al.* (2009) compared the state point analysis model with 2-D CFD models and field results obtained by following the WERF protocols (Wahlberg, 2001). Sludge settling and flocculation characteristics were analyzed and incorporated in the model parameters. The results show that the state point analysis consistently overestimated the secondary clarifier capacity of the five WWTPs tested, whereas the 2D CFD model underestimates it slightly, when compared to the results from stress tests.

With increased capacities in computing technology, CFD models in 3-D can now also be solved, something that was not possible just a few years ago. Computation in 3-D enables incorporation of secondary clarifier geometry, which is sometimes critical, as many tanks have asymmetry in devices such as energy dissipating inlets and scrapers for sludge removal. Secondary clarifiers are extremely complex: many processes occur simultaneously. Ramalingam *et al.* (2012) introduced the first 3-D model that includes more realistic flocculation and floc break-up in the supernatant zone. They developed a new discrete particle settling measurement technique when the sludge flocs were fractionated based on settling velocity. Similar to the 2DC model, parameters for floc break-up (K_B) and floc aggregation (K_A) were applied in a flocculation sub-model of the CFD model modified from Parker's (1970) floc kinetic equations. Continued advances in the development of CFD models are expected.

The widening use of CFD models on an international basis means that the models have become an important accumulation point for the determination of influence of tank design features on performance, a trend quite analogous to the international adoption of the IWA AS models (Chapter 14). Field experience will still be the final determinant of success of new developments, as it is doubtful that all factors can be included in the models. Unexpected outcomes are still possible. An example is the use of models to design concentric/slotted ring inlets, where CFD models predicted improved efficiency, but experience showed the novel inlet trapped nuisance foams and prompted a redesign to eliminate the problem in at least one plant.

11.6.3 Possibilities to increase capacity

Many treatment plants face immediate needs for increased secondary clarifier capacity. One interesting relatively new technology is degassing of the mixed liquor prior to sedimentation. By removing dissolved gasses, mainly the N_2 gas that is formed during denitrification, settleability and compressibility improve. The process is comprised of a vacuum tower situated between the AS tanks and the secondary clarifiers (Maciejewski *et al.*, 2011). To reach an under-pressure of 0.04 to 0.05 bar, the vacuum tower has to be placed 10 m above the water level in the AS tank. Based on Henry's Law, the saturation concentration of N_2 gas is approximately 1 mg N_2/L at 0.05 bar and 20°C, compared to approximately 15 mg N_2/L at 1 bar and 20°C. The energy input of the degasification process is 0.02–0.03 kWh/m^3. Plants equipped with degassing towers operate their secondary clarifiers at a very high surface solids loading rate of 7.5–10 kg/m^2/h; values in the same range as in plants that are not nitrifying and denitrifying (compare to Parker *et al.* 1996). Uncertainties with this process relate to the fact that the flocs are broken up just prior to secondary clarification and it is unknown how much a flocculation step can fully mitigate the impact on effluent quality.

Another new technology for upgrading of existing plants is to add magnetite (Fe_3O_4) to AS to increase density, which enables increased sludge concentrations in the AS tanks to very high values (e.g. 8 g/L) and SLRs to be increased three or four times (Catlow & Woodard, 2009). The magnetite particles are recovered from the waste AS by exposing the sludge to high shear mixing and magnetic drum separation, and subsequently returning them to the AS tank.

Finally, while studies in North America showed the equivalency of circular and rectangular clarifiers to handle similar SORs, it was found that differences in tank internals impacted perceived differences in allowable SLRs (Wahlberg *et al.*, 1994). Indeed, when data from a broader database are included, circular tanks of optimized designs have actually shown higher SLR capabilities for circular compared to rectangular tanks (Parker *et al.* 2001). There is a need to undertake more comprehensive research in this area, especially concerning the influences of sludge properties on tank design. Inlet and outlet structures, as well as sludge removal and the operation of the secondary clarifier and aeration tank as a whole, still offer opportunities for optimization.

11.7 REFERENCES

Anderson N. E. (1945). Design of final settling tanks for activated sludge. *Sewage Works J.*, **17**(1), 50–65.

Armbruster M. and Pahl K. (2003). *Absetzbecken* (Settling tank), European Patent Application 1354614.

ATV (1991). Dimensioning of single stage activated sludge plants upwards from 5000 total inhabitants and population equivalents. ATV Rules and Standards, Wastewater-waste, UDC 628.356.628.32.001.2(083), A 131, 11/92.

ATV-DVWK (2000). STANDARD ATV-DVWK-A 131E Dimensioning of Single-Stage Activated Sludge Plants ATV DVWK Rules and Standards, ATV-DVWK, Hennef, 2000.

Baumer P., Volkart P. and Krebs P. (1996). Dynamic loading tests for final settling tanks. *Water Sci. Technol.*, **34**(3–4), 267–274.

Billmeier E. (1978). *Verbesserte Bemessungsvorschläge für horizontal durchströmte Nachklärbecken* (Improved suggestions for the dimensioning of horizontal flow secondary settling tanks), Berichte aus Wassergütewirtschaft und Gesundheitsingenieurwesen der TU München, **21**.

Born W. (1997). *Phosphatrücklösungen im Nachklärbecken – Auswirkungen der Schlammeinschichtung in Abhängigkeit von Beckengeometrie und Betrieb* (Phosphate release in secondary settling tanks – Influence of sludge layering in dependence of tank geometry and operation). *Korresp. Abwasser*, **44**(9), 1589–1595.

Born W., Grünebaum T., Schmitt F., Thöle D. and Wilke A. (1999). *Betriebsbeobachtungen an Nachklärbecken von Belebungsanlagen* (Operational observations of secondary settling tanks of activated sludge plants). *Gas-und Wasserfach, Wasser, Abwasser*, **140**(15), 5–10.

Camp T. R. (1936). A study of rational design of settling tanks. *Sewage Works J.*, **8**(5), 742–758.

Camp T. R. (1945). Sedimentation and the design of settling tanks. *Trans. Amer. Soc. Civil Eng.*, **111**, Paper 2285, 895–958.

Catlow I. and Woodard S. (2009). Ballasted Biological Treatment Process Removes Nutrients and Doubles Plant Capacity. Proc. of WEFTEC, 2009, *Water Environ. Fedn.*, **12**, Orlando, FL, USA, October 10–14, pp. 3999–4021.

Clayfield G. W. (1974). Respiration and denitrification studies on laboratory and works activated sludges. *Water Pollut. Control*, **78**(1), 51–76.

Crosby R. M. (1984). Evaluation of the Hydraulic Characteristics of Activated Sludge Secondary Clarifiers. EPA/68-03-2782, USEPA.

Deininger A., Günthert F. W. and Wilderer P. A. (1996). The influence of currents on circular secondary clarifier performance and design. *Water Sci. Technol.*, **34**(3–4), 405–412.

Dick R. I. (1970). Role of activated sludge final settling tanks. *J.. Sanit. Eng. Div.*, Amer. Soc. Civil Eng, **96**(2), 423–435.

Dick R. I. and Ewing B. B. (1967). Evaluation of activated sludge thickening theories. *J. Sanit. Eng. Div.*, Proc. Amer. Soc. Civil Eng., **93**(4), 9–30.

Dunbar W. P. (1907). *Leitfaden für die Abwasserreinigungsfrage* (Guideline for the subject of wastewater treatment). R. Oldenbourg publishing, Munich, Germany.

DWA (2007a). *Betriebsempfehlungen bei Problemen mit Nachklärbecken von belebungsanlagen – Teil 1: Problemdeutung und Betriebsbeobachtung* (Operational recommendations for problems with secondary settling tanks in activated sludge systems – Part 1: Interpretation and observation), Working report of the DWA Specialist Committee 2.5 Sedimentation processes, KA 2007, **54**(9), 896–901.

DWA (2007b). *Betriebsempfehlungen bei Problemen mit Nachklärbecken von belebungsanlagen – Teil 2: Ursachen und Maßnahmen* (Operational recommendations for problems with secondary settling tanks in activated sludge systems – Part 2: origin and counter actions), Working report of the DWA Specialist Committee 2.5 Sedimentation processes, KA 2007, **54**(11), 1136–1141.

DWA (2013). *Einlaufbauwerke von Nachklärbecken* (Inlet structures of secondary settling tanks), Draft of working report of the DWA Specialist Committee 2.5 Sedimentation processes.

Eckenfelder W. W. Jr. and Melbinger N. (1957). Settling and compaction characteristics of biological sludges: I. General Considerations, *Sewage Industrl Wastes*, **29**(10), 1114–1122.

Ekama G. A. and Marais G. v. R. (1986). Sludge settleability and secondary settling tank design procedures. *Water Pollut. Control*, **85**(1), 101–113.

Ekama G. A. and Marias, P. (2004) Assessing the applicability of the 1D flux theory to full-scale secondary settling tank design with a 2D hydrodynamic model. *Water Res.*, **38**(3), 495–506.

Ekama G. A., Barnard J. L., Günthert F. W., Krebs P., McCorquodale J. A., Parker D. S. and Wahlberg E. J. (1997). IWA Scientific and Technical Report No. 6, Secondary Settling Tanks: Theory, Modelling, Design and Operation, IWA Publishing, London, UK.

Fitch E. B. (1961). Method and Means for Converting the Kinetic Energy of a Fluid Stream into Random Turbulence, US Patent 3,006,474.

Fuchs A. and Staudinger, G. (1999). Characterising the clarification of the supernatant of activated sludge. *Water Res.*, **33**(11), 2527–2534.

Gould R. H. (1943). Improved final settling tanks at Bowery Bay. *Civil Eng.*, **13**(6), 279.

Gould R. H. (1950). Wards Island Plant capacity increased by structural changes. *Sew. Industr. Wastes*, **22**(8), 997–1003.

Günthert F. W. (1984a). Thickening zone and sludge removal in circular final settling tanks. *Water Sci. Technol.*, **16**(10–11), 303–316.

Günthert F. W. (1984b). *Ein Beitrag zur Bemessung von Schlammräumung und Eindickzone in horizontal durchströmten runden Nachklärbecken von Belebungsanlagen* (A contribution to the dimensioning of sludge removal systems and thickening zone in circular horizontal flow secondary settling tanks of activated sludge systems). Berichte aus Wassergütewirtschaft und Gesundheitsingenieurwesen der TU München, **49**.

Günthert F. W. (1985). *Entwicklung und Anwendung der Schlammräumsysteme für Absetzbecken in der Abwassertechnik* (Development and application of sludge removal systems for sedimentation tanks in waste water treatment). Berichte aus Wassergütewirtschaft und Gesundheitsingenieurwesen der TU München, 60.

Günthert F. W. and Deininger A. (1995). Final settling tanks-New aspects and design approaches. In: Proc. 7th IAWQ Int. Conf. on Design Operation of Large WWTPs, A. Franz and H. Kroiss (eds), **1**, Vienna, Austria, August 27–September 1, pp. 265–274.

Jardin N. and Pöpel H. J. (1996). Behavior of waste activated sludge from enhanced biological phosphorus removal during sludge treatment. *Water Environ. Res.*, **68**(6), 965–973.

Kalbskopf K. H. and Londong D. (1970). *Entwicklungstendenzen bei Kläranlagen nach dem Belebungsverfahren für kleine Gemeinden.* (Trends of development for wastewater treatment plants with activated sludge system in small communities). *Gas-und Wasserfach, Wasser, Abwasser,* **111**(8), 455–459.

Kalbskopf K. H. and Herter H. (1984). Betriebserfahrungen mit den Absetzbecken der mechanischen und biologischen Stufe des Klärwerks Emschermündung (Operational experiences with sedimentation tanks in the mechanical and biological step of the Emscher Mouth treatment plant). *Gas-und Wasserfach, Wasser, Abwasser,* **125**(4), 199–206.

Keinath T. M., Hofer D. A., Dana C. H. and Ryckman M. D. (1977). Activated sludge-unified system design and operation. *J. Environ. Eng. Div., Amer. Soc. Civil Eng.*, **103**(5), 829–849.

Kelly R., Jimenez J., Klein A. and Merlo R. (2009). A tale of two methods: comparing state point analysis and computational fluid dynamic modeling for evaluating secondary clarifier performance. Proc. WEFTEC, 2009, Water Environ. Fedn, 2009, **14**, Orlando, FL, USA, October 10–14, pp. 2954–2967.

Krebs P. (1991). The hydraulics of final settling tanks. *Water Sci. Technol.*, **23**(4–6), 1037–1046.

Krebs P., Vischer D. and Gujer W. (1995). Inlet-structure design for final clarifiers. *J. Environ. Eng.*, **121**(8), 558–564.

Landon S. K., Freeman, D. B., Pincince, A. B., and Otoski, R. M. (1996). Forecasting full-scale plant performance through testing at the Deer Island pilot plant. Proc. 69th Ann. Water Environ. Fedn Conf., **1**, Chicago, IL, USA, October 5–9, pp. 519–529.

Lumley D. J. and Balmér P. (1991). Solids transport in rectangular secondary settlers. *Water Supply*, **9**(1), 123–132.

Maciejewski M., Golcz A., Budzisz G. and Oleszkiewicz J. A. (2011). Evaluation of capacity expansion and treatment improvement of wastewater treatment plant in Częstochowa, Poland using vacuum mixed-liquor degassing system. Proc. WEFTEC, 2011, Water Environ. Fedn, 2011, **11**, Los Angeles, CA, USA, October 15–19, pp. 5280–5297.

McCorquodale A., Griborio A. and Georgiou I. (2005). A public domain settling tank model. Proc. WEFTEC, 2005, Water Environment Federation, 2005, **13**, Washington, D. C., United States, October 29–November 2, pp. 2546–2561.

Merkel W. (1971). *Untersuchungen über das Verhalten des belebten Schlammes im System Belebungsbecken-Nachklärbecken.* (Research on the Behavior of Activated Sludge in the Aeration Basin-Secondary Clarifier System). *Gewässerschutz, Wasser, Abwasser*, **5**, D82, Doctoral dissertation, (Ges. zur Förderung der Siedlungswirtschaft an der Rheinisch-Westfälischen Techn. Hochschule).

Merrill M. S., Tetreault M., Parker D. S., Vitasovic Z., McCorquodale J. A. and Ji Z. (1992). Mathematical simulation of secondary clarifiers coupled with activated sludge reactors. Proc. WEFTEC, 1992, Water Environ. Fedn., 1992, New Orleans, LA, USA, September. 20–24, pp. 229–240.

Mikola A., Rautiainen J. and Vahala R. (2009). Secondary clarifier conditions conducting to secondary phosphorus release in a BNR plant. *Water Sci. Technol.*, **60**(9), 2413–2418.

Mohlman F. W. (1934). The sludge index. *Sew. Wks J.*, **6**(1), 119–122.

Moreno P. A. and Reed G. D. (2007). Evaluation of the operating performance of conventional versus flocculator secondary clarifiers at the Kuwahee Wastewater Treatment Plant, Knoxville, TN. *Water Environ. Res.*, **79**(5), 547–553.

Norris D. P., Parker D. S., Daniels M. L. and Owens E. L. (1982). High-quality trickling filter effluent without tertiary treatment. *J. Water Pollut. Control Fedn*, **54**(7), 1087–1098.

Parker D. S. (1970). Characteristics of Biological Flocs in Turbulent Regimes, Doctoral thesis, University of California, Berkeley, http://disexpress.umi.com/dxweb, UMI publication number 7109887

Parker D. S., Kaufman W. J. and Jenkins D. (1971) Physical conditioning of activated sludge floc. *J. Water Pollut. Control Fedn.*, **43**(9), 1817–1833.

Parker D. S. (1983). Assessment of secondary clarification design concepts. *J. Water Pollut. Control Fedn.*, **55**(4), 349–359.

Parker D. S. and Stenquist R. J. (1986). Flocculator-clarifier performance. *J. Water Pollut. Control Fedn.*, **58**(3), 214–219

Parker D., Butler R., Finger R., Fisher R., Fox W., Kido W., Merrill S., Newman G., Pope R., Slapper J. and Wahlberg E. (1996). Design and operations experience with flocculator-clarifiers in large plants. *Water Sci. Technol.*, **33**(12), 163–170.

Parker D. S., Kinnear D. J., and Wahlberg E. J. (2001). Review of folklore in design and operation of secondary clarifiers. *J. Environ. Eng.*, **127**(6), 476–484.

Pflanz P. (1969). Performance of (activated sludge) secondary sedimentation basins. IAWPR, Adv. Water Pollut. Res., Proc. 4th Intl Conf., Prague, Czechoslovakia, S. H. Jenkins (ed.), Pergamon Press, Oxford, pp. 569–593.

Ramalingam K., Xanthos S., Gong M., Fillos J., Beckmann K., Deur A. and McCorquodale J. (2012). Critical modeling parameters identified for 3D CFD modeling of rectangular settling tanks for New York City wastewater treatment plants. *Water Sci. Technol.*, **65**(6), 1087–1094.

Rankin R. S. (1958). Treatment of Sewage in Liquid Form and Similar Materials, US Patent 2,820,758.

Renner H. (1978). *Untersuchungen zur diskontinuierlichen Schlammräumung in rechteckigen, längsdurchströmten Nachklärbecken von Belebungsanlagen* (Investigations on discontinuous sludge removal in rectangular longitudinal flow secondary settling tanks). Institut für Siedlungs- und Industriewasserwirtschaft TU Graz Heft 4.

Resch H. (1980). Investigations on the performance of upflow final settling tanks in activated sludge plants. *Progress Water Technol.*, **12**(3), 201–211.

Rössle W. H. and Pretorius W. A. (2008). Batch and automated SVI measurements based on short-term temperature variations. *Water SA*, **34**(2), 237–243.

Schmidt-Bregas, F. (1958). *Über die Ausbildung von rechteckigen Absetzbecken für häusliche Abwässer* (On the design of rectangular settling tanks for domestic wastewater), Institute Series 3, Insitute for Sanitary Engineering and Waste Management, Leibniz Universität Hannover, Hannover, Germany.

Townsend D. W. and Brower J. (1930). New sludge-removal apparatus developed at Milwaukee. *Eng. News Record*, July 10.

Townsend D. W. and Brower J. (1934). Sludge Remover, US Patent 1,947,429.

Vesilind P. A. (1968). Theoretical considerations: design of prototype thickeners from batch settling tests. *Water and Sewage Works*, **115**(7), 302–307.

Wahlberg E. J. (2001). WERF/CRTC Protocols for Evaluating Secondary Clarifier Performance, Alexandria, VA.

Wahlberg E. J., Augustus M., Chapman D. T., Chen C., Esler J., Keinath T. M., Parker D. S., Tekippe R. J. and Wilson T. E. (1994). Evaluating activated sludge performance using the CRTC Protocol: four case studies. *Proceedings of the 67th Ann. Water Environ. Fedn Conf.*, **1**, Chicago, IL, USA, Oct 15–19, pp. 1–12.

Walker J. D. (1953). Water Clarification Apparatus and Method, US Patent 2,635,757.

Water Environment Federation (2006). Guide to Managing Peak Wet Weather Flows in Municipal Wastewater Collection and Treatment Systems. Alexandria VA.

Weidner J. (1967). *Zufluss, Durchfluss und Absetzwirkung zweckmäßig gestalteter Rechteckbecken* (Inflow, flow rate and sedimentation properties of rectangular settling tanks with functional design) Stuttgarter Bericht Nr. 30 zur Siedlungswasserwirtschaft, Institut für Siedlungswasserbau und Wassergütewirtschaft.

Wilson T. E. and Keinath T. M. (1990). Secondary clarifier design standards: an issue paper of the Clarifier Research Technical Committee. Proc. 1990 Amer. Soc. Civil Eng, Environ. Eng. Specialty Conf., Arlington, VA, USA, pp. 254–261.

Chapter 12

Energy considerations

Helmut Kroiss (Austria), Yeshi Cao (Singapore)

12.1 HISTORICAL DEVELOPMENT AND SCIENTIFIC PROGRESS

12.1.1 Introduction

The advent of urban sanitation can be traced back to 18th century Europe. Sewage was first transferred and collected purely for disposal and/or nutrient recycling to agriculture, but not for treatment. During this period, rivers receiving wastewater discharge were considered as part of the wastewater treatment process. Due to the rapid urbanization and industrialization, especially in the 19th century, severe urban hygienic and receiving water pollution problems occurred and had to be solved. The first biological wastewater treatment processes in Europe and the United States were biofilm processes such as land application (e.g. intermittent soil filtration, bacterial beds). A next step was the development of trickling filter technology which is still applied today. In 1914 Ardern and Lockett reported on their development of the AS process which later became the dominant biological wastewater treatment process worldwide despite the fact that it needed more energy than the other processes. For many decades thereafter the development of the AS process was driven by improving the treatment efficiency and reliability and sometimes the construction costs. Energy minimization has only become a long-term global goal of research and development in recent decades.

12.1.2 Evolution of treatment efficiency from BOD removal only to nitrification, nutrient and micro-pollutant removal

Even though Ardern and Lockett (1914) claimed that nearly complete BOD removal and full nitrification should be achieved for optimum receiving water protection, most of the early AS treatment plants were constructed for BOD removal only. The goal was to markedly reduce the oxygen consumption for the degradation of organic pollutants in the receiving waters. With increasing urbanization and industrialization, oxygen depletion caused by nitrification in the receiving water became relevant in several river basins. Thus, nitrification was introduced in the process. Accommodation of nitrification in the AS process doubled or even tripled the aeration tank volume.

In the 1920s the two-stage AS process was developed by Imhoff (1953) in Germany. Böhnke (1977) further developed and enhanced its worldwide application as the AB process. The original idea was to separate carbon removal processes in a highly loaded 1st stage (MCRT < 1 d) and to nitrify in a 2nd stage (MCRT ~ 8 d). This resulted in a marked reduction of the aeration tank volume required for reliable nitrification. Also the energy requirements for carbon removal could be reduced because bacterial decay and particulate matter hydrolysis did not play an important role in the 1st stage. Most of the energy contained in the removed organic pollutants was transformed into excess sludge and then, during anaerobic sludge digestion, into methane.

Ludzack and Ettinger (1962) developed a pre-denitrification process (MLE process) in which the carbonaceous pollutants are used as carbon sources for denitrification. Matsché (1972 and 1997) reported full-scale results from the Vienna Blumental treatment plant (150,000 PE) showing that ~90% N removal was achieved by simultaneous nitrification/denitrification in a 'race track' flow aeration tank (oxidation ditch).

In South Africa, Barnard (1973) combined features of the pre- and post-denitrification processes with enhanced configurations for N removal, which were either combinations of pre-denitrification followed by simultaneous nitrification/or intermittent aeration systems with control of the duration of anaerobic, anoxic and aerobic periods. The latter method can be applied in SBR systems and especially in small continuous flow plants designed for simultaneous aerobic sludge stabilization. At nitrifying treatment plants the energy requirement for carbon removal can be reduced by efficient N removal by denitrification.

The two-stage AS process was further optimized for denitrification resulting in low energy requirements for aeration and increased biogas production from sludge digestion. The 2nd stage excess sludge is returned to the 1st stage and the 1st and 2nd stages are used for denitrification (Matsché & Moser, 1994; Wandl *et al.*, 2009).

Because receiving water eutrophication is usually controlled by the availability of P for algae growth, wastewater treatment plants with N and P removal have become common technology, especially in 'sensitive areas'. The energy requirements for this removal are similar for P removal by either enhanced biological or chemical precipitation.

Recently the removal of micro-pollutants has assumed importance and the first legal requirements for their removal are in the process of being implemented (Swiss Federal Government, 2013). When physical/chemical post-treatment processes are used to meet such requirements (Schaar *et al.*, 2010) there are additional energy demands (Schaar *et al.*, 2013). These will not be discussed here.

12.1.3 Recent development of legal requirements for treatment efficiency (in developed countries)

Since the 1980s, the implementation of water policies and regulations in Europe and the USA has been further enforced. In Europe the Urban Waste Water Treatment Directive (UWWD, 1991) marks a milestone of basin-based water environment policies. Where eutrophication problems are dominant ('sensitive areas') strict minimum requirements for nutrient removal were imposed based on 'Best Available Techniques'. According to the European Water Framework Directive (WFD, 2000) wastewater discharge quality requirements are based on the so-called 'combined approach'. Depending on the 'sensitivity' of the river basin, including the coastal waters within 1 nautical mile from the coast, the minimum UWWD effluent standards have to be met whatever the receiving water quality. If these minimum requirements do not result in 'good receiving water quality' even stricter effluent standards for wastewater treatment plants must be imposed, for example according to European legislation the whole Danube catchment discharging to the Black Sea is a 'sensitive area' where all treatment plants >10,000 PE must meet the strict EU nutrient removal requirements. If this does not result in 'good water quality' in the Western Black Sea coastal waters then even stricter standards for nutrient removal (mainly for P) will be imposed.

In the USA, the Total Maximum Daily Load (TMDL) program, which was issued in 1992, was further enforced for implementation (USFR, 1999). The program sets a limit to the amount of a pollutant that can be discharged into a water body and sets up Waste-Load Allocations (WLAs) for discharges. As a consequence, water-quality, instead of technology-based, discharge standards are imposed and become extremely strict in some 'sensitive' areas where the wastewater discharge is a determining factor on the receiving water quality. All these developments drive wastewater treatment to be innovative in process configuration, robustness and increased efficiency of process equipment and process control. Significant progress has been achieved in developing and applying innovative processes and control procedures to reduce energy consumption, and increase energy recovery in wastewater treatment.

12.2 ENERGY CONTENT OF WASTEWATER

The organic pollutants in wastewater represent a potential energy input to the plant. A COD mass balance is technically the best method for assessing all types of energy considerations at treatment plants (Svardal & Kroiss, 2011). One kilogram of COD has an energy equivalent of ~14MWs (Heidrich *et al.*, 2011). All modern mathematical models of AS are based on COD balances. Assuming that wastewater contains 220 g biodegradable COD/inhabitant/d (120 g COD/PE/d, 2 PE/inhabitant, 92% COD removal) the mean 'power input' from municipal wastewater is approximately 35 W/inhabitant. Efficient nutrient removal plants have a mean power consumption in the range of 5–20 W/inhabitant indicating that energy self-sufficient treatment plants are possible if some of the COD can be converted to methane as a source of low entropy energy by for example, anaerobic sludge digestion.

The mean total primary power input (mainly from fossil energy sources) per inhabitant is between <3000 W in developing countries and up to >10,000 W in highly developed countries. Power for wastewater treatment therefore is far below 1% of the total power consumption. Nevertheless wastewater treatment plants consume a very significant proportion of total public power consumption of municipalities (WERF, 2009) and, in many cases it is the largest single consumer.

The nutrient content of wastewater can also be expressed as an energy equivalent (carbon footprint) if it is assumed that it replaces market fertilizers which need low entropy energy for production in the range of ~10 kWh/kg of N and P. However this energy cannot be recovered at the treatment plant!

Construction of sewer systems and wastewater treatment plants needs energy which can be converted into a consumption rate by taking their lifetime into consideration. A rough estimate by Svardal and Kroiss (2011) gave a range from 5 to 15 W/inhabitant, which is in the same order of magnitude as for the operation of wastewater treatment plants.

12.3 ENERGY CONSUMPTION OF WASTEWATER TREATMENT PLANTS

12.3.1 Introduction

The importance of energy in liquid and solid transformations, biological conversions, and solid treatment processes in any wastewater treatment plant (WWTP) is indisputable. Biological wastewater treatment with the AS process is an energy-consuming process but at wastewater treatment plants there are other net energy-consuming processes (e.g. pumping, sludge thickening and dewatering, ICA, lighting, heating, odor control, etc.).

Figure 12.1 shows a typical energy consumption distribution profile for a large municipal wastewater treatment plant: aeration is the largest portion (~60%), followed by pumping (~12%), anaerobic digestion (~11%), thickening and dewatering (4%), preliminary treatment, clarifiers and chlorination, and so on (7%); the remainder (~6%) is for non-process energy usages such as lighting and for heating, A/C, and so on in buildings.

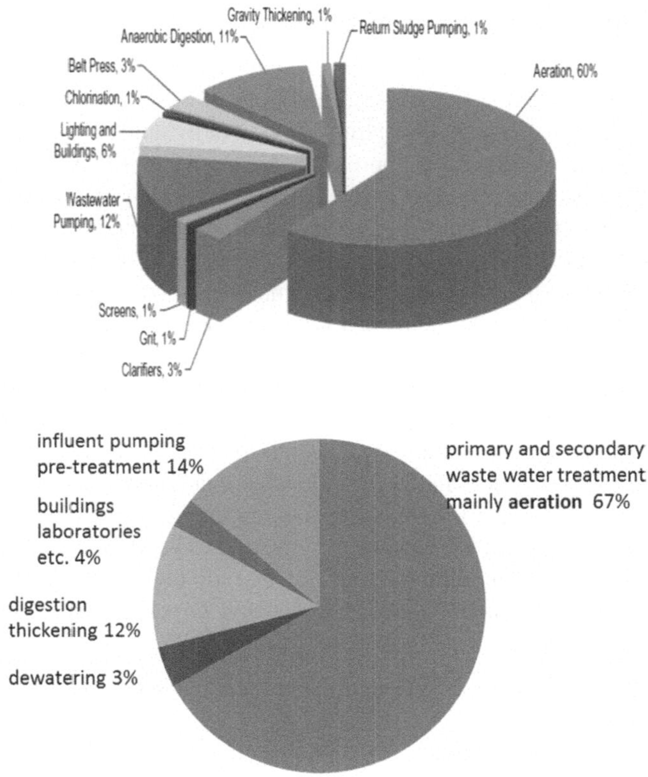

Figure 12.1 (top) Energy consumption distribution profile of conventional municipal. wastewater treatment plants (WERF, 2009); (bottom) data from Austria for nutrient removal plants.

Site specific situations and plant design considerations influence energy requirements. The energy consumption and distribution at a wastewater treatment plant depend on: plant size (design load), effluent standards, treatment processes, degree of capacity utilization, influent flow and pumping head, influent characteristics (mean concentration, C/N ratio, load variation over time, etc.), type and age of mechanical and electrical equipment, instrumentation and control systems used. The climate (mean temperature and annual variation) as well as other specific local circumstances (e.g. space availability, odor control requirements) can have an important influence. For comparing energy data for AS plants the mean yearly energy consumption should be related to the mean pollution load (e.g. kg COD in the influent or to population equivalent of 120 g COD/d) as it is much more relevant for energy consumption than the wastewater flow (m^3/d), which depends greatly on the water consumption infiltration, and on inflow and infiltration.

12.3.2 Auditing and benchmarking

In the early 1990s several Central European countries such as Switzerland, Austria, Sweden and Germany initiated the energy auditing and benchmarking of wastewater treatment plants. Starting from the optimization of individual units, and moving on to the development of systematic approaches for the whole treatment process, the auditing process became very well developed and was widely applied in practice.

Auditing manuals were published in several European countries, for example Switzerland (Anon, 1994), Germany (MURL, 1999) and Austria (LFUW, 2001). Similar performance assessments have been made in Australia (Krampe, 2013).

Energy consumption and production targets for the important units/processes of municipal wastewater treatment plants have been established and adopted as key benchmarking indicators in these countries. For comparison between countries, it is necessary to have comparable performance indicators. When using kWh/m^3 as an indicator, the differences in raw sewage COD, N, P and TSS concentrations have to be considered. Differences in the definitions of population equivalents must also be accounted for. To make the literature data on energy consumption indicators comparable, in this paper they have been converted to yearly mean kWh/kg COD and where necessary, by assuming a COD/BOD$_5$ ratio of 2.

The Swiss Energy Manual (Anon, 1994) has five targeted indicators for energy consumption assessment:

(i) Energy consumption by the aeration process in kWh/kg COD (Table 12.1).
(ii) Biogas reused ≥95%.
(iii) Biogas energy content converted into electrical energy ≥27%.
(iv) Electric energy supplied by biogas reuse: dependent upon plant size.
(v) Thermal energy supplied from biogas reuse: dependent upon plant size.

Table 12.1 Assessment of aeration energy using the Swiss Manual of Energy at WWTPs.

Treatment Target	Aeration energy (kWh/kg COD)*	Relevance to assess energy consumption
Carbon removal only	<0.27	Low
	0.27–0.40	Average
	>0.40	High
Nutrient removal	<0.44	Low
Nitrification at $T > 10$°C,	0.44–0.66	Average
No denitrification	>0.66	High

*COD = COD of aeration basin influent.
Source: Anon, (1994).

Benchmarking data from Austria show that the energy consumption of nutrient removal plants with anaerobic sludge stabilization varies between 0.50 and 1.25 kWh/kg COD (Lindtner *et al.*, 2008). The targets for total mean electric energy consumption for large (>50,000 PE) nutrient removal municipal treatment plants are 0.6 kWh/kg COD in Austria (Kroiss, 2012) and 0.45–0.6 kWh/kg COD in Germany (MURL, 1999). These targets do not consider the energy recovery from biogas.

One of the main problems encountered in energy assessment is the poor quality of the available data. The best tool for overcoming this is to conduct mass balances for COD, N and P for both the whole plant as well as for individual processes (Spindler & Vanrolleghem, 2012).

In Central Europe, after more than a decade's worth of effort on energy auditing and benchmarking, the energy cost of two thirds of all WWTPs, which have already undergone energy auditing, has been reduced by an astounding average of 38% in Switzerland, 50% in 344 wastewater treatment plants in Germany, and about 30% in Austria (Wett *et al.*, 2007).

Resource recovery is also emerging as another important theme. Current municipal wastewater treatment plants are being criticized for being environmentally unfriendly and unsustainable (GWRC, 2008; WERF, 2009); and there are calls for a 'paradigm shift' for current wastewater treatment processes. As a consequence, the pollutant-load specific external power consumption has become one of the indicators of wastewater treatment plant performance and sustainability. The total specific GHG emissions converted into CO_2 equivalents is likely to become a sustainability indicator. The recovery of P, a limited resource, and N have become topics of global research activity. The actual processes developed for N and P recovery as fertilizers are also energy consuming and thus competing with mineral fertilizer production (Siegrist, 2011; Scholz, 2013).

12.3.3 Economic considerations

The actual economic analysis of operational costs shows that, for large plants, the energy costs are between 15 and 20% of the total operating costs, ranking second after staffing and followed by sludge disposal (Figure 12.2).

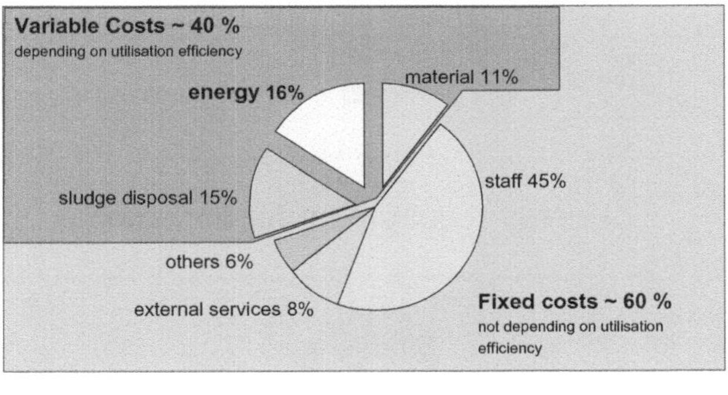

Figure 12.2 Distribution of operational costs for nutrient removal treatment plants >50,000 PE in Austria derived from extensive process benchmarking data (Lindtner & Svardal, 2008).

In most cases, the yearly capital recovery cost is higher than operating cost of wastewater treatment plants. As a consequence, the energy cost is less than 10% of the total cost of wastewater treatment. Since wastewater fees must cover the capital cost of sewer construction and the operating cost of sewer system maintenance and pumping, the energy costs are still a relatively small fraction of the overall wastewater fee (5 to 10%). In developing countries this fraction can be significantly higher. In addition to economic considerations, in recent years, the focus of discussion for the WWTP performance has turned to the carbon footprint (CF) and GHG emissions CO_2, CH_4 and N_2O. These emissions are also influenced by energy minimization and nutrient removal process development.

12.3.4 Energy consumption of AS process

12.3.4.1 Aeration

Aeration energy, the largest energy consumer in the AS process, is primarily determined by the oxygen uptake (OU) of the bacteria in the aeration tanks for carbon oxidation (OUC) and for nitrification (OUN). N removal by denitrification (or deammonification) exerts an oxygen demand (OUDN) for the oxidation of NH_4-N to N_2 gas. As a consequence, the aeration energy consumption strongly depends on the influent pollutant load, the treatment efficiency and the treatment process technology. Figure 12.3 shows this relationship for the removal of carbonaceous pollutants expressed as COD (Svardal & Kroiss, 2011). Treatment efficiency depends on SRT and temperature.

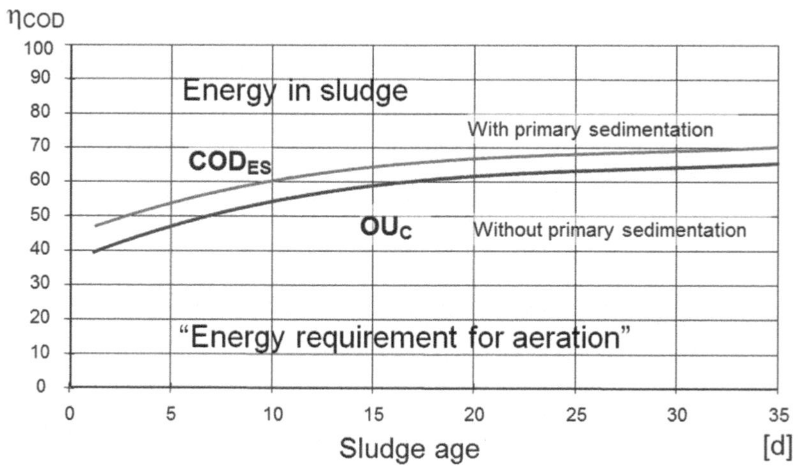

Figure 12.3 COD (Energy) balance: removed COD load (100%) = oxygen uptake for carbon removal + COD in the excess sludge; this figure is valid for a mean yearly temperature of ~15°C.

To relate aeration energy requirements to pollutant load and treatment efficiency, the following considerations can be used: the oxygen uptake for carbon removal (OU_C) primarily depends on SRT and temperature. Figure 12.3 shows the relationship between SRT and the COD balance regarding sludge production (in terms of COD) and the OU_C based on ASM 1 and extensive full-scale experience (Svardal & Kroiss, 2011). Rough estimates can be obtained for other temperatures by using a conversion factor for the SRT of $1.07^{(T-15)}$. From Figure 12.3 it can be seen that OU_C increases with increasing SRT, that is, with higher treatment efficiency. The OU_C of nitrifying treatment plants can be reduced by using a two-stage AS process with very short SRT (<1 d) 1st stage with relatively low COD-removal efficiency (~80%) and a 2nd stage with high SRT for nitrification and excellent overall COD removal efficiency (90 to 94 %). This results in more energy from wastewater pollutants being transferred into the excess sludge, which can then be converted to methane by anaerobic digestion.

Nitrification requires an OU_N of 4.6 g O_2/g effluent NO_3-N. The lower the nitrate effluent loads, the lower will be the aeration energy requirements of a nutrient removal plant. The removal of NH_4-N causes an OU_{DN} of 1.7 g O_2/g N_2 released to the atmosphere irrespective of the biological processes involved. For denitrification, an OU_C of 2.9 g O_2/gN_2 in the anoxic zones is necessary.

12.3.4.2 Aeration control

The aeration system consists of motors, aerators (surface aerators or blowers), sometimes mixers, instrumentation, and control and automation equipment. To minimize the aeration energy it is essential that, at all times, the oxygen required by the bacteria is available, because without it, treatment efficiency will be compromised. The pollutant load and composition can vary widely over a day (and also weekly and sometimes seasonally). In most cases the temperature varies seasonally. Variations of the actual oxygen consumption can have a ratio between minimum to maximum of ≥1:10. In nitrifying plants the instantaneous NH_4-N load has to be immediately oxidized to NO_3-N, because the nitrifying bacteria cannot store NH_4-N or NO_2-N. In addition, nitrifying bacteria cannot grow in the absence of DO and do not reach their maximum oxygen uptake until the DO concentration is ≥2 mg/L. At NH_4-N concentrations below about 1 mg /L, NH_4-N becomes limiting for the substrate utilization rate of nitrifying bacteria, following the Michaelis and Menten (1913) relationship shown in Figure 12.4. Because bacteria react immediately to changes in limiting substrate concentration (mainly DO and NH_4-N for nitrifiers), the aeration system has to be very flexible in adapting to load variations.

The dynamic mathematical AS models describing the oxygen uptake of the bacteria are based on Monod (1942) kinetics describing the relationship between the limiting substrate concentration and the growth rate of the bacteria. Assuming that the yield coefficient Y remains independent of substrate concentrations in the aeration tank, the Michaelis-Menten equation has the same form as the Monod equation so that the determination of the Monod constants (μ_{max} and k_s) in an activated sludge model for a real plant can be based on respiration rate measurements.

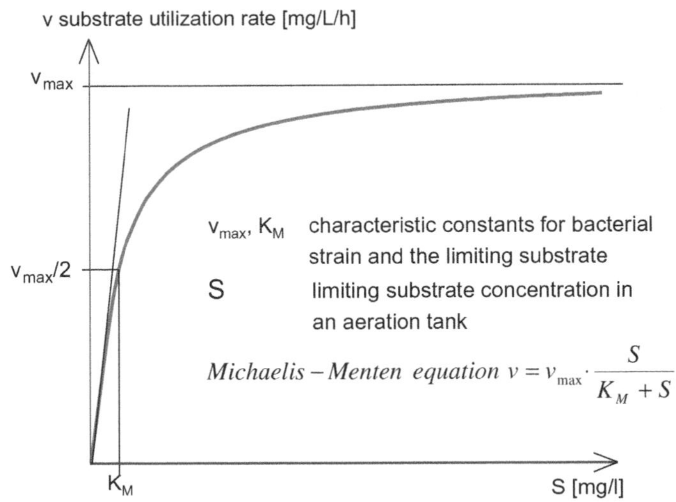

Figure 12.4 Michaelis–Menten relationship for substrate utilization rate (Kroiss, 2012).

Heterotrophic bacteria that remove COD can grow using DO as electron acceptor even at DO concentrations of very close to zero. They can also store dissolved substrate. Most of the carbonaceous pollutants in municipal wastewater are particulate so that a rather slow hydrolysis process controls the decay. There are indications that the hydrolysis rate is lower at low DO concentrations and under anoxic conditions than at higher DO levels. Most of these bacteria start to use NO_3-N or NO_2-N as electron acceptors under anoxic conditions (DO = 0, NO_3-N or NO_2-N available).

As a consequence of the above factors, an aeration control system has to continuously adapt aeration capacity to the changing conditions by adjusting the required aerobic aeration tank volume as well as the DO level so that NH_4-N is nearly completely oxidized and all of the volume that is not required for nitrification can be used for denitrification. In well-designed nutrient removal plants the variation of aeration capacity requirements mainly depends on the NH_4-N load variation. The variation of aeration capacity required for carbon removal is normally low, since denitrification and the storage capacity of the heterotrophs can cope with the load variation.

The DO level directly influences the energy requirement for aeration because the oxygen transfer efficiency is directly related to the oxygen saturation deficit, which becomes a maximum at a DO of zero. The higher the temperature, the lower is the oxygen saturation concentration and the greater is the influence of DO concentration on energy efficiency for aeration. For example, at 30°C an increase of

DO from 1.5 to 2 mg/L increases the aeration energy requirement by ~10% while at 15°C the increase is only ~6%. Since AS is a suspension of bacterial flocs, the DO (and all other substrate) concentrations measured in the bulk liquid are not the same as those at the bacterial surfaces in the flocs because of diffusion limitations. This effect increases with decreasing SRT and is also responsible for simultaneous denitrification even at DO levels in the bulk >0 mg/L and reduced activity of nitrifiers. Control of DO for energy minimization therefore is quite complex.

Over the last two decades, membrane bioreactors (MBRs) have been placed into operation to produce an effectively suspended solids-free effluent and to reduce the area requirements for AS treatment plants. Experience in Singapore for an MBR with a capacity of 23,000 m^3/d and with COD and N removals of 75% indicates that the process energy consumption for COD and N removal are the same order of magnitude (0.27 kWh/kg COD (~0.15 kWh/m^3) as that required for maintaining the membrane permeability. Membrane manufacturers are focusing their efforts on improving MBR operating modes and aeration systems to reduce this energy consumption.

12.3.4.3 Aeration system hardware

Fine-bubble diffusers were used in Ardern and Lockett's original AS experiments (Ardern & Lockett, 1914). Coarse bubble and surface aeration systems were used in the AS process almost immediately following its discovery. These systems are still used today since they provide both oxygen supply and complete mixing in the aeration tanks. For a long time, energy minimization played a minor role but since the 1970s many efforts have been made to develop high efficiency aeration systems sometimes to provide oxygen for nitrification/denitrification systems. For energy minimization, only the aeration efficiency under real operational conditions (kgO_2/kWh) is relevant. This must be clearly distinguished from aeration efficiency data produced under test conditions in clean water, which is used in much of the literature and in manufacturers' information. Comparable aeration system energy consumption and efficiency data should include the total energy withdrawn from the electrical grid, including aeration and mixing over the lifetime of the system.

Aeration efficiency is strongly dependent on the variation of oxygen requirements over time. For energy minimization, it is essential that the maximum efficiency of the mechanical and electrical equipment is achieved at the most frequent operational conditions. It is less relevant especially at the minimum oxygen requirements. The aeration system should therefore be adaptable to the capacity utilization development over time (the difference between design and actual loading).

Today, membrane fine bubble diffusers in tanks with an immersion depth between 5 and 7 m with separate mixing devices and an appropriate diffuser layout can achieve an optimum aeration energy efficiency of 2.5 to 3.5 kgO_2/kWh under real operating conditions. This is almost double the specific oxygen supply capacity of 1.2–2.4 kg O_2/kWh of traditional diffused air and surface aeration systems.

For energy minimization it is also essential to optimize aeration control for denitrification. One advantage of membrane diffusers over 'exposed' diffusers is that they can meet this goal by switching on and off without clogging. Disadvantages of the membrane diffusers are their tendency for scaling and fouling, both of which can cause reduced energy efficiency and lifetime of the membranes.

12.3.4.4 Dynamic model simulation for energy minimization

The IWA AS model (ASM) No.1 (Henze *et al.*, 1987), has served as the basis of newer iterations of IWA ASM models, has consistently shown its high applicability and has been widely used in the simulation of the nitrifying AS process. These dynamic models allow the comparison of different process configurations as well as the optimization of design and control concepts under real loading conditions. The reliability of the modeling results greatly depends on the quality of the database and the parameter adaptation to the specific local circumstances. It is often easier to adapt model parameters (which have not been determined by experiments) to existing data, than to develop a model which correctly represents the processes involved. Modeling the operation of existing plants is an excellent tool for minimizing aeration energy requirements by different operation modes and instrumentation, control and automation strategies.

On the anaerobic digestion front, the IWA model (ADM 1) (Batstone *et al.*, 2002) has provided a tool for calculating solids destruction and biogas generation in anaerobic digestion. Software packages based on these models such BioWin (EnviroSim, 2013) and GPS-x (Hydromantis, 2013) have been applied in plant level simulation, optimization, mass balances and energy recovery calculations. For municipal sewage sludge digestion design and operation they have been much less successful than the ASM models.

In order to minimize the carbon footprint of an entire treatment plant by making optimum use of the wastewater energy content, it is very useful to couple the two dynamic models; again this needs excellent full-scale data and for large plants even long-term pilot scale investigations to develop model parameters.

12.3.5 Pre-treatment by upflow anaerobic sludge blanket (UASB) reactors

Anaerobic treatment is characterized by the conversion of the organic pollutants into CH_4 (energy source) and low excess sludge production. Anaerobic treatment of many industrial wastewaters has been regarded as a sustainable and reliable technology. The Upflow Anaerobic Sludge Blanket (UASB) reactor was developed by Lettinga *et al.* (1980) from the full-scale experience with a Clarigester type of anaerobic reactor for treating winery and maize wastewaters in South Africa (Ross, 1984). It became a successful anaerobic treatment method for many industrial wastewaters. Since the 1980s UASB reactors coupled with polishing aerobic processes (AS, trickling filter, rotating disks and ponds) have been widely used in full-scale plants for municipal sewage treatment in regions with hot climates such as Brazil, Mexico, Columbia and India.

Despite the fact that UASB treatment of municipal sewage is cost-effective and relatively easy to operate when only partial COD removal is required, it has several shortcomings when used for municipal wastewater treatment (Cao & Ang, 2009) These include: (i) the downstream biological process may be short of carbon for nutrient removal; (ii) 10% to 30% of the CH_4 generated is dissolved in the effluent. This reduces the energy recovery efficiency and ultimately produces emissions of CH_4, a potent GHG, to the atmosphere; and (iii) the excess sludge production may be not much different from that of the conventional aerobic AS process with anaerobic sludge stabilization since particulate COD is dominant in municipal wastewater. The issues concerned with N removal may be less relevant in the future if main stream deammonification becomes a viable solution for anaerobic effluents.

12.3.6 Other energy consumers (Hardware)

Figure 12.1 shows that about 30 to 35% of the required electrical energy is not related to the mechanical biological wastewater treatment steps. Many of the mechanical and electrical facilities at treatment plants often do not get as much attention as the aeration system. The optimization of their power efficiency and their control can play important roles in energy consumption.

In most cases, cost minimization has priority over energy minimization. This may be reasonable because existing equipment replacement with more energy efficient equipment will consume energy at the equipment production site and this will be included in the cost of the new equipment. Whenever equipment has reached the end of its useful life the improvement potential for energy efficiency has to be considered for the replacement. A special challenge is to optimize the adaptation of all electrical equipment to changing conditions. This problem is discussed extensively in Chapter 13.

12.3.7 Wastewater treatment process developments for reduction of energy consumption

12.3.7.1 Introduction

Aerobic biological processes transform the energy contained in the organic pollutants to low temperature energy (14 kJ/g OU_c). The goal of the new process developments is to minimize oxygen consumption without compromising treatment efficiency. The processes can be physical/chemical, biological or combinations of both.

12.3.7.2 Chemically enhanced primary treatment (CEPT)

Primary Settling Tank (PST) removal efficiencies vary from 40% to 60% for TSS and from 25% to 40% for COD. By adding chemicals (flocculants, precipitants) to the influent these removal efficiencies can be enhanced to about 80 to 90 for TSS, from 50 to 70% for COD removal and 60 to 90% for P removal depending on the chemical addition and wastewater characteristics (e.g. seawater content). This type of treatment process is often used for wastewater treatment in coastal areas that have long sea outfalls (e.g. Hong Kong and even Shanghai several years ago) (Harleman & Murcott, 1992). As a consequence, biogas production from anaerobic digestion of primary sludge can be markedly increased. Thus the Sanitation Districts of Orange County CA, USA showed that biogas production could be increased by 12 to 18% by the increased TSS removal efficiency (from 65% to 80%) obtained by dosing $FeCl_3$ and polymer (EPA & USDE, 1995). CEPT can be advantageous for overcoming seasonal peak loadings such as in tourist and wine growing areas and low temperature peaks. When applied to municipal wastewater, CEPT increases the N/COD ratio since it largely increases particulate removal efficiency and a large part of the nitrogenous pollutants are dissolved. This can have negative consequences for N removal by denitrification. For the treatment of low alkalinity wastewaters, where low NO_3-N concentrations are essential to maintain an appropriate pH, the external addition of alkalinity may be necessary to avoid nitrification inhibition (Cao *et al.*, 2008).

12.3.7.3 Two-stage AS processes

The major objective for energy requirement reduction in the 1st stage is to maximize the adsorption of particulate material to the bacterial flocs, to minimize the hydrolysis of this adsorbed COD and to minimize the decay of bacterial mass produced from dissolved COD. All these factors result in low oxygen consumption and an excess sludge rich in biodegradable carbon compounds. This in turn results in more biogas from the anaerobic digesters. SRT is typically below 1 d. The 2nd stage AS plant has an MCRT of 5 to 10 d so as to ensure full nitrification and denitrification. The ability to achieve high N removals by denitrification depends on the influent N/COD ratio. Process schematics and operating results for two full-scale Austrian plants, both removing >80% N have been reported by Wett (2007) for the Strass plant and Wandl *et al.* (2009) for the Vienna main treatment plant.

The Strass plant, attained energy self-sufficiency in 2005 and has maintained it until the present date (2013). This plant, which has no primary sedimentation, is favored by a very low influent N/COD ratio (0.06) which results in a relatively low OU_N and excellent denitrification conditions. The plant is also equipped with a deammonification step for the sludge dewatering reject water which contains about 20% of the influent N.

The Vienna Main Treatment Plant (Wandl *et al.*, 2009) has primary sedimentation. All of the excess sludge from both stages is removed from the 1st stage, so that the excess sludge from the 2nd stage is used for adsorption, reaches maximum growth rate and enables partial nitrification/denitrification also in the 1st stage. It also can be used for denitrification by returning the effluent of the 2nd stage to the 1st stage (only used during dry weather and at low temperatures). Currently at this plant the dewatered raw sludge is incinerated but when the plant is upgraded (2020), anaerobic sludge digestion and energy generation from the biogas will allow it to achieve greater than 100% electric energy self-sufficiency (Wandl and Papp, 2012).

This process, like CEPT, typically increases the N/COD ratio in the feed to the nitrifying/denitrifying AS process. Several essential factors need to be considered when applying this process (or CEPT): (i) lower N/COD ratios are more advantageous to the process; (ii) if very low nitrate or TN concentrations (e.g. as in Germany: >85% removal on any grab sample) have to be met, supplementation with an external carbon source (ethanol, acetic acid) might be necessary to consistently meet the discharge standards. The two stage process is not recommended for small treatment plants.

12.3.7.4 The deammonification process

Unlike denitrification, deammonification does not require a carbon source for N removal. The existence of the anammox reaction in nature was first postulated from theoretical considerations (Broda, 1977) and the process was developed at Delft University of Technology (TUD) in the 1990s (van de Graaf *et al.*, 1996).

Deammonification is a two-step process. The 1st step is the conversion of half of the NH_4-N to NO_2-N (Sharon process). The 2nd step is the ANAMMOX (ANaerobic AMMonium OXidation) process in which NH_4-N is oxidized to N_2 by NO_2-N using CO_2as a carbon source. Because both processes are autotrophic the sludge production is low. The two processes can be performed in a one-step SBR system fitted with controls for pH, NH_4, NO_2, DO and reject water addition (Wett *et al.*, 2007). Currently more than 40 full-scale facilities for the treatment of reject water from digested sludge dewatering are operating in Europe, Asia and North America.

The oxygen demand for converting NH_4-N to N_2 is equal for denitrification and deammonification. The reduction of aeration energy achieved by applying deammonification is a consequence of the limited availability of the carbon source for denitrification, which results in higher effluent NO_3-N and NO_2-N concentrations. Each gram of additional nitrate in the effluent causes an additional demand of 2.9 g O_2 as compared to 1 g of N removed by denitrification.

In the two-stage AS process, N removal from reject water can be achieved almost as efficiently by converting the ammonia in it to nitrite (nitritation) and then adding it to the 1st step for denitrification. In this fashion the 1st stage could be operated even completely anoxically. By controlled addition of the nitrite-containing reject water, the energy requirements for aeration in the 1st step for carbon removal can be strongly reduced by denitrification.

12.4 ENERGY PRODUCTION AT WWTPS

12.4.1 Anaerobic sludge digestion

The formation of CH_4 during the microbiological decomposition of organic matter under anaerobic conditions was first reported in 1868 by Bechamp. In the 1890s Omelianski isolated microbes responsible for the release of hydrogen, acetic acid, and butyric acid during methane fermentation of cellulose. In 1881 Mouras was the first to apply anaerobic digestion to treat wastewater in his invention of the 'automatic scavenger' – a crude version of the septic tank (McCarty *et al.*, 1982). The Imhoff tank, a primary sedimentation tank combined with an anaerobic sludge stabilization tank at ambient temperature without biogas utilization was patented in Germany (German Patent DRP Nr. 187723) in 1907. During the 1920s, mesophilic sludge digesters were developed with the main goal of reliable sludge stabilization at larger plants, just as they are used today (Pruess, 1928). The biogas, consisting of CH_4 (~65%), CO_2 (34%), H_2S and NH_4 (<1%), can be used for sludge and digester heating as well as other heating purposes. Since the 1960s biogas driven engines have become common technology for the conversion of gas energy to electrical (and/or mechanical) power. The use of digester gas for power generation was spurred by the sharp price rise of fossil fuel in the early 1970s and by the increasingly stringent pollution control regulations resulting in increased power requirements. During the last several decades, the conversion efficiency of biogas to electrical energy has been improved from ~2 to 3 kWh/Nm^3 of biogas, which allows a marked increase in the renewable electric energy produced from anaerobic sludge digestion. The excess heat from the gas engines is sufficient for digester heating even in cold climates if the feed sludge can be thickened to at least 4.5% dry solids.

The calorific value of the raw sludge solids depends on the wastewater treatment processes that generate it. This value can be calculated using a COD mass balance for primary sedimentation (primary sludge) and for the excess biological sludge (Figure 12.3). One gram VSS corresponds to ~1.45 g COD or ~20.5 kJ. One gram COD degraded in a digester results in a CH_4 production of 0.35 Nm^3. One metre cubed CH_4 has a calorific value of ~40 MJ. Using these theoretical relationships it is possible to calculate the energy content in the biogas if the specific COD load in the digested sludge is known. From investigations at many municipal treatment plants with mesophilic sludge digestion it can be concluded that well stabilized sludge contains about 20 to 25% of the wastewater influent COD. Biogas composition depends on the composition of the wet feed sludge (COD and/or VSS, TKN, total sulfur, alkalinity, siloxanes). For most of the gas engines operated as Combined

Heat and Power (CHP) systems the biogas has to be pre-treated (to remove H_2S and siloxanes). Biogas can also be converted directly to electrical energy by fuel cells, but this is still not competitive. At actual energy prices the combination of anaerobic sludge digestion with electric power generation from biogas will only be economically competitive for plants with a capacity of >50.000 PE.

The CH_4 production from sewage sludge can be increased by increasing the energy content of the raw sludge and by reducing the energy content of the digested sludge (Section 12.4.2). Adding co-substrates to sludge digestion is another approach, but it will not be further discussed as there is no direct link to the AS process.

For decision making in design and operation, all energy considerations have to be based on realistic operational data rather than on design data. Therefore coupled wastewater treatment and sludge digestion models become more relevant, because the two process schemes are linked with regard to COD for sludge and biogas production.

12.4.2 Increase of energy recovery from sludge digestion by enhanced solids degradation

The efforts to increase energy recovery from biogas of anaerobic digestion mentioned above are:

(i) Maximization of the amount of biodegradable COD in primary and secondary sludge fed to digesters by alternative wastewater treatment process configurations.
(ii) The use of high efficiency gas engines for more electrical power production.

Other efforts also can contribute to the same goal:

(i) Optimization of digester performance through constant feeding and temperature, control of scaling and prevention of foaming by adjusting mixing to digester shape and equipment, reduction of grit and screenings accumulation by improved wastewater pre-treatment.
(ii) The increase of solids retention time by better thickening to achieve lower heat requirements for digester heating and lower reject water flows from dewatering.

Other methods of increasing solids biodegradability to produce higher biogas yields and lower amounts to be disposed of use additional sludge treatment steps in which chemical, thermal or electrical (mechanical) energy is used to increase the availability of the organic matter in the anaerobic decomposition process. These processes may aim at decreasing solids particle size thereby increasing the amount of free surface for bacterial attack and destroying bacterial cell walls and organic particles so as to make their contents available for biogas production.

At full-scale, most of these processes do not produce an overall increase in energy recovery compared to the results from well-designed and operated digestion plants. The situation can be different if the existing digestion facilities are overloaded and the additional sludge treatment steps take the place of increased digester volume or additional equipment. From this experience it can be concluded that biogas production is strongly influenced by the particulate material hydrolysis rate. Cell walls in organic particles are very resistant to enzymatic attack so their destruction needs large amounts of chemical, thermal or mechanical energy. The application of these processes may be economically successful for specific local situations but not for increased overall energy recovery.

The increased NH_4-N release to the liquid phase (reject water) due to the increased conversion of organic material into biogas and the NH_4-N inhibition of methanogenic bacteria have to be considered for application of these methods. The EU FP7, 'ROUTES' (2011) research project is currently investigating methods for enhancing energy recovery from sewage sludge.

12.4.3 Thermal sludge treatment

Thermal sludge treatment processes include: (i) incineration – dewatered sludge is burned in a chamber supplied with air. The excessive heat produced during combustion is converted to electricity by steam turbines; (ii) gasification – dewatered sludge is decomposed in a chemically reactive environment; the products are primarily fuel gasses; (iii) pyrolysis – dewatered sludge is decomposed in a primarily nonreactive environment. The products are primarily CO, H_2 and a solid residue (char). Gasification and pyrolysis are emerging viable bioenergy technologies with reduced N_2O emissions that produce energy using a modified combustion processes. With the exception of Japan it is uncommon to find full-scale facilities of this type. Compared to the biogas approach, thermal treatment allows the conversion of sludge to electrical energy (generated by steam and excess heat) together with maximum mass and volume reduction to producing an inert residue (ash), which reduces the cost of transportation and final disposal. Few reports are available that compare the electricity generation of full-scale anaerobic digesters and thermal treatment processes (Dauthuille, 2008).

12.4.4 Heat recovery and utilization

Up to 50% of the energy contained in the biogas from anaerobic digestion will be converted to heat even when highly efficient gas engines are applied. In addition to heating digesters and dryers, external utilization by nearby communities is also practiced. Efficient recovery of low temperature thermal energy from cogeneration of heat and power systems and side streams needs to be developed using technologies based on, for example, the Organic Rankine Cycle (ORC) (Hung *et al.*, 1997).

Another route of thermal energy recovery is from the low temperature heat of the wastewater itself. In western countries the energy consumption for warm water production (approximately 100 W/inhabitant) is much more than the energy equivalent of the pollutants. Even though its entropy level is much higher, there is much literature and many full-scale applications using the wastewater for heating and cooling (Müller & Kobel, 2001). Heat transfer from the wastewater reduces its temperature and therefore has to be considered for design and operation of biological treatment processes. If heat is recovered from the treatment plant effluent, this problem does not exist.

12.5 SHOWCASE OF LOW ENERGY MUNICIPAL NUTRIENT REMOVAL PLANT: STRASS, AUSTRIA (90,000–200,000 PE)

The four approaches used to reach energy sufficiency were: (i) use of two-stage AS to maximize COD fed to the anaerobic digesters to increase biogas production; (ii) adoption of on-line sensor based dynamic control of intermittent aeration; (iii) adoption of high efficiency electricity generators, and (iv) application of anammox for side-stream ammonium removal. Table 12.2 shows that an energy recovery efficiency of 108% was achieved in 2005 (Wett *et al.*, 2007). This performance is favored by the low influent N/COD = 0.067 ratio and the small diurnal COD and ammonia load variations due to the extraordinarily long trunk sewers (40 and 60 km). On the other hand the plant has to cope with significant variation in the pollution load caused by intensive winter and summer tourism in the catchment.

Table 12.2 Strass treatment plant energy balance for 2005.

Specific biogas production	24.5	l/PE/d1	
Biogas production	3648 m^3/d		
CO_2-content	38	%	
Methane production	2261 m^3/d		
Energy content	22,615	kWh/d	
Electrical efficiency of gas engines	37	%	
Electrical energy production.	8440	kWh/d (0.48 kWh/kgCOB)	108%
Electrical energy consumption	7910	kWh/d (0.45 kWh/kg COB)	100%

[1]PE is based on 120g COD/d.

12.6 FUTURE DEVELOPMENTS

12.6.1 Introduction

The scope of future wastewater treatment will not only cover water but will also include air emissions, solids disposal and resource recovery. These tasks will drive the development of new innovative processes, technology and management because they will be the indicators of the plant performance and sustainability. This section will highlight some of the foreseeable new processes for energy recovery in future wastewater treatment plant development.

12.6.2 Mainstream anammox

The application of anammox in mainstream wastewater treatment processes was first proposed by Jetten *et al.* in 1997 following its discovery and application in side-streams. Siegrist *et al.* (2008) conducted an energy balance study of mainstream anammox. The main challenges are stable partial nitritation, repression of nitrite oxidizing bacteria, retention of anammox bacteria and maintenance of a balanced population of aerobic ammonium oxidation and anammox bacteria at low temperature and low NH_4 concentrations. Some progress has been achieved in Europe and the USA (Wett *et al.*, 2012). Cao *et al.* (2013) have recently shown significant autotrophic N removal in one of the 200 000 m^3/d trains of the Changi (Singapore) Water Reclamation Plant.

12.6.3 Energy management tools

A real-time plant-level energy management system is essential for effective energy management. An Energy management system should have three components: (i) online sensors/meters; (ii) reliable equipment and interfaces between sensor and equipment; and (iii) software for data logging, transfer, and calculation. An example of this at the process level is an aeration system with off-gas monitoring to monitor transfer efficiency, auto-calibrating/operating and data transfer in real-time. Future Supervisory Control and Data Acquisition (SCADA) systems should ideally be able to monitor and control energy consumption and production based on on-line data monitoring and analysis at the plant level. This will require development of new capacities in sensor, equipment and automation in the future plants.

Energy is related to many aspects of plant operation and performance: effluent quality, GHG emissions, hazardous wastes and odor removal, and so on. Therefore, in addition to monitoring and control, the future energy management system should be able to diagnose, optimize and make selections between different operating strategies.

12.7 FINAL STATEMENT REGARDING ENERGY CONSIDERATIONS

Compared to current plants, future wastewater treatment plants will need to meet much broader and stricter legal requirements, such as pathogen disinfection, micropollutant discharges, greenhouse gas emission, and so on. These new developments will bring about new concerns and discussions: (i) more energy will be needed to perform such functions; and (ii) a new consensus will be necessary to establish the new requirements and indicators, which may have competing demands. For example, a low energy consumption N removal process may cause NO_2 concentrations and N_2O emissions. Balancing the advantages and disadvantages of all relevant alternatives will be necessary. For biological nutrient removal plants there are four highly relevant considerations for energy minimization, independent of specific local characteristics and process selection:

- Type of sludge stabilization (aerobic, anaerobic, thermal).
- High aeration system efficiency under real operating conditions (kg O_2/kWh).
- High conversion efficiency of electric power generation from biogas (kWh/ Nm^3 CH_4).
- Optimum aeration control.

Minimization of energy consumption and increase of energy recovery should be pursued because there is no justification for wasting low entropy energy. However this *is not the primary goal for WWTP design and operation* and, thus, should not be overstated. Reliable and highly efficient wastewater treatment should always be the basis for successful water protection and conservation at reasonable cost.

12.8 REFERENCES

Anon (1994). Energy in WWT in Germany, BUWAL. Swiss Federal Ministry for Environment, Forest & Landscape, Bern, Switzerland, ISBN: 3-905232-49-9.

Ardern E. and Lockett W. T. (1914). Experiments on the oxidation of sewage without the aid of filters. *J. Soc. Chem. Ind.*, **33**, 523.

Barnard J. (1973). Biological denitrification. *Water Pollut. Control,* **72**, 705–717.

Batstone D. J., Keller J., Angelidaki I., Kalyuzhnyi S. V., Pavlostathis S. G., Rozzi A., Sanders W. T .M., Siegrist H. and Vavilin V. A. (2002). The IWA anaerobic digestion model No. 1 (ADM1). *Water Sci. Technol.*, **45**(10), 65–73.

Böhnke B. (1977). *Das Adsorptions Belebungsverfahren* (the AB-type activated sludge treatment). *Korrespondenz Abwasser* (Wastewater Correspondence), **24**(7), 622.

Broda E. (1977). Two kinds of lithotrophs missing in nature. *Zeit. Allgemeine Mikrobiol.*, **17**(6), 491–493.

Cao Y. S. and Ang C. M. (2009). Coupled UASB – Activated sludge process for COD and nitrogen removals in municipal sewage treatment in warm climates. *Water Sci. Technol.*, **60**(11), 2829–2839.

Cao Y. S., Ang C. M., Raajeevan K. S., and Wah Y. L. (2008). Biological Nitrogen Removal Activated Sludge Process in Warm Climates: Full-Scale Process Investigation, Scaled-Down Laboratory Experimentation and Mathematical Modelling. IWA Publishing, London, UK, ISBN: 1843391872, 168.

Cao Y. S., Kwok B. H., Yong W. H., Chua S. C., Wah Y. L. and Yahya A. G. (2013). The main stream partial nitritation-anammox nitrogen removal in the largest full-scale activated sludge process in Singapore. WEF-IWA Nutrient Removal and Recovery, 2013, Tendency and Future, July 28–31, Vancouver, Canada.

Clean Water Act (1972). http://www.epa.gov/region5/water/cwa.htm/ (accessed 22 June 2013)

Dauthuille P. (2008). Concept of the 'plant+energy'. Lecture at PUB, Nov. 8, 2008, Singapore.

EnviroSim (2013). BioWin. http://envirosim.com/products/biowin/ (accessed 10 August 2013)

EPA and NRCS (1999). Comprehensive nutrient management planning, http://www.epa.gov/agriculture/ag101/impactcnmp.html/ (accessed 22 June 2013)

EPA and USDE (1995). Case studies in residual use and energy conservation at wastewater treatment plants, http://www.nrel.gov/docs/legosti/old/7974.pdf/ (accessed 22 June 2013).

GWRC (2008). State of Science Rept: Energy and Resource Recovery from Sludge. In: Y. Kalogo and H. Monteith (eds), IWA Publishing, London, p. 192, ISBN: 9781780404653.

Harleman D. R. F. and Murcott S. E. (1992). Upgrading and multi-stage development of municipal waste water treatment plants: Applicability of chemically enhanced primary treatment. Technical Report, World Bank.

Heidrich E. S., Curtis T. P. and Dolfing J. (2011). Determination of the internal chemical energy of wastewater. *Environ. Sci. Technol.*, **45**(2), 827–832.

Henze M., Grady C. P. L., Gujer W., Marais G. v. R. and Matsuo T. (1987). Activated Sludge Model No. 1, IAWPRC Sci. and Tech. Report No. 1, IAWPRC, London.

Hung T. C., Shai T. Y. and Wang S. K. (1997). A review of organic rankine cycles (ORCs) for the recovery of low-grade waste heat. *Energy*, **22**(7), 661–667.

Hydromantis (2013). GPS-X. http://www.hydromantis.com/GPS-X.html (accessed 10 August 2013).

Imhoff K. (1953). *Geschichte der biologischen Abwasserreinigung.* (History of biological waste water treatment) *GWF,* **94**, 344.

Jetten M. S. M., Horn S. J. and van Loosdrecht M. C. M. (1997). Towards a more sustainable municipal wastewater treatment system. *Water Sci. Technol.*, **35**(9), 171–180.

Krampe J. (2013). Energy benchmarking of South Australian WWTPs. *Water Sci. Technol.*, **6**(7), 2059–2066.

Kroiss H. (2011). Energy and Wastewater Treatment (Keynote address). 11th IWA Conf. on Design, Operation and Economics of Large Wastewater Treatment Plants, 4–8 September, Budapest, Hungary.

Kroiss H. (2012). Energy Efficiency in Wastewater Treatment Design, Lecture, Civil Engineering Department, National University of Singapore, March 22.

Lettinga G., van Velsen A. F. M., Horbna S. W., de Zeeuw W. J. and Klapwijk A. (1980) Use of upflow sludge blanket (USB) reactor concept for biological wastewater treatment. *Biotech. Bioeng.*, **22**, 699–734.

LFUW (Austrian Federal Ministry for Environment) (2001) Benchmarking in Water Management – Acquisition and Comparison of Technological and Economical Key Figures (in German), Final Report Vienna, Austria.

Lindtner S. and Svardal K. (2008). *Energieverbrauch von Kläranlagen – Schlüsse aus der österreichischen Benchmarking – Studie* (Energy requirements of waste water treatment plants – Consequences of the Austrian benchmarking study report). KAN Brochure 2009, Austrian Water and Waste Association, ÖWAV, 1010 Wien, Marc Aurel Strasse 5.

Lindtner S., Schaar H. and Kroiss H. (2008). Benchmarking of large municipal wastewater treatment plants treating over 100,000 PE in Austria. *Water Sci.Technol.*, **57**(10), 1487–1493.

Ludzack F. G. and Ettinger M. B. (1962). Controlling operation to minimize activated sludge effluent nitrogen. *J. Water Pollut. Control Fedn.*, **34**, 920–931.

Matsché N. (1972). The elimination of nitrogen at the treatment plant of Vienna Blumental. *Water Res.***6**,485–486.

Matsché N. (1977). Removal of nitrogen by simultaneous nitrification/denitrification in an activated sludge plant with mammoth rotor aeration. Proc. Conf. on Nitrogen as a Water Pollutant, Pergamon Press, ISBN/EAN: 978-0-08-020900-5.

Matsché N. and Moser D. (1994). Operation of a two-stage activated sludge package plant for high efficiency treatment. *Water Sci. Technol.*, **28**(10), 299–307.

McCarty P. L. (1982). One hundred years of anaerobic treatment. In: Anaerobic Digestion, 1981: Proceedings of the 2nd International Symposium on Anaerobic Digestion, Travemunde, Germany, 1981, D. E. Hughes, D. A. Stafford, B. I. Wheatley, W. Baader, G. Lettinga, J. NynsE, W. Verstraete, R. L. Wentworth (eds), Elsevier Biomedical Press, Amsterdam, The Netherlands, pp. 3–22.

Michaelis L. and Menten M. (1913). Die Kinetik der Invertinwirkung, *Biochem. Z.*, **49**, 333–369.

Monod J. (1942). The growth of bacterial cultures. *Ann. Rev. Microbiol.*, **3**, 371–394.

Müller E. A. and Kobel B. (2001). *Nutzung von Wärme aus Abwasser mit Wärmepumpen, Erfahrungen in der Schweiz, Potenziale in Deutschland* (Heat recovery from waste water, experiences from Switzerland and potential in Germany). *Korrespondenz Abwasser*, **8**, 1074–1080.

MURL (1999). Ministry for Environment, Nature Protection, Agriculture & Consumer Protection in the Germany State of North Rhine Westphalia. Energy in WWTPs (in Germany). Düsseldorf, Germany.

Pruess M. (1928) *Eine neue Frischwasserkläranlage für getrennte Schlammfaulung mit künstlicher Schlammumwälzung und künstlicher Beheizung* (A new wastewater treatment plant with mechanical mixing and heating equipment). *Gesundheitsingenieur*, 7.

Ross W. R. (1984). The phenomenon of sludge pelletisation in the anaerobic treatment of a maize starch process-ing waste. *Water S. A.*, **10**(4), 197–204.

ROUTES (2011). Novel processing routes for effective sewage sludge management. FP7 EU research project. http://www.eu-routes.org/ (accessed 25 June 2013).

Schaar H., Clara M., Gans O. and Kreuzinger N. (2010). Micropollutant removal during biological wastewater treatment and a subsequent ozonation step. *Environ. Pollut.*, **158**(5), 1399–1404.

Scholz W. (2013). Global transdisciplinary process on sustainable phosphorus management. *GWA Gewässerschutz-Wasser-Abwasser*, **232**, 13/1–13/16.

Siegrist H. (2011). Producing Fertilizer from Waste Water. EAWAG Ann. Rept 2011, 20–21.

Siegrist H., Salzgeber D., Eugster J. and Joss A. (2008). Anammox brings WWTP closer to energy autarky due to increased biogas production and reduced aeration energy for N-removal. *Water Sci. Technol.*, **57**(3), 383–388.

Spindler A. and Vanrolleghem P. A. (2012). Dynamic mass balancing for wastewater treatment data quality control using CUSUM charts. *Water Sci. Technol.*, **65**(12), 2148–2153.

Svardal K. and Kroiss H. (2011). Energy requirements for waste water treatment. *Water Sci. Technol.*, **64**(6) 1355–1361.

Swiss Federal Government (2013). *Botschaft zur Änderung des Gewässerschutzgesetzes* (Announcement on a change of the Swiss water protection act) *Verursachergerechte Finanzierung der Elimination von Spurenstoffen im Abwasser* (Application of the polluter pays principle to the elimination of micropollutants from municipal waste water). http://www.admin.ch/opc/de/federal-gazette/2013/5549.pdf (accessed February 2014)

USFR (1999). CFR 40 Part 130 – water quality planning and management. www.epa.gov/fedrgstr/EPA-WATER/1999/October/Day-01/w25307.htm/ (accessed 22 June 2013)

UWWD (1991). Directive on urban wastewater treatment (91/271/EEC). http://europa.eu.int/comm/ (environment/water/water-urbanwaste/directiv.html (accessed 25 June 2013)

van de Graaf A. A., Mulder A., de Bruijn P., Jetten M. S. M., Robertson L. A. and Kuenen J. G. (1996). Autotrophic growth of anaerobic ammonium oxidizing micro-organisms in a fluidized bed reactor. *J. Med. Microbiol.*, **142**(8), 2187–2196.

Wandl G. and Papp M. (2012). *Das Energieoptimierungskonzept der Hauptkläranlage Wien* (The energy optimization concept for the main treatment plant of Vienna). *Wiener Mitteilungen,* Inst. Water Quality Resource and Waste Management, Vienna University of Technology, **226**, pp. 217–230, ISBN: 978-3-85234-119-4.

Wandl G., Schaar H., Papp M. and Svardal K. (2009). The first two years of full scale operation of the two-stage main wastewater treatment plant of Vienna. Water Practice Technol. IWA Pub., London. doi: 10.2166/wpt.2009.005.

WERF (2009). Technology roadmap for sustainable wastewater plants in a carbon-constrained world. WERF Workshop, May 20–21, Chicago, IL, USA.

Wett B., Buchauer K. and Fimml C. (2007). Energy self-sufficiency as a feasible concept for wastewater treatment systems, Leading-Edge Conf., June 4–6, Singapore.

Wett B., Podmirseg S. M., Hell M., Nyhuis G., Bott C. and Murthy S. (2012). Expanding DEMON sidestream deammonification technology towards mainstream application. Intl IWA Conf. on Autotrophic Nitrogen Removal: from Research to Application, June 29, Milan, Italy.

WFD (2000). Directive 2000/60/EC of the European Parliament and of the Council establishing a framework for the community action in the field of water policy 'European Water Framework Directive'.

Chapter 13

Automation and control

Gustaf Olsson (Sweden), Zhiguo Yuan (Australia), Changwon Kim (Republic of Korea)

13.1 INTRODUCTION

Instrumentation, control and automation (ICA) of AS processes have attracted much attention since the early 1970s. Even earlier it was recognized that the process was subject to large load variations and that this called for some kind of control. In the 1930s the first operating requirements were the need to maintain DO concentration in the aeration tanks. Several proposals for the minimum DO concentration were recommended. The sludge activity, measured as the amount of DO required/g dry solid, h was considered a good index of operation for determining the rate of oxygen demand of AS. A number of control variables were defined, such as quantity of air used, aeration period, concentration of MLSS, but actuators were not flexible and the control methods were poorly developed. The developments during the last four decades in on-line instrumentation, computer technology, process understanding and subsequent model development, and control methods have been powerful driving forces for advanced control. At the same time regulatory requirements and increased design complexity have pushed the ICA development further and today ICA is a vital part of most water and wastewater systems.

13.2 THE ROLE OF CONTROL AND AUTOMATION

ICA attracted the attention of the water and wastewater industry in the early 1970s. Many operators recognized that the plants were designed for average loads, while the true load varied significantly during the day. The method of accommodating these variations was to design the wastewater treatment plant with a sufficiently large volume. Progress in control methodology suggested that there were other means of attenuating the consequences of load variations. In the 1980s biological nutrient removal (BNR) processes became standard for the wastewater treatment and resulted in much more complex unit processes that require advanced ICA technologies.

Feedback control requires measurements. Instrumentation was the weak link in the 1970s. Some progress had been made to develop DO sensors, COD analyzers and TSS meters and these instruments were tested in some pioneering control efforts. Still, instrumentation reliability was called into question. There were also important developments in computers, and real-time computing for on-line control was becoming affordable.

As early as the 1930s the control of operation of AS plants attracted attention. Palmer (1939) indicated a number of variables which influenced control. These were: quantity of air used, aeration period, concentration of MLSS, settling period, rate of sludge return and sludge condition as indicated by a determination of its oxygen consumption.

ICA is a hidden technology – it is not noticed as long as it works! Even though ICA is becoming ubiquitous in most water systems and nowadays almost every water and wastewater treatment plant is supplied with a computer control system, dynamic systems and process control are seldom part of the general civil/ environmental engineering curriculum. Consequently many water and wastewater system designers are unaware of the potential of ICA. This was recognized early on and in 1974 the late Professor John F. Andrews (1930–2011), then at Clemson University, South Carolina, USA recognized the need for such education at all levels when he noted: 'A course in *Process Dynamics and Control* is commonly found in most chemical engineering curricula. We would be well advised to include a course in *Dynamics and Control of Wastewater Treatment Systems* in environmental engineering curricula.' (Buhr *et al.*, 1974). Today there are such courses, but it has taken a long time to realize John Andrews' vision.

It has been demonstrated that ICA may increase the capacity of BNR plants by 10–30%. The advanced knowledge of the mechanisms involved in BNR that is being gained today is producing an increased understanding of the processes and the possibilities for control. With further understanding and exploitation of the relationship between operational parameters and the microbial population dynamics and biochemical reactions the improvements in BNR afforded by ICA may reach 20–50% of the total system investments within the next ten to twenty years. Various case studies of advanced control in water and wastewater treatment systems have shown significant savings in operating costs and remarkably short payback times (Olsson *et al.*, 2005).

ICA includes all the information that is generated and used in a system. With today's computer technology and instrumentation we take it for granted that we can generate a lot of data. However, data rich is not the same as information rich. Data has to be interpreted: signal patterns must be analyzed and various signals can be combined and compared to expected behaviour. This can form the basis for some action.

We often describe ICA has having three major purposes:

- Keep the plant running;
- Satisfy the effluent requirements;
- Maximize the efficiency.

A large number of equipment units are needed to keep the plant running. Levels, flow rates, pressures and temperatures are mostly controlled automatically by pumps, compressors and valves. This level of control is similar in all kinds of process industries and is usually highly standardized. Control at this level is a crucial pre-requisite for good operation of AS plants, but does not require specific process knowledge of the wastewater treatment system.

To satisfy the effluent requirements, many concentrations have to be controlled, such as DO, MLSS concentrations, SRT and RAS flow. Controlling these variables will help maintain the effluent quality, allow unmanned operation at night and on weekends, and save energy and other operational costs. Even at this level however many control systems fail, not because control is difficult or sensors are not sufficiently robust, but because many of the personnel implementing the control procedures do not have sufficient knowledge of the process dynamics or do not efficiently communicate this knowledge to their colleagues. Sensors are located in the wrong positions, data analysis is not adequate, sampling frequencies are often unrealistic (mostly too frequent), or the controller settings are not adequate.

13.3 DISTURBANCES

A major incentive for control is the presence of disturbances whose impact has to be compensated. Disturbances to a wastewater treatment plant are significant and quite different from most other process industries. The wastewater influent typically varies substantially in its concentration, composition and flow rate, with time scales ranging from a fraction of an hour to months. Discrete events such as rainstorms, toxic spills and peak loads may also occur from time to time. As a result, the plant is hardly ever in steady state, but is subject to transient behavior all the time (Olsson & Newell, 1999).

Too often, unnecessary disturbances are created within the plant itself. For example, if the influent pumps are operated in an on/off mode the plant will be subject to sudden flow rate changes which particularly affect clarifier operation. In the last few decades, there has been a revolution in the development of power electronics, and variable speed control of electric motors is now both affordable and reliable for everything from milliwatt scale motors to megawatt drives. Variable speed control has a huge influence both on wastewater flow rate and air flow controls.

Recycling of water and sludge in a wastewater treatment plant creates apparent couplings between various unit processes. If these interactions are not considered, then plant operation will suffer. For example, if anaerobically digested sludge supernatant is recycled to the plant influent during a high load period the N load to the plant may be very large and can cause a significant increase in the DO uptake rate.

Further *internal* disturbances can be generated due to inadequate or inappropriate operations including human errors, unsuitable or malfunctioning actuators and/or sensor breakdowns. These can cause major operational problems.

13.4 THE EARLY YEARS OF AUTOMATION AND CONTROL

Following the discovery of AS in 1913 (Ardern & Lockett, 1914) much attention was paid to the theory and limitations of the process. Very little was published by plant operators on the control of the process. This subject only attracted attention in the late 1930s and a comprehensive committee report on plant operation was published in the USA in 1941 (Pearse *et al.*, 1941).

The development of the simple concepts of AS into a large scale operating plant was a process of evolution carried out by practicing engineers in various parts of the world. The earliest tests were on a fill-and-draw basis. However, for large-scale operations such procedures became impracticable. Haseltine (1937) found that the first operating requirement of the AS process was the maintenance of DO in the aeration tanks. Heukelekian (1934) suggested that the minimum DO concentration should be >0.5–1 mg/L, while Collier (1939) suggested that 2 mg/L should be the minimum. This was confirmed by practical experiences from the North Side Works in Chicago, IL, USA. It was also found that sludge activity, measured in mgDO/gMLSS, h was a good index of operation. However, the use of an apparatus for quickly determining the DO demand rate of AS did not gain favor in the late 1930s. Many operators were convinced 'that the plant itself is its own best odeometer' (Pearse *et al.*, 1941).

Process control received more attention in the 1960s as being 'no less applicable in theory to sewage treatment as it is to any other microbiological process' (Jones *et al.*, 1969). In practice, however, it was considered more difficult, since 'both the quantity and the quality of the incoming sewage are variable and it is essential that the process operates continuously and efficiently at all times of the year if receiving waters are to be protected' (Jones *et al.*, 1969). The AS process was studied in particular, since between one third and one half of the total cost of sewage treatment was related to this process. In 1969 the SS concentration was one of the few variables that were considered controllable. It was recognized that the respiration, the sludge density index, the filterability of the sludge and the rate of nitrification all depended on SS concentration.

While it had been realized that DO concentration was a key variable Jones *et al.* (1969) indicated that to save energy the (DO) 'concentration should not be maintained at a level any higher than necessary'. As early as the 1950s a semi-continuous colorimeter in conjunction with the Winkler method was being used to measure DO concentration (Briggs *et al.*, 1954). Later this sensor was replaced by the dropping-mercury electrode (Briggs *et al.*, 1957). By the early 1970s the use of on-line DO sensors was well established in many wastewater treatment plants (Briggs *et al.*, 1967; Briggs, 1973; Brouzes, 1969; Roesler, 1974). Direct digital control of DO was performed in Palo Alto, California, USA (Petersack & Stepner, 1973), in Bridgeport, Connecticut, USA (Dobbins, 1974) and in Stockholm, Sweden (Olsson & Hansson, 1976) and very soon significant savings of air could be reported. For example, Guarino and Radziul (1977) reported a 22% reduction in air demand per unit of BOD removed.

Usually, in a plug flow AS basin, the air distribution along the reactor could only be controlled manually. Even at steady state – if it ever occurred – it was very difficult to change the valve openings along the aeration tank to obtain the desired air flow distribution. A change of one valve opening along the reactor immediately resulted in a new airflow distribution that was not necessarily the desired one. Most often, the aeration rate near the inlet of the aeration tank was insufficient to satisfy the DO requirement, while near the reactor outlet it was in excess of what was required (Figure 13.1).

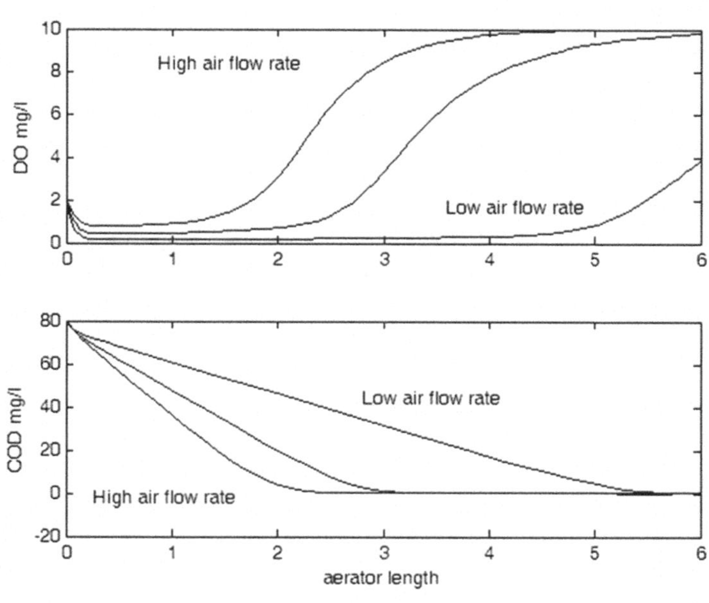

Figure 13.1 A typical DO profile in a plug flow reactor for carbon removal with uniform airflow distribution. The upper figure shows the DO profile (mg/L) for different air flows. The lower figure depicts the corresponding carbon concentration (mg COD/L) decrease along the plug flow reactor (from Olsson, 2012).

Haseltine (1937) used tapered aeration at the Salinas, California, USA plant from 1930 to 1933, applying 55–70% of the total air in the first half of the aeration tanks. In the 1940s Torpey (1948) found by experience that the DO profile levelled off at around 75% of the reactor length, indicating that the uptake reaction had gone to completion. Without any control, the profile was pushed towards the outlet of the aeration tank at high organic loads, while it receded towards the inlet at low loads. The interpretation of this type of profile was discussed in detail in Olsson and Andrews (1978). It was apparent that there was not one position in the tank which could be representative of the whole tank at all times. Instead the slope of the DO concentration towards the outlet could serve as an indicator of whether the reaction had gone to completion. Various attempts were made to use two DO sensors, located along the aeration tank towards the outlet, to find the most efficient air flow (Gillblad & Olsson, 1977; Olsson, 2012).

To keep the DO concentration at a desired value along the reactor requires control both along the length of the aeration tank and as a function of time. Today it is well established that the problem of DO profiles can be addressed better by providing independent air supply systems to different zones of the aerated reactor. The DO profile analysis, however, gave considerable insight since it is a direct mirror of the DO uptake.

Instrumentation and control in wastewater treatment received much attention in the early 1970s. A literature survey for the period 1971–1975 (WRC, 1975) found 52 publications that dealt with automation in sewage treatment plants. This included the volume of 70 papers presented at the first 1973 conference on ICA (Andrews *et al.*, 1974) organized in London by IAWPR (The International Association on Water Pollution Research), the original predecessor organization of IWA. During the forty years since then there have been another 10 ICA conferences – the latest one in 2013. In the early 1970s some people involved in operations realized that designers had been working on wastewater treatment plant automation without being completely aware of what was going on elsewhere. The following statement at the London conference expressed the general feeling of the participants: 'We accept that variability is of great importance, but being faced with the task of expressing its effect on performance, had to conclude that at the present time there are no data on which to assess its effects'. Details of the ICA conferences are described in Olsson (2012), but some reflections are worthy of being cited here: 'Even if we had the most sophisticated, automated plant in existence, it still would not be able to operate at maximum efficiency, because the designs of wastewater treatment plants are based on uniform combined sewer flow with consideration for periodic intensity due to storm flow or periodic lows during dry weather spells or hours of least demand. So, much of the time the flow into the plant is either above or below the maximum efficiency level' (Kukudis, 1973).

Presentations at the conferences of the 1970s were mostly from plant operators and instrument manufacturers. At the 1973 conference only 10% of the contributions were presented by academic researchers. Since then there has been a dramatic change in participation at the ICA conferences. At the last five conferences, since 1997, academic researchers have contributed almost two-thirds of the papers. Unfortunately the participation of the plant owners has decreased dramatically. It is apparent that the focus has shifted from problem driven (as defined by the 'end users') to methodology driven research. (Olsson, 2002). The progress of knowledge, as presented at the IWA conferences, has been described in Olsson *et al.* (2013). It is obvious that a closer cooperation between the end users and the researchers is highly desirable.

13.5 THE DEMAND

In the 1970s, the AS process was used mainly for organic matter removal and effluent requirements were mostly for BOD and SS. Regulatory requirements and water quality are important drivers for plant design and increasingly stringent effluent requirements, including nutrient removal, have required more sophisticated designs. The drivers for control and automation, however, are economics, efficiency and consistent operation. Often ICA is introduced as a second step after design while it would be more desirable to integrate it with the design in a so-called control-integrated design. Inflexible or under-dimensioned designs cannot be improved by control.

Energy is now the single largest operating expense at a treatment plant so it makes economic sense, where possible, to reduce this cost through good control. The vision of zero or even positive energy plants has already been realized in some cases (Nowak *et al.*, 2011). Different energy forms must be carefully defined, since electrical and thermal energy are not equivalent. While the traditional focus has been on the wastewater treatment process, a shift in emphasis may take place towards sludge treatment and waste-to-value conversion processes, leading to the renaming of wastewater treatment plants (WWTPs) as 'water resource recovery facilities' (WRRFs).

To achieve the best possible operation, many criteria have to be weighed together. Some applications are found in water supply systems (Baur *et al.*, 2003), and stormwater management (Bertrand-Krajewski *et al.*, 2007; Urrutiaguer *et al.*, 2010). In wastewater treatment systems, multi-criteria decisions have been presented by Flores-Alsina *et al.* (2008) on how to design the most efficient operation and control strategies. A new IWA Working Group on Life Cycle Assessment of Water and Wastewater Treatment is indicative that an understanding of efficient plant design and operation is the future (see: www.iwahq.org). In addition to technical and economic constraints, human behavior and motivation to operate at the highest possible efficiency must be considered (Rieger & Olsson, 2012).

13.6 COMPUTERS AND INFORMATION TECHNOLOGY

In the early 1970s a typical process computer was the Digital Equipment Corporation PDP-8 which, in its basic configuration, had a main memory of 4096

12-bit words. A common configuration was 28 kB of memory, supplied typically with eight 100 analog inputs, up to 200 digital outputs, and 4–16 analog outputs. The word size of 12 bits was large enough to handle unsigned integers from 0 to 4095 (or signed numbers from −2048 to +2047). This was sufficient for controlling simple machinery. Twelve bits could also store two 6-bit ASCII subset characters. More than 50,000 PDP-8 systems were sold at a typical price of about US$ 30,000. It was about the size of a small household refrigerator, the operator interface was usually a black and white cathode ray tube (CRT) and paper tape was the software medium. In the wastewater industry these types of computers were mostly used for data acquisition and monitoring; very few closed loops were found in a large US survey (Babcock, 1974) and most of these were for level control.

A few very large wastewater treatment plants had real-time computers. The computer for the Philadelphia, Pennsylvania, USA, Southwest Water Pollution Control Plant that treated 135 MGD (5.9 m^3/s) had around 2000 digital inputs, 600 digital outputs, 600 analog inputs, 40 analog outputs and 300 computer-controlled loops mostly for control of basic equipment operation (Ballotti *et al.*, 1974). The term direct digital control (DDC) had been defined as a feedback system where all the controllers were implemented in a time-shared manner in a digital computer.

In the 1970s, cost alone prohibited complete automation in many process industries and in particular the wastewater industry. Consequently there was much resistance to the use of computers for control. The economic return from process improvement had to exceed the investment in computer control equipment. This meant that the complexity was substantial. Several obstacles had to be overcome: instrumentation was expensive, the process dynamics were not well known, the actuators had to be sufficiently flexible, the computer had to be reliable, and adequate control theory had to be developed. Even so, process control steps were developed by computers that could perform modeling and simulation and later on, control (Busby & Andrews, 1975). It should be noted that automation was not expected to save manpower but to improve operational reliability and facility capacity.

Recent decades have witnessed extensive discussions on the attractiveness of treating wastewater with decentralized systems and of source separation (Larsen *et al.*, 2013). Today, very small processes can have real-time computer control for monitoring and control and this should be a necessary feature of such systems (Olsson, 2013).

13.7 OBSERVING THE PROCESS-MEASURING AND MONITORING

Many sensors are implemented simply to keep the process running. These are the so-called 'low level' sensors that check levels, pressures, temperatures, flow rates, and other physical variables. Like in any process industry, the equipment in the wastewater treatment processes has to operate satisfactorily. Any deviations from 'normal' performance of equipment and physical variables should be detected automatically in order to guarantee safe operation of the plant. Automation in the process industry has developed in parallel with computer development and many wastewater systems are in their second or third generation control systems. Standardization of the automation of physical variables has developed for both the control of continuous and of discrete signals and events for example ISO 61131 (IEC, 2013).

Reliable measurements are the fundamental requirement for successful feedback control. It was recognized early on that the main obstacle to developing control and automation in wastewater systems was the lack of adequate sensors, especially field instrumentation (Jones *et al.*, 1969). Many instruments had been tested only in lab or pilot scale environments and lacked the necessary robustness. Much attention was given to on-line instrument development during the ICA conferences in the 1970s. Since aeration is a key operation, DO sensors were considered to be particularly important (Briggs, 1973). The need for efficient and dependable sensors was discussed at length at a workshop in 1974 at Clemson University, Clemson, South Carolina, USA, (Buhr *et al.*, 1974). Flow rate, DO concentration, sludge blanket level, settling velocity, respiration rate, SS concentration, short-term BOD, ammonia, nitrate and phosphate concentrations were considered key variables. A central location for gathering and dispensing information on instrumentation testing was recognized as being 'of considerable assistance'.

Another obstacle was the level of process understanding. Many operators had only vague ideas of how to use the information from a complex sensor like a COD instrument, so many people gave up the idea of applying these kinds of analytical instruments. If the purpose of an instrument is not recognized by the operating and maintenance personnel, then the interest in keeping the sensor calibrated and running fades away quickly. These types of difficulties with the advanced instruments created a lot of suspicion about such equipment and may have caused an unnecessary delay in the application of advanced devices, when improved instruments became available in the market.

It took at least two decades to make the instrumentation wish list from the Clemson workshop in 1974 a reality (Olsson & Newell, 1999; Olsson *et al.*, 2005, 2013). While commercial DO sensors have been used routinely in control systems since the early 1980s (Olsson *et al.*, 1985), DO control was far from being universally utilized in the early 2000s (Ingildsen *et al.*, 2002a; Jeppsson *et al.*, 2002). The instruments required to measure nutrient concentrations were either unavailable or considered too unreliable to be used in practical applications.

Today, sensors are no longer the bottleneck for ICA in practice. The increased confidence in instrumentation is now driven by the fact that clear definitions of performance characteristics and standardized tests for instrumentation have become available. Nutrient analyzers have developed into *in situ* sensors, and luminescent DO sensors that do not need membranes. Knowledge of sensor dynamics is critical since they usually do not give an immediate response and this must be considered in any control system (Rieger *et al.*, 2003).

The most common online instrumentation is shown in Table 13.1. Further developments of instrumentation are described in Rieger *et al.* (2003), Vanrolleghem and Lee (2003), Rieger *et al.* (2008), and Plósz *et al.* (2007).

Table 13.1 Commonly used online instrumentation.

Physical variables	Primary waste water properties	Advanced online sensors
Flow rate	DO	Organic content (COD, TOC)
Pressure level	Turbidity	Respiration rate
Temperature	SS	Ammonia
Conductivity	Sludge concentration	Nitrate/nitrite
pH	Sludge blanket level	Phosphate
ORP	Sludge settling velocity	

There is a huge potential for using *sensor networks*. A network consists of a group of sensors with a communications infrastructure that can monitor variables at diverse locations. Today there are several applications of networks measuring variables such as temperature, rainfall intensity, chemical concentrations and pollutant levels.

The Internet is now ubiquitous and is slowly becoming utilized for remote monitoring in wastewater treatment systems. This possibility was mentioned in Olsson *et al.* (2005) and an actual application of a centralized control system using the Internet to remotely control small decentralized plants in South Korean rural communities is described in Lee *et al.* (2004).

Special attention needs to be paid to measurement quality checking since automation of advanced nutrient removal depends profoundly on on-line instruments. Decisions and control actions based on on-line sensors have to be made automatically and reliably. All signals must be screened and tested. Signal amplitude and rate of change need to be checked. High-frequency noise should be eliminated, outliers detected and missing data compensated for. Most instruments are robust today but they still become fouled or fail during operation, which can have serious consequences if the instrumentation is used in closed loop control. Tools are needed to work around these disturbances and to correct the sensor signals. This also requires safety nets for the control as well as fault-tolerant algorithms and methodologies. Irizar *et al.* (2008) give some examples of advanced signal processing for control loops and ways to estimate the DO uptake rate, the oxygen transfer rate (K_La) as well as the detection of bending points in online signals from sequential batch reactors.

The screened measurement can then be used as a basis for *monitoring* (tracking the process operational state) or directly for *control*. If confidence in a measurement decreases, it might be possible (on a short-term basis) to use an estimated value, but eventually control must be set to a default scheme until confidence in the measurement has been restored.

Many process variables are not directly measureable, but are still interesting to monitor or use as a basis for control. For example, DO uptake rate can be calculated from DO measurements in the aeration tank, using a DO mass balance (Olsson & Newell, 1999). Sludge concentration instruments can be intercalibrated by using sludge mass balances. The sludge blanket level can be estimated from flow rate and sludge concentration measurements in and out of the settler. Effluent total P is sometimes estimated from the effluent turbidity. Flow rates can be calculated from the rate of change of levels. As on-line instrumentation becomes increasingly important in the operation of modern wastewater treatment plants it will be necessary to have diagnostic methods that can indicate sensor deviation at an early stage. At the Rya WWTP, Göteborg, Sweden (Lumley, 2002), soft sensors have been used to verify instrument readings. On-line mass balance calculations are used to estimate a measurement that can be compared with the real sensor signal.

In a sophisticated treatment plant there is a huge data flow from the process and additional instrumentation will provide even more data. Unlike humans, computers are infinitely attentive and can detect abnormal patterns in plant data. Multivariate analysis has been used for many years in the chemical process industry, but was only introduced into the wastewater industry in the late 1990s (Rosen & Olsson, 1998). They have been successful in many applications, but have been much less useful in others. Rosen *et al.* (2003) have analyzed some of the shortcomings of these methods and also present guidance on how to adapt the use of them for wastewater treatment operations. Wastewater treatment data quality and reliability are often poor and present specific challenges. Multivariate methods have often been applied as if the system is static, but in reality the data vary significantly. Principal component analysis (PCA) has been applied in both continuous flow and in SBRs for monitoring (Lee & Vanrolleghem, 2003; Villez *et al.*, 2008; Ruiz *et al.*, 2011) as well as a basis for control of the phase length (Villez *et al.*, 2010).

13.8 CONTROLLABILITY – MANIPULATING THE PROCESS

Since the AS process is hardly ever at steady state with a constant load the process needs to be manipulated in real time to dampen the influence of disturbances. For many years, the obvious method to obtain this was to design the system with sufficient volume that it would attenuate load changes and other disturbances. This is an expensive solution and does not provide sufficient protection against all changes. The process needs 'steering wheels' to adapt it to varying conditions on many time scales. Thus, the process can behave differently in summer and winter, on weekdays and over weekends, and during the day and night. Furthermore some disturbances appear suddenly, such as an industrial spill or a short rain storm. All of these require a flexible operation and adequate 'control handles'.

13.8.1 Control variables

In 1939 Palmer (1939) noted that there were some key variables available for manipulating the AS process. Even in the 1970s there were still relatively few variables available for this purpose, viz.:

- The DO concentration could be controlled to some extent with the aeration rate. However, the compressors had limited controllability and the air flow distribution along the aeration tank could seldom be controlled automatically.
- The RAS was mostly controlled to be proportional to the influent flow rate.
- The waste sludge flow rate was controlled manually.

A nutrient removal plant of today has many more control variables:

- A bioreactor has several zones; anaerobic, anoxic and aerobic. Some of them – the swing zones – can be either aerobic or anoxic.
- Air supply systems are much more sophisticated. Aeration zones can be controlled separately, pressure losses can be minimized by variable pressure control and variable speed compressor control.
- More intermittent systems, such as SBRs are being used and these are more flexible for control.
- Control systems have been developed where a portion of the aerated part of the plant has been used as a settler during high load situations (aerated tank settling operation).
- More recirculation streams are present, for example nitrate recirculation.
- Chemicals can be added for enhanced primary clarification as well as for chemical P removal.
- Volatile fatty acids can be added from the primary settler for enhanced biological P removal.
- External carbon can be added to anoxic zones to enhance denitrification.

13.8.2 Actuators

Today it is widely recognized that the variable speed control of motors can provide smooth pumping and easily adjustable compressors. During and prior to the 1970s, variable speed pumping was a luxury and on/off pumping often resulted in severe disturbances of the secondary settler, causing unnecessary increases of the effluent SS content. For air compressors, guide vane control could be offered, but this provided only limited control authority. As a result, it was difficult to provide accurate DO control. Olsson (1985) pointed out that valves were (and still are) often not designed for good control and the need for smooth control action was questioned.

During the last few decades we have witnessed a power electronics revolution that has made variable speed control both affordable and reliable. Power electronic devices such as IBGT (Insulated-Gate Bipolar Transistors) are now generally available for currents up to 1200 A and voltages up to 3000 V with switching frequencies of more than 1 MHz. DO control is no longer restricted by insufficient control authority of the blowers and pumping can deliver smooth changes of the flow rates.

In some cases it is a great advantage to measure the actuator movements. If the opening of an air valve is measured, then it is possible to control the air pressure according to the most-open-valve control method. This will save energy for aeration (Olsson & Newell, 1999; Åmand *et al.*, 2013). Moreover, the actuator condition can be monitored. By comparing the air or liquid flow rate with the valve opening, it is possible to detect pipe clogging or increased friction in the valve operation.

13.9 DYNAMIC MODELING AND SIMULATION

13.9.1 The importance of dynamics

Because wastewater treatment plants are dynamic systems the result of a corrective action will take some time to produce a response. Therefore the time scales of process changes are important.

Typical time scales in a wastewater treatment plant are:

Fast (min–h):

- Hydraulics and flow dynamics,
- Oxygen mass transfer,
- Chemical precipitation,
- DO concentration dynamics,
- Solids–liquid separation.

Medium fast (h–several h):

- Concentration dynamics,
- Nutrient removal.

Slow (d–months):

- Biomass growth.

The time scale influences the design of the control strategy. One way to express this control task is to supply the right amount of air, or add the correct amount of chemicals, or move the sludge to the right place to match the substrate load, *at the right time.*

Since there is such a wide difference between the *fast* and the *slow* time scales it is often possible to separate the various control actions into different time domains. In the *fast* time scale the variables that change very slowly can be considered constant, for example in DO control the biomass concentration can be considered constant, and; in the *slow* time scale, for example for total sludge inventory control, the DO concentration can be considered to change instantaneously.

It is always important to consider the dynamics when closing the loop. Sometimes controllers are tuned to be too rapid in their response, for example while a DO sensor can show a new DO concentration value every 10 sec, the airflow rate response has a response time of 15–30 min, so that changing the airflow more often than every minute will only produce meaningless control actions and wear out the actuators. A control action frequency of one every 5–12 min is adequate. Thus typically the DO measurement fed to the controller is an averaged (and possibly further filtered) value taken over a 5 to 12 min time period.

SBR systems are purposefully operated in a transient state. The oxidation phase will continue until oxidation is completed then a reduction phase (such as denitrification) will follow and will finish when the reduction is completed.

13.9.2 Modeling

The importance of modeling and simulation was recognized already in the early 1970s. A typical statement of the time was made by Bowden and Wright (1974): 'We accept that variability is of great importance, but being faced with the task of expressing its effect on performance had to conclude that at the present time there are no data on which to assess its effects.' A literature survey (WRC, 1975) lists 12 publications on modeling and simulation of the AS process. A paper by Andrews (1974) reviewed some of the methods available for studying the dynamic behaviour of the AS process, including mathematical models, transient response analysis and techniques for evaluation of process stability. This paper included a bibliography of 49 papers. Jones (1973) presented a comprehensive description of bacterial growth kinetics.

The understanding of nutrient removal kinetics has been manifested in the splendid Activated Sludge Models 1, 2 and 3 (Henze *et al.*, 2000). Several commercial simulation software packages have implemented the ASM models for various reactor configurations together with dynamic models of the settler and clarifier. An early example of a model library was described in Olsson *et al.* (1985). Early simulators for model development were developed, such as ASIM (Gujer & Larsen, 1995) and SSSP (Bidstrup & Grady, 1988). Research at McMaster University in Hamilton, Ontario, Canada led to the commercial package GPS-X from Hydromantis (Patry & Takács, 1990; GPS-X, 2013). Several AS simulator packages have emerged, such as Aquasim (Reichert, 1994; Aquasim, 2013), BioWin (Dold, 1990, 1992; Biowin, 2013), Simba (Simba, 2013), STOAT (Stoat, 2013) and WEST (Vanhooren *et al.*, 2003; West, 2013). Takács, one of the key developers of both GPS-X and Biowin, is now developing the Sumo simulator (Dynamita, 2013). General purpose platforms such as Matlab/Simulink are frequently used for simulation of the AS process.

Some of the simulators can combine a process model with real time online modules, data filtering, sensor fault detection, parameter estimation, model parameter extraction from respirograms, uncertainty analysis, decision support modules and the software to make all these modules work together. Given almost unlimited computing power, we can easily simulate complicated non-linear models, but the challenge is still the verification and validation of the models and the underlying database (Hauduc *et al.*, 2010).

It is important to realize that the ASM models are not meant to be the basis for controller design. Rather they represent detailed descriptions of the way we understand the mechanisms of the biological processes. For control design, one has to identify certain key parameters that are crucial for the operation of the plant such as oxygen uptake rate (OUR), and rates of BOD removal, nitrification or denitrification. Consequently, estimation of dynamic parameters is an important part of the modeling that can form the basis for more advanced control.

13.10 UNIT PROCESS CONTROL

The control of the equipment and physical variables such as flow rates, levels, and pressures is a pre-requisite for 'keeping the plant running'. Variables directly related to the effluent quality must be controlled. An overview of control loops is found in Chapter 6 of Olsson *et al.* (2005):

- DO control with a constant or variable DO setpoint;
- Nitrate recirculation control in a pre-denitrification plant based on nitrate and DO measurements in the aeration tank and in the anoxic zone;
- External carbon dosage;
- RAS control based on sludge blanket measurements in the clarifiers;
- Chemical precipitation control;
- Aeration phase duration control in alternately fed plants based on nutrient sensors;
- Advanced SRT control based on local measurements of effluent ammonia concentration and estimates of nitrification capacity;
- Aeration tank settling (ATS) as a method for temporarily increasing plant capacity under storm conditions (Nielsen *et al.*, 2000; Gernaey *et al.*, 2004).

13.10.1 DO control

DO control is of primary importance and today, the control of the DO to a fixed setpoint is a mature technology. However, even correct DO control cannot compensate for insufficient air supply at extremely high loads. With the development of nutrient sensors in recent years, the primary focus of aeration control has been the on-line adjustment of the DO set point. Aeration control is discussed in detail in Chapter 5 of Olsson *et al.* (2005) while Åmand *et al.* (2013) present a comprehensive review of aeration control after the year 2000.

Many different kinds of DO controllers have been tried and reported over the years. In 1984, experiments were performed on the use of a more complex self-tuning regulator of higher order in a full-scale process (Olsson *et al.*, 1985). The controller parameters were continuously and automatically updated by estimating the measured concentration and the controller output. After some time, the controller parameters converged towards a proportional plus integral (PI) controller performance. In other words, the controller did not need to be more complex than a PI controller to perform well. Despite this long-standing finding, attempts are still published to control the DO concentration with advanced control algorithms that provide little additional benefit in the actual implementation (Olsson, 2012).

Adequate design of the aeration system is the pre-requisite for energy efficient aeration and the two principally important parts are the compressor and the diffuser system. The compressor has to allow for variable air flow rates, which is crucial for any DO control.

The real challenge today, is to decide what DO concentration is required for the process, given disturbances and uncertainties. There are several aspects to this challenge including the legislative framework with which the plant has to comply. For example, with respect to N, if there is only a limit on total N concentration and not on ammonia concentration, then the best strategy – from an energy point of view – is to maximize the ammonia concentration while satisfying the total N. For example, if the effluent criteria are defined as averages over a day or over a week, then it is not crucial to maintain the DO very close to the set point. On the other hand, if the effluent criteria have to be met at all times, then the control has to be stricter. In some countries, there are effluent fees based on the plant performance. Under such a criterion, operational costs can be compared with effluent costs, and the best strategy sought.

A further possibility for energy saving is to allow the *air pressure set point* to be variable. There is room for energy savings by minimizing the air pressure at all times using a 'most open valve' strategy (Olsson & Newell, 1999, Chapter 17; Åmand *et al.*, 2013). Assume that there are two or more parallel aeration tanks and the DO control uses the air flow valves to supply the aeration tank zones with adequate air flow. These valves are most probably not fully opened so by gradually reducing the air pressure, all of the air valves must open up more. The air pressure reduction can be continued until the most open valve is almost fully open. In this way, the pressure drop over the air valves is minimized and energy is saved.

13.10.2 Recycle flow controls

A number of recycle flow rates in an AS process must be controlled:

- *Nitrate recirculation flow control* must recycle sufficient nitrate for reduction in the denitrification zone. If the flow rate is too low, then the denitrification zone is not fully utilized. On the other hand, if the flow rate is too high, then there is not enough capacity in the anoxic zone to denitrify all of the nitrate. There are various ways to determine whether the nitrate has been reduced and denitrification has been completed. Nitrate measurement is the obvious way (Yuan *et al.*, 2002a). The importance of ORP (redox) measurements was recognized early on (Briggs & Jones, 1968). Redox measurements have also been used for denitrification control (Cecil, 2003, 2008).

- *External carbon dosage* is needed if there is insufficient COD in the anoxic zone; various carbon dose control methods have been presented. Carbon dose must always be minimized to save costs. Lindberg and Carlsson (1996) presented an adaptive carbon flow rate controller for a pre-denitrifying pilot-scale plant. Yuan *et al.* (1997) achieved similar results on a full-scale wastewater treatment plant. Ingildsen (2002, Section 8.3) proposed a control structure for external carbon dosage control. The control of external carbon should be coordinated with the internal nitrate recirculation. Yuan and Keller (2004) studied the integrated control of nitrate recirculation and external carbon addition and proposed a structure with four feedback control loops. They manipulated the nitrate recirculation and the carbon dosage flows in a highly coordinated manner so that external carbon consumption was minimized while the nitrate effluent concentration was met.
- *RAS flow* rate can seldom be used for control on an hourly basis. There are two common practices for the RAS flow rate, constant flow rate or ratio control, that is the ratio between the RAS flow rate and the influent flow rate is kept constant. Neither of these control methods calculates how the aeration tank SS concentration varies or how the AS settling conditions change. Olsson and Newell (1999), in their detailed discussion of RAS control, indicate the many constraints that have to be considered. The settler sludge blanket level needs to be controlled dynamically; the MLVSS concentration in the aeration tank should be maintained at a level consistent with the target SRT. RAS flows differ during dry weather and wet weather. N loading is another important factor that determines how to operate the sludge blanket. Yuan *et al.* (2002b) describe a two-level controller that takes all the various aspects of the sludge blanket control into consideration.

13.10.3 Hedging point strategies

The typical criterion for comparing various control strategies includes a weighted sum of the cost of operation and effluent quality. Since all control is related to risk, large safety margins mean less efficient and more costly operation. In DO control, the DO set point can be calculated on-line based on ammonia measurements. If the effluent requirement is based on the total N with no specific limitation of ammonia N, then the most cost-effective control is to maximize the ammonia N concentration. This strategy will save aeration energy but may lead to a smaller nitrifier population since ammonia removal is not maximized. This will make the plant more vulnerable to disturbances, such as increased ammonia load, lower temperatures or the presence of inhibitory compounds. This has led to the development of hedging point strategies (Ingildsen *et al.*, 2002b). By defining the 'safety margins' of control the sensitivity for future disturbances can be decreased.

Knowledge-based representation techniques also complement standard deterministic models for the risk assessment of microbial-related operational problems (e.g. filamentous bulking). These issues cannot be described with standard deterministic models due to the lack of fundamental knowledge precisely describing the mechanisms for the phenomenon, for example the growth or death of filamentous organisms with relationship to the plant operational parameters. In some of these cases, only cause–effect relationships are known (Comas *et al.*, 2008).

13.10.4 Chemical precipitation control

The most important aspect of chemical precipitation is to achieve rapid and complete mixing. Once the chemicals are mixed with the wastewater, the dynamics are very fast and responses occur in a matter of seconds. Chemical precipitation is either used as the only mechanism for P removal or as a complement to the BioP removal. In any case, the P concentration towards the outlet is an indicator of whether the removal mechanism has worked or not.

A number of process schemes have been developed for chemical dosage. Post-precipitation was commonly used in the 1970s and pre-precipitation was shown to be an alternative. Apart from removing P, more organic matter is also removed by pre-precipitation in the primary clarifier. This in turn decreases the load on the biological treatment step and can theoretically decrease the amount of energy needed for aeration while increasing the gas production in the digester. A combination of pre-precipitation and post-precipitation has been common used. Most of the P is removed by the pre-precipitation and simultaneous precipitation, while the post-precipitation serves as a polishing stage.

Traditionally, the simplest dosage controllers have been based on flow rate and turbidity measurements but in the early 1980s another approach was tried at the Himmerfjärden, Sweden, treatment plant (Hellström *et al.*, 1984). Influent flow rate measurements were supplemented with information on the historic records of daily and hourly variations of P concentration. The dosage was adjusted for wet weather flows. With an on-line *in situ* P sensor in place, excellent control performance and significant chemical savings was achieved using a simple feedback controller (Ingildsen, 2002). The sensor was located in the effluent of the flocculation chamber. Also Devisscher *et al.* (2002) reported successful control of the chemical dosage for P removal using on-line P measurements. *In-situ* P sensors used nowadays need to be more robust for field use.

13.11 FROM UNIT PROCESS TO PLANT-WIDE

Individual unit processes cannot be controlled as separate and isolated units. The sequential influence from the sewer all the way through the primary treatment, aeration tanks and settlers on the effluent is widely recognized. The numerous recycle flows further increase the complexity of the situation. The return sludge flow rate creates close coupling between the aeration tank and the settler. The nitrate recycle requires careful DO control. At the outlet of the aeration tank the DO concentration should be sufficiently high to achieve nitrification, but at the same time the concentration low enough that the anoxic reactions in denitrification are not hindered. Similarly, the recycle of the supernatant from the anaerobic digester to the wastewater influent should be synchronized with the influent load.

To maximize the efficiency, for example by minimizing the electrical power requirement, means that the couplings between the various unit processes have to be considered. As stated by Kukudis (1973): 'We must speak of automation in the entire system – the network of sewers and the plants'. Sewer control and flow equalization were used in the early 1970s (Kukudis, 1973). The presence of recycles makes it necessary to consider the overall effects of the control strategies (Olsson & Newell, 1999), much like the hedging point strategies mentioned above.

The ultimate goal of system-wide control is the protection of receiving water quality. This objective was first formulated by Young and Beck (1974) and later expressed by Harremoës *et al.* (1993) and should still be a guiding principle for control: 'Wastewater management must be looked at in its totality and in close combination with the processes and quality aspects of the receiving waters. The system from the sink … to the ultimate consequential water quality in the environment has to be regarded as an entity.' System-wide control is still a topic for advanced research almost forty years after these ideas were first expressed.

Knowledge-based systems and other artificial intelligence techniques have been applied to systematically make use of heuristics, experience of practitioners and existing databases (Rodríguez-Roda *et al.*, 2002). Another possibility to support the operator in decision making is to use data mining techniques for knowledge extraction from a historical database containing the disturbances and control actions and to match patterns to recognize the shape of the sensor profiles (Kim *et al.*, 2012).

The complexity of urban water systems requires an integrated perspective. The performance indices have to include not only technical, environmental and economic aspects but, though more difficult to deal with, social aspects, for scenario assessment. This demands comprehensive understanding of Life Cycle Analysis.

13.12 CONCLUSIONS

Process control is a necessary part of any AS system, from the individual equipment operation to the plant-wide perspective. The need for instrumentation, control and automation of water and wastewater systems is pressing. A recent study (ARC, 2013) concluded that the water and wastewater industry represents one of the greatest opportunities for the automation business in the next twenty years.

As mentioned in the introduction, the lack of understanding of control and instrumentation was recognized as early as 1974 by John Andrews. Today the need for ICA is seldom questioned. However, it is essential that the control challenges in wastewater treatment should engage not only water professionals but also specialists from control engineering, electrical, mechanical and chemical engineering, as well as professionals having a comprehensive knowledge of environmental issues (Hug *et al.*, 2009). The generation shift among plant operators and engineers in many countries provides a great opportunity to realize this approach. The new generation joining the water industry may have less practical process experience but generally have much more experience and interest in computers.

The purpose of a good process control system is to make complexity simpler. Education and training are important components and some aspects should be particularly highlighted:

- Control and operation must be an integral part of the design process;
- Many new sensors and instruments are being developed and the best use of them should be evaluated. This also includes adequate maintenance plans and on-site and standard operation procedures;
- Maximum use should be made of the enormous computing and storage capacity in real time computers in modern industrial control systems;
- Data validation tools and monitoring, detection and diagnosis methods should be used to obtain the maximum use of the information from available measurements;
- Appropriate process control methods should be employed.

13.13 REFERENCES

Åmand L., Olsson G. and Carlsson B. (2013). Aeration control – a review. *Water Sci. Technol.*, **67**(11), 2374–2398.

Andrews, J. F. (1974). Dynamic models and control strategies for waste-water treatment processes. *Water Res.*, **8**(5), 261–289.

Andrews J. F., Briggs, R., and Jenkins S. H. (1974) Instrumentation, Control and Automation for Wastewater Treatment Systems. Pergamon Press, Oxford, UK. Also in *Prog. Water Technol.*, **6**(5–6), IAWPR (Intl Assoc. Water Pollut. Res.), London.

Aquasim (2013). www.eawag.ch/forschung/siam/software/aquasim/index (Retrieved 2 August 2013).

ARC (2013). http://www.arcweb.com/market-studies/pages/automation-expenditures-for-water-wastewater.aspx (Retrieved 22 June 2013).

Ardern E. and Lockett W. T. (1914). Experiments on oxidation of sewage without the aid of filters. *J. Soc. Chem. Ind.*, **33**, 523.

Babcock R. H. (1974). Commercially available proven measuring and control systems – a critical review of U. S. practice. In: Andrews *et al.* (1974).

Ballotti E. F., Guarino C. F., Edwards A. B. and Nelson M. D. (1974). Activated sludge wastewater treatment plant control by instrumentation and computer. In: Andrews *et al.* (1974).

Baur R., Le Gauffre P., Saegrov S. (2003). Multi-criteria decision support for annual rehabilitation programmes in drinking water networks. *Water Supply,* **3**(1–2), 43–50.

Bertrand-Krajewski J.-L., Barraud S., Le Gauffre P., and Baer E. (2007). Infiltration in sewer systems: multi-criteria comparison of investment/rehabilitation strategies. *Water Practice Technol.*, **2**(1), IWA Publishing, doi:10.2166/wpt.2007.027.

Bidstrup S. M. and Grady C. P. L. Jr. (1988). SSSP-simulation of single-sludge processes. *J. Water Pollut. Control Fedn.*, **60**, 351–361.

Biowin (2013). www.environmental-expert.com/software/biowin-simulator-for-municipal-wastewater-treat-ment- systems-17197 (Retrieved 15 June 2013).

Bowden K. and Wright D. E. (1974). The CIRIA optimization study of sewage treatment: its significance for the control of treatment works. In: Andrews *et al.* (1974).

Briggs R. (1973). Instrumentation and control in sewage treatment. In: Andrews *et al.* (1974).

Briggs R. and Jones K. (1968). The significance of redox potential measurements in sewage treatment. The 1968 Effluent & Water Treatment Manual, Water Pollution Research Laboratory, Stevenage, UK.

Briggs R., Knowles G. and Scragg L. J. (1954). A continuous recorder to dissolved oxygen in water. *Analyst,* **79,** 744.

Briggs R., Davies F. S., Dyke G. V. and Knowles G. (1957). Use of wide-bore dropping-mercury electrode and zinc reference electrode for continuous polarography. *Chem. Ind.*, Part 8, 223–224.

Briggs R., Jones K. and Oaten A. B. (1967). Monitoring and automatic control of dissolved oxygen level in activated sludge plants. Effluent and Water Treatment Convention, Thunderbird Enterprises, London, UK, 11.

Brouzes P. (1969). Automated activated sludge plants with respiratory metabolism control. In: Adv. Water Pollut., S. H. Jenkins (ed.), Pergamon Press, New York.

Buhr H. O., Andrews J. F. and Keinath T. M. (1974). Research needs for automation of wastewater treatment systems. In: Proc. USEPA Workshop, Clemson Univ., SC, USA, September 23–25.

Busby J. B. and Andrews J. F. (1975). Dynamic modelling and control strategies for the activated sludge process. *J. Water Pollut. Control Fedn.,* **47**(5), 1055–1080.

Cecil D. (2003). Controlling nitrogen removal using redox and ammonium sensors. *Water Sci. Technol.*, **47**(11), 109–114.

Cecil D. (2008). The control of denitrification time in full scale by the automatic detection of the low nitrate bend in the redox curve. *Water Sci. Technol.*, **57**(7), 1095–1101.

Collier J. R. (1939). Activated sludge sewage treatment, *Ohio Conf. Sewage Treatment*, **13**, 124–126.

Comas J., Rodríguez-Roda I., Gernaey K. V., Rosen C., Jeppsson U. and Poch M. (2008). Risk assessment modelling of microbiology-related solids separation problems in activated sludge systems. *Environ. Model. Software,* **23**(10–11), 1250–1261.

Devisscher M., Bogaert H., Bixio D., Van de Velde J. and Thoeye C. (2002). Feasibility of automatic chemicals dosage control – a full-scale evaluation. *Water Sci. Technol.*, **45**(4–5), 445–452.

Dobbins W. E. (1974). Design of activated sludge plants for computer control. In: Andrews *et al.* (1974).

Dold P. L. (1990). A general activated sludge model incorporating biological excess phosphorus removal. In: Proc. Canadian Soc. Civil Eng. Ann. Conf., Hamilton, ONT, Canada.

Dold P. L. (1992). Activated Sludge System Model Incorporating Biological Nutrient (N & P) removal. Tech. Rept, Dept. of Civil Eng. & Eng. Mech., McMaster Univ., Hamilton, Ontario, Canada.

Dynamita (2013). http://www.dynamita.com (Retrieved 9 September 2013).

Flores-Alsina X., Sin G., Rodriguez-Roda I. and Gernaey K. V. (2008). Multicriteria evaluation of wastewater treatment plant control strategies under uncertainty. *Water Res.*, **42**(17), 4485–4497.

Gernaey K. V., Nielsen M. K., Thornberg D., Höök B., Munk-Nielsen T., Ingildsen P. and Jørgensen S. B. (2004). Conservation principles suspended solids distribution modeling to support ATS introduction on a recirculating WWTP. *Water Sci. Technol.*, **50**(11), 179–188.

Gillblad T. and Olsson G. (1977). Computer control of a medium sized activated sludge plant. *Prog. Water and Technol.*, **9**(5–6), 427–433. Pergamon Press, Oxford, UK.

GPS-X (2013). http://www.hydromantis.com/GPS-X.html (Retrieved 2 August 2013)

Guarino C. F. and Radziul J. V. (1977). Water–wastewater, instrumentation and automation (I & A), USA. *Prog. Water Technol.*, **9**(5–6), 35–39.

Gujer W. and Larsen T. A. (1995). The implementation of biokinetics and conservation principles in ASIM. *Water Sci. Technol.*, **31**(2), 257–266.

Harremoës P., Capodaglio A. G., Hellström B. G., Henze M., Jensen K. N., Lynggaard-Jensen A., Otterpohl R. and Soeberg H. (1993). Wastewater treatment plants under transient loading-performance, modeling and control. *Water. Sci. Technol.*, **27**(12), 71–115.

Haseltine T. R. (1937). Operating control tests for the activated sludge process. *Waterworks and Sewerage,* **84**, 121–123.

Hauduc H., Rieger L., Takács I., Héduit A., Vanrolleghem P. A. and Gillot S. (2010). Systematic approach for model verification–application on seven published activated sludge models. *Water Sci. Technol.*, **61**(4), 825–839.

Hellström B. G., Vopatek P. and Österman, A. (1984). Ferrous sulphate – dissolution tanks and a computer application for controling the dosage at the Himmerfjärden sewage treatment plant. *Vatten*, **40**, 40–45.

Henze M., Gujer W., Mino T. and van Loosdrecht M. C. M. (2000). Activated Sludge Models ASM1, ASM2, ASM2d and ASM3. IWA Sci. Technol. Rept No. 9, IWA Publ., London, UK.

Heukelekian H. (1934). Some biochemical indices of the condition of activated sludge. *N.J. Sewage Works Assoc. Proc.*, **19**, 50–55.

Hug T., Benedetti L., Hall, E. R., Johnson B. R., Morgenroth E. F., Nopens I., Rieger L., Shaw A. R. and Vanrolleghem P. A. (2009). Mathematical models in teaching and training: mismatch between education and requirements for jobs. *Water Sci.Technol.*, **59**(4), 745–753.

IEC (2013). International Electrotechnical Commission, IEC 61131 standard for programmable logic controllers (PLCs). www. http://www.iec.ch (Retrieved 15 February 2014)

Ingildsen P. (2002). Realising Full-Scale Control in Wastewater Treatment Systems Using *in situ* Nutrient Sensors. PhD thesis, Dept. of Ind. Electrical Engineering and Automation (IEA), Lund Univ., Lund, available at www.iea.lth.se. (Retrieved 26 February 2014)

Ingildsen P., Lant P. and Olsson G. (2002a). Benchmarking plant operation and instrumentation, control and automation in the wastewater industry. *Water Sci. Technol., Water Supply,* **2**(4), 163–171.

Ingildsen P., Olsson G. and Yuan Z. (2002b). A hedging point strategy – balancing effluent quality, economy and robustness in the control of wastewater treatment. *Water Sci. Technol.*, **45**(4–5), 317–32.

Irizar I., Alferes J., Larrea L., and Ayesa E. (2008). Standard signal processing using enriched sensor information for WWTP monitoring and control. *Water Sci. Technol.*, **57**(7), 1053–1060.

Jeppsson U., Alex J., Pons M. N., Spanjers H. and Vanrolleghem P. A. (2002). Status and future trends of ICA in wastewater treatment – a European perspective. *Water Sci. Technol.*, **45**(4–5), 485–494.

Jones G. L. (1973). Bacterial growth kinetics: measurement and significance in the activated sludge process. *Water Res.*, **7**, 1475–1492.

Jones K., Briggs R., Carr J. G. and Potter A. H. (1969). Automatic control of aeration in a fully nitrifying activated-sludge plant. *Inst. Pub. Hlth Engr. J.*, **68**(4), 271–295.

Kim Y., Kim H. S., Kim M. S., Piao W., Kang D. and Kim C. (2012). Knowledge extraction from the historical database of wastewater treatment plant operation using various data mining techniques. Proc. New Developments in IT & Water, November 4–6, Amsterdam, The Netherlands.

Kukudis R. (1973). Problems involved in automating the wastewater treatment plant. In: Andrews et al. (1974).

Larsen T. A., Udert K. M. and Lienert J. (eds) (2013). Source Separation and Decentralization for Wastewater Management. IWA Publishing, London, ISBN: 9781843393481.

Lee D. S. and Vanrolleghem P. A. (2003). Monitoring of a sequencing batch reactor using adaptive multiblock principal component analysis. *Biotechnol. Bioeng.*, **82**, 489–497.

Lee H., Min Y. M., Park C. H. and Park Y. H. (2004). Automatic control and remote monitoring system for biological nutrient removal on small wastewater treatment plants in Korea. *Water Sci. Technol.*, **50**(6), 199–206.

Lindberg C.-F. and Carlsson B. (1996). Adaptive control of external carbon flow rate in an activated sludge process. *Water Sci. and Technol.*, **34**(3–4), 173–180.

Lumley D. (2002). On-line instrument confirmation: how can we check that our instruments are working? *Wat. Sci. Technol.*, **45**(4–5), 469–476.

Nielsen M. K., Bechmann H., and Henze M. (2000). Modeling and test of aeration tank settling (ATS). *Water Sci. Technol.*, **41**(9), 179–184.

Nowak O., Keil S. and Fimml C. (2011). Examples of energy self-sufficient municipal nutrient removal plants. *Water Sci. Technol.*, **64**(1), 1–6.

Olsson G. (1985). Dynamics and control of wastewater treatment systems. Final address, 4th IAWPRC Workshop on Instrumentation and Control of Water and Wastewater Treatment and Transport Systems, Denver, CO, USA, May 3.

Olsson G. (2002). Lessons learnt at ICA 2001. *Water Sci. Technol.*, **45**(4–5), 1–8.

Olsson G. (2012). ICA and me – a subjective review. *Water Res.*, **46**(6), 1585–1624, Available online at doi:10.1016/j.watres.2011.12.054.

Olsson G. (2013). The potential of control and monitoring. Chapter 12 in Larsen *et al.* (2013).

Olsson G. and Andrews J. F. (1978). The dissolved oxygen profile – a valuable tool for the control of the activated sludge process. *Water Res.*, **12**, 985–1004.

Olsson G. and Hansson O. (1976). Stochastic modeling and computer control of a full scale wastewater treatment plant. Proc. Symp. Systems and Models in Air and Water Pollut., Inst. Measurement and Control, London, England.

Olsson G. and Newell B. (1999). Wastewater Treatment Systems. Modeling, Diagnosis and Control. IWA Publishing, London.

Olsson G., Rundqwist L., Eriksson L. and Hall L. (1985). Self-tuning control of the dissolved oxygen concentration in activated sludge systems. In: Adv. Water Pollut. Control, R. A. R. Drake (ed.), Intl. Assoc. Water Pollut. Res. and Control, London, UK, pp. 473–480.

Olsson G., Nielsen M. K., Yuan Z., Lynggaard-Jensen A. and Steyer J. P. (2005). Instrumentation, Control and Automation in Wastewater Treatment Systems. Sci. Tech. Rept No. 15, IWA Publishing, London, UK.

Olsson G., Carlsson B., Comas J., Copp J., Gernaey K. V., Ingildsen P., Jeppsson U., Kim C., Rieger L., Rodríguez-Roda I., Steyer J.-P., Takács I., Vanrolleghem P. A., Vargas Casillas A., Yuan Z. and Åmand L. (2013). Instrumentation, Control and Automation in Wastewater–from London 1973 to Narbonne 2013. IWA Conference on Instrumentation, Control and Automation (ICA-2013), Narbonne, France.

Palmer J. R. (1939). Activated sludge control. *Ohio Conf. Sewage Treat.*, **13**, 21.

Patry G. G., and Takács I. (1990). Modular/multi-purpose modelling system for the simulation and control of wastewater treatment plants: an innovative approach. In: R. Briggs (ed.), Adv. Water Pollut. Control, Pergamon Press, Oxford, UK, pp. 385–392.

Pearse L. (ed.) (1941). The operation and control of activated sludge sewage treatment works. Report of Committee on Sewage Disposal. American Public Health Association (APHA), Atlantic City meeting of the APHA, October 14. *Sewage Works J.*, **14**(1), 3–69.

Petersack J. R. and Stepner D. E. (1973). Computerized data management and control of a secondary wastewater treatment plant. In: Andrews et al. (1974).

Plósz B., Weiss M., Printemps C., Essemiani K. and Meinhold J. (2007). One-dimensional modeling of the secondary clarifier-factors affecting simulation in the clarification zone and the assessment of the thickening flow dependence. *Water Res.*, **41**(15), 3359–3371.

Reichert P. (1994). Aquasim – a tool for simulation and data analysis of aquatic systems. Water Quality International 94, IAWQ 17th Biennial Int. Conf., **2**, 21–30.

Rieger L., Alex J., Winkler S., Boehle, M., Thomann M. and Siegrist H. (2003). Progress in sensor technology–progress in process control? Part I: Sensor property investigation and classification. *Water Sci. Technol.*, **47**(2), 103–112.

Rieger L., Vanrolleghem P. A., Langergraber G., Kaelin D. and Siegrist H. (2008). Long-term evaluation of a spectral sensor for nitrite and nitrate. *Water Sci. Technol.*, **57**(10), 1563–1569.

Rieger L. and Olsson G. (2012). Why many control systems fail. *Water Environ. Technol., Water Environ. Fedn.*, June 43–45.

Rodríguez-Roda I., Sànchez-Marrè M., Comas J., Baeza J., Colprim J., Lafuente J., Cortés U. and Poch M. (2002). A hybrid supervisory system to support WWTP operation: implementation and validation. *Water Sci. Technol.*, **45**(4–5), 289–297.

Roesler J. F. (1974). Plant performance using dissolved oxygen control. *J. Environ. Eng. Div., Amer. Soc. Civil Eng.*, **100**, *EE5*, 1069–1076.

Rosen C. and Olsson G. (1998). Disturbance detection in wastewater treatment systems. *Water Sci. Technol.*, **37**(12), 197–205.

Rosen C., Röttorp J. and Jeppsson U. (2003). Multivariate on-line monitoring: challenges and solutions for modern wastewater treatment operation. *Water Sci. Technol.*, **47**(2), 171–179.

Ruiz M., Sin G., Berjaga X., Colprim J., Puig S. and Colomer J. (2011). Multivariate principal component analysis and case based reasoning for monitoring, fault detection and diagnosis in a WWTP. *Water Sci. Technol.*, **64**(8), 1661–1667.

Simba (2013). www.ifak-system.com/en/environmental-simulation/simba/ (Retrieved 2 August 2013)

Stoat (2013). www.wateronline.com/doc/STOAT-ASM-Simulation-Model-Of-Wastewater-Trea-0001 (Retrieved 2 August 2013)

Torpey W. N. (1948). Practical results of step aeration. *Sewage Works J.*, **20**, 781.

Urrutiaguer M., Lloyd S. and Lamshed S. (2010). Determining water sensitive urban design project benefits using a multi-criteria assessment tool. *Water Sci. Technol.*, **61**(9), 2333–2341.

Vanhooren H., Meirlaen, J., Amerlinck Y., Claeys F., Vangheluwe H. and Vanrolleghem P. A. (2003). WEST: Modeling biological wastewater treatment. *J. Hydroinformatics*, **5**, 27–50.

Vanrolleghem P. A. and Lee D. S. (2003). On-line monitoring equipment for wastewater treatment processes: state of the art. *Water Sci. Technol.*, **47**(2), 1–34.

Villez K., Ruiz M., Sin G., Colomer J., Rosen C. and Vanrolleghem P. A. (2008). Combining multiway principal component analysis (MPCA) and clustering for efficient data mining of historical data sets of SBR processes. *Water Sci. Technol.*, **57**(10), 1659–1666.

Villez K., Rosen C., D'hooge E. and Vanrolleghem P. A. (2010). Online phase length optimization for a sequencing batch reactor by means of the Hotelling's T2 statistic. *Ind. Eng. Chem. Res.*, **49**, 180–188.

West (2013). www.dhisoftware.com/Products/Cities/WEST.aspx. (Retrieved 2 August 2013).

WRC (1975). Instrumentation and control in waste water treatment. A literature survey covering the period 1971 to 1975. Tech. Rept, Water Res. Centre, Stevenage Laboratory, UK, September 1975.

Yuan Z. and Keller J. (2004). Integrated control of nitrate recirculation and external carbon addition in a predenitrification system. *Water Sci. Technol.*, **48**(11), 345–354.

Yuan Z., Bogaert H., Vanrolleghem P. A., Theoye C., Vansteenkiste G. and Verstraete W. (1997). Control of external carbon addition to pre-denitrification systems. *J. Environ. Eng.*, **123**(11), 1080–1086.

Yuan Z., Oehmen A. and Ingildsen, P. (2002a). Control of nitrate recirculation flow in predenitrification systems. *Water Sci. Technol.*, **45**(4–5), 29–36.

Yuan Z., Bogaert H., Rosen C. and Verstraete W. (2002b). Sludge blanket height control in secondary clarifiers. Proc. 1st IWA Conf. on Instrumentation, Control and Automation, Malmö, Sweden, June 2001, 81–88. Published in *Water Intelligence Online,* ISSN online: 1476–1777, IWA Publishing.

Young P. and Beck B. (1974). The modelling and control of river quality in a river system. *Automatica*, **10**, 455–468.

Chapter 14

Modeling

George A. Ekama (South Africa), Imre Takács (France)

14.1 INTRODUCTION

Modeling the AS system really started in the 1950s after the years of legal controversy (1915–1935) were settled and several technical issues were resolved. Alleman and Prakasam (1983) have discussed important lawsuits that led to some plants in the US being shut down and have cited the chemical vs. biological and diffused air vs. mechanical aeration debates as major technical issues that needed resolution for widespread application of the system. Sawyer (1965) listed 12 important milestones that advanced the application of AS, including the resolution of whether the process was chemical or biological, the development of the Kraus process (1945) to overcome bulking problems, the development of tapered aeration, step aeration/feeding, modified aeration, high rate, contact-stabilization systems, which led to the definition of SRT and the development of extended aeration, aerated lagoons and oxidation ditch plants. None of these milestones advanced mathematical modeling of the AS system in a generalized way.

Mathematical modeling of the AS system in the generalized way that we know it today required resolution of the following six fundamental issues:

(1) Characterization and measurement of influent wastewater.
(2) Finding a consistent measure that allowed mass balances to be made over the system.
(3) Integration of biomass gains and losses and oxygen utilization in the bioprocesses.
(4) Elimination of system specific sludge production and oxygen utilization parameters.

(5) Integration of denitrification with the aerobic bioprocesses.
(6) System operation and control.

These six issues are interwoven and the research groups that spearheaded the development of both steady state and dynamic mathematical AS system models, touched on several at the same time in their research, advancing some issues further than others. However, the inter-relatedness of the issues demanded that all six needed to be identified, unraveled from the each other and resolved before re-integration into generalized steady state and dynamic models could be achieved. The six issues were largely resolved between 1950 and 1980, laying the foundation for the development of ASM1 (Henze *et al.*, 1987). The modeling advances that led to the resolution of these 6 issues are presented in the first part of this paper. The second part of the paper presents the developments in AS modeling since publication of ASM1.

14.2 FUNDAMENTALS

14.2.1 Growth – Monod kinetics

All AS process models have their roots in pure culture kinetic studies. The definitive initiation of this was from batch studies on soluble substrates (S_B) by Monod (1942) from which the following two important relationships for ordinary heterotrophic microorganism (X_{OHO}) growth were proposed:

(1) The biomass generated ($\Delta M_{X_{OHO}}$,mg) is proportional to the mass of substrate utilized (ΔM_{S_B}, mg), that is

$$\Delta M_{X_{OHO}} = Y_{OHO} * \Delta M_{S_B} \tag{14.1}$$

where Y_{OHO}, the specific yield coefficient, is constant for a particular substrate and organism, and

(2) The specific growth rate, $\mu_{OHO} = dX_{OHO}/(dt\ X_{OHO})$ is linked to the concentration of the growth limiting substrate, S_B, surrounding the organism (X_{OHO}) by the (now famous) Monod expression, that is

$$\mu_{OHO} = dX_{OHO}/dt * 1/X_{OHO} = \mu_{OHO,max} S_B/(K_S + S_B) \tag{14.2}$$

where $\mu_{OHO,max}$ is the maximum specific growth rate and K_S = half saturation constant, the concentration of S_B at which $\mu_{OHO} = \mu_{OHO,max}/2$.

Equation 14.2 is empirically analogous to the Michaelis–Menten (1913) enzymatic rate equation and Monod suggested that this was because μ_{OHO} was controlled by enzymatic reactions. If K_M is defined as $\mu_{OHO,max}/Y_H$, then Equations 14.1 and 14.2 can be written in terms of dS_B/dt, which is the form that Lawrence and McCarty (1970) used to develop their basic process equations for AS. Today, the Monod equation is used in numerous ways in bioprocess models.

The kinetic and stoichiometric constants Y_{OHO}, $\mu_{OHO,max}$ and K_S were obtained from experimental measurements in continuous flow chemostats. Both Monod (1950) and Novick and Szilard (1950) derived the kinetic equations for chemostat effluent biomass (X_{OHO}) and substrate (S_B) concentrations which constrain the growth equations with the dilution rate (D = 1/HRT or = 1/SRT:

$$dX_{OHO}/dt = (\mu_{OHO,max} * S_B/(K_S + S_B) - D)X_{OHO} \tag{14.3}$$

$$dS_B/dt = D(S_{B,in} - S_B) - X_{OHO}/Y_{OHO} * \mu_{OHO,max} * S_B/(K_S + S_B)) \tag{14.4}$$

At steady state, that is constant influent substrate ($S_{B,in}$), flow (Q), volume (V), the biomass (X_{OHO}) and substrate (S_B) concentrations exiting the chemostat reduce to

$$X_{OHO} = Y_{OHO} * (S_B, \text{in-}S_B) \quad \text{and} \quad S_B = K_S/(\mu_{OHO,max}/D - 1) \tag{14.5}$$

The Monod approach was extensively investigated using single and multiple soluble substrates and organism types. These indicated numerous behavioral responses that deviated from those described by the Monod relationship (e.g. diauxic, sequential utilization of substrates, internal substrate storage, cryptic growth, growth rate hysteresis, byproduct formation). These deviations were evident at high dilution rates (short SRTs) but at low dilution rates (long SRTs) they were insignificant. One further observation of relevance to the AS process is that, at SRTs greater than about 7 h, if a multiple component substrate is fed to an organism mass, the substrate components are not utilized sequentially, but simultaneously, although the rates of utilization of the individual components may differ (Harder & Dijkhuizen, 1976). The influence of byproduct formation on the Monod relationship is not yet clear because even though the chemical composition of the effluent organics is not the same as that of the influent, this concentration often is utilized to calibrate the Monod rate equation. The saving grace here is that normally in aerobic studies on the AS process, the soluble substrates are utilized at such high rates that their filtered effluent concentrations are so low that usually it is not necessary to know K_M and K_S to high degrees of accuracy, provided their magnitudes are appropriate to the behavior observed. These remarks only apply to soluble readily biodegradable substrates; colloidal and some chemically complex substrates do not behave according to the Monod rate equation; these will be discussed in the 'extended models' section of this paper.

14.2.2 Reduced yield

At long SRTs, an effect not reflected in the Monod equations became apparent – the specific yield (Y_{OHO}) decreased. Two hypotheses were advanced to explain this behavior: endogenous metabolism (Herbert, 1958); and maintenance energy (Pirt, 1965). Herbert assigns the substrate utilization totally to active biomass generation and thereafter organism biomass loss occurs with time due to oxidation of the metabolic material. Pirt assigns the substrate first to maintenance of essential organism functions with the fraction assigned to essential functions depending on the mass of organisms already present, and with the balance being assigned to growth. Both concepts were proposed on the basis of what the respective authors surmised as a logical explanation for the observed reduced yield. Each enables the observed biomass concentration to be more accurately predicted than by Monod's hypotheses. However, both approaches essentially are still lumped parameters and they find acceptance mainly through the fact that they each mimic the observed behavior satisfactorily within the restrictive range of conditions to which they have been applied.

Including Herbert's endogenous mass loss, Monod's net biomass growth equation gives

$$dX_{OHO}/dt = (\mu_{OHO} - b_{OHO})X_{OHO} = (\mu_{HOHO,max} * S_B/(K_S + S_B) - b_{OHO})X_{OHO} \quad (14.6)$$

Herbert's endogenous biomass loss model was universally adopted into AS models. This was partly because it separated the growth and endogenous processes, which made developing AS models easier, but mainly because it could be easily modified to resolve a second associated deviation from Monod's equations – inert endogenous residue generation.

14.2.3 Yield coefficient and endogenous respiration rate

Between 1952 and 1955, the Porges, Jasewicz and Hoover group applied mass balanced stoichiometry to the growth and endogenous processes to assist in interpreting their experiments with skim milk powder (a mixture of lactose and casein) as substrate (Porges *et al.*, 1955). Without using the words 'electron donation', 'anabolism' and 'catabolism' that are in use today and using the 'total oxygen demand' (actually the COD) to define the concentration of substrate, Hoover and Porges (1952) made the important observation that some of the skim milk COD was conserved as biomass (anabolism) and the balance was the oxygen utilized (catabolism), thereby balancing the COD over the growth process. They measured the composition of the biomass formed and determined the classic $C_5H_7O_2N$ composition. They also found that the biomass had an ash content (ISS) of 0.10 mgISS/mgVSS. From their batch test data, they found a yield coefficient $Y_{OHO} = 0.44$ gVSS/gCOD and a biomass COD/VSS ratio (i_{CV}) = 1.42 gCOD/gVSS, which meant that of the substrate COD, 63% ($i_{CV}Y_{OHO}$) was conserved as biomass and 37% ($1 - i_{CV} * Y_{OHO}$) was oxidized by oxygen.

From the measured composition of the biomass, Porges *et al.* (1955) also applied stoichiometry and a COD mass balance to the endogenous process. From oxygen consumption rate data, they concluded that the endogenous respiration rate was variable, but that at a rate of 14.3 $mgO_2/(gVSS * h)$, the endogenous mass loss takes place at 1%/h, or 0.24/d.

The Porges, Jasewicz and Hoover group made an early and major contribution to AS modeling with concepts that have stood the test of time, and measured constants remarkably close to those used in ASMs today. However, their assumption that the biomass was completely biodegradable, from which they concluded that if the correct SRT could be found there would be no net biomass production, needed to be revised.

14.2.4 Inert endogenous residue generation

Symons and McKinney (1958) showed that total oxidation of biomass does not take place, and that a small inert organic residue remains. Kountz and Fourney (1959) endeavored to quantify the production of inert material and came to the conclusion that 20 to 25% of the active organism mass generated is nonbiodegradable inert organic material. Washington and Symons (1962) found that in fill-and-draw experiments where the same amount of substrate was fed each day and where the sludge mass was retained, a constant rate of increase in the sludge mass was observed. They confirmed Kountz and Fourney's result that about 20% of the active mass generated was unbiodegradable. Substantially the same conclusions were formed by McCarty and Brodersen (1962) and Washington and Hetling (1965).

McKinney (1962) and McKinney and Ooten (1969) were the first to incorporate explicitly endogenous residue generation (X_E) into an AS model, as follows:

$$dX_{OHO}/dt = -b_{OHO}X_{OHO} \tag{14.7}$$

$$dX_E/dt = +f_E\, b_{OHO}X_{OHO} \tag{14.8}$$

where b_{OHO} = specific endogenous mass loss rate (d^{-1})
f_E = fraction of biomass that is unbiodegradable = 0.20.

14.2.5 Substrate description – BOD, COD or TOC

While useful for defined soluble substrates because the substrate concentration could be measured directly by tests for known organic types, Monod's Eq. (14.2) was of limited value in wastewater treatment, where the substrate comprised a mixture of organics of unknown types and concentrations and BOD_5 was used to measure the 'substrate' concentration. During the 1950s and 1960s the defined substrate and 'real' (often using complex organics like skim milk powder) wastewater studies using the BOD_5 progressed along parallel paths, each looking for a bridging parameter with which to express the yield coefficient in the Monod–Herbert growth kinetics. From the literature between 1950 and 1970, four parameters emerged, that is the yield based on the substrate's (1) electron donation capacity (EDC), (2) total oxidative energy (kJ), (3) ATP production (Y_{ATP}) and (4) carbon content (Y_C). Payne (1970) published an incisive review of yield characteristics of heterotrophs (X_{OHO}), which sheds considerable light on the respective utility of the four above parameters. The work reviewed by Payne is profound and extensive, touching on issues of thermodynamics, bioenergetics and stoichiometry that are too detailed to consider here.

Evaluating the four approaches, the Y_{ATP} was important in unraveling the internal mechanisms of growth, but had little value in the practical modeling of growth. The

Y_{kJ} presented practical measurement problems, but did demonstrate that the energy changes in growth were closely proportional to the changes in EDC. The Y_C fell into disuse not so much because the kJ/C for organics is not constant, but it was (and still is) very difficult to measure the CO_2 produced by aerobic systems to make carbon balances. So the COD ($Y_{av\ e-}$ or Y_{COD}) became the preferred approach because it (i) is closely related to substrate available energy (kJ), a primary driver for biological growth (BOD and TOC are not), (ii) gives a close approximation to the calculated theoretical oxygen demand (ThOD) of known organics, (iii) allows COD mass balances to be made over aerobic systems because it is easier to measure oxygen utilization rates than CO_2 production in biological systems (BOD and TOC do not balance), (iv) it includes unbiodegradable particulate organics (X_U from the influent and X_E), which affect sludge production in the AS systems (BOD does not), (v) can be easily and quickly done with simple wet chemical methods (BOD takes 5 d), (vi) oxidizes practically all organics with suitable catalysts, and (vii) does not oxidize ammonia and so gives the EDC of the organics only (provided sulfide and nitrite are low). Thus, the description of the AS process in so far as it affects organics removal became a description in terms of electron donor and acceptor behavior. That this description is also closely linked with the energy behavior is of theoretical interest, but was not directly of crucial or practical importance in modeling the AS process.

14.2.6 Wastewater COD fractions

Accepting the COD as the measure of substrate, from the work of McKinney and Ooten (1969) the influent carbonaceous material to a wastewater treatment plant can be viewed as being composed of three fractions: (1) unbiodegradable particulate organics (X_U), (2) unbiodegradable soluble organics (S_U) and (3) biodegradable organics (X_B and S_B).

In the reactor, X_U is enmeshed in the AS flocs, but does not undergo transformation. It accumulates as part of the sludge mass, attaining some steady state concentration, depending upon the SRT and its mass flux (gCOD/d) in the influent. S_U is not affected by the biochemical or flocculation action of the sludge and so is discharged with the effluent. The X_B and S_B undergo biochemical transformation in the OHO growth process which, during the 1960s, was modeled with Monod kinetics as modified by Herbert's endogenous mass loss and the endogenous residue generation hypotheses.

14.3 THE FIRST AS MODELS

14.3.1 Empirical models

With the continued use of the BOD_5 as a measure of wastewater organic strength, empirical AS models were developed during the 1950s and 1960s based on BOD_5. Heukelekian *et al.* (1951) fed real settled wastewater to intermittently-fed fill and draw systems and wasted measured volumes of mixed liquor from the reactor to control the SRT. From their data they concluded (i) that AS and excess sludge when measured as VSS were better correlated to BOD_5 load than sludge measured as TSS to TSS load, (ii) SRT should be defined in terms of TSS and (iii) for a given SRT sludge production was independent of the reactor TSS concentration.

Later, Gellman and Heukelekian (1953) provided plots of F_{VSS}/M_{VSS} (=1/SRT) and F_{O2} (flux oxygen utilized, gO_2/d) per M_{VSS} vs. F_{BOD}/M_{VSS}, where the F_{BOD}/M_{VSS} (gBOD load/d per g VSS in reactor) became known as the Food-to-Microorganism (F/M) ratio, the Load Factor (LF) or the specific substrate utilization rate (SUR). From these plots the specific sludge growth rate (SSGR) and specific oxygen utilization rate (SOUR) could be determined for a particular wastewater and F/M. It was found (e.g. Gellman & Heukelelian, 1953; Eckenfelder & Weston, 1956) that experimental data plotted this way produced straight lines (deviations were considered experimental error) with general equations:

$$F_{VSS}/M_{VSS}\ (SSGR) = a'\ F_{BOD}/M_{VSS}\ (F/M) - b'$$
$$\text{gVSS produced/gVSS in reactor} \tag{14.9}$$

$$F_{O2}/M_{VSS}\ (SOUR) = a''\ F_{BOD}/M_{VSS}\ (F/M) + b''$$
$$gO_2 \text{ utilized/d per gVSS in reactor} \tag{14.10}$$

where a' and a'' are the specific sludge growth rate (gVSS/gBOD) and b' and b'' the specific endogenous respiration rate (/d). The fact that the signs on the endogenous respiration rate (b) were correct (decreasing sludge growth and increasing oxygen demand) confirmed the approach. These plots became the basis for the empirical model for the AS system and were used for over twenty years and the '*a*' and '*b*' constants were deemed to be the yield coefficient (Y) and endogenous respiration rate (b). Four to six AS systems, each at a different *F/M*, would be operated on a wastewater and the steady state VSS and OUR measured. The data were plotted in the SSGR and SOUR vs. F/M domain and the '*a*' and '*b*' constants determined by least squares regression. Different wastewaters and different ranges of F/M produced different '*Y*' and '*b*' values. This led to doubt about the applicability of the pure culture models to real wastewater and the conclusion that the endogenous respiration rate decreased with decreasing *F/M* (increasing SRT). The seeds of this problem had already been observed by Heukelekian *et al.* (1951) who noted 'Not all organic materials in sewage are equally available as bacterial food. A portion is not available at all and their removal by AS is not dependent on their availability as bacterial food' and by Symons and McKinney (1958) who noted that not all of the biomass organics was biodegradable.

14.3.2 Kinetic models

14.3.2.1 Eckenfelder model

In 1955, Eckenfelder and O'Connor proposed a mathematical model, which was modified and expanded in the 1960s (Eckenfelder & O'Connor, 1961; Eckenfelder,

1970). This model was based on Equations 14.9 and 14.10 in the SSGR and SOUR vs. F/M plots and was widely accepted and used.

14.3.2.2 McKinney model

In 1962 McKinney proposed a kinetic model that integrated many of the developments above. His model (i) balanced COD over the growth process where the COD conserved in biomass was double that passed to oxygen ($k_1 = 0.5$, equivalent to $Y_{COD} = 0.67$), (ii) set the substrate utilization rate proportional to growth rate (dX_{BH}/dt) rather than to specific growth rate ($dX_{BH}/dt.\ 1/X_{BH}$) as in Monod kinetics, (iii) based the endogenous respiration process on (active) biomass and determined the rate (k_2) to be 0.01 $gO_2/(gVSS * h)$ (equivalent to $b_H = 0.24/d$), (iv) incorporated endogenous residue generation at a rate (k_8) of 0.036/d at 20°C (with k_2 equivalent to a biomass unbiodegradable fraction (f_E) = 0.23), (v) divided reactor VSS into biomass (X_{OHO}), endogenous residue (X_E) and unbiodegradable particulate organics (X_U) from the influent, and (vi) included a 'biomass' ISS (inorganic solids associated with biomass growth) at 0.10 gISS/gVSS in the biomass (X_{OHO}) and endogenous residue (X_E) which, with the ISS from the influent, made up the reactor TSS concentration. While connecting substrate utilization, sludge production and oxygen utilization in the COD-balanced bioprocesses of growth and endogenous respiration, this connection was not followed through to application of the model because (i) the BOD_5 was used for wastewater organic strength measurement and (ii) different constants for sludge production and oxygen utilization were used with the result that COD balances could not be made over the system. Also, model calibration focused on predicting the effluent BOD_5 concentration, sludge production and oxygen utilization separately, not as an integrated model where all three model outputs were accurately predicted with the same set of constants. Like the Eckenfelder model, the McKinney model was also expanded and modified by McKinney during the 1960s and it was also widely accepted and used.

The different appearance of the kinetic equations and their 'kinetic' constants in the Eckenfelder and McKinney models created confusion and led to the impression that the models were different. However, Goodman and Englande (1974) showed that the equations in each model could be cast into a similar form and gave the same results when calibrated against the same experimental data.

14.3.2.3 Lawrence and McCarty model

Lawrence and McCarty (1972) published a model which was based on (i) specific growth rate and the Monod–Herbert kinetics of growth and endogenous respiration, (ii) SRT defined as mass of sludge in system divided by flux of sludge wasted, that is SRT = $M_{TSS}/F_{TSS,WAS}$/d, (iii) formal definitions of *F/M*, Loading Factor and substrate utilization rate as gCOD or BOD_5 removed/d per g VSS or TSS in the reactor, (iv) defined a washout (minimum) SRT in terms of the OHO biomass maximum specific growth rate ($\mu_{OHO,max}$) and endogenous respiration rate (b_{OHO}), (v) emphasized the importance of considering the live OHO biomass as part of the VSS and (v) considered nitrification by autotrophic nitrifying organisms (ANO). In fact, in the McCarty model, it is hard not to recognize the Monod approach to modeling nitrification by ANO of Downing *et al.* (1964) for modeling organics removal by OHO, except that endogenous respiration of OHO was added. While the model recognizes OHO biomass, its kinetic constants Y_{OHO}, b_{OHO} and substrate removal rate q_B (=$\mu_{OHO,max}/Y_{OHO}$) are determined by calibration against experimental data in which it is implicitly accepted that VSS represents the OHO biomass concentration. The model also does not include equations for calculating oxygen demand.

14.3.2.4 Marais and Ekama model

The steady-state model of Marais and Ekama (1976) was based on the strengths of the McKinney and McCarty models, viz. they (i) accepted specific growth rate and the Monod–Herbert kinetics of growth and endogenous respiration, (ii) balanced the COD over the growth process where the e^- of the organics (donor) are conserved in biomass (anabolism, $i_{CV}Y_{OHO}$= 0.66) and passed to oxygen (acceptor, catabolism, $1 - i_{CV}Y_{OHO} = 0.34$), (iii) balanced COD over the endogenous process where the e^- of the biomass (donor) are conserved in endogenous residue ($f_E = 0.20$) and passed to oxygen (acceptor, catabolism, $1 - f_E = 0.80$), (iv) integrated the equations for biomass gain and loss and oxygen utilization so that these are based on the same kinetic constants Y_{OHO} and b_{OHO}, (v) based the endogenous respiration process on (active) biomass and determined the rate (b_{OHO}) to be 0.24/d at 20°C, (vi) incorporated endogenous residue generation with a biomass unbiodegradable fraction (f_E) of 0.20, (vii) included unbiodegradable particulate organics (X_U) from the influent so the reactor VSS comprises biomass (X_{OHO}), endogenous residue (X_E) and unbiodegradable particulate organics (X_U) from the influent, (viii) defined SRT as mass of sludge in system divided by flux of sludge wasted, that is SRT = $M_{TSS}/F_{TSS,WAS}$/d and (ix) considered nitrification by autotrophic nitrifying organisms (ANO).

By retaining the active biomass (X_{OHO}) and accepting that the measured VSS (X_{VSS}) included endogenous residue (X_E), Marais and Ekama (1976) found that their model could satisfactorily predict their experimentally measured AS system reactor VSS concentration, sludge production and OUR over a range of SRTs and *F/M* ratios for a range of Y_{OHO} and b_{OHO} pairs, for example Y_{OHO} = 0.43 gVSS/gCOD and b_{OHO}= 0.24/d or Y_{OHO}= 0.32 gVSS/gCOD and b_{OHO}= 0.10/d. So while their method yielded a consistent set of Y_{OHO} and b_{OHO} values that could predict sludge production and oxygen demand, data from continuous systems alone could not identify the correct Y_{OHO} and b_{OHO} pair. They measured b_{OHO} independently in batch aerobic digestion tests on sludge harvested from the continuous systems. Contrary to the 'b' determined from the SSGR or SOUR vs. *F/M* plots, they found b_{OHO} to be constant with SRT at 0.24/d and mildly temperature sensitive ($\theta_{bOHO} = 1.029$). With b_{OHO} known, Y_{OHO} was determined to be 0.43 gVSS/gCOD and also found to be constant with SRT. By including endogenous residue in the VSS, they showed, from their data and from literature data, that the lines in the SSGR and SOUR vs. *F/M* are not straight but curved starting from the origin. Later, Ekama and Wentzel (2004) added a biomass ISS of 0.15 gISS/gOHOVSS to the steady-state model which, together with the measured influent ISS, allowed prediction of the reactor ISS and TSS concentrations. This steady-state model's predictions from the Heukelekian *et al.* (1951) and Goodman and Englande (1974) reactor TSS data are shown in Figure 14.1 and Figure 14.2 for a COD/BOD_5 ratio = 2 and a settled wastewater unbiodegradable particulate COD fraction (X_U) of 0.

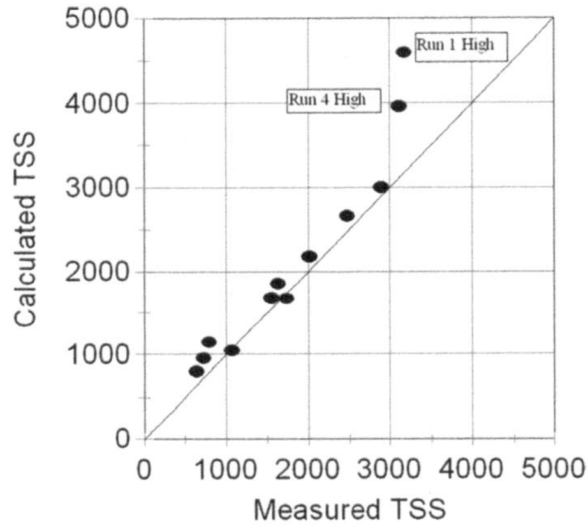

Figure 14.1 Marais and Ekama (1976) steady model: calculated reactor TSS *vs.* measured Heulelekian *et al.* (1951) TSS data.

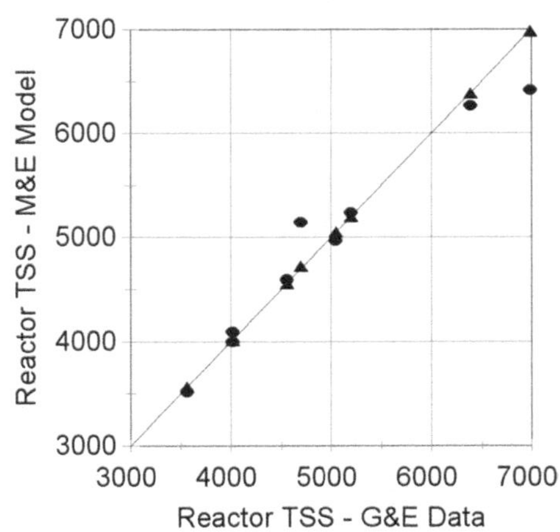

Figure 14.2 Marais and Ekama (1976) steady model: calculated reactor TSS *vs.* measured Goodman & Englande (1974) full scale plant reactor TSS data.

The independence of Y_{OHO} and b_{OHO} to the system constraints SRT and HRT indicated that the actual OHO yield and endogenous respiration rate had been determined from real wastewater systems. This paved the way for developing dynamic models for AS systems.

14.3.2.5 ASM1

Based on Marais and Ekama's steady-state model, Dold *et al.* (1980) and van Haandel *et al.* (1981) published a dynamic model which included all the elements of the Marais-Ekama model, extended by OHO denitrification, ANO (one-step) nitrification, and alkalinity use and production calculations. One element of the model was the separation of influent biodegradable substrate into readily and slowly biodegradable substrate (S_B and X_B), both being used directly but with different kinetics (dual substrate model). This was based on a square wave (12 h feeding, 12 h without feeding) loading test which showed an immediate drop of OUR after feed stopped and an initial rapid rate of denitrification in the primary anoxic reactor (Figure 14.3 and Figure 14.4). In 1982 IWA (at that time IAWPRC) formed a Task Group (Henze, Marais, Grady, Gujer, Mino) which in 1987 published a modified, simplified and updated form of the Dold/van Haandel model converted to single substrate use (Henze *et al.*, 1987). This model became widely known as ASM1. The original matrix format was extended by COD, charge and elemental balancing (known now as the Gujer matrix). The international consensus on which this model was based had an immense effect on practically all model development efforts after 1987. It has truly stood the test of time since it was able to accurately predict electron acceptor demand and sludge production in a wide variety of system configurations.

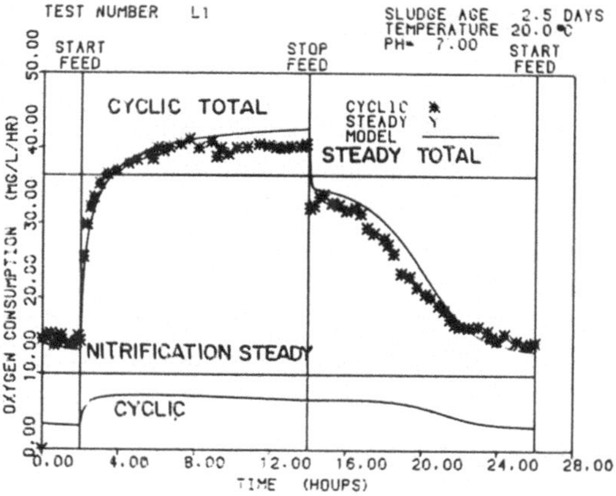

Figure 14.3 Precipitous decrease in OUR at STOP FEED (1400 h) indicating presence of readily biodegradable organics (RBCOD) in influent wastewater – the magnitude of the decrease is proportional to the influent rbCOD concentration (data from Ekama & Marais, 1978).

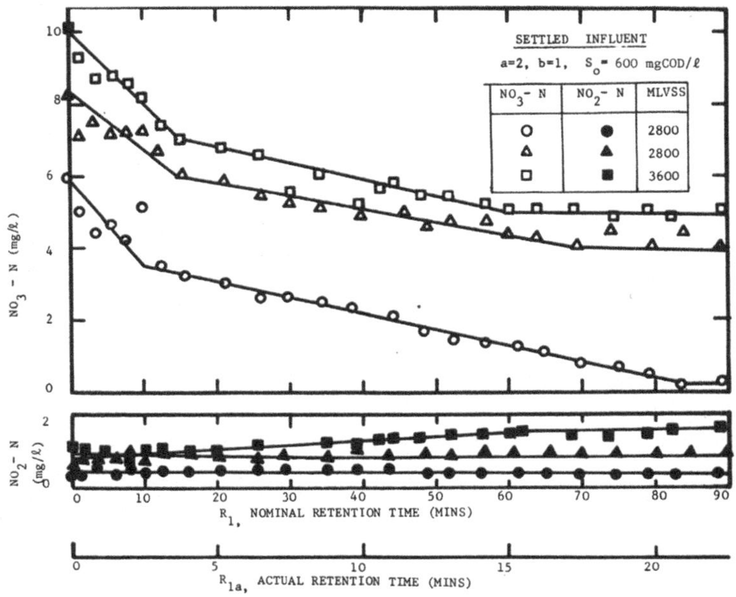

Figure 14.4 Initial (0–10 min) rapid rate of denitrification in a plug flow primary anoxic reactor indicating presence of readily biodegradable organics (rbCOD) in influent wastewater – the magnitude of the initial decrease is proportional to the influent rbCOD concentration (data from Stern & Marais, 1974).

14.4 EXTENDED AS MODELS

14.4.1 Anoxic yield

As nutrient removal progressed in the late 20th century, unaerated (anoxic and anaerobic) zones became more and more common in AS plants, and it became apparent that ASM1 over predicted sludge production and underpredicted denitrification proportionally to the unaerated volume fraction. Copp and Dold (1998) published an extensive set of data showing that Y_{OHO} decreases under anoxic/ anaerobic conditions. A value of $Y_{OHO,anox} = 0.54$ was proposed based on best fit. Muller *et al.* (2003) reviewed and experimentally confirmed this value.

14.4.2 Substrate storage

ASM1 is based on direct utilization of readily biodegradable substrate, leading to a direct proportionality between electron acceptor demand and the use of electron donor (e.g. OUR and filtered COD). It can be observed, particularly in highly loaded systems, that substrate disappears from solution without a proportional use of oxygen, and that this oxygen deficit is later recovered. This phenomenon is based on storage of substrate for later use, and has practical significance in designing the layout of tapered aeration systems. The IWA Task Group published ASM3 (Gujer *et al.*, 1999) which is based on storage and use of stored substrate exclusively. In reality, depending on the composition of influent and even more characteristically for specific organics (e.g. methanol or glycol), both direct use and post-storage use of substrate occurs, but the complexity of models increases exponentially with each new substrate use pathway, and a model that uses both mechanisms concurrently has not yet found its way into practical use.

14.4.3 Influent colloidal material

ASM1 contained two types of influent substrate, readily biodegradable (S_B) and slowly biodegradable (X_B) and differentiation between these was based on biokinetic tests. However, the fact that *most* S_B is soluble (as measured by paper or glass fiber filtration) and *most* X_B is particulate, and the expense of biodegradability testing, led to the use of filtration to approximate slowly and readily biodegradable substrate. This simplification degrades the accuracy of the model to predict electron acceptor use patterns in plug-flow reactors. In addition, it was difficult to reconcile the original biokinetic test results with typical measurements such as influent TSS, since part of the slowly degradable material is soluble by the filtration test. Melcer *et al.* (2003) proposed filtration followed by flocculation/filtration to define one more influent constituent-colloidal biodegradable material (C_B). In this concept C_B is 'soluble' by a simple filtration test, but is slowly degradable, since before utilization it must first be converted to X_B in the reactor. This approach makes the influent TSS measurements usable because they are only connected to the particulate fraction of X_B. The IWA Nomenclature Task Group (Corominas *et al.*, 2010) systemized the symbols for these constituents so that *S*, *C* and *X* refer to soluble, colloidal and particulate material, respectively. Thus S_B is truly soluble and readily biodegradable, C_B is colloidal and slowly biodegradable, and X_B is particulate and slowly biodegradable. The former 'slowly degradable material' is indicated as $CX_B = C_B + X_B$. This nomenclature guideline is followed in this chapter as far as possible without altering the original intent of the early modeling pioneers.

14.4.4 Specific substrates and biomasses

External carbon addition to remove excess nitrate is a widespread practice in certain areas where extremely low TN limits are in place (e.g. 2 mgTN/L in some areas around Chesapeake Bay, USA). When investigating methanol (and later other carbon sources), Purtschert *et al.* (1996) and later Mokhayeri *et al.* (2009) determined that adaptation to the new substrate is required, particularly for methanol. This can be modeled using a specific biomass (e.g. anoxic methylotrophs) which requires time to establish itself in the reactor. Kinetics (growth and decay rates, half saturations) and stoichiometry (particularly yield) proved to be specific to each specific biomass and different from values used for typical wastewater. Thus the lower yield (0.4) of anoxic methylotrophs utilizing methanol, the use of OHOs and typical wastewater kinetics/stoichiometry for determining methanol dosage can lead to undersized anoxic zones and overestimated methanol doses.

14.4.5 Nitrification

The conversion of ammonia to nitrate is a multi-step, biologically mediated process with several intermediates (of which nitrite is the most important). However, from the perspective of the oxygen demand and alkalinity consumption, the first models considered nitrification as a one-step aerobic process in which ammonia was oxidized to nitrate mediated by autotrophic nitrifying organisms (ANOs). This simplification provided a more manageable mathematical structure and, in typical AS systems where SRT was above a critical value, the results obtained were almost identical. The maximum specific nitrifier growth rate (typically assumed to be 0.9/d), the decay rate (0.17/d) and temperature sensitivity in the 12°C to 26°C region (doubles every 10°C; Arrhenius coefficient = 1.072) are key design parameters. If the aerobic SRT falls below the critical SRT for nitrification, ANOs may wash out of the system and nitrification stops.

In the past decade nitrification pathways specifically and N pathways in general have been investigated in detail, and 2, 4 or 5 step nitrification and autotrophic denitrification models have been proposed to describe the relationships between ammonia (NH_3 and NH_4^+), nitrite (NO_2^-), nitrate (NO_3^-) hydroxylamine (NH_2OH), nitric oxide (NO) and nitrous oxide (N_2O) (Chandran *et al.*, 2011).

14.4.6 P removal

In wastewater treatment plants, P is removed by cell synthesis, inorganic polyphosphate synthesis and chemical addition. Wentzel *et al.* (1989) developed a model for strictly aerobic P uptake behavior in enhanced cultures of P accumulating organisms (PAOs) and integrated it into the Dold/van Haandel ND model (UCTOLD) to form UCTPHO (Wentzel *et al.*, 1992). An adapted version of this model together with chemical precipitation by metal salt addition was later included in a new version of the ASM Task Group model (ASM2), which described all three P removal mechanisms. Around this time, denitrification by PAOs was observed and ASM2 was modified to ASM2d to include this phenomenon. An overview of AS models available at the time is given by Hu *et al.* (2003). Developments to include denitrification by PAOs are recorded by Barker and Dold (1997) and Hu *et al.* (2007a, b).

An alternative, metabolic biological P removal model that included glycogen storage was published by Smolders *et al.* (1994). A rigorous scientific comparison of the two model approaches has never been published.

14.4.7 pH

Weak acid/base chemistry plays an important role in many aspects of AS, particularly in nitrification/denitrification. Most dynamic models to date use alkalinity as a reliable indicator of pH stability. However, pH has an effect on process rates, gas transfer (e.g. CO_2 stripping) and chemical precipitation reactions which cannot be linked directly to alkalinity. Fairlamb *et al.* (2003) presented a comprehensive pH model linked to an extended AS model (as implemented in BioWin^{TM}) and later Sötemann *et al.* (2005a, b) published an AS (and AD) model with complete elemental mass balances and integrated, mixed, weak acid/base chemistry (with ion pairing) for pH calculation. While the fundamentals of acid/base reactions and the dissociation constants are well known, the complexity of ionic species in wastewater and the necessary corrections for ionic strength and ion pairing are still under investigation (Tait *et al.*, 2012). The effect of pH on biological rates is usually modeled as a bell-shaped curve (Batstone *et al.*, 2002). pH also allows the quantification of chemical species such as free ammonia and nitrous acid which can be inhibitory or toxic to various microorganisms. A pH model coupled with an AS model enables estimation of stripping effects for H_2CO_3 and $NH_{3,,}$ which can be especially significant when treating high strength wastewaters.

14.4.8 Gas Transfer

One of the main engineering tasks for which AS models were developed is to estimate electron acceptor and specifically actual oxygen requirement (AOR). AOR varies along the length of a plug-flow reactor with time, with loading and through other dynamic effects such as intermittent sidestream loads. A steady-state model cannot calculate peak oxygen demand during high loading periods if it is used with peak loading data, it will overestimate reactor sizing and aeration equipment requirements. Traditionally an empirical safety factor was used on top of the AOR value calculated using average loading conditions. One of the most important benefits of dynamic models are their ability to estimate peak AOR and to properly size facilities and equipment.modeling.

14.4.9 Precipitation

In a model fitted with pH and gas transfer sub-models, precipitate formation, both natural and engineered, can be estimated. There is evidence (Maurer *et al.*, 1999) that calcium or magnesium phosphates can precipitate under typical AS conditions, particularly in anoxic zones where the pH may increase due to alkalinity production during denitrification. If a purely biological model (e.g. ASM2D) that does not contain Ca or Mg P precipitation processes is calibrated to data from a system where significant amounts of P were in fact precipitated, the calibration will lead to erroneous kinetic or stoichiometric parameters. This forced calibration attributes changes in P concentration exclusively to biological processes, and the model's prediction power is compromised for changes in system configuration or operational performance.

14.1 MODELING IN PRACTICE

14.5.1 Whole plant models

AS models still dominate our field even after sixty plus years of development. A recent overview of the application and limitations of the publicly available ASM models can be found in Hauduc *et al.* (2013). However the focus is slowly shifting to models that can describe the whole plant. ADM1 (Batstone *et al.*, 2002) can be considered as the first internationally-used model extension for describing anaerobic digestion. Ikumi *et al.* (2011) published a comprehensive stoichiometric model linking AS to anaerobic and/or aerobic digestion that includes elemental balances and mixed weak acid/base chemistry for pH calculations.

Whole plant models can be constructed from existing process unit models using interfaces, for example ASM1 interfaced to ADM1 for the excess sludge line and vice versa for return lines (e.g. the continuity based interface method (CBIM) approach of Vanrolleghem *et al.*, 2005). Whole plant models can also be constructed using one model with the same set of components and describing the behavior of all components under all conditions (aerobic, anoxic, anaerobic, highly concentrated warm sidestream) in the treatment plant. Grau *et al.* (2007) developed a 'supermodel' approach which used the same (large) model component set in all biological reactors and analyzed the different approaches, their benefits and disadvantages. These models, in addition to the OHO, AOB, NOB, anaerobic fermenting and methanogenic biomasses also included slow growing ($\mu_{max} = 0.1/d$) anaerobic ammonia oxidizers (e.g. anammox bacteria).

14.5.2 Engineering use

While the original models consisted of only a few equations, ASM1, as applied to a complex nitrifying/denitrifying plant configuration consists of thousands of lines of calculations, and many, typically dynamically changing input parameters. Other models and their extensions are often many times larger than ASM1. To be usable in practice these models have to accept parameters that can be measured in practice and produce relevant results quickly. These demands led to the development of many commercial or research software packages starting in 1987, with SSSP, and later, in alphabetic order, BioWin©, EFOR©, GPS-X©, Simba©, Stoat©, West©, and others. The availability of user-friendly engineering software has facilitated the development of more efficient and complex better performing plant configurations, reduced the risk of permit violations in the design and operation phases, and allowed countless young professionals to understand the inner workings of AS by starting from the knowledge that the modeling community collected and stored in the form of these mathematical equations.

14.5.3 Research

Concerted research effort by thousands of researchers over many decades has led to the development of AS models as we know them today; conversely however research itself has derived a huge benefit from dynamic modeling. A mechanistically-based model is a hypothesis of what key components are present in the system and what is their fate (as described in process rates). The model can be tested with much less effort than that required for collecting experimental data, and relevant experiments can be identified to help focus a research effort. Once the model has been calibrated, it can be used to predict the performance of systems that have not been extensively tested experimentally. And where a model fails, a new research issue arises. The pursuit of these questions is what has led to today's extensive, detailed models and this pursuit will continue for the next 100 years and beyond.

14.6 ACKNOWLEDGEMENTS

George Ekama: I am indebted to Professor Gerrit van Rooyen Marais for starting the first half of this review of AS modeling up to ASM1, which he worked on in his retirement after 1992. Unfortunately he did not complete it before succumbing to lung cancer on December 9, 2005. By following the ideas and thoughts in his drafts and notes, I have attempted to complete it without the benefit of personally knowing the early model developers and their work as well as he did.

Imre Takács: I am grateful for the direct and indirect guidance and inspiration I have received over the years primarily from the ASM Task Group and Dr Peter Dold, as well as from many younger (by years or spirit) wastewater process modelers from all over the world who I have been lucky enough to meet over the past 30+ years.

14.7 REFERENCES

Following is a small selection out of tens of thousands of articles published that have influenced AS modeling in the past 100 years.

Alleman J. E. and Prakasam T. B. S. (1983). Reflections on seven decades of activated sludge history. *J. Water Pollut. Control Fedn.*, **55**(5) 436–443.

Barker P. S. and Dold P. L. (1997). General model for biological nutrient removal activated-sludge systems: Model presentation. *Water Environ. Res.*, **69**(5), 969–984.

Batstone D. J., Keller J., Angelidaki I., Kalyuzhnyi S. V., Pavlostathis S. G., Rozzi A., Sanders W. T. M, Siegrist H. and Vavilin V. A. (2002). The IWA anaerobic digestion model no. 1 (ADM1). *Water Sci. Technol.*, **45**(10) 65–73.

Chandran K., Stein L., Klotz M. and van Loosdrecht M. C. M. (2011). Nitrous oxide production by lithotrophic ammonia-oxidizing bacteria and implications for engineered nitrogen-removal systems. *Biochem. Soc. Trans.*, **39**, 1832–1837.

Copp J. B. and Dold P. L. (1998). Comparing sludge production under aerobic and anoxic conditions. *Water Sci Technol.*, **38**(1) 285–294.

Corominas L., Rieger L., Takács I. Ekama G., Hauduc H., Vanrolleghem P. A., Oehmen A., Gernaey K. V., van Loosdrecht M. C. M. and Comeau Y. (2010). New framework for standardized notation in wastewater treatment modeling. *Water Sci. Technol.*, **61**(4) 841–857.

Dold P. L., Ekama G. A. and Marais G. v. R. (1980) A general model for the activated sludge process. *Prog. Water Technol.*, **12**, 47–77.

Downing A. L., Painter H. A. and Knowles G. (1964). Nitrification in the activated sludge process. *J. Proc. Inst. Sewage Purif.*, Part 2, 130–153.

Eckenfelder W. W. Jr. (1970). Water Quality Engineering for Practicing Engineers. Barnes & Noble, New York, NY.

Eckenfelder W. W. Jr. and O'Connor D. J. (1955). The aerobic treatment of organic wastes. Proc. 9th Industr. Waste Conf., Purdue Univ., W. Lafayette, IN, Extn Series 89, **39**(2), 512.

Eckenfelder W. W. Jr. and O'Connor D. J. (1961). Biological Waste Treatment. Pergamon Press, Oxford, England.

Eckenfelder W. W. Jr. and Weston R. F. (1956). Kinetics of biological oxidation. In: Biological Treatment of Sewage and Industrial Wastes: Vol. 1 – Aerobic Oxidation. Bro. J. McCabe and W. W. Eckenfelder, Jr. (eds), Reinhold Publishing Corp., New York, pp. 18–34.

Ekama G. A. and Marais G. v. R. (1978). The Dynamic Behaviour of the Activated Sludge Process. Res. Rept W27, Dept. of Civil Eng., Univ of Cape Town, Rondebosch, RSA.

Ekama G. A. and Wentzel M. C. (2004). A predictive model for the reactor inorganic suspended solids concentration in activated sludge systems. *Water Res.*, **38**(19) 4093–4106.

Gellman I. and Heukelekian H. (1953). Studies of biochemical oxidation by direct methods: III Oxidation and purification of industrial wastes by activated sludge. *Sewage Industr. Wastes*, **25**(10), 1196–1209.

Fairlamb M., Jones R., Takács I. and Bye C. (2003). Formulation of a general model for simulation of pH in wastewater treatment processes. Proc. Water Environ. Fedn., WEFTEC 2003 pp. 511–528(18).

Goodman B. L. and Englande A. J. Jr. (1974). A unified model for the activated sludge process. *J. Water Pollut. Control Fedn.*, **46**(2), 312–332.

Grau P., de Gracia M., Vanrolleghem P. A. and Ayesa E. (2007). A new plant wide methodology for WWTPs. *Water Res.*, **41**, 4357–4372.

Gujer W., Henze M., Mino T., and van Loosdrecht M. C. M. (1999). Activated sludge model no. 3. *Water Sci. Technol.*, **39**(1), 183–193.

Harder W. and Dijkhuizen L. (1976). Mixed substrate utilization,. In: Continuous culture 6: Applications and new fields. A. C. R. Dean, D. C. Ellwood, C. G. T. Evans and J. Melling (eds), Ellis Horwood, Chichester, UK, pp. 297–314.

Hauduc H., Rieger L., Oehmen A., van Loosdrecht M. C. M., Comeau Y., Héduit A., Vanrolleghem P. A. and Gillot S. (2013). Critical review of activated sludge modeling:

state of process knowledge, modeling concepts, and limitations. *Biotechnol. Bioeng.,* **110**(1), 24–46.

Henze M., Grady C. P. L., Gujer W., Marais G. v. R. and Matsuo T. (1987). Activated sludge model no.1. IAWPRC Scientific and Technical Report, No. 1, Internatl Assoc. Water Pollut. Res. and Control, London, p. 33, ISSN: 1010-707X.

Herbert D. (1958). Recent progress in microbiology. In: VII International Congress for Microbiology, G. Tunevall (ed.), Almquist and Wiksel, Stockholm, pp. 381–396.

Heukelekian H., Orford H. E. and Manganelli R. (1951). Factors affecting the quantity of sludge production in the activated sludge process. *Sewage Industr. Wastes,* **23**(8), 945–958.

Hoover S. R. and Porges N. (1952). Assimilation of dairy wastes by activated sludge II – The equation of synthesis and rate of oxygen utilization. *Sewage Industr. Wastes,* **24**(3), 306–312.

Hu Z.-R., Wentzel M. C. and Ekama G. A. (2003). Modeling biological nutrient removal activated sludge systems – a review. *Water Res.,* **37**(14), 3430–3444.

Hu Z.-R., Wentzel M. C. and Ekama G. A. (2007a). A general model for biological nutrient removal activated sludge systems-model development. *Biotechnol. Bioeng.,* **98**(6), 1242–1258.

Hu Z.-R., Wentzel M. C. and Ekama G. A. (2007b). A general model for biological nutrient removal activated sludge systems-model evaluation. *Biotechnol. Bioeng.,* **98**(6), 1259–1275.

Ikumi D. S., Harding T. H., Brouckaert C. J. and Ekama G. A. (2011). Plant wide integrated biological, chemical and physical bioprocesses modeling of wastewater treatment plants in 3 phases (aqueous-gas-solid). Research Report W136, Dept of Civil Eng., Univ. of Cape Town, Rondebosch, Cape, RSA.

Kountz R. R. and Forney C. (1959) Metabolic energy balances in a total oxidation activated sludge system. *Sewage Industr. Wastes,* **31**, 819.

Kraus L. S. (1945). The use of digested sludge and digester overflow to control bulking of activated sludge. *Sewage Works J.*, **17**(6) 1177–1190 and 1946, **18**(6) 1099–1112.

Lawrence A. and McCarty P. L. (1972). Unified basis for biological treatment design and operation. *J. Sanit. Eng. Div., Amer. Soc. Civil Eng.,* **96**(SA3), 757–778.

Marais G. v. R. and Ekama G. A. (1976) The activated sludge process part 1 – steady state behavior. *Water SA,* **2**(4) 163–200.

Maurer M., Abramovich D., Siegrist H. and Gujer W., (1999). Kinetics of biologically induced phosphorus precipitation in waste-water treatment. *Water Res.*, **33**(2), 484–493.

McCarty P. L. and Brodersen C. F. (1962). Theory of extended aeration activated sludge. *J. Water Pollut. Control Fedn.,* **34**(11), 1095–1103.

McKinney R. E. (1962). Mathematics of complete-mixing activated sludge. *J. Sanit. Eng. Div., Amer. Soc. Civil Eng.,* **88**(SA3), 87–113.

McKinney R. E. and Ooten R. J. (1969). Concepts of complete-mixing activated sludge. Trans. 19th Eng. Conf., Univ of Kansas, 32.

Melcer H., Dold P. L., Jones R. M., Bye C. M., Takács I., Stensel D. H., Wilson A. W., Sun P., and Bury S. (2003). Methods for Wastewater Characterization in Activated Sludge Modeling. *Water Environ. Res. Foundn.*, Alexandria, VA.

Michaelis M. and Menten L. (1913). Die Kinetik der Invertinwirkung, *Biochem. Z.,* **49**, 333–369.

Mokhayeri Y., Riffat R., Murthy S., Bailey W., Takacs I. and Bott C. (2009). Balancing yield, kinetics and cost for three external carbon sources used for suspended growth post-denitrification. *Water Sci. Technol.,* **60**(10), 2485–2491.

Monod J. L. (1942). *Recherches sur la Croissance des Cultures Bacteriennes* (Research on the Growth of Bacterial Cultures). Hermann et cie, Paris. 1942. Actualite´s scientifiques et industrielles; 911. Microbiologie; no. 1.

Monod J. L. (1950). La technique de culture continue – theorie et applications. *Ann. Inst. Pasteur,* **79**, 390–410.

Muller A., Wentzel M. C., Ekama G. A. and Loewenthal R. E. (2003). Heterotroph anoxic yield in anoxic aerobic activated sludge systems treating municipal wastewater. *Water Res.,* **37**(10), 2435–2441.

Novick A. and Szilard L. (1950). Description of the chemostat. *Sci.,* **112,** 715–716.

Payne W. J. (1970). Energy yields and growth of heterotrophs. *Ann. Rev. Microbiol.,* **24**, 17–52.

Pirt S. J. (1965). The maintenance energy of bacteria in growing cultures. *Proc. Royal Soc., London, Series B*, **163**, 224.

Porges N., Jasewicz L. and Hoover S. R. (1955). Principles of Biological Oxidation. In: Biological Treatment of Sewage and Industrial Wastes: Vol. 1 – Aerobic Oxidation, Bro. J. McCabe and Eckenfelder W. W. Jr. (eds), Reinhold Publishing Corp., New York, NY, 35–48.

Purtschert I., Siegrist H. and Gujer W. (1996) Enhanced denitrification with methanol at WWTP Zürich-Werdhölzli. *Water Sci. Technol.,* **33**(12), 117–126.

Sawyer C. N. (1965). Milestones in the development of the activated sludge process. *J. Water Pollut. Control Fedn,* **37**(2), 151–162.

Smolders G. J. F., van der Meij J., van Loosdrecht M. C. M. and Heijnen J. J. (1994). A structured metabolic model for anaerobic and aerobic stoichiometry and kinetics of the biological phosphorus removal process. *Biotechnol. Bioeng.,* **47**(3), 277–287.

Sötemann S. W., Musvoto E. V., Wentzel M. C. and Ekama G. A. (2005a). Integrated chemical, physical and biological processes kinetic modelling Part 1 – Anoxic and aerobic processes of carbon and nitrogen removal in the activated sludge system. *Water SA*, **31**(4), 529–544.

Sötemann S. W., van Rensburg P., Ristow N. E., Wentzel M. C., Loewenthal R. E. and Ekama G. A. (2005b). Integrated chemical, physical and biological processes kinetic modeling; Part 2 – Anaerobic digestion of sewage sludges. *Water SA*, **31**(4) 545–568.

Stern L. and Marais G. v. R. (1974). Raw Sewage as Electron Donor in Biological Denitrification. Res. Rept W7, Dept. of Civil Eng., Univ of Cape Town, Rondebosch, RSA.

Symons J. M. and McKinney R. E. (1958). Biochemistry of nitrogen in the synthesis of activated sludge. *Sewage Industr. Wastes,* **30**(7), 874–890.

Tait S., Solon K., Volcke E. I. P. and Batstone D. J. (2012) A unified approach to modeling wastewater chemistry: model corrections. Proc. WWTmod2012, Mont Saint-Anne, Quebec, Canada, Feburary 26–28.

van Haandel A. C., Ekama G. A. and Marais G. v. R. (1981) The activated sludge process. Part 3-Single sludge denitrification. *Water Res.*, **15**(10), 1135–1152.

Vanrolleghem P. A., Rosen C., Zaher U., Copp J., Benedetti L., Ayesa E. and Jeppsson U. (2005). Continuity-based interfacing of models for wastewater systems described by Petersen matrices. *Water Sci. Technol.,* **52**(1–2), 493–500.

Washington D. R. and Hetling L. J. (1965) Volatile sludge accumulation in activated sludge plants. *J. Water Pollut. Control Fedn.,* **37**4, 499–507.

Washington D. R. and Symons J. M. (1962). Volatile sludge accumulation in activated sludge systems. *J. Water Pollut. Control Fedn.,* **34**(8) 767–790.

Wentzel M. C., Dold P. L., Ekama G. A. and Marais G.v.R. (1989). Enhanced polyphosphate organism cultures in activated sludge systems. Part III-Kinetic model. *Water SA,* **15**(2), 89–102.

Wentzel M. C., Ekama G. A. and Marais G. v. R. (1992). Processes and modeling of nitrification denitrification biological excess phosphorus removal systems-a review. *Water Sci. Technol.,* **25**(6), 59–82.

Chapter 15

Hybrid systems

Hallvard Ødegaard (Norway), Magnus Christensson (Sweden), Kim Kelleshoj Sørensen (France)

15.1 INTRODUCTION

In the broadest sense, a hybrid system involves the combination of a suspended biomass (AS) and an attached biomass (biofilm) in any manner (Figure 15.1). One can differentiate between the systems where the biofilm system and the AS system are separated into different reactors (Figure 15.1a–15.1c) and those where the biofilm system is integrated into the AS system (Figures 15.1d–15.1f).

The purposes for choosing such combined (hybrid) processes may be several:

(1) To improve biomass separation
   (a) from a biofilm reactor by improving flocculation when passing the biofilm effluent through a suspended biomass (AS) reactor, such as in Figures 15.1a and 15.1b
   (b) from an AS reactor by mixing the heavier biofilm biomass with the lighter AS biomass, such as in Figures 15.1c–15.1f.

(2) To reduce the organic load on the AS reactor and hence the footprint of the plant by using a high-rate biofilm reactor ahead of it, such as in Figures 15.1b and 15.1c.

(3) To enhance nitrification (or other slow-growing microbial processes, that is in industrial wastewater treatment) in an AS reactor by increasing the biomass in the integrated biofilm/AS reactor, such as in Figures 15.1b–15.1f.

(4) To reduce the footprint of an AS plant – especially when nitrification (Figure 15.1f) or biological N- and P-removal (Figure 15.3e and 15.3f) is required.

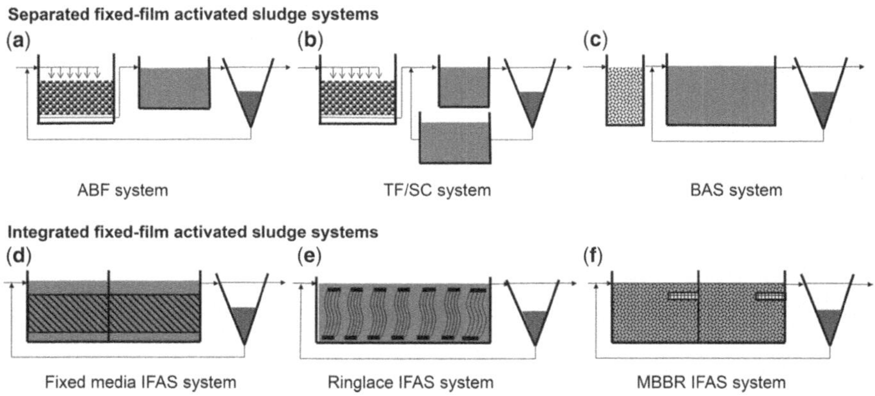

Figure 15.1 Different hybrid systems.

In recent years, the integrated fixed-film AS system (IFAS) based on moving bed biofilm carriers in the AS reactor has become the dominating hybrid system and this will be the main focus of this paper. However, we will first describe and discuss the other systems briefly.

15.2 AN OVERVIEW OF HYBRID SYSTEMS

Before 1980 Atkinson *et al.* (1980) pointed out the possibility of intensifying AS processes using cell support systems.

15.2.1 Separated fixed-film, AS systems

These systems include a biofilm reactor preceding an AS plant either for the main purpose of improving operational stability, performance reliability and energy savings, as in the ABF process (Figure 15.1a) (Rakness *et al.*, 1982), improving the settling properties of the biomass as in the TF/SC-process (Figure 15.1b) (Norris *et al.*, 1980, 1982; Hui *et al.*, 1983), or removing the bulk of the organic matter through a high-rate biofilm process as in the BAS process (Figure 15.1c) (Ødegaard, 2006).

The activated biofilter (ABF) system (see Figure 15.1a) was introduced in the USA in the early 1980s. An EPA-report at the time (Rakness *et al.*, 1982) concluded that there was a potential for energy savings, better process stability and better settling characteristics than for a trickling filter or an AS process separately.

The trickling filter solids contact (TF/SC) process was demonstrated in 1979 (Norris *et al.*, 1980, 1982). It has mostly been used to upgrade or enhance trickling filter process efficiency. The primary function of the aerated solids contact tank(s) is to increase solids capture by flocculation and thus reduce particulate BOD in the. The process may use one or two contact tanks, the first placed in between the trickling filter and the settling tank and possibly a second tank placed on the RAS line, as in contact stabilization systems. The TF/SC system has primarily been used in the USA. Parker and Bratby (2001) referred to 98 US applications of the process in their review of the TF/SC process.

The high-rate MBBR/AS (BAS) system was introduced after the development of the moving bed biofilm reactor (Ødegaard *et al.*, 1994; Ødegaard, 2006). The MBBR was developed in Norway as a pure biofilm reactor using plastic carriers made of high density polyethylene that were kept within the reactor by a screen. The advantage over an AS reactor is primarily its compactness and no need for recycle.

Helness *et al.* (2005) demonstrated that very high rates of organic matter degradation (35 g SCOD/m^{2}d) could be achieved at high organic loads in an MBBR. The idea behind the BAS process is to have the MBBR reactor remove the easily biodegradable, soluble organic matter and leave the particulate organic matter for the AS with its superior flocculation characteristics. The process has been used for municipal wastewater treatment to increase capacity without increasing footprint (Rusten *et al.*, 1998) but has been especially successful for various industrial treatment purposes (Ødegaard, 2008a).

15.2.2 Integrated fixed-film AS system (IFAS)

Processes using the carrier integrated process principle have been given many names, such as Bio-2, Linpor® , Captor®, Hybas™, and so on. The current most commonly accepted acronym for the general process principle seems to be IFAS (Randall & Sen, 1996).

The biofilm support media in IFAS systems may be introduced in the AS tank in various ways (Figure 15.2):

(a) fixed in the tank in the form of modular plastic media, for example AccuFAS or submerged plastic nets (Lang, 1981; Hirose, 1983; Eberhardt *et al.*, 1984; Müller, 1998; Gebara, 1999) (Figures 15.2a–b).
(b) partly submerged RBC in the AS tank (SURFACT) (Guarino *et al.*, 1977).
(c) cords hanging into the AS tank, also referred to as looped-cord or strand media, in the form of a woven rope with protruding loops (e.g. Ringlace), which provide a surface for growth of biomass (Lessel, 1991, 1994; Jones *et al.*, 1998) (Figure 15.2d).
(d) suspended carriers that move freely with the turbulence that is set up in the AS tank (Walker & Austin, 1981; Hegemann, 1984; Müller, 1998; Bretscher, 2005; Ødegaard, 2008a).

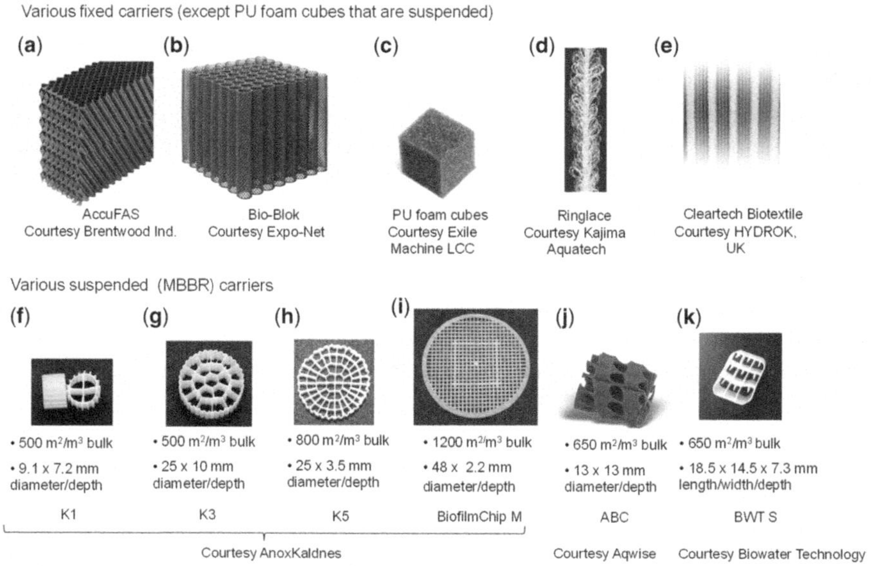

Figure 15.2 Various biofilm carriers used in IFAS systems.

Each type carrier system has its advantages and disadvantages. The Bio-2 sludge system that was introduced in Germany in 1982 (Müller, 1998) used a fixed bed carrier material submerged in the aeration tank. Müller (1998) referred to six German plants using four carrier media types including plastic structure blocs (ACCU-FAS) (Figure 15.2a) and vertical PE tubes with a net-like wall structure welded into cubic blocs (Bio-Blok) (Figure 15.2b). Serious media clogging can occur with fixed modular plastic media and nets (Müller, 1998).

The partially submerged RBC system (SURFACT) (Guarino *et al.*, 1977) has not been used widely. The HYBACS® system with an RBC preceding the AS and with recycle to the RBC seems to be more successful. The loop-cord media system (Ringlace, Figure 15.2d) (Lessel, 1991, 1994) has worked well in some instances, but has eventually collapsed at many plants under the weight of the biomass on them.

Several investigators observed a very high amount of ciliates and rotifers in the biofilm in the fixed bed and Ringlace IFAS systems (Lang, 1981; Eberhardt *et al.*, 1984), which may explain the low effluent SS concentration from several of these systems. Poor nitrification was often observed. Müller (1998) suggested that this might be attributed to uneven distribution of air to the fixed biomass, especially when using Flocor E carrier material in which there are closed channels. The higher organisms (worms) may be grazing on the nitrifiers in the biofilm (Lessel, 1994).

Most success has been experienced with systems based on suspended biofilm carriers (Figures 15.1f and 15.2d–e). In Germany, the Linpor® system (Linde) was developed in the early 1980s (Hegemann, 1984; Reimann, 1990; Morper & Wildmoser, 1990; Morper, 1994). It used 1 cm^3 polyurethane foam cubes as biofilm carriers (Figure 15.2c), added to a normally operated conventional AS system. This was intended to increase the amount of biomass so that a higher volumetric COD turnover could be achieved. This direct effect was not always apparent (Wang *et al.*, 2000), but the SVI decreased after addition of the carriers. This made it possible to increase the MLSS concentration (due to the better performance of the clarifier), which led to an increase of volumetric turnover.

Biomass grew on the outside of the plastic foam as well as in its pores giving a considerable growing surface. The system had some operational challenges, however, mainly because the carriers were so light that they were difficult to distribute throughout the entire AS reactor volume. Frequently the carriers floated and accumulated at the tail end of the normally rectangular reactors against the screen used to keep them in the reactor.

Sen *et al.* (1994) demonstrated N-removal in bench-scale with various media including fixed sponge, free floating sponge and Ringlace.

A special version of the Linpor® system – the so-called Captor® system (Simon-Hartley) (Golla *et al.*, 1994; Reddy *et al.*, 1994) – was developed in parallel in the UK. The sponge cube carriers were larger (2.5 × 2.5 × 1.25 cm) and the removal of excess sludge was carried out by squeezing the carriers off-line.

The breakthrough for IFAS systems came after the introduction of the moving bed biofilm reactor (MBBR) (Ødegaard *et al.*, 1994). The suspended carriers (Figures 15.2f–k) move freely in the water body by the movement set up by aeration in the aerobic reactors and by mixers in the anoxic reactors. One or several usually horizontally mounted cylindrical screens retain the carriers in the reactor (Figure 15.1f). Since this is currently the dominant system, it will be the subject of the remainder of this paper.

15.3 THE MBBR IFAS SYSTEM

15.3.1 Objectives and applications

MBBR IFAS systems are used for a variety of reasons, including enhanced BOD removal, enhanced nitrification, enhanced N removal, enhanced biological P removal, improved settling, reduced footprint, improved operational stability, and so on (Christensson & Welander, 2004) (Figure 15.3). The most frequently used objective is to enhance nitrification, with the goal of being able to fully nitrify at a lower suspended biomass aerobic SRT than in a conventional AS system. This results in a reduced footprint for nitrifying, N removal and BNR plants. Side-effects are improved biomass settling and operational stability. Therefore IFAS systems are frequently used for upgrading conventional AS (CAS) plants.

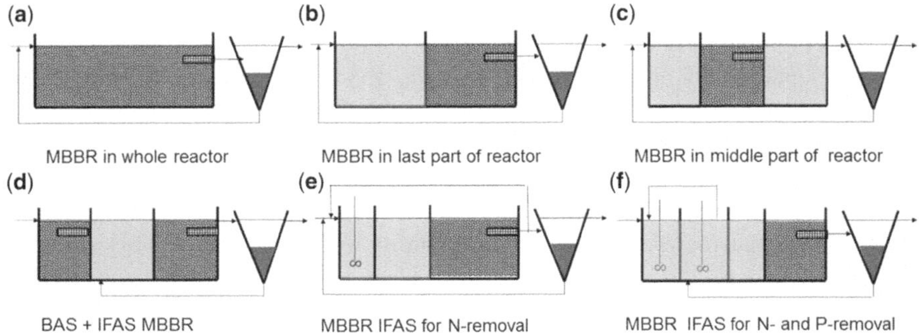

Figure 15.3 Various IFAS systems for different objectives based on moving bed biofilm carriers.

One of the benefits of the MBBR IFAS system is that one need only add the amount of attached growth area (i.e. the amount of carriers) that is necessary. This feature makes the process ideal for upgrading CAS plants. The maximum recommended filling fraction in a pure MBBR system is 67% (Ødegaard *et al.*, 1994). In MBBR IFAS systems, however, filling fractions >60% are seldom used, and one may adapt the biofilm growth area necessary to the volume available. For example, the first AS plant to be upgraded to an MBBR IFAS-plant (Broomfield, Colorado, USA) had a filling fraction of 30% in both aerobic reactors (McQuarrie *et al.*, 2004).

As shown in Figure 15.3 one may add carriers to the entire reactor volume or to one or more segments of the reactor. Figure 15.3 only shows applications where carriers are used in the aerobic part of the reactor train. Carriers can also be used in anoxic and anaerobic reactors, although this is currently uncommon. All of the carrier types shown in Figures 15.2f–h have been used in full-scale MBBR IFAS applications.

Several investigations (Rusten *et al.*, 2003; Germain *et al.*, 2007; Di Trapani *et al.*, 2010) have demonstrated through parallel studies on the same wastewater that an MBBR IFAS system was superior to a conventional AS system. Rusten *et al.* (2003) demonstrated that for a three-reactor pre-denitrification MBBR IFAS pilot plant at Tau WWTP, Norway, an 18% filling fraction (64 m^2/m^3) of the Kaldnes K2 carrier in the two nitrification reactors, increased their capacity 40% over that of an AS system.

15.3.2 Nitrification

Nitrification in MBBR IFAS systems is mainly influenced by the aerobic SRT, the organic carrier area load (relative to ammonium load), the reactor DO concentration, the ammonium concentration and the temperature. Figure 15.4 shows data from the Tau, Norway pilot plant and from the Broomfield, Colorado, USA WWTP (Rusten *et al.*, 2003). The Tau pilot plant, with an 18% filling fraction of 64 m^2/m^3 carriers, achieved complete nitrification with about a 40% lower suspended biomass aerobic SRT than without carriers (Figure 15.4a). The Broomfield pilot plant demonstrated that a DO concentration of 4 mg/L was needed to achieve an effluent of 1 mg NH_4/L (Figure 15.4b). Figure 15.4b also shows that the carrier biomass nitrification rate agreed well the nitrification design model (section 15.5.1) published by Rusten *et al.* (1995) for pure MBBR systems, indicating that this model is also useful in MBBR IFAS systems. Figure 15.4a demonstrates that complete nitrification in the MBBR IFAS pilot plant at the Tau plant was achieved at mixed liquor aerobic SRTs as low as 2.2 d at 15°C.

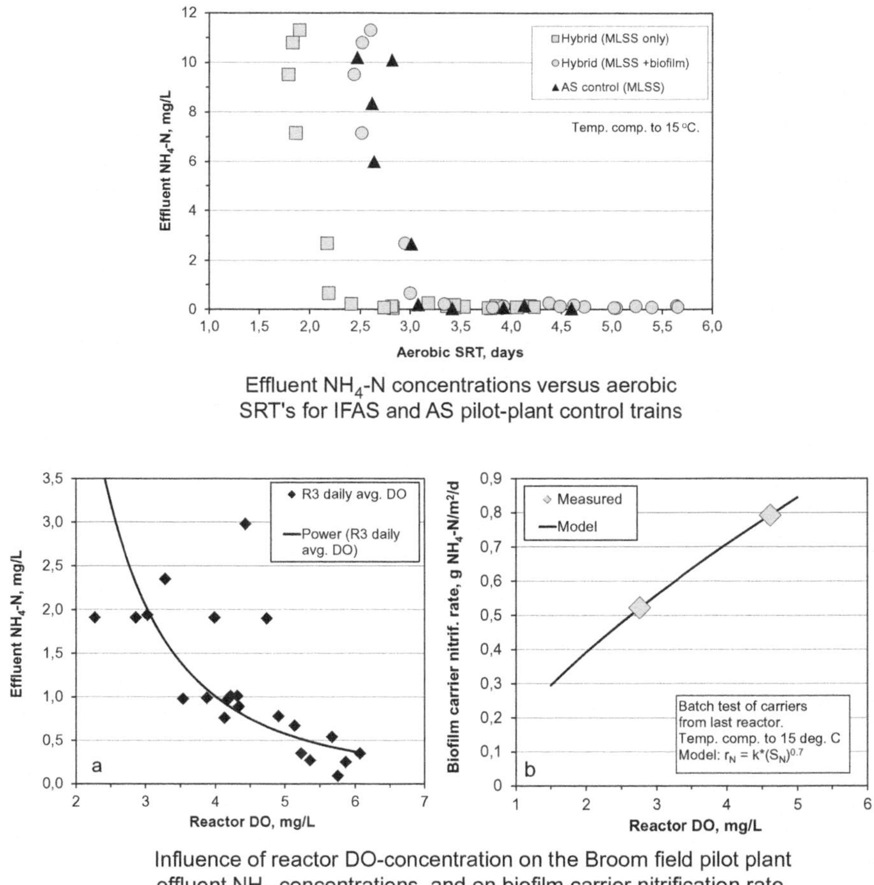

Figure 15.4 Influence of $SRT_{aerobic}$ and DO in MBBR IFAS pilot plants (Rusten *et al.*, 2003).

Figure 15.5 shows the relationship between design SRT and temperature for various US-plants designed for nitrification (Johnson, 2009) together with the curves for nitrifier growth rate and the ATV design curve for nitrification in AS plants. With one exception, the SRTs for these full-scale MBBR IFAS systems are far lower than the German design curve.

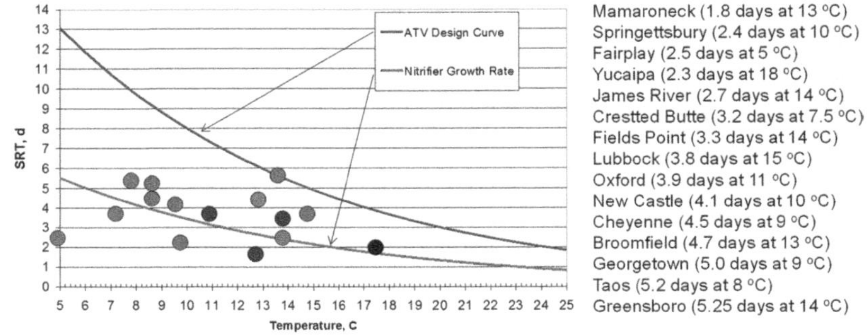

Figure 15.5 Design SRT versus temperature for full scale IFAS systems (Johnson, 2009).

Different models have been used to explain the enhancement of nitrification in IFAS systems; (a) the high attached biomass SRT, (b) an increased nitrifier biomass per unit volume, (c) an increased nitrifier activity in the attached biomass compared to the suspended biomass, (d) the seeding of nitrifiers into the suspended biomass from the attached biomass. McQuarrie (2009) studied the nitrification activity in the suspended biomass and the attached biofilm in an MBBR IFAS system based on data from Thomas (2009). He found that the specific nitrification activity (mgNH_4-N/gSS, h) was much higher in the attached biomass than in the suspended biomass (Figure 15.6a).

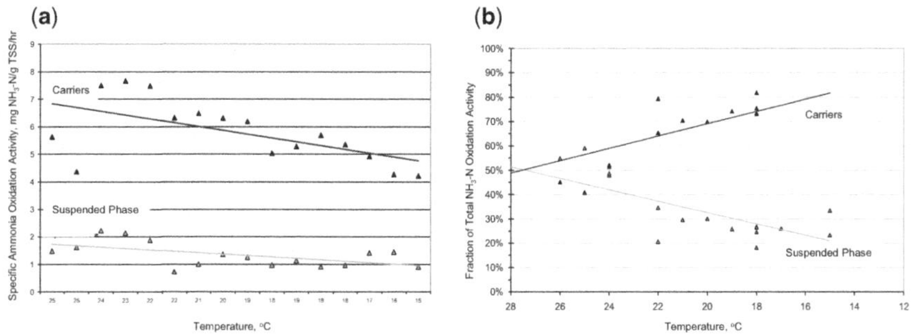

Specific nitrification activity in attached and suspended biomass versus temperature

Fraction of NH_3-N oxidation activity in attached and suspended biomass versus temperature

Figure 15.6 Distribution of nitrification activity in attached and suspended biomass as a function of temperature (McQuarrie, 2009; Thomas, 2009).

A probably explanation for this finding is that the nitrifiers in the attached biomass are always exposed to low soluble biodegradable carbon to ammonia ratio, resulting in an attached biofilm enriched with nitrifying bacteria. McQuarrie *et al.* (2009) also demonstrated that the fraction of NH_4-N oxidation activity taking place on the carriers, increased with decreasing temperature while the fraction in the suspended biomass decreased with decreasing temperature. This demonstrates that the benefit of the carriers in the overall process is greater at low temperatures.

15.3.3 Denitrification

Denitrification (DN) in IFAS systems may be by pre-DN, post-DN and combined pre- and post-DN (Ødegaard, 2006). Pre-DN-systems, in which the anoxic reactor does not contain carriers, are most frequent, but there also is an interesting potential for using carriers in anoxic reactors (Choi *et al.*, 2007) – as in pure MBBR-systems (Ødegaard *et al.*, 1994; Ødegaard, 2006). Post anoxic IFAS reactors are now installed at the Mamaroneck, New York, USA WWTP and at the Lundtofte, Denmark WWTP.

In the previously mentioned Tau pilot plant, Rusten *et al.* (2003) found a >50% higher DN rate in the IFAS pre-anoxic reactor biomass than in the CAS pre-anoxic reactor biomass (3.15 mg NO_x-N/g MLSS, h and 2.00 mg NO_x-N/g MLSS, h respectively). This is an indication that the low aerobic SRT of hybrid processes leads to higher DN-rates than in CAS plants. In systems with higher MLSS and SRT, much of the carbon is lost during the aerobic phase; hence, the shorter the aerobic SRT, the more carbon is available for DN.

15.3.4 Biological P removal

It is quite seldom that plants are upgraded to an IFAS system EBPR. This is possibly because the plants have been operated at low mixed liquor and low SRTs, and EBPR bacteria might wash out of them (especially at low temperatures), making it difficult to maintain EBPR.

Sriwiriyarat and Randall (2005) demonstrated in a pilot-plant investigation with Biowebs® as biofilm carrier that EBPR could be accomplished in three-stage IFAS BNR systems. Reasonably robust EBPR was accomplished in systems with fixed film media installed in the anoxic reactors only, in the aerobic reactors only, and in both the anoxic and aerobic reactors. Even though the net P release and uptake were similar in the IFAS and in the control system (suspended sludge only), the IFAS process tended to have higher P release in the anoxic zones with fixed film media. It appeared that the changes in the COD/TP ratio had a greater impact on EBPR than changes in SRT or the integrations of media into the anoxic zones. The P removal decreased by only 10% when the SRT was decreased from 10 to 6 d, while it decreased by 50% when the COD/TP ratio was increased from 20 to 52. It was confirmed that EBPR could be maintained in the IFAS systems without serious complications.

Onnis-Hayden *et al.* (2011) studied the PAOs and EBPR activity at Broomfield, Colorado, USA plant and found that the PAO activity was predominantly associated with the mixed liquor rather than the biofilm media. The relative abundance of

PAOs and *Accumulibacter*-like PAOs was estimated to be 20–30% and 15.8 ± 1.4% in the mixed liquor and 3–8% and 4 ± 1% in the biofilm, respectively. These results demonstrated that in the IFAS-EBPR process, the N-removing and P-removing populations that require or prefer conflicting SRT values (e.g. >15 d for nitrifiers and <5 d for PAOs) can be decoupled, therefore allowing for separate SRT control and overall optimization of both N and P removal.

15.3.5 Biomass separation in IFAS systems

Several authors (Fouad & Barghava, 2005b; Blank *et al.*, 2007; Briggs, 2009; Kim *et al.*, 2010; Wilson *et al.*, 2012) have reported better AS settling properties after introducing IFAS into an existing AS plant. Figure 15.7 (Briggs, 2009) and Figure 15.8 (Wilson *et al.*, 2012) both show improvements of SVI after introduction of IFAS.

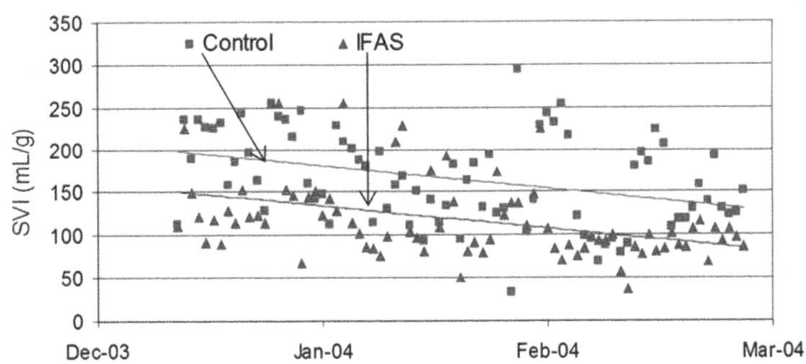

Figure 15.7 Comparison of SVI in two parallel lines, at Lakeview WWTP, Ontario, Canada (Briggs, 2009).

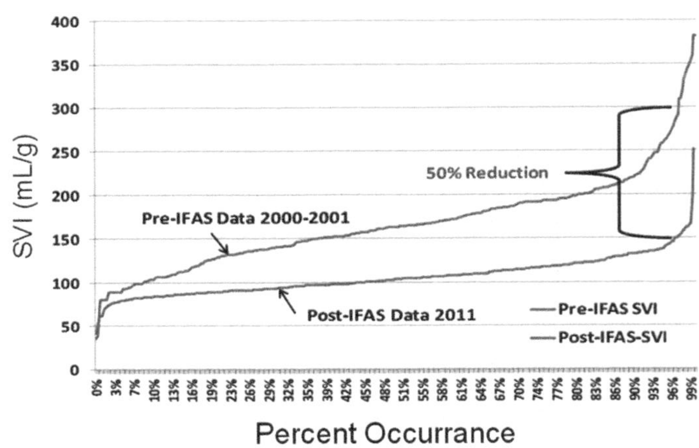

Figure 15.8 Pre- and post-IFAS SVIs at Field's Point WWTP, USA (Wilson *et al.*, 2012).

15.3.6 New applications of IFAS

A new application for the MBBR IFAS is the IFAS ANITA MOX (Christensson *et al.*, 2013; Zhao *et al.*, 2013), a deammonification process, in which the aerobic ammonium-oxidizing bacteria are enriched in the suspended AS and the anaerobic ammonium-oxidizing (anammox) bacteria are enriched in the biofilm on suspended carriers. This partition of the two groups of bacteria involved in deammonification, ensures that the anammox bacteria are not substrate-limited by nitrite. The process has recently been used at the Sjölunda, Sweden WWTP to treat reject water from sludge digestion, with a removal capacity exceeding 2 kg N/m^3, d. In the future, IFAS ANITA MOX may be used as a mainstream anammox technology.

15.4 MODELING OF IFAS SYSTEMS

The mathematical description of MBBR IFAS systems is complex because one must account for transformation processes occurring in both the suspended and biofilm biomass and for the interaction between these two (Albizuri *et al.*, 2009). Even though a wide range of simulation methods are available for both AS (Henze *et al.*, 2000) and biofilm systems (Wanner *et al.*, 2006) only a few models have been developed to simulate hybrid systems. These models are mainly for steady state conditions, to reduce the computational burden of resolving the 1st order differential equation involved in such systems (Sen & Randall, 2008a). Currently modeling is not usually used for design, but for dynamic simulation it may become an important tool for understanding the kinetic behavior of these systems.

Fouad and Bhargava (2005a) proposed a simple mathematical model based on Monod kinetic expressions, for both suspended and attached growth, and the Fickian diffusion law for biofilm. The model aimed at investigating a simple expression for substrate flux into the biofilm. Since the latter was the main parameter in the model and its value changed with HRT, the model is able to accurately describe the hybrid system if an accurate value for the flux is established.

The model proposed by Sriwiriyarat *et al.* (2005) was able to predict carbonaceous removal, nitrification and denitrification at steady state conditions in AS systems with integrated sponge media. Rate expressions and substrate fluxes were used to establish mass balance equations, and the mass balance equations were solved simultaneously with a system of nonlinear equations, using Newton modified techniques. These model predictions agreed very well with experimental data from pilot studies, except for mixed liquor volatile suspended solid (MLVSS) concentrations and denitrification in aerobic reactors.

Takács *et al.* (2007) introduced a modified 1-D dynamic biofilm model for process engineering purposes that could be used with one parameter set for a large range of process situations. The biofilm model is integrated with a general AS/ anaerobic digestion model combined with a chemical equilibrium, precipitation and pH module. This allows the model to simulate the complex interactions that occur in the aerobic, anoxic and anaerobic layers of the biofilm. Both MBBR and IFAS systems were simulated using the same model and parameter set.

Sen and Randall (2008a, b) presented a model for IFAS and MBBR systems that embeds a biofilm model into a multi-cell AS model. The model can operate with up to 12 cells in series and with biofilm media incorporated into one or more cells except for the anaerobic cells. The process configuration can be any combination of anaerobic, anoxic, aerobic, post-anoxic with or without supplemental carbon, and re-aeration; it can also include any combination of step feed and recycles, including recycles for mixed liquor, RAS and nitrate. The biofilm flux rates for organics, nutrients, and biomass can be computed either by using a simple semi-empirical biofilm model or a computationally intensive diffusion model.

Boltz *et al.* (2009a, b) proposed a mathematical model for steady-state IFAS and MBBR simulation. The model was an extension of the IWA ASM2d (Henze *et al.*, 2000) and biofilm modeling techniques presented by Wanner *et al.* (2006). The model included simultaneous diffusion and Monod-type reaction kinetics inside the biofilm, competition between aerobic autotrophic nitrifiers, non-methanol-degrading facultative heterotrophs, methanol degrading heterotrophs, slowly biodegradable chemical oxygen demand, and was inert for substrate and space inside the biofilm. The model assumed that the reaction kinetics was identical for bacteria within suspended biomass and biofilm.

Mannina *et al.* (2011) developed a simplified model in which the computational burden was reduced by replacing the 2nd order differential equations typical of some of the existing models. The model comprises two connected Monod-type models for the simulation of suspended biomass and biofilm, which are in competition for electron donors and acceptors. In addition to biochemical conversions, the model contains a simulation of particulate detachment from biofilm into the bulk liquid, which represents the connection between suspended biomass and biofilm compartments, since detached biofilm is converted to active suspended biomass.

15.5 DESIGN OF IFAS SYSTEMS

15.5.1 Design procedures

Three types of design procedures for IFAS systems have evolved over time:

(a) Purely empirical experience based methods that normally use AS design procedures as a base and then compensates for the fact that an attached biomass is included, by for instance using an 'equivalent sludge age' approach.
(b) Semi-empirical methods based on steady-state rate expressions developed for AS and MBBR processes respectively and combined in such a way

that the results fit full scale MBBR IFAS data. For such methods, spreadsheets have been developed and are used by manufacturers, process suppliers and consultants, allowing for a certain degree of analysis and optimization.

(c) Methods based on dynamic, process kinetic models with different levels of integration of AS and biofilm kinetics.

Currently it is fair to say that the semi-empirical procedures are dominant for design but that dynamic models are increasingly being used for checking the design result and analyzing operational options for the MBBR IFAS system. The steady state calculations may be entirely SRT based (similar to several AS procedures) or entirely process rate based (similar to biofilm processes) or a combination of both. The approach may be based on whether or not one is upgrading an existing AS-plant or designing a greenfield plant.

Most focus has been given to the design of the nitrification step of the IFAS since BOD removal, denitrification and possibly EPBR is often carried out in reactors without carriers and design of these reactors is similar to that used for CAS. Different approaches are used to determine the nitrification rate of the suspended biomass that will depend on the relative amount of autotrophic and heterotrophic biomass.

One approach when upgrading an existing plant is to use CAS-design to determine the amount of biomass that is needed to ensure nitrification. Based on this, the available reactor volume and the MLSS, the specific nitrification rate (g NH_4-N/kg MLSS, d) is determined and this rate is later used for the suspended biomass in the hybrid reactor, while the nitrification capacity of the attached biomass is determined in the same way as for a pure MBBR-process (see model below). Another approach is based on the availability of organic matter relative to NH_4-N, that is the C/N-ratio (gBOD/gNH_4-N) in the influent to the reactor. Figure 15.9 shows the relationship between MLSS, nitrification rate and C/N-ratio proposed by Ødegaard (2008b).

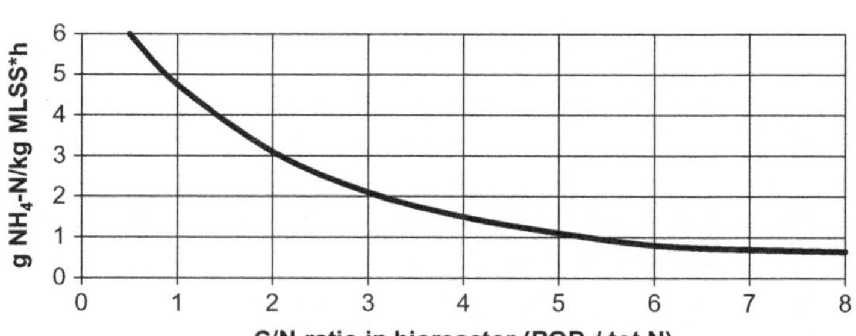

Figure 15.9 Nitrification rate in MLSS *vs.* C/N of influent to the bioreactor (15°C) (Ødegaard, 2008b).

Determination of the nitrification rate in the attached biomass may be based on the rate equation developed for the pure MBBR-process (Hem *et al.*, 1994; Rusten *et al.*, 1995):

$r_N = k(S_n)^n$
r_N = nitrification rate (g NH_4-N/m^2, d)
k = reaction rate coefficient
n = reaction order constant, can be estimated at $n = 0.7$
S_n = rate-determining ammonium concentration, mg NH_4-N/L (can be estimated from $S_n = (DO_{bulk} - 0.5)/3.2$

k depends on the bioreactor influent composition (the C/N-ratio) and can be determined from Figure 15.10. If there is more than one hybrid aerobic reactor or if the hybrid reactor follows suspended reactors that are designed to remove essentially all BOD, then (i) the influent BOD to the final hybrid reactor may be assumed to be zero, and (ii) the nitrification rate coefficient for these reactors may be set at 0.75 (Ødegaard, 2008b).

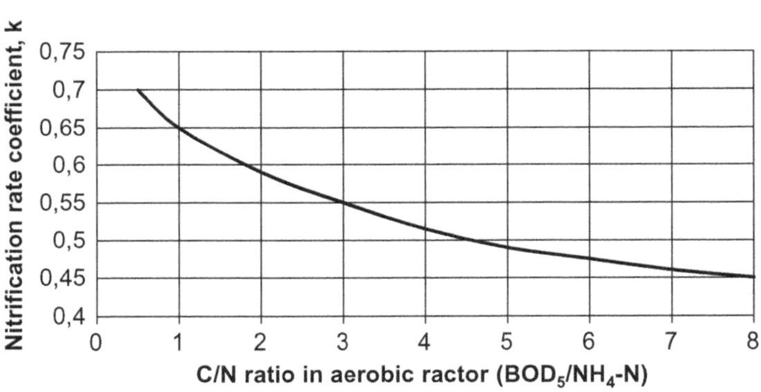

Figure 15.10 Reaction rate coefficient in the attached biomass *versus* C/N of influent to the bioreactor (15°C) (Ødegaard, 2008b).

It has been shown in practice that there is a relationship between the nitrification taking place in the attached and the suspended biomass, which depends on the bioreactor load and can be described by either the aerobic SRT_{MLSS} or the F/M-ratio. The higher the SRT_{MLSS} the lower is the fraction of nitrification taking place in the biofilm. This has to be corrected for when determining the actual nitrification rate of the attached biomass. The maximum rate is multiplied by a correction factor (K) to obtain the actual rate. Ødegaard (2008b) proposed that the correction factor decreased linearly from 1 to 0.2 when the aerobic SRT_{MLSS} increased from 2 to 8 d.

McQuarrie *et al.* (2010) proposed a design procedure based on the use of a nitrification safety factor ($SF_{Nitrification}$) defined as the ratio of aerobic SRT attained with the suspended growth component of the IFAS system and the calculated nitrifier minimum SRT. They claimed that CAS systems that operate at sufficient

$SF_{Nitrification}$ (e.g. >2.0) do not benefit from the addition of plastic carriers since the SRT is high enough to retain a robust mass fraction of autotrophic organisms. For SF < 2.0 McQuarrie *et al.* (2010) proposed four conceptual approaches for IFAS system design based on the typical flow diagram shown in Figure 15.11.

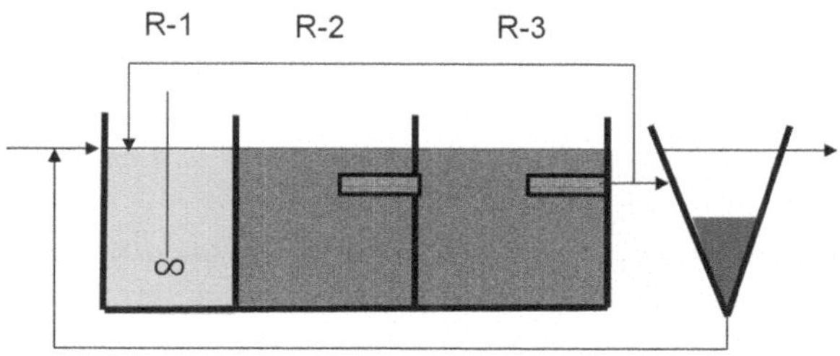

Figure 15.11 Schematic of the typical IFAS system for which the $SF_{Nitrification}$ design approach is used (McQuarrie *et al.*, 2010).

Table 15.1 indicates whether carriers are recommended in one or both aerated reactors and the substrate that governs selection of ammonia flux ($J_{F,\ NH3\text{-}N}$) based on an analysis of the results of four IFAS plants.

Table 15.1 Four approaches for IFAS system design as proposed by McQuarrie *et al.* (2010).

Design approach	$SF_{Nitrification}$	Effluent ammonia target, mg NH_3-N/L	Install media in reactor?		Parameter for determining ($J_{F,\ NH3\text{-}N}$)	
			R-2	R-3	R-2	R-3
A	0.5–1.0	<2.0	Yes	No	S_{O2}[2]	–
B	0.5–1.0	<1.0	Yes	Yes	S_{O2}	S_N
C	1.0–1.5	<1.0	Yes	Option[1]	S_{O2}	S_N
D	1.5–2.0	<1.0	Option[1]	No	S_{O2}	–

[1]Plastic carriers added to reactor to increase reliability against system perturbations.
[2]Contribution of nitrification achieved by suspended growth component of the IFAS system is ignored.

Various consulting companies and suppliers of IFAS systems have developed their own design models and spreadsheet design procedures, some of which are based on experience and some on dynamic modeling.

15.5.2 Compartment partition

There are different views on where to place the biofilm carriers in the process train. If needed, carriers may be placed in the entire aerated AS reactor. In most cases, however, they are placed in the last compartment of the aerobic reactor.

Here, however, the ammonium concentration is at its lowest, and does not favor the higher nitrification capacity of the attached biomass.

Placing the media in an early stage of a nitrification reactor may have several advantages:

- Here the ammonium concentration is highest, favoring the nitrification capacity of the attached biomass.
- The DO concentration may be reduced in the later compartment since the ammonium concentration here will already be low. Less oxygen is returned to the anoxic reactor (in the case of N-removal).
- The reduced intensity of mixing (because of less intensive aeration) improves flocculation and hence subsequent biomass separation.
- 'Seeding' of nitrifiers from the attached to the suspended biomass may be favored.

Di Trapani *et al.* (2011) demonstrated in a pilot-plant operating at low temperature (11.5°C), the effectiveness of the design shown in Figure 15.3c. A much higher attached biomass activity (average 4.9 mgNH_4-N/gVSS, h) than in the suspended biomass (2.2 mg NH_4-N/gVSS, h) was found. The use of at least two aerobic reactors in series is highly recommended since in this way the ammonium concentration in the 1st aerobic reactor will always be high enough and its biofilm will not be limited by a lack ammonium. Furthermore, the use of two reactors enhances operational flexibility with respect to automatic control (Albizuri *et al.*, 2010).

15.5.3 Oxygen transfer

Normally, one designs for a higher DO and aeration capacity in IFAS systems than in CAS systems because nitrification in biofilms may be governed by DO-diffusion limitation. Normally one would choose an average DO of approximately 3 mg/L and a peak design DO of 5 mg/L.

There is some debate about what type of aeration system to choose. Some suppliers of IFAS systems recommend medium-sized bubble aerators while others recommend fine bubble diffusers. It has been demonstrated that the oxygen transfer in a tank filled with carriers is considerably higher than without carriers and that it increases with carrier filling fraction up to approximately 60%, where values above 15 gO_2/Nm^3, m have been measured in clean water tests (Christensson, 2011).

At the Field's Point, Providence, RI WWTP (USA) a full scale clean water oxygen transfer test was performed (Wilson *et al.*, 2012) using the medium-sized bubble aeration system of AnoxKaldnes. The bubble pattern was uniform throughout each of the three tests that showed an SOTR of approximately 12 g O_2/Nm^3, m in the first two tests and approximately 14 g O_2/Nm^3, m in the third test. These are quite high OTR values and in good agreement with tests made in Sweden with the same aeration system (10–15 g O_2/Nm^3, m) depending on the filling, carrier type and air flow (Christensson, 2011).

15.5.4 Approach velocity and screen design

The screens are normally built from cylindrical wedge wire or perforated plates. A screen aerator is often installed to prevent excessive head-loss caused by media build-up at the screen. The screen velocity is normally designed at 50–60 m/h for the screen types mentioned above. Biofilm carrier approach velocity is an important design parameter for minimizing hydraulic accumulation of carrier media at the screens and the possibility of structural failure of the screens or overflowing the reactor. It is normally recommended that the approach velocity be <30 m/h at peak flow at a length/width-ratio of 1.0 decreasing linearly to 15 m/h at a length/width-ratio of 3.0. A comprehensive discussion on screens and hydraulic problems has been given by Leaf *et al.* (2011).

The approach velocity may be challenging when converting an existing plug flow CAS to IFAS. An example is the design proposed for the Sha Tin, Hong Kong WWTP (AECOM, 2013) (Figure 15.12).

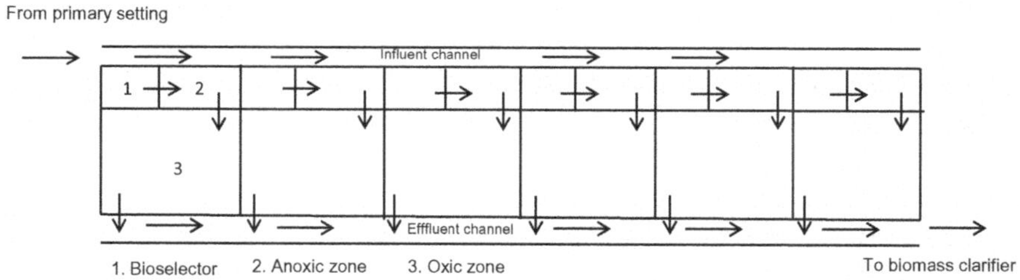

Figure 15.12 Potential bioreactor configuration for the IFAS option feasibility study for the relocation of Sha Tin WWTP to caverns in Hong Kong (AECOM, 2013). Courtesy of Drainage Services Department, Government of Hong Kong Special Administrative Region.

15.6 OPERATION OF MBBR IFAS SYSTEMS

Operation of MBBR IFAS systems is similar to CAS plant operation, but there appears to be an improvement in robustness and resistance to load variations (Maas *et al.*, 2007). Since the nitrification rate is dependent on DO, it is useful to control nitrification rate by measuring ammonium (or nitrate) out of the IFAS and let this concentration control the aeration blowers. Operational challenges that are frequently mentioned are foaming, especially at start-up of plants and foam accumulating in the anoxic reactors (Wilson *et al.*, 2012).

15.6.1 Full-scale MBBR IFAS examples

At the time of writing, there are approximately 200 IFAS plants operating worldwide that range in size from very small to quite large. Many more are under planning and construction. The first large plant was at Broomfield, Colorado, USA (30,000 m^3/d), which was retrofitted from a traditional BNR plant to an IFAS BNR plant in 2003. Considerable testing and follow-up has taken place at this plant. The James River WWTP, Virginia, USA , which was designed with low SRT (2.7 d) has been in operation for approximately five years. Very large plants in operation today are Field's Point, Rhode Island, USA (291,000 m^3/d max flow and 189,000 m^3/d ave. flow) and the Buji, China WWTP (200,000 m^3/d). The largest plant in Europe is the Marquette-Lez-Lille, France WWTP (242,000 m^3/d). Table 15.2 presents some design and operation features of selected plants.

The *Broomfield, Colorado, USA WWTP* was upgraded in 2003 to one of the first IFAS BNR plants with both anaerobic and anoxic IFAS reactors. Rutt *et al.* (2006) reported that this allowed the existing aerobic reactors to maintain year-round nitrification at close to the conventional AS SRTs typical of carbonaceous treatment plants.

The *James River, HRSD, Virginia, USA WWTP* consists of pretreatment and primary sedimentation, followed by nine plug-flow aeration tanks, five secondary clarifiers and chlorine disinfection. After upgrading to IFAS, substantially improved nitrification performance at cold temperature conditions (McQuarrie *et al.*, 2009) has been demonstrated. Interestingly, low effluent TP has been observed despite (a) no formal anaerobic zone, and (b) no metal salt addition. It has been speculated that the low SRT possible with an IFAS system may in fact make it easier to maintain viable PAOs while out-selecting GAOs.

The *Field's Point, Providence, Rhode Island, USA WWTP* is to be upgraded to achieve a monthly average seasonal effluent permit limit of 5 mg/L total N. Six of the ten reactors are currently converted for IFAS operation. When completed in 2014, it is expected to be one of the largest IFAS facilities in the world.

The *Marquette-Lez-Lille, France WWTP* is a greenfield Hybas™ plant placed in operation in 2013. After pre-treatment in Multiflo, biological treatment takes place in three parallel trains combining pre- and post-denitrification. P is removed with EBPR and tertiary treatment is provided by Hydrotech filters. Combining a suspended carrier with high protective surface area (AnoxKaldnes BiofilmChip M, 1200 m^2/m^3) with a high filling degree (52%) gives a very compact plant. The footprint of the new plant, including enhanced N and P removal is only half of the old plant that was previously used for carbonaceous removal.

The Sharjah, UAE WWTP, has one line (37,500 m^3/d) which is a highly loaded CAS-plant retrofitted to be a nitrifying plant (effluent standard 1 mg NH_4-N/L) at the double the organic load without additional tankage. The existing single stage aeration tank (13,000 m^3) was converted into a three-stage process, as in Figure 15.2.d, with a highly loaded, pure MBBR (BAS) step (20% of tank volume, 50% fill fraction of AnoxKaldnes K3 carrier) followed by an IFAS in two steps, the 1st step without carriers (50% of tank volume) and the 2nd step with carriers (30% of the tank volume, at a 50% K3 fill fraction). A performance test carried out in April and May 2010, showed averages of 6.4 mg BOD_5/L (98% removal), 0.92 mgNH_4-N/L (97% removal) and 98.3% TKN-removal.

Table 15.2 Design and operation values for some IFAS plants.

Plant	Broomfield, CO, USA (Phase 1)	James river, VA, USA (Demonstration)	Field's point, RI, USA	Marquette-Lez-Lille, France
Reference	Rutt (2011); Rutt *et al.* (2006)	McQuarrie *et al.* (2010)	Wilson *et al.* (2012)	
Type of plant				
– Primary settlers	Yes	Yes	Yes	Yes (Multiflo)
– Upgrading CAS/Greenfield	Upgrading	Upgrading	Upgrading	Greenfield
Design flows				
– Ave. daily design flow (m^3/d)	25,400	6820	189,000	136,000
– Max. design capacity (m^3/d)	30,320	7580	291,000	242,000
Average daily loads (kg/d)				
– BOD_5	4410	953	19,833	33,320
– Tot N	1236	270	3430	7860
– C/N-ratio (BOD_5/total N)	3.6	3.5	5.8	4.2
Effluent requirements				
– $SBOD_5$	10	–	10	15
– Nitrification (mg NH_4-N/L)	1.5	–	1.0	2
– N-removal (mg total N/L)	10	12	5.0	10
– P-removal	1	1	–	1

(Continued)

Table 15.2 Design and operation values for some IFAS plants (*Continued*).

Plant	Broomfield, CO, USA (Phase 1)	James river, VA, USA (Demonstration)	Field's point, RI, USA	Marquette-Lez-Lille, France
Bioreactor				
– Total volume (m^3)	6833	1906	36,000	57,800
– MLSS SRT (d)	7	4.5	4.0	8
– Max MLSS/(mg/L)	3500	2500	2750	4000
– Aerobic reactor				
– Aerobic volume	4546	918	15,000	20,050
∘ Carrier and fill fraction	K1	K3	K3	BiofilmChipM
∘ Fill fraction	30%	50%	45%	52%
∘ DO (mg O_2/L)	>3.0	4.0	4.0	4.0
∘ Nitrific. rate in biofilm, (g/m^2,d)	–	0.844	0.78	0.6
– Anoxic reactor				
∘ Pre-anoxic volume	1319	686	3400	18,800
∘ DN-rate MLSS (g/kg,h)	–	–	5.5	3.0
∘ Post-anoxic volume	None	None	9000	8400
∘ DN-rate MLSS (g/kg,h)	–	–	3.6	5.0
– Anaerobic reactor volume (m^3)	968	None	None	10,550
– Design temperature, °C	14	14	14	12
Plant operation				Start-up during
– Average effluent conc. (mg/L)				2012
∘ BOD_5	3	8	9	No data available
∘ SS	4	6	8	
∘ NH_4-N	0.4	1.3	1.2	
∘ Tot-N	15	10.1	4.0	
∘ Tot P	0.9	1.1	–	
– Average RAS (%)	35	80	25	
– Average SVI (mL/g)	120	120	120	

15.7 CONCLUSIONS

Hybrid systems (combination of biofilm and AS systems) have been used in various forms since the 1980s, but the breakthrough for the so called IFAS (Integrated Fixed Film Activated Sludge) systems came after the turn of the century with the introduction of the combination of the moving bed biofilm reactor (MBBR) and an AS reactor. Plastic carriers with a specific surface areas in the range of 300–1200 m^2/m^3 are introduced to the whole, or part of an AS reactor and retained there by screens.

(1) The IFAS systems have proven especially suitable when an AS plant is to be upgraded to accommodate nitrification or nutrient removal, where the existing plant does not have sufficient reactor volume to achieve this. IFAS systems have, however, also been used in a number of greenfield plants where space is a limiting factor. The space needed for an IFAS system is typically 40–60% of that for an equivalent conventional AS system.

(2) When used for enhancing nitrification, it is found that:

   (a) Nitrification is essentially independent of the aerobic MLSS SRT and can be achieved at SRTs in the suspended biomass as low as approximately 2 d, because it is governed by the nitrification taking place in the biofilm biomass when the loading is high (low MLSS SRT). At lower organic loadings more nitrification takes place in the suspended biomass.

   (b) The nitrification activity of the biomass (g NH_4-$N_{removed}$/kg VSS, h) is far higher in the biofilm biomass (up to 8–9 g NH_4-$N_{removed}$/kg VSS, h at 20°C) than in the suspended biomass (normally 2–3 g NH_4-$N_{removed}$/kg VSS, h at 20°C) at a typical influent C/N-ratio of 3–4 and DO of 4 mg O_2/L.

   (c) As in every biofilm system, the nitrification rate is strongly influenced by reactor DO concentration. It is not advisable to design for a DO below 4 mg O_2/L. Oxygen transfer is high (freshwater SOTR of around 12 g O_2/Nm^3, m provided the filling fraction is >50%) even with the use of medium-bubble diffusers, because of the presence of the carriers.

   (d) Design of nitrification reactors may be based on the design in pure biofilm systems with an additional effect from the suspended biomass – depending on the load (C/N-ratio or SRT).

   (e) As the temperature decreases, the fraction of the total nitrification that takes place in the biofilm biomass increases. This makes the IFAS resilient to temperature variations.

(3) Most commonly the anaerobic and anoxic reactors are designed without carriers, but there is an increasing interest in using carriers also in stirred reactors to reduce footprint even more. Since application of carriers means that the $SRT_{aerobic}$ will be lower, more carbon will be available in the anaerobic and anoxic reactors leading to higher reaction rates.

(4) Knowledge of the design of IFAS reactors has not yet reached the same level as that for AS systems. Most designs are made on basis of design rules of pure biofilm systems and AS systems respectively. Modeling of IFAS systems can help in evaluating designs.
(5) Experience shows that the operation of IFAS plants is simpler and more robust than for conventional AS plants. The SVI decreases after introduction of the carriers.

15.8 ACKNOWLEDGEMENTS

The authors would like to acknowledge Joshua P. Boltz, CH2M HILL, Tampa, Florida, USA, James P. McQuarrie Metro Wastewater Reclamation District, Denver, Colorado, USA, Chandler Johnson, World Water Works, Oklahoma City, Oklahoma, USA, Mark Steichen, Black and Veatch, Kansas City, Missouri, USA, Daniele di Trapani, Università degli Studi di Palermo, Italy and Bjørn Rusten, Aquateam, Norway for their valuable support and comments during the preparation of this chapter.

15.9 REFERENCES

AECOM (2013). Relocation of Sha Tin Sewage Treatment Works to Caverns – Feasibility Study. Report on Sewage and Sludge Treatment Facilities for relocated STSTW (Draft). AECOM, Asia Co., Ltd, Hong Kong.

Albizuri J., Santamaria A. and Larrea L. (2010). IFAS process for nitrogen removal with an HRT of 4 hours. Proc. WEF/IWA Biofilm Reactor Technology Conf., 2010, Portland, Oregon, USA, August 15–18.

Atkinson B., Black G. M. and Pinches A. (1980). Process intensification using cell support systems. *Process Biochem.*, **15**(4), 24–32.

Blank S. A., Hoffmann H. and Hahn H. H. (2007). *Integration von schwimmenden Auf-wuchskörpern in Belebungsanlagen zur Erhöung der Sedimentationsleistung von Nachklärbecken* (Integration of floating biofilm carriers in activated sludge plants for the improvement of the settling capacity in post sedimentation tanks). *KA- Abwasser Abfall*, **54**(5), 484–487 (in German).

Boltz J. P., Johnson B. R., Daigger G. T. and Sandino J. (2009a). Modeling integrated fixed-film activated sludge and moving-bed biofilm reactor systems I: mathematical treatment and model development. *Water Environ. Res.*, **81**(6), 555–575.

Boltz J. P., Johnson B. R., Daigger G. T., Sandino J. and Elenter D. (2009b). Modeling integrated fixed-film activated sludge and moving-bed biofilm reactor systems II: evaluation. *Water Environ. Res.*, **81**(6), 576–586.

Bretscher U. (2005). Enhancing the Performance of the Activated Sludge Process. Article published by KUSTER + HAGER Civil Engineers Ltd, CH-8730 Uznach, Switzerland.

Briggs T. (2009). Full-Scale Testing of IFAS at G.E. Booth (Lakeview) WWTP – Lessons learned. Presentation at workshop on biofilm reactors, *WEFTEC*, Orlando, October 11.

Choi N., Phillips H., Johnson T. L., Butler R. and Lauro T. (2007). Pilot testing of MBBR and IFAS treatment processes for nitrification and denitrification at the Mamaroneck WWTP. Proc. 80th WEFTEC – National Conference of the Water Environ. Fedn, San Diego, CA., **80,** October 13–17.

Christensson M. (2011). Moving on with MBBR. Proc. WEF/IWA Conf. on Nutrient Recovery and Management, Workshop A. Miami, Florida, USA, January 9–12 2011.

Christensson M. and Welander T. (2004). Treatment of municipal wastewater in a hybrid process using a new suspended carrier with large surface area. *Water Sci. Technol.,* **49**(11–12), 207–214.

Christensson M., Ekström S., Andersson Chan A., Le Vaillant, E. and Lemaire, R. (2013). Experience from start-ups of the first ANITA Mox plants. *Water Sci. Technol.,* **67**(12), 2677–2684.

Di Trapani D., Mannina G., Torregrossa M. and Viviani G. (2010). Comparison between hybrid moving bed biofilm reactor and activated sludge system: a pilot plant experiment. *Water Sci. Technol.,* **61**(4), 891–902.

Di Trapani D., Christensson M. and Ødegaard, H. (2011). Hybrid activated sludge/biofilm process for the treatment of municipal wastewater in a cold climate region: a case study. *Water Sci. Technol.*, **63**(6), 1121–1129.

Eberhardt H., Klee O. and Weber W. (1984). Increase in efficiency of an overloaded activated sludge plant by installation of submerged solid blocks. *Wasserwirtschaft,* **74**, 47–53 (in German).

Fouad M. and Bhargava R. (2005a). A simplified model for the steady-state biofilm-activated sludge reactor. *J. Environ. Management,* **74**, 245–253.

Fouad M. and Bhargava R. (2005b). Sludge production and settleability in biofilm-activated sludge process. *J. Environ. Eng.,* **131**(3), 417–424.

Gebara F. (1999). Activated sludge biofilm wastewater treatment system. *Water Res.,* **33**(1), 230–238.

Germain E., Bancroft L., Dawson A., Hinrichs C., Fricker L. and Pearce P. (2007). Evaluation of hybrid processes for nitrification by comparing MBBR/AS and IFAS configurations *Water Sci. Technol.,* **55**(8–9), 43–49.

Golla P. S., Reddy M. P., Simms M. K. and Laken T. J. (1994). Three years of full-scale Captor® process operation at Moundsville WWTP. *Water Sci. Technol.,* **29**, 175–181.

Guarino C. F., Nelson M. D. and Marrazzo W. J. (1977). Philadelphia develops SURFACT. Paper presented at 50th Ann. WPCF Conf., Philadelphia. October 2–7.

Hegemann W. (1984). A combination of the activated sludge process with fixed film biomass to increase the capacity of waste water treatment plants. *Water Sci. Technol.,* **16**, 119–130.

Helness H., Melin E., Ulgenes Y., Järvinen P., Rasmussen V. and Ødegaard H. (2005). High-rate wastewater treatment combining a moving bed biofilm reactor and enhanced particle separation. *Water Sci. Technol.*, **52**(10–11), 117–127.

Hem L., Rusten B. and Ødegaard H. (1994) Nitrification in a moving bed reactor. *Water Res.*, **28,** 1425–1433.

Henze M., Gujer W., Mino T., Matsuo T., Wentzel M., Marais G. v. R. and van Loosdrecht M. C. M. (2000). Activated sludge model No. 2d, In: Activated Sludge Models ASM1, ASM2, ASM2d and ASM3, M. Henze, W. Gujer, T. Mino, T. Matsuo, M. Wentzel,

G. v. R. Marais and M. C. M. van Loosdrecht (eds), Scientific and Technical Report No. 9, IWA Publishing, London, UK, pp. 75–98.

Hirose M. (1983). *Ein neues System einer Kombination des Belebungsverfahrens mit sessilen Organismen auf Aufwuchsflächen* (A new system for combining activated sludge systems with sessile organisms on growth carriers). *GWF-Wasser Abwasser.*, **124**, 239–242 (in German).

Hui A. M., Michalczyk B. L., Tekippe R. J., Coe D. A. and Stallings R. B. (1983). Pilot study investigation of TF/SC and RBC/SC processes. Paper presented at the *55th Ann. WPCF Conf.*, Atlanta, October 4.

Johnson C. H. (2009). Personal communication.

Jones R. M., Sen D. and Lambert R. (1998). Full scale evaluation of nitrification performance in an integrated fixed film activated sludge process. *Water Sci. Technol.*, **38**(1), 71–78.

Kim H., Gellner J. W., Boltz J. P., Freudenberg R. G., Gunsch C. K. and Schuler A. J. (2010). Effects of integrated fixed film activated sludge media on activated sludge settling in biological nutrient removal systems. *Water Res.*, **44**(5), 1553–1561.

Lang H. (1981). Nitrification in biological treatment with the aid of 'BIO-2-SCHLAMM processing'. *Wasserwirtschaft,* **71**, 166–169 (in German).

Leaf W. R., Boltz J. P., McQuarrie J. P., Menniti A. and Daigger G. T. (2011). Overcoming hydraulic limitations of the integrated fixed-film activated sludge (IFAS) process. Proc. 84th Annual WEFTEC, Los Angeles, CA, USA, pp. 5236–5256.

Lessel T. H. (1991). First practical experience with submerged rope-type (Ringlace) biofilm reactors for upgrading and nitrification, *Water Sci. Technol.,* **23,** 825–834.

Lessel T. H. (1994). Upgrading and nitrification by submerged bio-film reactor – experiences from a large scale plant. *Water Sci.Technol.,* **29**(10/11), 167–174.

Maas C. L. A., Parker W. J. and Legge R. L. (2007). Detachment of Solids and Nitrifiers in IFAS Systems, Proc. 80th Annual WEFTEC, San Diego, CA.

Mannina G., diTrapani D., Viviani G. and Ødegaard H. (2011). Modeling and dynamic simulation of hybrid moving bed biofilm reactors: model concepts and application to a pilot plant. *Biochem. Eng. J.,* **56**, 23–36.

McQuarrie J. P. (2009). Integrated fixed film activated sludge (IFAS). Proc. WEFTECH, 2009, Workshop W213 Biofilm Reactors: Application to Today's Global Wastewater Challenges, Orlando, FL.

McQuarrie J., Rutt K., Seda J., Haegh M. (2004). Observations from the First Year of Full-scale Operation – The IFAS/BNR Process at the Broomfield Wastewater Reclamation Facility. Proceedings of the Water Environment Federation, WEFTEC 2004: Session 91 through Session 95, Broomfield, CO, pp. 274–285.

McQuarrie J. P., Waltrip D., Rutherford R., Thomas W., Bott C., Baumler R. and Katehis D. (2009). Full-scale design challenges for the James River Treatment Plant IFAS improvements project: Using a demonstration scale project to optimize final design. Proc. WEF/IWA Nutrient Removal Conf., Baltimore, MD, USA.

McQuarrie J. P., Boltz J. P. and Daigger G. T. (2010). Interactions between suspended biomass and biofilm in integrated fixed-film activated sludge (IFAS) bioreactors: Process design implications for IFAS systems. Proc. WEF/IWA Biofilm Reactor Technology Conf., Portland, OR.

Morper M. (1994). Upgrading of activated sludge systems for nitrogen removal by application of the Linpor-CN process. *Water Sci. Technol.,* **29,** 167–176.

Morper M. and Wildmoser A. (1990). Improvement of existing wastewater treatment plant efficiencies without enlargement of tankage by application of the Linpor process – case studies. *Water Sci. Technol.,* **22**(7/8), 207–215.

Müller N. (1998). Implementing biofilm carriers into activated sludge process – 15 years of experience. *Water Sci. Technol.,* **37**(9), 167–184.

Norris D. P., Parker D. S. and Daniels M. L. (1980). Efficiencies of advanced waste treatment obtained with upgraded trickling filters. *J. Civ. Eng Amer. Soc. Civil Eng.,* **50**(9), 78–81.

Norris D. P., Parker D. S., Daniels M. L. and Owens E. L. (1982). High-quality trickling filter effluent without tertiary treatment. *J. Water Pollut. Control Fedn.,* **54**(7), 1087–1098.

Ødegaard H. (2006). Innovations in wastewater treatment: the moving bed biofilm process. *Water Sci. Technol.,* **53**(9), 17–33.

Ødegaard H. (2008a). The use of the moving bed biofilm reactor (MBBR) technology for industrial wastewater treatment. Proc. IWA Specialized Conf. on Industrial Water Treatment Systems, Amsterdam, October.

Ødegaard H. (2008b). Design procedures for HYBAS. Internal working document. Scandinavian Environmental Technology AS, Trondheim, Norway.

Ødegaard H., Rusten B. and Westrum T. (1994). A new moving bed biofilm reactor – applications and results. *Water Sci. Technol.,* **29**(10–11), 157–165.

Onnis-Hayden A., Majed N., Schramm A. and Gu A. Z. (2011). Process optimization by decoupled control of key microbial populations: distribution of activity and abundance of polyphosphate-accumulating organisms and nitrifying populations in a full-scale IFAS-EBPR plant. *Water Res.,* **45**, 3845–3854.

Parker D. S. and Bratby J. R. (2001). Review of two decades of experience with TF/SC process. *J. Environ. Eng.,* **127**, 380–387.

Rakness K. L., Schultz J. R., Hegg R. A., Cranor J. C. and Nisbet R. A. (1982). Full Scale Evaluation of Activated Bio-Filter Wastewater Treatment Process. USEPA Rept 600/S2-82-067.

Randall C. W. and Sen D. (1996). Full-scale evaluation of an integrated fixed-film activated sludge (IFAS) process for enhanced nitrogen removal. *Water Sci. Technol.,* **33**(12), 152–162.

Reddy M. P., Pagilla K. R., Senthilnathan P. R., Johnson H. W. and Golla, P. S. (1994). Estimation of biomass concentration and population dynamics in a Captor® activated sludge system. *Water Sci. Technol.,* **29**(7), 149–152.

Reimann H. (1990). The Linpor process for nitrification and denitrification. *Water Sci. Technol.,* **22**, 297–298.

Rusten B., Hem L. and Ødegaard H. (1995). Nitrification of municipal wastewater in moving bed biofilm reactors. *Water Environ. Res.,* **67**(1), 75–86.

Rusten B., McCoy M., Proctor R. and Siljudalen J. G. (1998). The innovative moving bed biofilm reactor/solids contact re-aeration process for secondary treatment of municipal wastewater. *Water Environ. Res.,* **70**(4), 1083–1089.

Rusten B., Nielsen M., Welander T. and Rasmussen, V. (2003). Increasing the capacity of activated sludge plants by using AS/ MBBR hybrid process with Kaldnes biofilm carriers. Proc. 5th IWA Conf. on Biofilms Systems, Cape Town, South Africa.

Rutt K. (2011). IFAS – A Broomfield Colorado Success Story with Lessons Learned. 16th Ann. Central States Water Environ. Assoc. Education Seminar on Emerging Activated Sludge Technologies and Operational Strategies, Madison, WI, USA.

Rutt K., Sedal J. and Johnson C. H. (2006). Two year case study of integrated fixed film activated sludge (IFAS) at Broomfield, CO WWTP. Proc. WEFTEC, Dallas, TX, USA.

Sen D., Mitta P. and Randall C. W. (1994). Performance of fixed film media integrated in activated sludge reactors to enhance nitrogen removal. *Water Sci. Technol.,* **30**(11), 13–24.

Sen D. and Randall C. W. (2008a). Improved computational model (Aquifas) for activated sludge, integrated fixed-film activated sludge, and moving-bed biofilm reactor systems, part I: semi-empirical model development. *Water Environ. Res.*, **80**, 439–453.

Sen D. and Randall C. W. (2008b). Improved computational model (Aquifas) for activated sludge, integrated fixed-film activated sludge, and moving-bed biofilm reactor systems, part II: multilayer biofilm diffusional model. *Water Environ. Res.*, **80**, 624–632.

Sriwiriyarat T. and Randall C. W. (2005). Performance of IFAS wastewater treatment processes for biological phosphorus removal. *Water Res.,* **39**, 3873–3884.

Sriwiriyarat T., Randall C. W. and Sen D. (2005).Computer program development for the design of integrated fixed film activated sludge wastewater treatment processes. *J. Environ. Eng., Amer. Soc Civil Eng.,* **131**(11), 1540–1549.

Takács I., Bye C. M., Chapman K., Dold P. L., Fairlamb P. M. and Jones R. M. (2007). A biofilm model for engineering design. *Water Sci. Technol.,* **55**(8–9), 329–336.

Thomas W. A. (2009). Evaluation of Nitrification Kinetics for a 2.0 MGD IFAS Process Demonstration. MS. Thesis, Environmental Engineering Department; Virginia Polytechnic Institute and State University: Blacksburg, VA, USA.

Walker I. and Austin E. P. (1981). The use of plastic porous biomass supports in a pseudo-fluidised bed for effluent treatment. In: Biological Fluidised Bed Treatment of Water and Wastewater. P. F. Cooper and B. (eds), Atkinson, Ellis Horwood, Chichester.

Wang J., Shi H. and Qian Y. (2000). Wastewater treatment in a hybrid biological reactor (HBR): effect of organic loading rates. *Process Biochem.*, **36**, 297–303.

Wanner O., Eberl H., Morgenroth E., Noguera D., Picioreanu C., Rittmann B. and van Loosdrecht M. C. M. (2006). Mathematical Modeling of Biofilms. Scientific and Technical Report No. 18; IWA Publishing, London, UK.

Wilson M., Sandino J., Cote T., Brueckner T., Boltz J. P. and Michelsen D. (2012). Design and operational insights from the world's largest integrated fixed-film activated sludge (IFAS) process for low nitrogen limits. Proc. 85th WEFTEC, New Orleans, LA, USA.

Zhao H., Lemaire R., Christensson M., Thesing G., Veuillet F., Ochoa J., Lamarre D. and Gadbois A. (2013). Single-stage deammonification process perfomance – MBBR versus IFAS configurations. Proc. WEF/IWA Nutrient Removal and Recovery Conference, 2013, Vancouver, BC, Canada.

Chapter 16

Membrane bioreactors

George V. Crawford (Canada), Simon Judd (UK), Tamas Zsirai (UK, Hungary)

16.1 INTRODUCTION

16.1.1 Definition

The term 'membrane bioreactor' (MBR) is generally used to define those wastewater treatment processes where a perm-selective membrane is integrated with a biological process. This is to be distinguished from a 'polishing' process where the membrane is employed as a discrete tertiary treatment step with no return of the active biomass to the biological process. Whereas conventional AS (CAS) systems employ a clarifier (or settlement tank) for solid/liquid separation, an MBR employs a membrane for this function. All currently available commercial MBR technologies use the membrane ostensibly as a filter, rejecting the solid materials developed by the biological process to provide a product of high clarity and reduced pathogen concentration. The technological development of MBRs can be appraised with reference to their historical development and market penetration with respect to the market value and installed capacity.

16.1.2 History

The first commercial MBRs were introduced by Dorr–Oliver in the early 1970s (Bemberis *et al.,* 1971). This technology combined ultrafiltration (UF) membranes with a CAS) process, and was originally intended for ship wastewaters (Bailey *et al.,* 1971). Early MBRs were designed with the membrane placed outside the biotank, a configuration now termed 'sidestream' or 'pumped' (sMBR, Figure 16.1a), as opposed to the now more commercially significant 'immersed' configuration (iMBR, Figure 16.1b). The relatively high pressures demanded (~3.5 bar) combined with the low membrane fluxes attained (17 L/(m^2h)) meant that sidestream MBRs were relatively high in energy demand. Nonetheless, the Dorr–Oliver system successfully demonstrated the key MBR principle of achieving biotreatment and clarification in a single step, and it was marketed in Japan under license to Sanki Engineering with some success, up until the early 1990s. Similar sidestream MBR technologies were subsequently developed by Rhône Poulenc (the PLEIADE® technology) in France and Weir Envig in South Africa, the latter for anaerobic treatment (ADUF), as well as Thetford Systems in the USA (the Cycle–Let process).

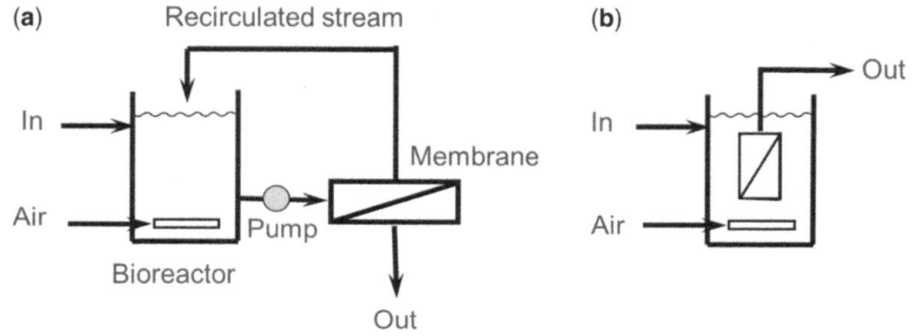

Figure 16.1 Configurations of a membrane bioreactor: (a) sidestream, (b) immersed (adapted from Judd & Judd, 2010).

A significant technological development took place in the late 1980s which was instrumental in commercializing the immersed (iMBR) configuration. In Japan, the government-instigated water recycling program led to pioneering work at Tokyo University (Yamamoto *et al.,* 1989) in developing a hollow fiber (HF) UF iMBR process, as well as the development of a flat sheet (FS) microfiltration (MF) iMBR by the agricultural machinery company Kubota. The first FS iMBR demonstration plant (25 m^3/d capacity) was installed at Hiroshima in 1990. Also in Japan, Mitsubishi Rayon introduced its *SUR* MBR membrane module, based on its Sterapore product, in 1993.

Parallel developments also took place in North America in the early 1990s. Zenon Environmental, a company formed in 1980 and who subsequently acquired Thetford Systems, were developing their own iMBR system. By the early 1990s, the ZenoGem® immersed HF UF MBR process had been patented (Tonelli & Canning, 1993; Tonelli & Behmann, 1996).The first immersed HF ZeeWeed module, the ZW145 which provided 145 ft^2 of membrane area, was introduced to the market in 1993.

Whereas the relatively high costs, and energy demand in particular, associated with the sMBRs meant that their market penetration was largely limited to niche industrial applications, the implementation of the immersed configuration was more widespread and rapid. The first Kubota municipal wastewater treatment works installed outside Japan was at Porlock in the UK in 1997, following successful UK pilot trials conducted in the mid-1990s. The first Zenon (later acquired by GE) membrane-based plant of similar size installed outside of the USA was the Veolia (then Vivendi) Biosep® plant at Perthes en Gatinais in France in 1999, although around a dozen Zeeweed-based installations had been previously commissioned between 1995 and 1999. Both Porlock and Perthes en Gatinais have a peak flow capacity just below 2 MLD, and represent landmark plants in the development and implementation of immersed MBR technology.

By the late 1990s many other MBR membrane products and systems were under development, leading to an explosion of commercial activity from the turn of the Millennium which continued up to the end of the decade. Whereas the first half of the 1990s saw the launch of only three major immersed MBR membrane products, originating from just two countries (Canada and Japan), the first five years of the following decade saw the launch of at least 10 major products originating from 7 countries, coupled with some significant acquisitions. For 12 major suppliers, as of 2010, there were either existing or planned MBR installations of more than 10 MLD capacity (Table 16.1). In addition to those products listed for which there are 'flagship' large plants, there are a number of other MBR membrane products – a large number originating from China – all of which have come to the market since about 2000. There are additionally a number of proprietary MBR technologies based on a few of the membrane products, for example the sidestream MBR technology based on the multitube membrane modules.

Table 16.1 Key MBR membrane products, bulk municipal market.

Supplier	Country	Date launched	Acquired	Date, first >10 MLD plant
Asahi Kasei	Japan	2004	–	2007
GE-*ZeeWeed*®	US	1993	Jun-06	2002
Econity[1]-*KSMBR*®	Korea	2000	–	2008
Huber	Germany	2003	–	2014
Hyflux (*Porosep*)	Singapore	2012		2013
Koch Membrane Systems – *PURON*®	US	2001	Nov-04	2011

(*Continued*)

Table 16.1 Key MBR membrane products, bulk municipal market (*Continued*).

Supplier	Country	Date launched	Acquired	Date, first >10 MLD plant
Kubota *EK*	Japan	1990	–	1999
Kubota *RW*	Japan	2009	–	2011
Memstar	Singapore	2005	–	2010
Microdyn Nadir	Germany	2005	–	–
Mitsubishi Rayon (*SADF*)	Japan	2005	–	2006
Mitsubishi Rayon (*SUR*)	Japan	1993	–	–
Motimo	China	2000	–	2007
Pentair-Norit	Netherlands	2002	–	2012
Siemens Water Tech. - *MEMCOR®*	Germany	2002	Jul-04	2008
Sumitomo	Japan	2010	–	–
Toray	Japan	2004	–	2011

[1]Formerly Korea Membrane Separations.
Source: Adapted from Judd and Judd (2010).

16.1.3 Commercial status

Growth rates and the extent of implementation of MBR technology vary regionally according to the state of economic development and infrastructure. Common to all regions, however, is more rapid growth in the MBR market compared with the GDPs of those countries installing them, significantly so in China. Global growth rates of 10–15% have been routinely quoted in reports produced by market analysis companies (BCC, 2008; Srinivasan, 2007; Markets & Markets, 2013), with most recent projections reporting an estimated CAGR of about 15%. The market value for the membrane equipment portion of MBR facilities was estimated at $0.75b in 2011, with about 40% of the revenue going to the Asia Pacific region. The total installed capacity of MBRs in 2017 is expected to exceed 12,000 MLD.

The commercial status of the technology is in part reflected in the ever increasing capacity of the individual installations (Figure 16.2). In 1999 the first MBR installation of more than 10 MLD capacity was commissioned in Swanage, UK, based on Kubota membranes. As of 2012, the 20 largest MBR plants worldwide all had a peak 24-h capacity in excess of 60 MLD (Table 16.2), with GE-Zenon membranes accounting for about two thirds of the installed capacity. Whilst some of these plants may never be challenged at peak flow, it is a testament to the increasing confidence in the technology that installations continue to increase both in size and number.

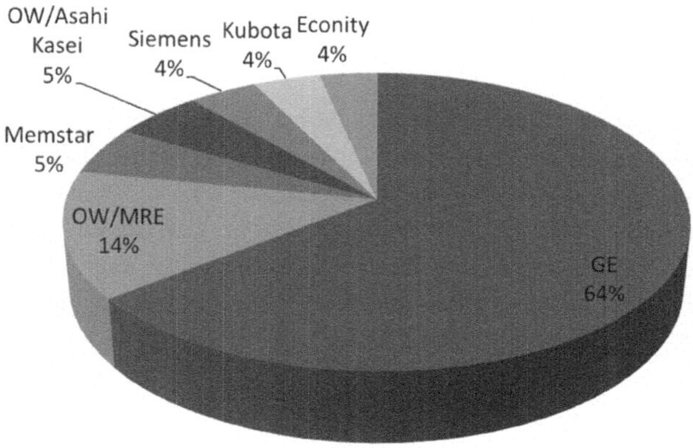

Figure 16.2 Installed capacities of 20 largest MBRs, by membrane supplier (The MBR Site, 2012).

Table 16.2 The 20 largest MBRs in the world.

Installation	Membrane Supplier	Date	PDF[1]	ADF[1]
Brightwater, WA, USA	GE	2011	170	117
Qinghe, China	OW/MRC	2011	150	150
North Las Vegas, NV, NV	GE	2011	133	95
Yellow River, GA, USA	GE	2011	111	69
Shiyan Shendinghe, China	OW/MRC	2009	110	110
Aquaviva, Cannes, France	GE	2012	106	59
Busan City, Korea	GE	2012	100	100
Guangzhou, China	Memstar	2010	100	100
Wenyuhe, Beijing, China	OW/Asahi Kasei	2007	100	100
Johns Creek, GA, USA	GE	2009	94	42
Awaza, Turkmenistan	GE	2011	87	69
Jordan Basin WRF, UT, USA	GE	2012	79	53
Beixiaohe, China	Siemens	2008	78	–
Al Ansab, Muscat, Oman	Kubota	2010	77	55
Cleveland Bay, Australia	GE	2009	75	29
Broad Run WRF, VA, USA	GE	2008	71	38
Christies Beach, Australia	GE	2011	68	27
Gongchan, Korea	Econity	2012	65	–
Lusail, Qatar	GE	2011	61	61
Ecosama, São Paulo, Brasil	Koch	2012	61	57

[1]Peak Daily Flow (PDF) and Average Daily Flow (ADF) in MLD.
Source: Adapted from The MBR Site (2012).

16.2 PROCESS DESCRIPTION

16.2.1 MBR design

The main components of all membrane bioreactor technology systems comprise screening, the process biotank, and the membrane or membrane tank. The membrane tank sludge is then recirculated from the membrane tank to the biotank(s).

Sidestream MBRs (sMBRs, Figure 16.1a) employ a pump to transfer the sludge through an external membrane. This is conventionally achieved by crossflow filtration: the sludge is pumped through narrow membrane channels – usually through horizontally-mounted membrane modules – at an elevated pressure (3–4 bar) and crossflow velocity (1.5–4 m/s). Such systems provide a relatively high flux – up to 200 LMH, depending on the application – but also generally the highest specific energy demand (SED) of between 1.5 and 3.5 kWh/m^3 permeate. A more recent development has been the introduction of the air-lift sidestream MBR, in which sludge is air-lifted using a combination of aeration and sludge pumping through vertical multitube membranes. These systems are capable of operating at SEDs of lower than 0.6 kWh/m^3.

The most commonly employed MBR process configuration is the submerged/ immersed (iMBR, Figure 16.1b). In this configuration the membrane is immersed in the AS and the permeate product removed under gravity and/or by applying a suction, of up to 0.5 bar, on the permeate side. Sludge is then transferred from the membrane tank to the biotank at 3–5 times the flow through the process – a much lower-energy operation than that for pumped sMBR systems. The immersed membranes are then kept clean by scouring with air delivered from an aerator positioned below the membranes and which may be separate from the membrane module or integrated with it.

16.2.2 Operation

The key operating determinant of any membrane separation process is the flux, which is expressed in SI units of m^3/s of flow/m^2 area (and hence as m/s overall). Units of convenience most commonly used for flux are L/m^2, h (LMH), m/d (for MBRs specifically) or, in North America, gal/ft^2, d (GFD). The flux then determines the rate of increase in the transmembrane pressure (TMP), with units of kilopascals (kPa), bars (bar) or pounds per square inch (psi), due to fouling of the membrane surface and/or clogging of the channels between the membranes (Figure 16.3). The ratio of the flux and pressure is then normally referred to the permeability, which then normally has units of LMH/bar or GFD/psi.

Loss of membrane permeability is ameliorated both by membrane scouring during the filtration cycle and periodic cleaning. Scouring is affected by the use of a crossflow of sludge and/or by the use of large air bubbles – as opposed to the much finer bubbles used for process aeration – both of which are designed to generate shear at the membrane surface and thus encourage permeate flow. In sMBRs a crossflow is employed to generate shear at the membrane surface, similarly to RO modules used for desalination. For the iMBRs, and also air-lift sMBRs, shear is generated by aerating the membrane to generate slug flow and/or agitating the membranes whilst also lifting the sludge through the membrane channels (Section 16.3.2.2).

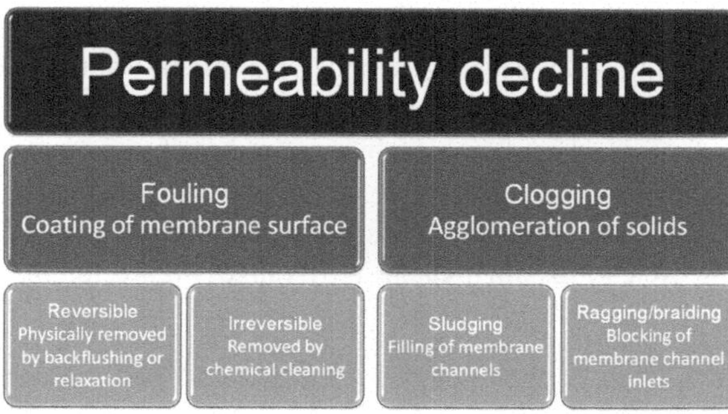

Figure 16.3 MBR membrane permeability decline mechanisms.

The main impact of the operating flux is on the period between cleanings, which may be by physical or chemical means and is usually both (Figure 16.4). Physical cleaning is normally achieved either by backflushing, that is reversing the flow, or relaxation, which is simply ceasing permeation whilst continuing to scour the membrane with air bubbles. These two techniques may be used in combination, and backflushing may be enhanced by combination with air. Chemical cleaning is carried out with mineral or organic acids, caustic soda or, more usually in MBRs, sodium hypochlorite solution, and can be performed either *in situ* ('cleaning in place' or CIP) or *ex situ*. It is common for HF systems to be regularly periodically cleaned by adding a relatively lower concentration (300–500 mg/L) of chemical cleaning agent to the backflush water to produce a 'chemically enhanced backflush' (CEB). This is sometimes referred to as 'maintenance cleaning', whereas cleaning that includes prolonged soaking and is in response to a loss of permeability is called 'recovery cleaning'.

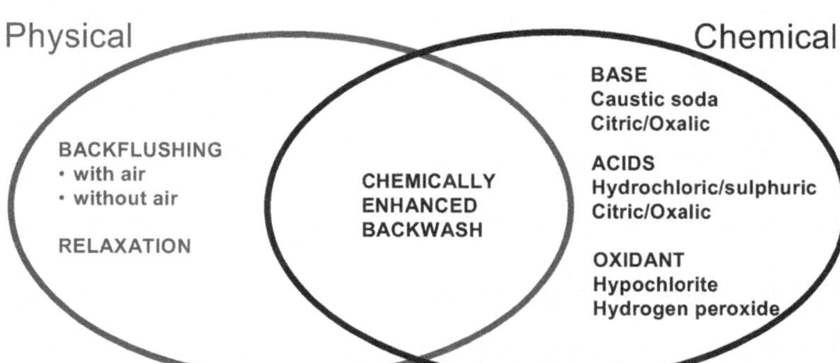

Figure 16.4 Membrane cleaning methods (from Judd & Judd, 2010).

Physical cleaning lasts no more than 2 min and is generally applied every 10–15 min. for an HF membrane, and less frequently for FS membranes. It can remove gross solids attached to the membrane surface, generally termed 'reversible' or 'temporary' fouling, whereas chemical cleaning removes more tenacious material often termed 'irreversible' or 'permanent' fouling (Figure 16.5a). Since the original virgin membrane permeability is never recovered once a membrane is fouled through normal operation, there remains a residual resistance which can be defined as 'irrecoverable fouling'. It is this fouling which builds up over a number of years and may ultimately determine membrane life. Operation of the membrane component of an MBR is thus cyclic in nature, with the short-term physical cleaning cycle supplemented with a chemical cleaning cycle to sustain a viable TMP range (Figure 16.5b).

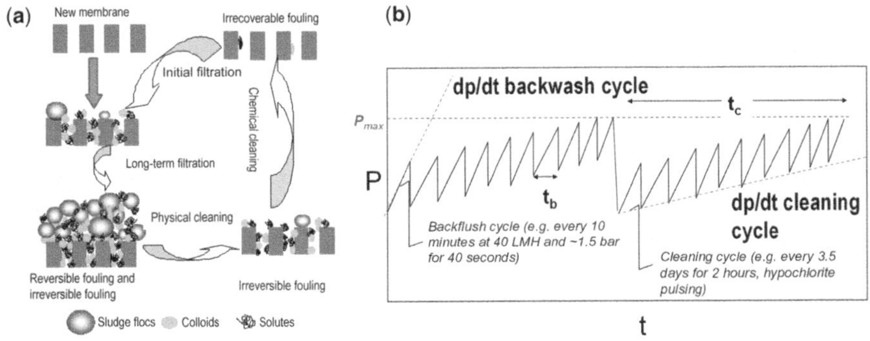

Figure 16.5 Physical and chemical cleaning: (a) foulant removal (Meng *et al.* 2009), and (b) pressure recovery.

Whilst chemical cleaning of the membrane is generally effective at removing foulant materials from the membrane surface, such mitigation methods have little or no effect on clogging. Clogging is characterized either by the filling of membrane channels with solids ('sludging') or the agglomeration of fibrous materials at the membrane channel inlet to form rags ('ragging' or 'braiding'). Whilst the latter has conventionally been attributed to poor pretreatment, it appears that rags form *in situ* from the agglomeration of small lengths of textile fibres (Stefanski *et al.*, 2011; Gabarrón *et al.*, 2013). Sludging, on the other hand, is generally promoted by a combination of aggressive fluxes, inadequate aeration and/or poor sludge quality.

16.3 PROCESS DEVELOPMENT

16.3.1 Biological treatment

MBRs were originally envisaged as offering intensive biological treatment through extended SRTs with the biological process still based upon the biological growth and treatment principles of the CAS process. Since the technology completely uncouples the HRT and the SRT, it offers the possibility of operation under conditions which encourage the development of slow-growing micro-organisms, such as the nitrifiers, without detriment to treated water quality through washout. Such long SRT conditions also lead to reduced sludge generation. The development of biological process designs for MBRs has been governed by two driving forces: the owner's increasing understanding of the MBR system resulting in their greater willingness to accept some of commercial and process performance risk, and recognition of the fundamental differences between MBR biological process design and that of the CAS.

Early MBR installations, up until the early 1990s, featured mixed MLSS concentrations of generally between 15 and 20 g/L in combination with extremely long SRTs. Because MBRs were an unproven process, commercial membrane vendors were asked by owners to take full system responsibility for the MBR performance, and the installations were therefore conservatively designed by the vendors. An early review of data reported for 20 industrial installations (Stephenson *et al.*, 2000) revealed the mean MLSS concentration to be 17 ± 7 g/L, with the highest peak value reported as being 40 g/L. These data are associated with sMBRs and corresponding values for permeability that are generally below 50 LMH/bar.

As more information became available on MBRs, their design evolved in stages (Crawford *et al.*, 2001). The first stage, having long SRT and high MLSS concentrations, was effective for ammonia removal but generally ineffective for Total N removal, and P removal was typically not required. The two most significant developments associated with what might be called the second generation of MBRs are related to the removal of total N and P. The MBRs in Ganges, BC, Canada and Arapahoe County, Colorado, USA included recirculation of mixed liquor to an anoxic inlet zone, thereby achieving nitrate reduction and the coincident recovery of alkalinity. Small demonstration systems at Milton, ON, Canada, and at the Arapahoe County installation, are early examples of the inclusion of chemical addition for P removal.

The third stage of the evolution of MBR system design was marked by efforts to increase membrane flux and capacity, to minimize the system SRT and MLSS, and to optimize the total cost of MBR systems. One indirect aspect was the increasing role of owners (utilities and consultants) in the development, understanding and risk management of MBR systems. Membrane equipment suppliers such as GE began challenging the limits of SRT and MLSS, in both cases by testing lower values while evaluating the effects on flux and performance. One objective was to reduce the SRT to levels closer to actual process requirements. This decreased the tank volume required, but increased the influence of peak flows and loads. Membrane flux was found to improve at lower MLSS, and the flux was less sensitive to reduced SRT than had previously been observed provided that a minimum SRT (approximately equal to the SRT required for nitrification) was maintained. A target MLSS of 10,000 mg/L and reduced SRTs closer to those needed to satisfy biological process kinetics typify third generation systems.

The fourth generation of MBR system design was driven by two new forces:

(1) The increased size and scale of MBR systems, and
(2) The increased number of membrane equipment manufacturers, creating additional competition in the marketplace.

These two aspects contrived to increase involvement of the owner and consultant in the facility planning, and design of the MBR installation itself and that of the associated plant utility and sludge management systems. This then provided owners greater flexibility in selecting the design SRT and MLSS, and therefore the total volume of MBRs including redundancy and for peak conditions. However, such flexibility inevitably creates trade-offs. For example, increasing the bioreactor volume to reduce the MLSS concentration increases membrane flux, but will also proportionally increase the waste sludge volume since sludge is wasted from the mixed liquor. Selection of MLSS had to be made in the context of total plant costs, including sludge stabilization, thickening and dewatering. Many equipment systems formerly provided by the membrane vendor in smaller systems became plant utilities that were designed and provided by the owner and general contractor. Some key examples included the MBR aeration and air scour blower systems, the backwash pumping systems, zone mixing systems, and RAS pumping systems. As owners and consultants took increasing responsibility for the plant and process design, the role of the equipment manufacturer focused to a greater degree on membrane equipment performance, reliability and cost. Equipment procurement specifications prepared by owners (certainly in North America, and to a lesser degree elsewhere) emphasized total flow capacity, membrane life, and cleaning systems performance, making them in many ways similar to that of more mature processes such as membrane equipment specifications for water treatment applications.

Throughout the last three of the four stages of MBR design evolution, researchers and practitioners had to modify and adapt the biological process designs to recognize some fundamental differences between the MBR and CAS processes. These differences include the high DO concentration in the RAS, the high RAS rate required to control the MLSS concentration at the membranes, the increased potential for short-circuiting within process zones due to the reduced tank volumes and high recirculation flow rates, the increased potential for foam generation, and the reduced aeration efficiency due to the increased mixed liquor viscosity. The development and evolution of the biological process design of MBRs for nutrient removal is exemplified by the process flow diagrams for some large operating MBR facilities (Figures 16.6–16.7).

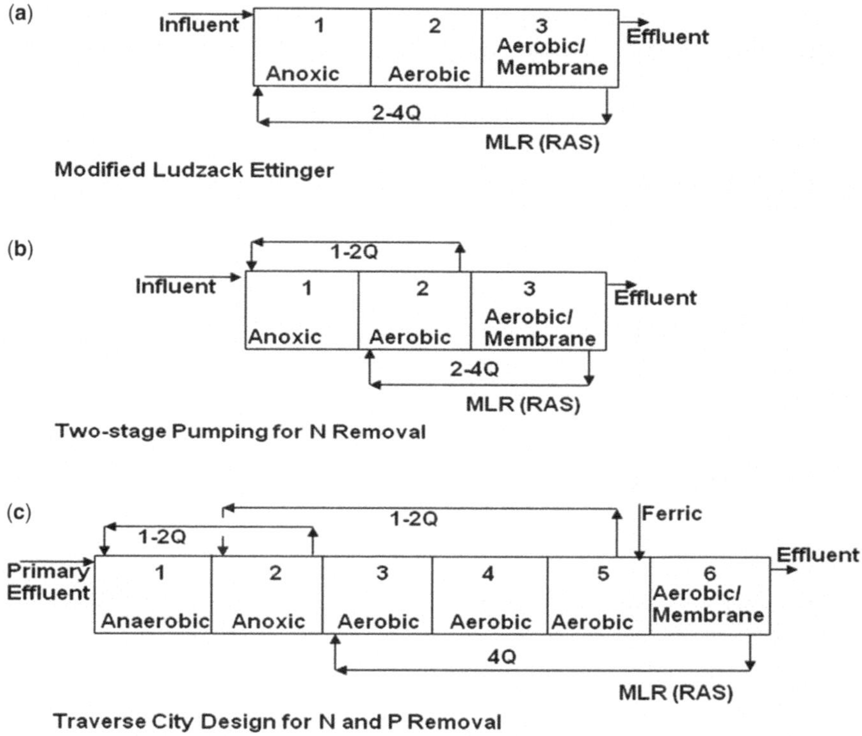

Figure 16.6 Early MBR biological process designs.

Some early MBR designs attempted to use a modified Lutzack Ettinger (MLE) type of process design (Figure 16.6a). It was soon discovered that two factors prevented success: the membrane air scour aeration system caused the RAS to be rich in DO, and the RAS rate needed to be high – 300% to 400% of the plant flow – to maintain an effective membrane operation. As a result, the anoxic zone operated with a measureable DO concentration, rarely achieving truly anoxic conditions, and denitrification did not occur. Designs then evolved to include two stage pumping (Figure 16.6b), wherein the DO in the RAS was used up in the aerobic zone, the DO in the aerobic zones was tapered if possible, and a new mixed liquor transfer occurred from the aerobic zone to the upstream anoxic zone. This new transfer stream was lower in DO than the RAS stream, and its flow rate could be set based on process requirements rather than membrane requirements. These same principles were applied to the design of the 32 MLD Traverse City, Michigan, USA MBR (Figure 16.6c) commissioned in 2004. In this system a third pumped transfer was located upstream of the anoxic zone so as to include an anaerobic zone for EBPR within the process design in a manner based upon the VIP process (Daigger *et al.,* 1988). The 71 MLD Broad Run, Loudoun County, Virginia, USA MBR process design (Figure 16.7a) uses EBPR to achieve nutrient limits of 3 mg/L TN and 0.1 mg/L TP. The process is based upon multiple transfer streams similar to the Traverse City, Michigan, USA plant but also includes methanol dosing as a supplemental carbon source to ensure sufficient denitrification in the second anoxic stage of the process.

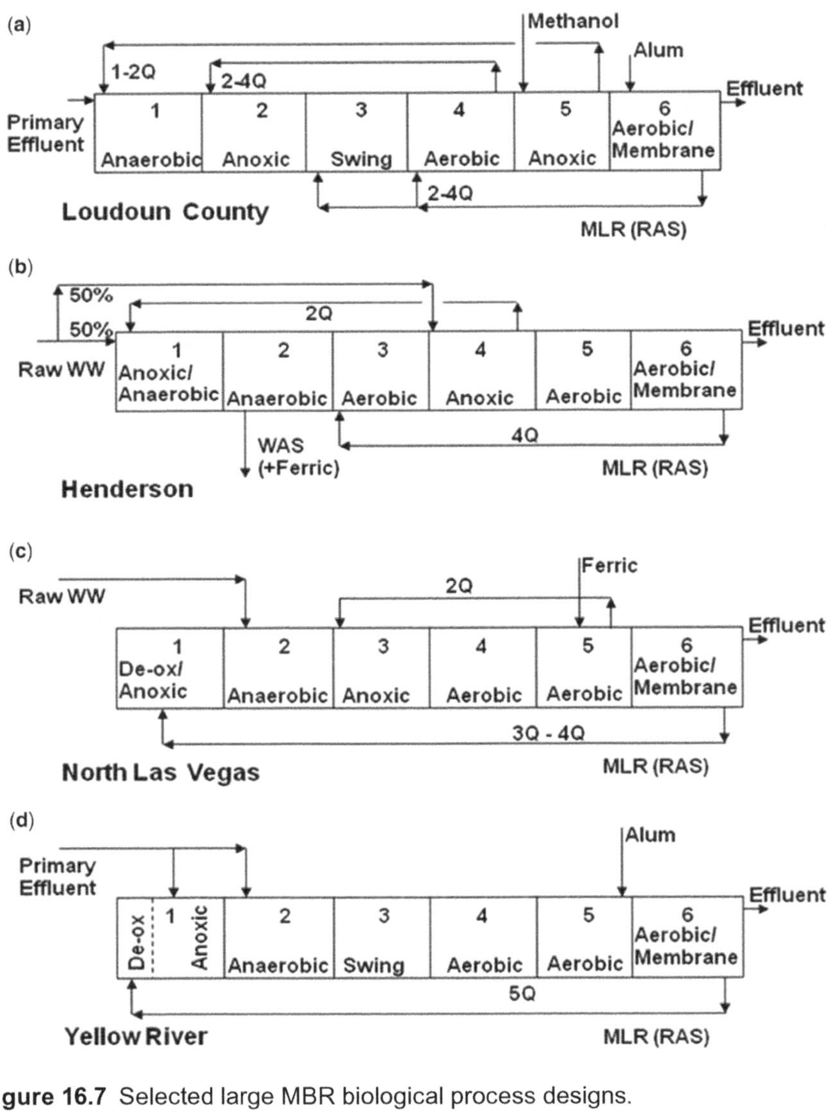

Figure 16.7 Selected large MBR biological process designs.

One drawback of the processes developed for such MBRs was the complexity and the operating cost associated with the multiple upstream pumping provisions. At the same time, the membrane air scour systems evolved so that the DO concentration in the RAS was reduced. Figure 16.7b illustrates the design for the Henderson, Nevada, USA MBR for the biological removal of N and P. In this design, one of the upstream transfers is eliminated, and instead the incoming wastewater flow is split in a step-feed arrangement so that half of the incoming flow is diverted to the second anoxic stage of the process. The stepped feed stream provides the carbon source for denitrification in the second stage. It was recognized that step feed could be used for MBRs required to remove N but not P by removing the Stage 1 anoxic/anaerobic zone and redirecting the upstream transfer from Stage 1 of to Stage 2 (Figure 16.7b), with Stage 2 becoming anoxic.

Figures 16.7c and 16.7d illustrate the biological process designs at two recently commissioned (2011) MBR facilities: the 133 MLD North Las Vegas, Nevada, USA MBR and the 111 MLD Yellow River, Georgia, USA MBR. These two designs adhere to the same process design principles already described for MBRs, but with less concern for DO in the RAS and with each being customized for the wastewater characteristics and discharge requirements of the specific site. The North Las Vegas plant employs a Johannesburg-type configuration with a large Stage 1 deoxygenation/anoxic zone that endogenously deoxygenates and denitrifies the RAS before it reaches the Stage 2 anaerobic zone, where the influent wastewater is introduced. A significant portion of the Stage 1 volume is provided by the RAS return channels that run from the membrane area to the front of this large plant. The Yellow River MBR plant employs step feed to manage the RAS denitrification prior to the Stage 2 anaerobic zone. Both of these facilities have been in successful compliance with their discharge requirements since commissioning.

16.4 MEMBRANE TECHNOLOGY

16.4.1 Membrane material and configuration

Membranes and membrane modules can generally be defined on two bases: the configuration and the material. The membrane configuration describes both the shape of the membrane and the direction of flow through it. The material includes both the type of polymeric (or ceramic) material of which the membrane is composed and the pore characteristics. In principle, any polymeric or ceramic material can be used as a membrane provided it can be manipulated to form pores of a uniform size. In practice the choice of material is limited both by the manufacturing method and the requirements of the MBR application. A high degree of permselectivity (permeation of the water and rejection of the SS and solutes) demands as high a degree as possible of isoporosity, that is all the membrane pores having the same shape and size.

There are currently only three membrane configurations employed for MBR applications, although the recent development of more robust and inexpensive ceramic membranes has extended the range slightly. Hollow fibre (HF) and flat sheet (FS), Figures 16.8a and 16.8b respectively, are currently the only configurations employed for immersed membranes. Pumped systems are predominantly multitube (MT) (Figure 16.8c), though there are still examples of the original flat sheet plate-and-frame configuration. For both the immersed configurations, flow is from the outside to the inside of the membrane element, whereas for the MT the flow is in the opposite direction.

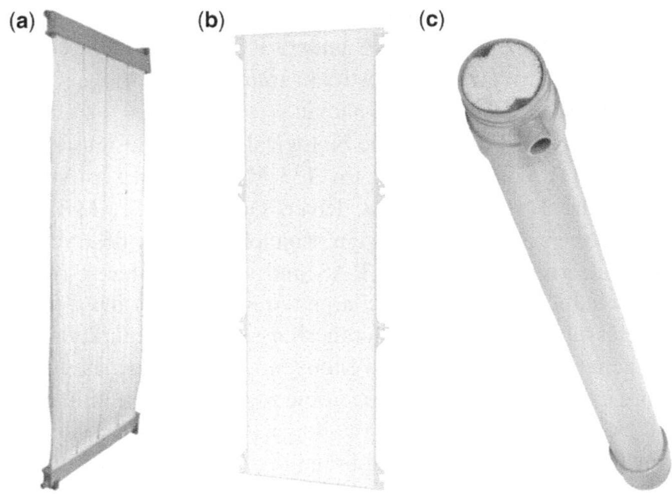

Figure 16.8 (a) Hollow fibre (HF), (b) flat sheet (FS) and (c) multitube (MT) (with kind permission of GE, Kubota and Pentair).

HF membranes have innate mechanical integrity that allows them to be backflushed, where backflushing is the periodic reversing of the flow in the opposite direction to that of the filtration flow to clean the membrane. Conventionally the FS and MT modules are thought of as not backflushable, since in both cases, reversing the flow can compromise the integrity of either the membrane layer or the seal between the membrane sheet and the rigid backing panel on which it sits. However, a number of more recent MBR FS and MT membrane products are backflushable. For FS membrane panels the mechanical integrity required for backflushing is achieved in some of the more recent products by basing the panel on a composite membrane material rather than attaching the membrane sheets to a backing plate.

A review of 48 commercially-available membrane products (as of 2010) each claiming to be appropriate for MBR applications, revealed a relatively narrow range of materials being employed (Figure 16.9). More than half the membranes surveyed at this time were based on polyvinylidene difluoride, or

PVDF, with the next most common material being polyethersulfone (PES). The combination of good chemical resistance and surface structure has meant that the sulfonated and fluorinated polymers, and PES and PVDF in particular, dominate in modern MBR membrane materials. However, the polyolefinic membranes (polypropylene, PP, and in particular polyethylene, PE) are also used – as one example, the FS membrane market leader (Kubota) employs a chemically modified PE material. The polyolefenic HF membranes are amongst the lowest in raw production cost of all MBR membrane materials. The pores are generated by extruding the material under controlled conditions (or 'dry spinning') to produce slit-like pores (Figure 16.10a). These tend to provide high packing densities (membrane surface area per bulk module volume) from relatively small (sub-1 mm) filament diameters. This slit-like pore structure is to be distinguished from the more complex thermally induced phase separation (TIPS) process for PVDF, which produces the more classic pseudo one-dimensional pores (Figure 16.10b). The remaining identified materials – polyacrylonitrile (PAN), polysulfone, polyvinyl alcohol (PVA) and polytetrafluorethane (PTFE) – are much less common.

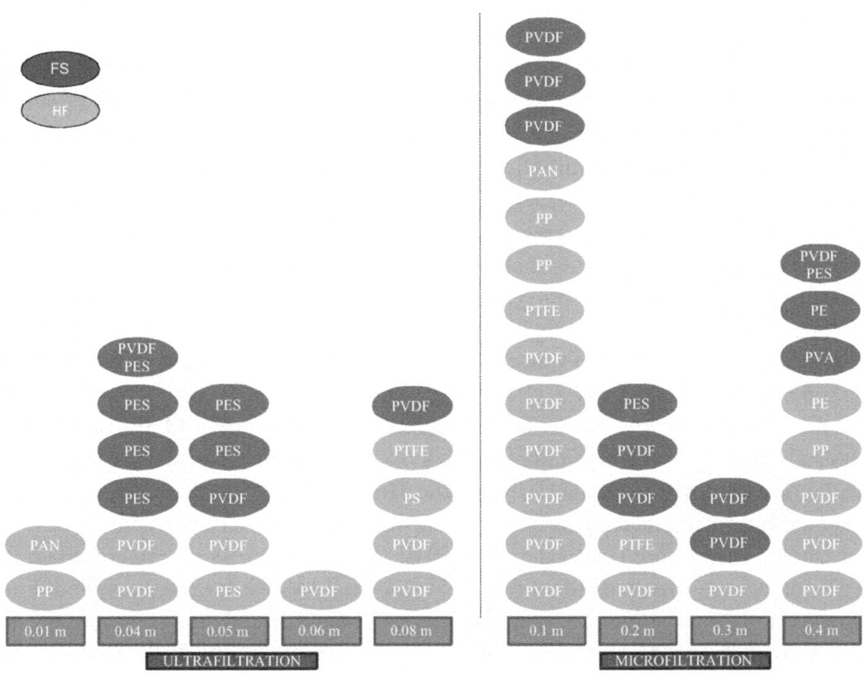

Figure 16.9 Review of 48 individual MBR membrane products (Santos & Judd, 2010).

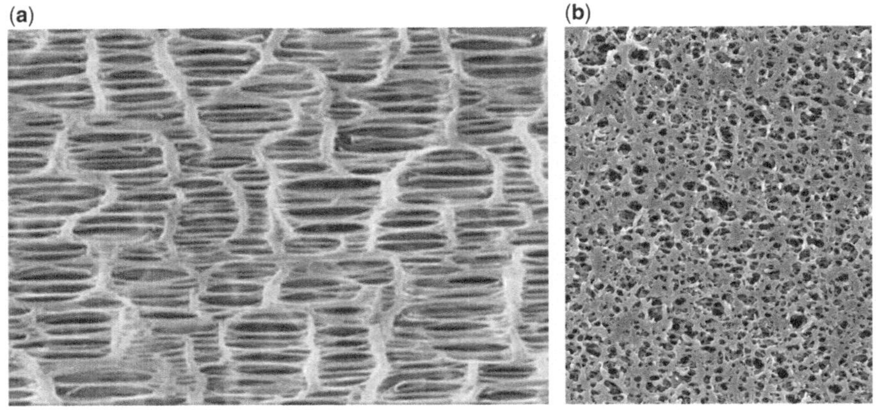

Figure 16.10 Membrane surface (a) dry-spun membrane, and (b) TIPS (by kind permission of Korea Membrane Separations and Asahi-Kasei respectively).

The hydrophobicity of most of these polymeric materials means that almost all polymers used for MBR membranes require wetting prior to use, to suppress fouling by naturally hydrophobic materials. Wetting is normally achieved simply by coating the membrane with a hydrophilic gel which is eluted when first used. Once wet, most membrane products cannot be allowed to dry out, since this denatures the material. An advantage of the more recently introduced ceramic membrane materials (such as alumina, titanium dioxide and silicon carbide) is that they are innately hydrophilic and therefore, in principle, less susceptible to fouling by the hydrophobic organic matter which predominates in the MBR mixed liquor. They are also chemically robust, allowing the application of more aggressive cleaning chemicals than those used for polymeric membranes, and are available across a range of pore sizes. FS configured ceramic membranes have recently been introduced, allowing them to be employed as immersed modules rather than being limited to sidestream processes. However, ceramic membranes remain comparatively costly.

Much of the development work conducted in polymeric MBR membrane materials relates to the reduction of the pore size distribution and assurance of mechanical integrity – particularly for the HF membranes which are exposed to considerable mechanical stress from flexing of the fibre in the region next to the pot. The quoted mean pore sizes range from 0.01 to 0.4 μm across all the commercial MBR membrane materials identified in Figure 16.9. Whilst larger pores would be expected to be more prone to pore plugging and breakthrough of smaller particles, with viruses being of particular concern, there appears to be little evidence to suggest that either of these issues are of significance in practice. Pore plugging, or internal fouling generally, is minimized by the somewhat benign operational conditions of an MBR, with fluxes being relatively low and/or shear rates (from air scouring of the membrane or crossflow operation) fairly high. Whilst it has been shown that breakthough of free-swimming viruses can potentially be significant (Hirani *et al.*, 2010), in practice viruses are predominantly associated with TSS and are thus rejected by the membrane.

16.4.2 The membrane technology

The efficacy of the MBR membrane technology design lies both with the membrane material and the design of the module and ancillaries. A key aspect of the latter, in the case of iMBRs or air-lift sMBRs, is the contacting between the air and the membrane. It is this which produces shear at the membrane surface, which in turn sustains a flux of permeate water through the membrane.

All FS membrane modules are designed as cassettes, with the individual membrane panels/cartridges spaced 6–9 mm apart within the cassette to produce a parallel flow channel for the sludge. The sludge is air-lifted through the channel by the application of coarse-bubble aeration-from aeration ports that may be several mm in size – producing 'slug' bubbles which are large enough to fill the channels. The rising bubble is thus in contact with the membrane and creates significant interfacial dynamic shear (changes in the shear force and direction at the membrane surface) as it passes up through the channel. A critical operational aspect for an FS iMBR is thus the balance between (i) the rate of accumulation of MLSS within the channel (or clogging), and (b) the shear imparted at the membrane surface and the flux this generates (in conjunction with the TMP). Since both of these phenomena directly relate to the membrane aeration, the balance between the membrane air scour rate applied and the flux generated is of critical importance.

A similar situation arises with the immersed hollow fibre modules. In this case, the action of the membrane aerator is two-fold, in that the air bubbles apply a degree of direct scouring of the membrane surface, but also agitate the fibres (from their lateral movement) creating shear and thus promoting fluxing of water through the membrane. In this case there is thus an additional balance between the direct shear (air scour) and the indirect shear (membrane lateral movement) generated by membrane aeration. Agitation of the fibres may be expected to ameliorate clogging to some extent.

The membrane aeration rate can be normalized against the membrane area (Nm^3/h, m^2), in which case it becomes the specific aeration demand (SAD_m). If this parameter is then divided by the flux in consistent units of m/h, then the parameter produced (the specific aeration demand normalized against permeate flow, SAD_p) takes units of Nm^3/m^3 and is the ratio of air scour applied to permeate generated. Since each Nm^3 incurs a fixed operational cost (according to the efficiency of the blower, the aerator depth and other system properties), SAD_p provides a reasonable indication of the energy efficiency of the membrane permeation process.

MBR technological developments since original commercialization of the immersed technology in the early 1990s have seen significant reductions in costs for two principal reasons. First, membrane SAD_p costs have decreased primarily due to the simple laws of supply and demand, and of competition in the market, suppressing the purchase cost. SAD_p costs have steadily decreased due to improvements in technology design, and in the air-to-membrane contacting specifically. Secondly, membrane life expectancy has generally increased for plants operating with a minimum of manual intervention, further decreasing life cycle costs.

The extent of improvement in membrane air scour efficiency is striking. Zenon (now GE) first commercialized the Zeeweed ZW150 immersed HF module in 1993. The module design has since been changed four or five times, primarily to allow improved aeration distribution within the fibre bundle, and improved aeration methods. Cyclic aeration was introduced by GE in 2001, in the ZW500C model, which equated to applying air intermittently for 10 s by switching the air between two trains of membranes. In 2005 GE demonstrated that aeration could be further reduced by a factor of two by distributing the air between four trains. In 2009, Siemens introduced a method to achieve even higher aeration efficiency by applying extremely large air bubbles very briefly at periods of about 10–30 s and, in 2012, GE introduced the LEAP system, which also incorporated intermittent large-bubble aeration. Overall, the technological changes developed have provided a more than ten-fold improvement in energy efficiency overall (Figure 16.11).

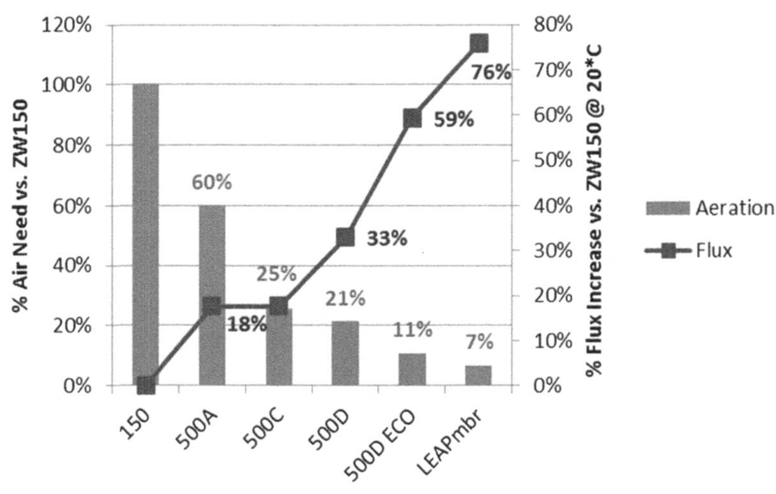

Figure 16.11 GE-Zenon MBR technological developments (by kind permission of GE).

Despite the significant improvements in technological design over the years, there remains the key issue of standardization. Standardization with respect to the membrane module size and connectivity would expedite interchangability of membrane products when change-out is required. Whilst the multitube sMBR products are largely standardized, normally being based on 200 mm diameter modules, the same is not necessarily true of the immersed modules where the submerged depth and the ratio of the module membrane area A_m to its footprint F (the cross-sectional area at the module base) vary widely.

The A_m:F ratio can be increased by stacking (placing one module deck on top of another in the membrane tank), which is now common in the case of large FS module based installations. However, because most FS panels are 4–7 mm thick (with the exception of some of the newer composite membrane modules) and generally separated by 6–9 mm in the stack, simple geometric considerations mean that the volumetric packing density (the ratio of effective membrane area to bulk module volume, or the A_m:F ratio divided by the height) cannot exceed ~250 m^2/m^3 within the module. In practice, almost all commercial FS technologies have volumetric packing densities below 140 m^2/m^3. Against this, a much wider variation in packing density arises for HF products, partly because of the commensurate variation in filament diameter of 0.3–2.8 mm across the entire range of available products. Packing densities for HF modules thus tend to be higher than for FS modules, with more than half being above 140 m^2/m^3. This therefore limits opportunities for interchanging of HF and FS modules, though it is clearly possible in principle to install HF modules in place of FS ones, notwithstanding the increased operational complexity of the former.

16.4.3 Other design aspects

Some aspects of MBR facility design are unique and the development of appropriate design practices has been required. In some cases such design issues have yet to be resolved and acceptable solutions developed.

All MBR facilities employ screens which are necessarily finer than the 6 mm-rated screens conventionally used for classical municipal wastewater treatment by the CAS process. In general sub-3 mm screens are required for FS and MT membranes, whereas sub-2 mm screens are needed for HF modules. One recent unresolved development has been the desire by vendors to use a test measuring the concentration of screenings in the MLSS, and to limit that concentration regardless of the influent screening size. The benefits are two-fold: the test would directly measure the parameter that affects membrane performance, maintenance and life, and the test method is simplified. One drawback is that the owner is asked to accept the risk and consequences of screenings re-forming within the process to form rags (Figure 16.3), and of debris such as leaves, paper and plastics entering the tanks. This may lead to owners putting covers on all plant tankage to block external debris, although ragging would still be a risk. An acceptable design resolution has yet to be found that satisfies both the vendors and owners' concerns.

One feature of MBRs is that the higher MLSS concentration allows for a more compact design. However the concentrations of surfactants and the mixed liquor viscosity are also higher, with the result that the typical foam that is associated with all AS processes is present to a greater degree in MBRs. Design practices have evolved: transfers to downstream zones within an MBR are typically achieved by surface overflow, and interferences where foam could become trapped are avoided, all with the intent of allowing the foam to move to a point in the process where it can be managed and/or removed.

Efficient grit removal is important to avoid abrasion of the membranes. Initially, most vendors requested that effective grit removal systems be provided, tested and proven upstream of the MBR. Coincidentally, foam management practices were being developed, and the practice of using surface overflow transfers was seen as an effective means to prevent grit from reaching the membrane tanks. The need to test and prove the effectiveness of the plant grit removal system is therefore no longer a priority when surface transfers are employed.

The design evolution of permeate pump priming systems for MBRs is one example of a support system technology advancement being developed by the major membrane vendors. In order to simplify the disconnection and removal of individual membrane cassettes and modules, permeate pump suction piping is typically located above the membrane tank liquid level. This, in combination with the high head loss associated with water passing through the membrane, results in a need to prime the permeate pump suction piping prior to operation. Early MBRs employed vacuum systems for this application, however by 2004 it became apparent that the difficulties associated with leak detection in vacuum piping were onerous and often could not be achieved without taking a significant portion of the facility out of operation. The industry then uniformly adopted a high pressure eductor permeate priming system, first introduced by Siemens.

The need to clean the membranes using strong acid and hypochlorite solutions has already been discussed. As the scale of MBR plants increased, *in-situ* membrane cleaning became the norm, and designers recognized the need to provide acid and chemical resistant coatings or liners on the membrane tank walls (or in some cases to fabricate the tanks using specialty steel). Such coatings are difficult to apply due to the controls required, for example, on concrete dryness and temperature. Liners are also challenged by several protrusions for piping, inlet and outlet gates, and pipe supports. This design issue is not yet resolved, and continues to be a critical design decision.

Finally, there is a drive to extend the degree of automation inherent in MBR design to permit continuous unstaffed operation, other than for scheduled equipment maintenance. MBRs are highly automated systems that monitor and control the air scour, membrane pumping, backwash, permeate priming and other key systems, while also managing the number of membranes in service to serve the current plant flow. A number of large, flagship plants have operated unstaffed overnight and during weekends, including the Traverse City, Michigan, USA MBR. In concept, an unstaffed MBR, monitored remotely by staff at a central or regional plant, would be an effective satellite water reuse production plant operated with a regulated constant flow and extracting only a portion of flow from a sewer that continues to a centralized plant. The water produced would then be reused locally, with the assumption that the sewer flow would continue if the MBR plant were automatically shut down under alarm conditions. Some MBRs such as those at Tempe, Arizona, USA, Henderson, Nevada, USA and Ulu Pandan, Singapore, have included many of these features, however design practices for such an application are still being developed.

16.5 CASE STUDY: TRAVERSE CITY

Client	Traverse City, MI, USA
Primary contractor/contractor	CH2M HILL
Primary consultant	CH2M HILL
Membrane Technology	GE
(Judd & Judd, 2010)	

Traverse City, Michigan, USA is a community of about 15,000, well known for the natural beauty surrounding the city which contributes to a well-developed round-the-year tourism and recreation industry. The wastewater treatment plant also serves parts of Grand Traverse and Leelanau Counties, such that the overall population equivalent is close to 50,000. MBR technology was selected for the site development plan following extensive public consultation which revealed widespread interest in providing effluent of as high a quality as reasonably possible. The public was also decidedly uninterested in increasing the footprint of the existing plant, which is bounded on two sides by Boardman Lake, on another side by the Pere Marquette railroad, and on the fourth side by a recently improved city park. The conversion of the plant to an MBR was part of a project which took place over a 2.5 years period at a cost of $31 m, increasing the plant capacity by 42%, with respect to BOD loading, while improving effluent quality. The 32 MLD maximum monthly flow (64 MLD peak hourly flow) plant began producing MBR permeate in July 2004.

The plant influent consists of wastewater from Traverse City and six surrounding townships, and treated effluent from Grand Traverse County Septage Treatment Facility. It passes through a 6 mm screen, then Detritor grit removal systems. The screenings and grit are disposed of in a sanitary landfill. Wastewater then flows through primary sedimentation tanks (PSTs). Sludge from the PSTs is directed to anaerobic digesters. A portion of the PST underflow can be returned to the plant influent to facilitate liberation of volatile fatty acids (VFAs) generated under the anaerobic conditions prevailing in the PST sludge blanket. The PST thus can also serve as a fermenter to promote biological P removal in the downstream aerobic process.

Following screening to 2 mm using perforated plate screens, the primary effluent is transferred by screw pumps to the bioreactor basins. These comprise anaerobic, anoxic and aerobic zones with recirculation of mixed liquor from the aerobic to the anoxic zones, at a recycle ratio of 1, and from the anoxic to the anaerobic zones, also at a recycle ratio of 1. Ferric chloride can be added to the mixed liquor to precipitate phosphate as necessary to supplement its biological removal and reduce plant effluent P levels. The mixed liquor flows from the aeration basins to the membrane vessels and is recycled from the downstream end of the membrane vessels back to the aerated zone of the aeration basins at a recycle ratio of 4 (Figure 16.12).

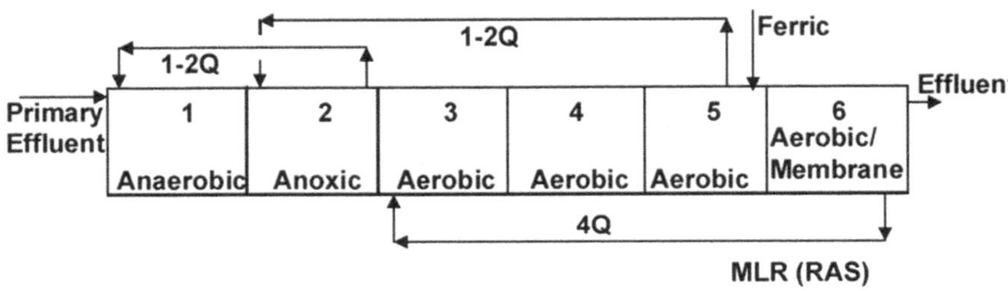

Figure 16.12 Recycle flows for the Traverse City plant (Schwarz *et al.*, 2006).

Figure 16.13 MBR permeate and air headers, Traverse City, MI, USA WWTP (by permission of CH2MHill).

The membrane vessels (Figure 16.13) comprise 8 trains each containing 13 cassettes of 32 ZW500c modules giving a total installed membrane area of 68,091 m^2. The membranes are typically operated on a standard 10 s on/30 s off cyclic aeration mode at a specific aeration demand of 0.18 m^3/hr, m^2, with relaxation for 40 s every 12 min. Maintenance cleaning using a 200 mg/L sodium hypochlorite solution is carried out every 10 d and is supplemented by a clean with 2000 mg/L citric acid every 20 d. This schedule maintains a net average flux of 19.2 LMH at an MLSS concentration of 4–5 g/L (6–10 g/L in the membrane basins). The permeate flows through a UV disinfection system (Infilco Degremont) prior to discharge to the Boardman River.

The plant has on-site solids processing, with half of the holding tank volume provided by conversion of the out-of-service secondary clarifiers. WAS is withdrawn at the point where the RAS exits the membrane vessels. It is dosed with polymer and concentrated in a gravity belt concentrator. Ferric chloride can be applied to the WAS to fix the P chemically, thus preventing its reintroduction into the wastewater in side-stream flows. It can be dosed either upstream or downstream of the concentrator. The concentrated WAS and primary sludge are anaerobically digested; the digested sludge is dosed with polymer and the solids further concentrated in sieve drum concentrators prior to storage in tanks with a total capacity exceeding 11,000 m^3. About 15,000 m^3 sludge (800 tonnes dry solids) is generated annually, and is sub-surface injected into local farmland.

The target effluent quality parameters are BOD_5 and TSS = 4 mg/L, 1 mg/L, NH_4-N = 1 mg/L, TP = 0.5 mg/L. These values are all significantly lower than the statutory requirements of 25, 30, 11 and 1 for these respective determinants. The actual effluent quality achieved during plant performance testing was BOD_5 = ND, TSS = <0.5 mg/L, NH_4-N = <0.5 mg/L and TP = 0.05–0.4 mg/L.

16.6 REFERENCES

Bailey J., Bemberis I. and Presti J. (1971). Phase I Final Report – Shipboard Sewage Treatment System. General Dynamics Electric Boat Division, Nov., NTIS.

BCC (2008). Membrane Bioreactors: Global Markets. BCC Report MST047B, June.

Bemberis I., Hubbard P. J. and Leonard F. B. (1971). Membrane sewage treatment systems-potential for complete wastewater treatment. *Amer. Soc. Agric. Eng. Winter Mtg.*, 71–878, 1–28.

Crawford G., Fleischer E., Lozier J. and Daigger G. (2001). The evolution of membrane bioreactor system designs for wastewater treatment. Presented at the Ben-Gurion University of the Negev Internl Conf. on Membrane Technology for Wastewater Reclamation and Reuse, Tel-Aviv, Israel, September 2001.

Daigger G. T., Waltrip G. D., Romm E. D. and Morales L. M. (1988). Enhanced secondary treatment incorporating biological nutrient removal. *J. Water Pollut. Control Fedn*, **60**, 1833–1842.

Gabarrón S., Gómez M., Monclús H., Rodríguez-Roda I. and Comas J. (2013). Ragging phenomenon characterisation and impact in a full-scale MBR. *Water Sci. Technol.*, **67**(4), 810–816.

Hirani Z. M., DeCarolis J. F., Adham S. S. and Jacangelo J. G. (2010). Peak flux performance and microbial rejection by selected membrane bioreactors. *Water Res.*, **44**(8), 2431–2340.

Judd S. and Judd C. (2010). The MBR Book, Elsevier, London/NY.

Markets and Markets (2013). Membrane Bioreactor Systems Market by Types, Configuration & Applications – Trends & Forecasts To 2017. marketsandmarkets.

Meng F., Chae S.-R., Drews A., Kraume M., Shin H.-S. and Yang F. (2009). Recent advances in membrane bioreactors (MBRs): membrane fouling and membrane material. *Water Res.*, **43**(6), 1489–1512.

Santos A., and Judd S. (2010). The commercial status of membrane bioreactors for municipal wastewater. *Separation Sci. Technol.*, **45**(7), 850–857.

Schwarz A. O., Rittmann B. E., Crawford G. V., Klein A. M., and Daigger G. T. (2006). Critical review on the effects of mixed liquor suspended solids on membrane bioreactor operation. *Sep. Sci. Technol.*, **41**(7), 1489–1511.

Srinivasan J. (2007). MBR still growing in EU wastewater treatment market. *Water and Waste Internatl.*, **22**(5), 43–44.

Stefanski M., Kennedy S., and Judd S. J. (2011). The determination and origin of fibre clogging in membrane bioreactors. *J. Membrane Sci.*, **375**, 198–203.

Stephenson T., Judd S., Jefferson B. and Brindle K. (2000). Membrane Bioreactors for Wastewater Treatment. IWA Publishing, London. Carriage return www.thembrsite.com/largest-mbr-plants/ (accessed February 2014)

Tonelli F. A. and Behmann H. (1996). Aerated Membrane Bioreactor Process for Treating Recalcitrant Compounds. US Pat. No. 410730.

Tonelli F. A. and Canning R. P. (1993). Membrane Bioreactor System for Treating Synthetic Metal-working Fluids and Oil Based Products. US Patent No. 5204001.

Yamamoto K., Hiasa M., Mahmood T. and Matsuo T. (1989). Direct solid-liquid separation using hollow fibre membrane in an activated sludge aeration tank. *Water Sci. Technol.*, **21**(10), 43–54.

Chapter 17

Industrial wastewater treatment

Karl-Heinz Rosenwinkel (Germany), Willy Verstraete (Belgium), Siegfried E. Vlaeminck (Germany), Martin Wagner (Germany), Sabrina Kipp (Germany), Nina Manig (Germany)

17.1 HISTORY OF INDUSTRIAL AS TREATMENT

In the 100 years since the introduction of the AS process (Ardern & Lockett, 1914), many researchers have made significant contributions to its application for industrial wastewater treatment. By improving and modifying the conventional AS process, a number of new process configurations and operational features have been developed to meet specific industrial treatment objectives.

The need for modifying the AS process to suit particular industrial wastewater circumstances was addressed by Lockett as a result of the experiences gained during the operation of one of the first full-scale AS installations at Davyhulme, Manchester, UK where a large proportion of the wastewater was industrial. Large quantities of oils created operational problems soon after start-up by clogging the diffuser media (von der Emde, 1964). For highly concentrated wastewater, Imhoff developed the two-stage AS process in the 1920s (Imhoff, 1955). Mechanical surface aeration was introduced at the same time (using Haworth paddles at Sheffield, UK, and Kessener brushes in The Netherlands) to overcome diffuser clogging in AS plants treating highly concentrated industrial wastewaters (von der Emde, 1964).

In the 1960s, the City of Nordhorn, Germany had problems treating textile wastewater effluent from three large factories in addition to its municipal wastewater because the industrial wastewater volume was 10 times higher than the municipal wastewater volume. In 1964, a wastewater treatment plant using the AS process preceded by a chemical precipitation step was built (Kehr, 1964). The first brewery using the AS process for on-site wastewater treatment was the Veltins brewery in Grevenstein, Germany. The plant started operation in 1966 (Seyfried, 1969).

In the 1970s, the deep shaft technique was introduced. Its invention arose from the Pressure Cyclone Contactor (PCF) reactor developed for producing protein animal feed additive from methanol (Rosenzweig & Ushio, 1974). In 1974, Imperial Chemical Industries (ICI) put the first deep shaft pilot plant into operation (Hines *et al.*, 1975) and in 1975, a large deep shaft reactor was built for treatment of wastewater from a potato starch factory in Emlichheim, Germany (Seyfried, 1978). One of the largest deep shaft processes was introduced in Japan for the Omatsu-Itagami Cardboard AG (Hitachi Puranto Kenetsu Co.). The first deep shaft system for wastewater pretreatment was at Molson's Brewery Ltd., Barrie, Canada in 1980 (Lom & Fedderson, 1981).

The use of pure oxygen instead of air for aeration of AS was also introduced during the 1970s. Pure oxygen systems are applied for the treatment of strong and difficult wastewaters, for example paper and pulp, pharmaceutical, food producing industry and for rendering plants because of its smaller volume (Tilley, 2011). Another process developed for treatment of industrial wastewaters during the 1970s was the 'Katox' (catalytic oxidation) process. Here, activated carbon is added to the AS treatment system to detain specialized microorganisms and exoenzymes on its surface. It was applied for treatment of wastewaters from textile, leather and pulp industry (Oehme, 1981). A process working on the same general principle developed at the same time is the PACT (Powdered Activated Carbon Treatment) process in which powdered activated carbon is added to an AS system. Especially complex organics, toxic or inhibitory substances could be removed by adsorption and biological degradation mechanisms. The PACT system is used, for example, in chemical, petrochemical and pharmaceutical industries (Meidl, 1999).

17.2 INDUSTRIAL AS TREATMENT – STATE OF THE ART

17.2.1 Influence of industrial wastewater characteristics and loading

Efficient AS treatment requires an understanding of the effects of the wastewater characteristics and loadings on the process. Industrial wastewaters can have great variability both in composition (Table 17.1) and flow rate with substantial differences between the various industrial sectors, and between plants in the same sector. The main reasons for these differences are different production methods, water supply and water processing, technical conditions and age of the production site, training and motivation of the employees, use of certain additives and cleaning agents, number of shifts and campaign operation, effluent requirements (direct or indirect discharger), extent of production-integrated environmental protection means, and the number of wastewater treatment facilities (Rosenwinkel *et al.*, 2005). Compared to municipal wastewater, industrial wastewaters can contain high organic loads and insufficient nutrients, and unique physical/chemical properties including, high salinity concentrations, high and low pH, and hazardous/inhibitory substances that may negatively affect the efficiency of the AS process. Problems may occur due to the following:

Table 17.1 Examples of wastewater composition from various industries.

	COD mg/L	**BOD$_5$ mg/L**	**Total N mg/L**	**Total P mg/L**	**pH**	**Other Characteristics**
Slaughterhouse ('red line', pig)[(1)]	2000–8000	1000–4000	150–500	15–50	7.15–8.4[(6)]	Fat: 500–2500 mg/L
Brewery[(2)]	1013–7775	1080–4270[(3)]	22–1069	0.8–113	1.4–14	
Yeast[(4)]	5000–25,000	3500–18,000	500–1200	10–50	4.8–6.5	SO_4^{2-}: 600–1200 mg/L
Textiles [(5)]	420–1400	80–500	<6.0 (Total $N_{mineral}$)	26–80	8.5–10.3	Conductivity: 420–1400 mS/m
Paper [(2)]	20–5500	10–3000				

Sources: [(1)]ATV (1995); [(2)]Bischofsberger *et al.* (2005); [(3)]Seyfried (1969); [(4)]Rüffer (1995); [(5)] Schulze-Rettmer and Marzinkowski (2001); [(6)]Total raw wastewater according to Jäppelt and Neumann (1985).

High temperatures and variable pH value. Many industries, such as chemical and pulp and paper, generate wastewaters with high and varying temperatures that can affect DO uptake rate and biodegradation kinetics. An increase in temperature increases the activity of microorganisms, and at the same time results in a lower oxygen saturation concentration that can limit the aeration system oxygenation capacity. To ensure stable mesophilic processes, the AS temperature should be <40°C. Self-heating up to thermophilic temperatures can also occur when treating highly concentrated wastewaters. Similarly to temperature, pH can impact microbial activity. Generally, influent pH should be in the range of 6.5–10 because extreme aeration basin pH values can upset the AS process. Low pH wastewater is a common problem in the fruit and vegetable processing industry; here a pH of 4.5 in the AS tank influent is possible due to rapid biological degradation and the production of high organic acid concentrations; requirements for protection of municipal sewers also have to be considered. High pH wastewaters, for example from textile processing industries, can, to some extent, be neutralized by the CO_2 produced by the AS during aerobic biodegradation.

Intermittent and shock loading, for example due to campaign operations. Many industries do not operate continuously so that hydraulic and organic loads can vary considerably with time. Intermittent loads can occur daily (e.g. dairies) or seasonally, (e.g. sugar and wine industry). Intermittent and shock loadings can significantly decrease the efficiency of AS process and may call for equalization.

Lack of N, P and trace elements. Many industrial wastewaters (e.g. brewery, pulp and paper, and food-processing industries with high-carbohydrate loads) lack sufficient N and/or P for growth and maintenance of active microbial populations. Trace elements such as Ca, Mg, Mn, Zn, Co, Mo, Ni, Cu, V, B, Fe, I and vitamins can also be lacking. N and P deficiencies can lead to the growth of filamentous organisms responsible for bulking sludge – one of the main problems in industrial AS plants. N and P requirements for biological treatment depend on the required influent BOD removal. Recommended values are in the ranges 2.2–7.4% N/BOD_5 and 0.3–1.25% P/BOD_5 (Kilgore & Sawyer, 1953; Wuhrmann, 1958; von der Emde *et al.*, 1973).

High pollutant loads, for example carbohydrate, fat or protein loads. In addition to the nutrient deficiency issues referred to above, high industrial organic loads can cause floating sludge (high oil and grease loads, e.g. in the petroleum industry) and difficulties with N removal (high protein loads, for example from slaughtering, meat processing and rendering). Hrudey (1981) suggested that a daily loading of less than 0.10 kg lipid/kg MLSS would ensure reasonable protection against lipid overloading of the AS process. A conservative value of 0.05 kg fat/kg MLSS/d is recommended by Rosenwinkel (1999).

Hazardous and inhibitory substances (including disinfectants). Industrial wastewaters may contain a wide variety of hazardous and inhibitory substances. They can be toxic, refractory, biomagnified, carcinogenic, teratogenic and mutagenic. They include absorbable organic halogens (AOX), halogenated hydrocarbons, hydrocarbons (benzene, phenols, etc.), heavy metals and cyanides (Rosenwinkel *et al.*, 2005). In the petrochemical industry, wastewaters often contain phenols, mercaptans, ammonium compounds and cyanide (Fitzke & Wienands, 2008). Peak toxic substance concentrations due to spills, may result in AS process failure. In the food-processing industry, the use of disinfectants such as chlorine, peroxide or quaternary ammonium compounds can cause both reduced performance as well as shock loads to the AS process.

17.2.2 Industrial AS process technologies

The AS process has been modified and optimized in many ways. These modifications can be classified by their flow regime into continuous-flow and discontinuous-flow technologies, and according to the method of separating the suspended biomass after biological treatment. Settling tanks are most commonly used but membrane separation is becoming more common.

Many AS technologies are well known in municipal wastewater treatment (e.g. conventional, two-stage and cascade AS plants, SBR, and membrane technologies). However, compared to municipal plants, the design and application of industrial AS plants requires more attention to meet the particular industrial wastewater circumstances and the highly variable wastewater composition.

17.2.2.1 Continuous-flow AS technologies for industrial wastewater treatment

Conventional AS treatment plants. The conventional AS process is used to treat some types of industrial wastewaters (for example food and beverage industry). It is also used in combination with various pretreatment steps, such as anaerobic pretreatment, floatation or separation of fat. For example anaerobic pretreatment is often used in pulp and paper or dairy industry and floatation and separation of fat are important for slaughterhouse and animal by-products wastewaters. In the petrochemical industry, AS treatment is typically preceded by oil removal devices such as API separators and dissolved or induced air flotation units.

Two-stage AS treatment and cascades. Two-stage AS processes consist of two complete AS systems (aeration basin, settling tanks and RAS) connected in series. The benefits of this modification compared to conventional one-stage processes include shorter aeration times, ability to increase 2nd stage sludge retention time with a comparatively small tank volume increase, removal of poorly degradable materials, more efficient use of plant capacity and often more stable operation. The main disadvantage is the need for two settling tanks.

An example of the use of a two-stage process is the Asta Medica pharmaceutical plant Bielefeld, Germany. This plant treats an average daily flow of approximately 490 m^3/d with a maximum COD load of 1240 kg/d and comprises two aerated equalization tanks each with a volume of 1100 m^3, a two-stage AS process and sand filtration. For denitrification, ethanol is added to the 2nd AS tank. The 1st AS stage has a volume of 500 m^3 and an F/M of 0.4 kg COD/kg/d; the 2nd stage AS volume is 1,000 m^3 and operates at an F/M of 0.08 kg COD/kg/d (Rosenwinkel, 1994, 1999).

AB processes are a special form of the two-step AS process in which the 1st (A) stage is very highly loaded (about 10 kg BOD5/m^3/d) and the 2nd (B) stage is a conventional low-loaded system. The AS from the first step is wasted from the intermediate clarifier. The City of Krefeld, Germany uses an AB process to treat a municipal wastewater containing a high proportion of wastewaters from the food, fertilizer, chemical and steel industries. Following the first experiments on the feasibility of this technology at Krefeld (Böhnke, 1977; Diering, 1980). The plant was expanded and modified several times and is now designed for 1.2 million PE. The A-stage has three parallel channels and an intermediate clarifier and the B-stage has four parallel channels, two of which have simultaneous nitrification/denitrification (NDN) and two of which have pre-denitrification, then a 2nd clarifier and finally flocculation and filtration for SS polishing (Hobus *et al.*, 2004).

AS tanks in series (plug-flow) have a single sludge recirculation from the clarifier after the last tank to the 1st tank. While this type of plant is less prone to sludge bulking than a single-stage well-mixed basin it is more sensitive to inhibitors because their concentration will be much higher in the smaller 1st tank than in a single, larger, totally mixed basin.

The fruit juice manufacturing company Ernteband Fruchtsaft GmbH used cascade technology for wastewater pretreatment before discharging into a municipal wastewater treatment plant (Rosenwinkel, 1984); a plant with two AS cascades is used for full treatment by the dairy industry in Heideblume Molkerei Elsdorf AG, Rotenburg, Germany. The company processes about 400,000 kg of milk/d and generates a wastewater flow of about 1500 m^3/d. The plant is designed for 28,300 PE and consists of mechanical pretreatment, fat separation, anaerobic pretreatment and two AS cascades with final clarifiers (DWA, 2011).

Pure oxygen processes. In pure oxygen AS (HPOAS) processes, 98% O_2 is used for aeration instead of air (21% O_2). This type of process is used for carbonaceous BOD removal from highly concentrated wastewaters because their high DO concentration allows them to meet very high DO demands. The low off-gas volume from HPOAS processes can also be advantageous if, because of odor or other off-gas problems, the off-gas must be treated. Because oxygen transfer from 98% O_2 requires approx. 0.25–0.3 kWh/kgO_2 and the energy required to produce 98% oxygen is approx. 0.4–0.45 kWh/kgO_2 the economic viability must be checked. HPOAS systems are often applied in the animal by-products industry. BASF in Minden, Germany, produces actives and additives for the pharmaceutical and cosmetics industry and uses an HPOAS plant for biological effluent treatment. The HPOAS plant, designed for 170,000 PE and a flow of 10,000 m³/d, is preceded by aerated equalization neutralization and primary clarification. The HPOAS consists of, a four-step cascade aerated with pure oxygen aeration and a variable tank volume of 4200 to 5100 m³ (Klinzmann & Stammann, 1981).

Reactors with unique shape and depth. Oxygen supply by diffusion to the sludge floc interior is largely influenced by the contact time between the flocs and gas bubbles and the oxygen partial pressure. To enhance oxygen supply, tall (normally 8–12 m depth) slim reactors were developed in which contact time and oxygen partial pressure are increased. Oxygen solubility in water increases with total pressure, which in turn, increases with tank depth. This allows the F/M to be increased and also results in reactors that have a small footprint. However with deep tanks and low air flows, CO_2 stripping is reduced and concrete corrosion may occur.

Processes with very deep tanks include the Bayer tower biology (12–30 m tall steel cylinder with externally attached circular final sedimentation tank, see Figure 17.1 and the tall UDHE bioreactor or BIOHOCH® reactor. Tower biology plants for example are used by two German breweries: Königsbacher Brauerei AG, Koblenz and Bitburger Brauerei, Th. Simon GmbH, Bitburg (Rosenwinkel & Brinkmeyer, 2004). An example of the industrial application of the BIOHOCH® reactor is the wastewater treatment plant of the chemical industrial park at Bitterfeld, Wolfen, Germany where four BIOHOCH® reactors with tank volumes of 13,000 m³ and 17,000 m³ are used to treat industrial wastewaters from more than 500 industrial companies together with the municipal wastewater of 83,000 inhabitants (Rudolph, 2013).

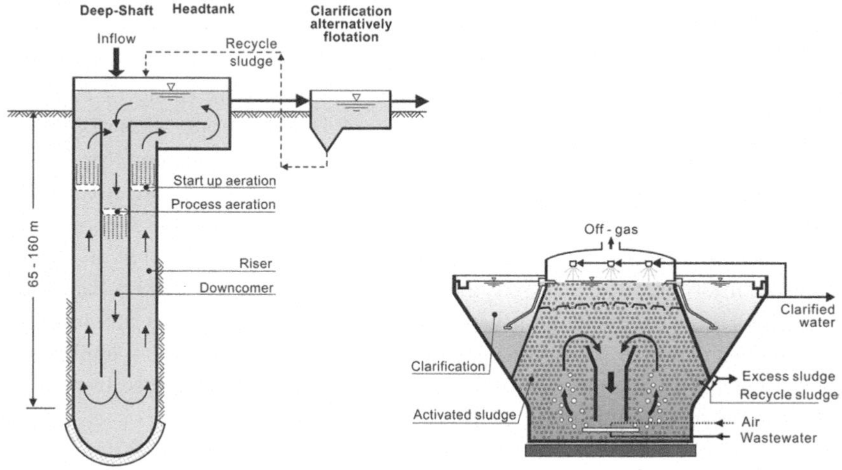

Figure 17.1 Layout of a deep-shaft reactor (left) and the scheme of a BIOHOCH-Reactor (right).

The deep shaft high intensity AS technology also uses tall, slim bioreactors (Figure 17.1). It comprises two concentric deep shafts (50–150 m) and an above-ground degassing tank. The wastewater–recycle–sludge mixture is fed down the inner shaft (downcomer) and flows back in the outer shaft (riser). The liquid circulation is generated by the buoyancy imparted by compressed addition air. During startup the downcomer is also aerated to induce circulation. Aeration at 30–50% of the tank depth and the high pressure in the downcomer contribute to excellent oxygen utilization. The main advantages of this process are its small footprint, low exhaust air volume and slow cooling of the wastewater. Deep shaft technology can only be used with favorable site and sub-surface conditions (Longdong & Rosenwinkel, 2007; Tilley, 2011).

The Omatsu-Itagami Cardboard AG, Japan, has built one of the largest deep shaft reactors with integrated vacuum degassing. It treats floatation-pretreated wastewater (20,000 m³/d, average BOD_5 concentration 160 mg/L, volumetric load 3.7 kg/m³/d). It has a reactor volume of 615 m³ (2.8 m diameter and 100 m deep) and two-step secondary clarification. The average effluent BOD_5 concentration is about 10 mg/L giving an approx. BOD_5 removal of 94% (Hitachi Puranto Kenetsu Co.).

Molson's Brewery Ltd. (Barrie, Ontario, Canada) pretreats their wastewater (2050 m³/d, 5000 kg/d BOD_5) with two deep shaft bioreactors (1.37 m diameter, 152 m deep). A BOD_5 removal of 95% is achieved at F/Ms of 0.33 –3.3 kg/kg/d and an MLSS of 10 g/L. The deep shaft technology has a very low footprint and can operate satisfactorily at winter temperatures of –40°C (Lom & Fedderson, 1981).

Membrane bioreactor technology in industrial wastewater treatment. In MBRs, AS separation is carried out by filtration with micro- or ultrafiltration membranes. Membranes (plate, tubular or capillary) are arranged in modules operated either outside the aeration tank (tubular) (sidestream configuration) or submerged in the aeration tank or a separate filtration tank (plate or capillary) (submerged configuration). The basic principle and mode of operation of MBRs is the same for industrial and municipal wastewater treatment (Chapter 16).

However, a number of special characteristics of the MBR make it especially interesting for industrial application: first, the <0.01 μm membrane pore size ensures complete retention of the AS so the discharge is virtually free of TSS and pathogenic microorganisms (DWA, 2007; Judd, 2011). Thus, a major advantage of using MBRs is the excellent quality of the discharge, especially for water reuse and recycling. For example MBRs are particularly suitable for industrial applications in which further treatment of wastewater by nanofiltration or reverse osmosis for water reuse or recycling is required (Cornel & Krause, 2006).

Furthermore, MBR systems allow operation with significantly higher sludge concentrations compared to the conventional AS process. Therefore, they require smaller AS tank volumes and have correspondingly higher SRTs (Stephenson *et al.*, 2000). Coupled with the fact that they need no additional sedimentation tank, MBRs have a low space requirement which is a significant advantage for industrial applications. The reduced tank volume for membrane systems also allows one to meet higher treatment standards or to increase treatment capacity by modifying existing conventional AS facilities. MBR systems are also an attractive option for industrial wastewater treatment since industrial wastewaters often have poor sludge settling characteristics.

Industrial wastewaters usually are more concentrated than municipal wastewaters and are often equalized. Thus the membrane area of MBRs required for industrial applications may be small compared to a similarly-sized municipal plant (DWA, 2008).

Even though MBRs are considered state-of-the-art, the biological degradation mechanisms in them are the same as in conventional AS plants and their sizing can be based on wastewater load, however, for unknown wastewater compositions, pilot testing is recommended. This is particularly important since it is not yet possible to predict the conditions under which membrane fouling occurs.

The energy requirements of membrane bioreactors are higher than for conventional AS plants due to the energy required for filtration through the membrane and to a lower α value for aeration. However, because industrial wastewaters are usually more concentrated they can require less membrane surface per kg of COD. Henkel *et al.* (2009) showed that α decreased almost linearly with the MLVSS concentration (Figure 17.2). Thus for high sludge concentrations the aeration energy requirements will increase. This has led to the practice of generally limiting the MLSS concentration in submerged MBR systems to 10–12 g/L (MLVSS = 6–9 g/L).

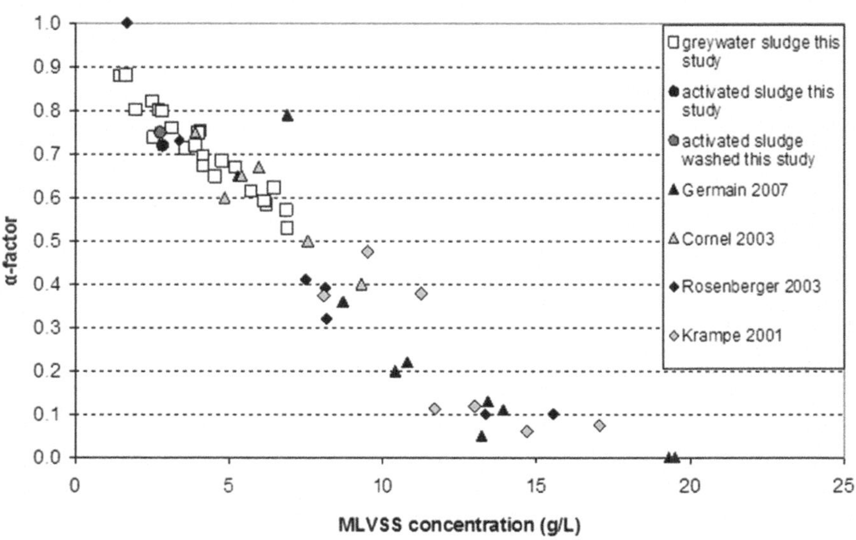

Figure 17.2 Effect of MLVSS concentration on the α-factor (Henkel *et al.*, 2009).

MBRs are applied in many industrial sectors. The German brewery Badische Staatsbrauerei Rothaus in Grafenhausen, Germany uses a combination of SBR and membrane filtration for wastewater treatment. The process consists of a mechanical pretreatment step, two equalization tanks with volumes of 1350 und 1600 m^3, 10 SBRs (each 400 m^3), an equalization tank (300 m^3) and four membrane filtration units, each with a membrane surface area of 1500 m^2. The treatment plant is designed for a wastewater load of 100,000 PE and an average flow of 1400 m^3/d. containing about 4500 mg/L COD, 70 mg/L total N and 17 mg/L total P. The effluent requirements of <45 mg/L COD, <1 mg/L total P and total N are met (DWA, 2010).

A German chemical company uses a three-step biological system with external membrane filtration for wastewater treatment. After primary clarification, neutralization and P precipitation the wastewater flows to an equalization tank and then to three AS basins followed by membrane filtration with four filtration units and a permeate storage tank. The plant is designed for a maximum flow of about 200 m^3/h or 3600 m^3/d. The influent has an average COD of about 3500 mg/L, NH_4-N of 100 mgN/L and <10 mgP/L. The treated wastewater contains an average COD of 300 mg/L and NH_4-N and PO_4-P concentrations <1 mg/L (DWA, 2008).

The wastewater treatment plant of Kellogg Limited in Manchester, UK is an example of MBR technology for treating food industry wastewater. The company produces cereals and its wastewater volume is about 1500 m^3/d with concentrations of 3000–4000 mg COD/L and 2000 mg BOD_5/L. An externally installed membrane filtration system consisting of four units with a membrane surface area of 864 m^2 is preceded by AS with two tanks each with a volume of 1050 m^3 and an equalization tank with a volume of 1200 m^3. The AS MLSS is about 20 g/L and the F/M is 0.14 kg COD/kg MLSS/d. The membrane filtration unit has a flux of 70–120 $L/m^2/h$. With this system about 98% COD removal results in effluent COD concentrations <120 mg/L. A special characteristic of the plant is aeration through injectors (Wagner, 2004).

Hybrid processes in industrial wastewater treatment. Hybrid processes combine suspended and attached biomass (on a carrier material) in an AS process flow sheet. Growth on the carrier surfaces increases the biomass concentration and can enhance biodegradation efficiency (Chapter 15). Their use for retrofitting overloaded plants and for the provision of N removal in a small space is making them an attractive option in the industrial wastewater treatment field (Schmitt-Schmieder, 2001). Suspended carriers are kept in suspension by aeration mixing and retained in the aeration basin with a screen. Industrial sectors using these types of system include paper, beverage, electronics, textile, coke oven and distillery (Strabag-Linpor®-process, www.strabag.de).

Besides common aerators, specially constructed disk wheels such as the Stählermatic system, are used both as growth supports and aerators (Görner & Hübner, 2001). The advantages of this system are its simple, robust configuration and its very low energy demand (DWA, 2004).

17.2.2.2 Discontinuous-flow AS technologies for industrial wastewater treatment

Because simple, easy to operate systems are very appealing, conventional SBRs have met with great success in industrial wastewater treatment. Their operational flexibility is particularly appropriate for industrial wastewater treatment because it can accommodate large variations in flow and pollutant concentrations. Mace & Mata-Alvarez (2002) performed an extensive review of important SBR applications in industries such as wineries/breweries, landfilling (leachate), paper/pulp industry, slaughterhouses and the food/dairy industry.

Typical SBR design parameters include a SRT of 20–40 d (although this can be considerably lower) and an MLSS concentration of 3–4 g/L (WEF & ASCE, 2005). The inclusion of anaerobic and anoxic periods in the feeding/reaction phases will depend on the need for EBPR and denitrification and for settleability control. An additional economic benefit of SBRs is that, because industrial wastewaters are usually more concentrated than municipal wastewater, the SBR requires smaller exchange volumes.

Several types of hybrid or modified SBR configurations exist that combine the advantages of continuous influent/effluent flow and constant liquid level of a CAS system with the properties of flexibility and compactness of SBRs (Wu *et al.*, 2001), for example the UNITANK® approach. This system consists of a tank divided into three equal-volume compartments ('1', '2' and '3') each equipped with aeration and mixing. The wastewater flow direction can be from '1' through '2' to '3', or vice versa. Thus compartments '1' and '3' can either be a reactor or a settler while tank '2' always functions as a reactor (Zhang *et al.*, 2007). This type of SBR has been used in some 200 industrial wastewater treatment plants (Keppel Seghers Belgium, Willebroek, Belgium, 2013). A second example is the Aqua MSBR®, which uses 3–7 reactor cells, two of which operate as SBRs with the following sequence: (i) feeding, anoxic mixing and MLSS recycle, (ii) feeding, aeration, mixing and MLSS recycle, (iii) settling, and (iv) clarification. While one of the SBR cells is in phases (i)–(iii), the other is in phase (iv), and vice versa. A dedicated 'air weir' device (Aqua-Aerobics, 2010) is installed at the water surface of these cells to control effluent flow without allowing MLSS or scum discharge. The recycled mixed liquor is concentrated in a phase separator cell. Further additions can include compartments for fermentation, anaerobic P release, anoxic denitrification, aerobic P uptake and carbon/nitrogen oxidation.

Currently, one Aqua MSBR® system is operating on industrial wastewater and four on combined industrial and domestic wastewater (Aqua Aerobics Systems Inc., 2013).

17.3 SPECIAL TOPICS IN INDUSTRIAL AS TREATMENT

Industrial wastewater composition variation can create significant difficulties that adversely affect AS process efficiency. The following topics are of particular interest to industrial AS treatment and must be addressed to ensure the successful operation of the above-described technologies: (1) bulking sludge control using selectors and chemicals, (2) special industrial wastewater aeration applications, and (3) biostimulation and bioaugmentation. Partial nitritation/anammox processes and the biological floc technology used in the aquaculture industry are examples of novel solutions that can be applied to process optimization, cost-effectiveness and energy-efficiency of AS technologies in the industrial sector.

17.3.1 Selector application and use of chemicals for bulking control in industrial AS plants

Adequate settling and growth of floc-forming bacteria is the primary prerequisite for retaining the microorganisms in an AS bioreactor. Even though industrial effluents can have highly variable flows and loads, some of these variations can be tackled in the design phase by providing for the appropriate F/M and SRT. An additional measure for preventing sludge bulking is to include a selector in the process scheme (Chapter 10). Selectors have been used to control AS settleability in plants treating wastewaters from the cotton fiber industry (Munirathinam, 2003), the pulp and paper industry (Keays *et al.*, 1996; Andreasen *et al.*, 1999; Melcer *et al.*, 2003) and from beet sugar mills (Prendl & Kroiss, 1998). Installation of a selector as an initial contact zone enhanced AS tolerance towards toxic compounds, such as phenolics (Abulnour *et al.*, 2008) and heavy metals Cr(VI) and Zn(II) (Alkan *et al.*, 2008). Selectors can also maintain high AS quality in municipal WWTPs receiving a considerable part of their COD load from industrial discharges (Duine & Kunst, 2002; Cao *et al.*, 2006).

Chemical addition is widely used in the industrial AS systems as a short-term solution for occasional rapid bulking control. Chemicals aid in compaction of highly filamentous flocs but generally they do not target the underlying causes of bulking sludge. Before using a specific chemical, it is recommended that the required dosing quantity be determined in laboratory tests to avoid an overdose that may adversely affect the biological activity of the AS. Flocculation aids (polymers, iron salts), adsorbents (activated carbon, lignite powder), and lime, activators (e.g. trace elements and vitamins), or oxidizing agents (hydrogen peroxide, ozone) are commonly used (Rosenwinkel, 1996) (Chapter 10).

17.3.2 Industrial wastewater aeration

The large variability in wastewater composition as well as in the design of AS plants for industrial wastewater treatment poses a challenge for the selection and dimensioning of aeration units. Some (but not all) of the factors that are rather uncommon with municipal AS plants are listed below since they can play a decisive role in designing industrial wastewater treatment plants:

(a) Large aeration tank water depth (up to 25 m).
(b) High water temperature (up to 35°C) leading to reduced oxygen solubility.
(c) Odorous wastewater requiring off-gas treatment AND minimization of off-gas volume.
(d) High dissolved salt concentration influencing bubble size and coalescence and floc dispersion.
(e) Constituents that can precipitate/form scale on/within diffusers, for example high calcium concentrations.
(f) AS downstream of anaerobic units in which CO_2 stripping can elevate pH and allow $CaCO_3$ precipitation
(g) High pH influents can be neutralized by the CO_2 formed during AS treatment.
(h) Upgrading of existing wastewater treatment plants, for example for increased N removal.

For large water depths (>20 m) such as in tower biology plants, fine-bubble diffusers should not be used to achieve a high oxygen transfer efficiency (kg O_2/kWh) since the energy demand for generating fine air bubbles increases considerably with water depth. Coarse or medium bubble aeration systems such as perforated pipes or injectors are more robust and economical for large water depths and can achieve oxygen transfer rates of >70% even with bubble diameters of several millimetres.

In Figure 17.3, total standard oxygen transfer efficiency (SOTE, %) and standard specific oxygen transfer efficiency (%/m) are plotted as a function of diffuser submergence depth (0–20 m) for standard specific oxygen transfer efficiency (SSOTE, η_{s0}) for 4, 6 and 9%/m. η_{s0} is the SOTE at a theoretical water depth of 0 m. The more efficient the aeration system, the higher is η_{s0} and the more pronounced is its decrease with water depth because of the decreasing oxygen content of the rising air bubbles. Thus, the SOTE curves at first rise steeply and almost linearly, then rise progressively less steeply. With an efficient aeration system in deep aeration tanks with depths of submergence of 20 m, the utilization of the oxygen transferred by the aeration system can be as high as 90%.

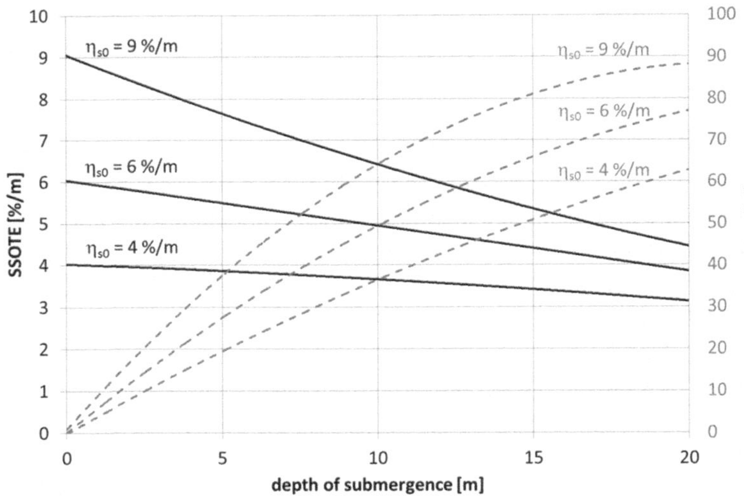

Figure 17.3 Influence of the depth of submergence on total (SOTE) and specific (SSOTE) oxygen transfer efficiencies for different basic-standard specific oxygen transfer efficiencies η_{s0}.

In some cases, aeration with high purity oxygen instead of air might offer advantages for:

- Improving N removal efficiency in non-nitrifying plants (e.g. at the BASF, Ludwigshafen plant).
- Minimizing off-gas volume.
- Increasing DO concentration.
- Lowering the pH of high pH wastewaters though reduced CO_2 stripping.
- Reducing foaming potential by reducing the amount of injected gas introduced.

High wastewater salt concentrations impede air bubble coalescence, and increase oxygen transfer rates and possibly α-factors (≫1). Wastewaters with high calcium concentrations can cause scaling on diffusers. Formic acid can be injected into the air distribution system as a preventive measure against this problem. In addition, diffuser membranes should be allowed to 'relax' at least once a day for about 15 min and subsequently should be operated at the maximum permissible air flow rate per diffuser.

17.3.3 Biostimulation and bioaugmentation

Biostimulation attempts to enhance process conditions so that they are favorable for the growth and activity of the existing AS microorganisms, for example the addition of the appropriate nutrients, in appropriate amounts, to nutrient-deficient wastewater (Burgess *et al.*, 1999).

Bioaugmentation is the addition of either an externally, or an internally cultivated microbial community to the AS reactor. Available products target wastewaters from specific industries such as petrochemical, coking, pesticide, pulp/paper, food/dairy (Jogdand, 2006) and focus mainly on the degradation of greases and fats, oil and petroleum hydrocarbons, lignin (Chen *et al.*, 2012), phenols (Watanabe *et al.*, 2002), chlorinated compounds, ketones/acetones. Other products target the enhancement of a sufficient nitrifying sludge inventory, for example under cold conditions (Head & Oleszkiewicz, 2004), the amelioration of toxicity (Boon *et al.*, 2003; Nejidat *et al.*, 2006), to compensate for unexpected hydraulic or pollutant loadings, and to improve the AS settling properties (Van Limbergen *et al.*, 1998).

Microorganisms can be purchased as liquid or freeze-dried biomass generated from pure cultures (e.g. Pseudomonas strains in Munox® series 1 or Bacillus strains in Bacilox® by Osprey Biotechnics, Sarasota, Florida, USA) or in the form of mixed communities (e.g. the nitrifying products ABIL and HANDS from Avecom nv, Wondelgem, Belgium). The use of enzymes is less common but they are also on the market, for example Bac-Zyme® by Microtack (Bangkok, Thailand). *In situ* bioaugmentation schemes can be implemented to cultivate the desirable microbial communities but the availability and suitability of wastewater sidestreams on an industrial site is often limited, so this practice is more useful in municipal wastewater treatment (Berends *et al.*, 2005).

17.3.4 Partial nitritation/anammox technology

Partial nitritation/anammox is a biological N removal process for wastewaters with low biodegradable COD (bCOD)/TKN ratios that can provide cost savings of about 30–40% compared to nitrification/denitrification with external carbon addition (Fux & Siegrist, 2004; Vlaeminck *et al.*, 2012). Most full-scale applications focus on the treatment of the reject water from dewatering anaerobically digested sewage sludge. The process is also called the OLAND process (Oxygen-Limited Autotrophic Nitrification/Denitrification) (Kuai & Verstraete, 1998). N removal without heterotrophic denitrification is gradually finding application in the treatment of the following industrial wastewaters: potato processing (Abma *et al.*, 2010; Desloover *et al.*, 2011), semi-conductors (Tokutomi *et al.*, 2011), glutamate and yeast production, landfilling leachate treatment, tanneries, dairies, meat processing, sweetener production, and distilleries (personal communication, ARAconsult GmbH, Innsbruck, Austria; Cyklar-Stulz GmbH, Gommiswald, Switzerland; Paques BV, Balk, the Netherlands). An overview of full-scale industrial partial nitritation/anammox installations is summarized in Table 17.2.

Table 17.2 Full-scale industrial partial nitritation/anammox installations.

Technology	Location	Operational since	Wastewater	Reactor volume	Design load	
				m^3	kg-N/d	kg $N/m^3/d$
***SBR* (*one stage*)**	***DEMON*® *process; Cyclar Stulz, ARA consult***					
	Alltech, RS	2011	yeast		2400	
	Louis, I		leachate		1800	
	Rickenbach, DE		dairy		50	
	Trento, IT	2011	leachate		700	
***AS* (*multiple stage*)**	***NAS*®; *Colsen***					
	Bergen op Zoom, NL	2009 (retrofitted from N/DN)	Digestate potato industry	7920 (3 stages + clarifier)	720	0.09
	Kruiningen, NL	2004 (retrofitted from N/DN)	Digestate potato industry	15300 (4 stages + clarifier)	325	0.02
***MBR* (*multiple stage*)**	***NAS*®; *Colsen***					
	Boxtel, NL	(in design)	Digestate slaughterhouse	5310 (3 stages, last stage as MBR)	888	0.17
	Budrio, IT	2010	Digestate potato industry	1509 (3 stages + MBR)	344	0.23

Source: Adapted from Lackner et al. (2014).

In these industrial applications, biofilm or granular sludge reactors are used to achieve high biomass retention so that the slowly growing anammox bacteria can be retained in the reactor. The DEMON® process (DeaMmONification) uses a combination of flocs (more aerobic activity) with granular particles (more anoxic activity) for N removal in an SBR configuration. As part of the control strategy, the small flocs are selectively washed out with a hydrocyclone; this allows the preferential wasting of the faster growing aerobic organisms, including the undesirable nitrite oxidizing bacteria (Wett *et al.*, 2010).

The New Activated Sludge (NAS®) process relies on completely flocculant sludge and the control of SRT and DO concentration. One plant treating digested potato factory wastewater has been studied in depth by Desloover *et al.* (2011) (Figure 17.4). At an N loading rate of 0.18–0.23 kg N/m³/d the batch-fed partial nitritation step oxidizes about half of the ammonia to nitrite at an SRT of 37 h and a DO of 0.9–1.0 mg/L. Serial anammox, denitrification and nitrification compartments are followed by a final clarifier; the overall SRT is 46 d and the SVI was 167 mL/g. In the anammox step, 0.26 kg N/m³/d was removed, with an estimated contribution of 71% removal by anammox; the anammox bacteria constituted 3.1% of the total sludge microorganisms. Overall, the NAS® plant obtained an N removal efficiency of 95% and demonstrated that cost-effective partial nitritation technology can be retrofitted to existing industrial AS systems removing N by nitrification/denitrification.

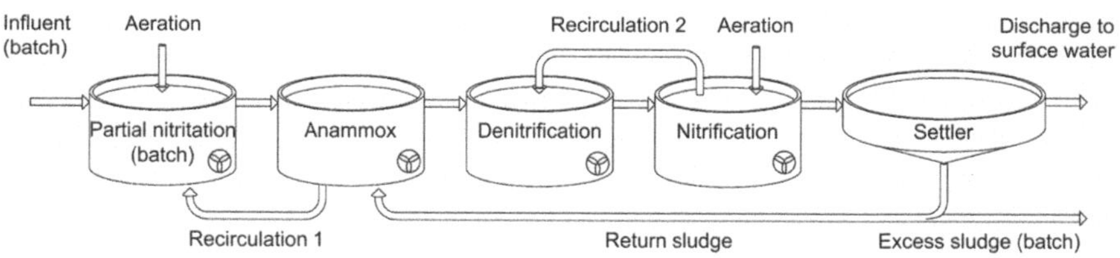

Figure 17.4 Schematic of a 4-stage NAS® process (Desloover et al., 2011).

17.3.5 AS in aquaculture industry: biofloc technology

An interesting advance in the industrial aquaculture sector is the development and implementation of the so-called biofloc technology (Avnimelech, 1999; Crab *et al.*, 2012). Aquaculture is a food production sector industry with a growth rate of about 9%/y since the 1970s (FAO, 2004). In aquaculture, typically about 75% of the feed N ends up in the cultivation water. As a means of recovering this N and converting it into microbial protein, a carbon source is added at about a C/N of 10 ratio to grow heterotrophic 'bio' flocs. Carbon sources range from low protein aquaculture feed to carbohydrates such as acetate, glycerol, glucose, molasses, and starch. Since the resulting 'bio' flocs contain 28–49% protein, they serve as excellent recycled feedstock for cultivated fish or shrimps (De Schryver *et al.*, 2012) with feed cost savings of up to 31–36% for fish and 50% for shrimp cultivation (Avnimelech & Taw, 2012). The aeration cost for supplying oxygen to the 'bio' flocs only consumes some 10–20% of these savings. Two typical 'bio' floc technology configurations are depicted in Figure 17.5.

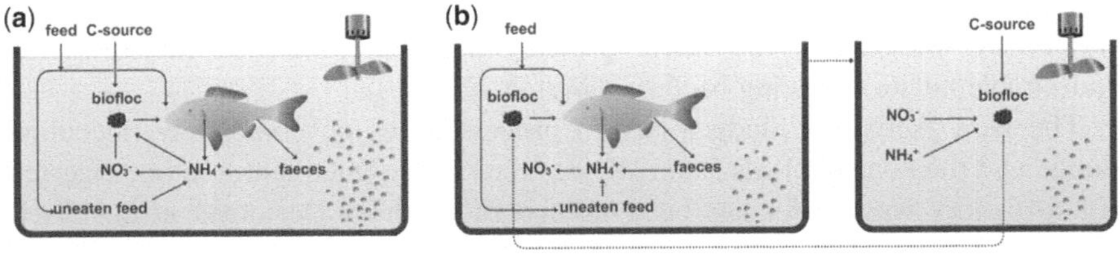

Figure 17.5 Schematic of two possible implementation strategies for 'bio' floc technology (Crab *et al.*, 2012): A. 'Bio' flocs integrated in the aquaculture tank, using a feed with a relatively low protein content and/or the addition of a carbon source, B. Use of a separate 'bio' floc reactor, where a carbon source is added, and the consecutively grown 'bio' flocs are recirculated to the aquaculture tank

17.4 INDUSTRIAL AS TREATMENT – FUTURE DEVELOPMENTS

A major challenge for industrial AS systems will be to evolve from an end-of-pipe treatment system to a highly reliable resource recovery system. In the decades to come, microbiome-based systems must be configured to produce valuable commodities. System biology based route maps will be of crucial value for this objective, such as being able to opt for the production of young nutritious biomass, as in the 'bio' floc technology, or for a biomass rich in valuable products, for example alginate-like polymers, polyhydroxybutyrate, polyphosphate, rare earth metals, and so on. It would be also quite intriguing to combine algae and bacteria in the same floc. This combination should be able to use light energy as a driver for immobilizing mineral nutrients and supplying oxygen while the heterotrophic biomass would supply carbon dioxide to the phototrophs (Van den Hende *et al.*, 2011), which would produce algal biomass with potential value as a feed or for producing biodiesel/oil.

AS systems currently are monitored and controlled using simple physical–chemical parameters (e.g. pH, Eh, ORP, rates and levels of COD, N and P conversion, sludge settling characteristics). Extensive documentation of the microbes present in industrial AS systems is still in its infancy and largely restricted to scientific studies (Bramucci & Nagarajan, 2006). Such data should be extended to improve operation of AS systems, for example by using the data to design early warning indicators of changes in critical conversions or of the need to re-stock the microbiome with specific process capacities, for example by bioaugmentation. To achieve this type of capability, the road for microbial resource management (MRM) must be further developed by calibrating parameters such as species diversity, rate of change of species populations, and intrinsic community organization (=adherence to the Pareto distribution pattern, as discussed by Marzorati *et al.*, 2008), and possibly extended with information on the spatial juxtaposition of certain functional organism groups. Such a set of parameters extending from metagenomics to treatment performance will be very helpful in preventing the failure of an SBR by, for instance, sludge wash-out or unsatisfactory conversion efficiencies. They will also be helpful for evaluating the effects of (cross) inoculation (Wittebolle *et al.*, 2009a), the evolution of specific microbiomes (Wittebolle *et al.*, 2009b) and the application of real-time controllers.

Overall, the future major challenge for industrial wastewater treatment by AS is to determine the most efficient route for maximum 'dissipation' of all pollutants necessary to obtain the most important end product of all, that is 'NEWater'. The future of AS is therefore to be found in its capacity to become part of the 'water factory' and in the ever increasing need to effectively deal with 'urban mining' (Verstraete *et al.*, 2009; Verstraete & Vlaeminck, 2011). Indeed, our societies are becoming more and more urbanized and 'hot spots' of human activity are integrating massive amounts of resources which need to be up-cycled. Harnessing the 'intelligence' of constantly evolving microbial communities to achieve such value recovery is a beautiful task for the next 100 years.

17.5 REFERENCES

Abma W. R., Driessen W., Haarhuis R. and van Loosdrecht M. C. M. (2010). Upgrading of sewage treatment plant by sustainable and cost-effective separate treatment of industrial wastewater. *Water Sci. Technol.*, **61**(7), 1715–1722.

Abulnour A. E.-G. G., Shaalan H. F., El-Razek T. M. A., Hassan A. N. and Ali S. S. (2008). Treatment of phenolic industrial wastewater using activated sludge process. *Catrina J.*, **3**(3).

Alkan U., Eleren S. C., Nalbur, B. E. and Odabas E. (2008). Influence of the activated sludge system configuration on heavy metal toxicity reduction. *World J. Microbiol. Biotechnol.*, **24**(8), 1435–1443.

Andreasen K., Agertved J., Petersen J. O. and Skaarup H. (1999). Improvement of sludge settleability in activated sludge plants treating effluent from pulp and paper industries. *Water Sci. Technol.*, **40**(11–12), 215–221.

Aqua-Aerobic Systems Inc (2010). Aqua MSBR Modified Sequencing Batch Reactor. Bulletin #M102 04/10, Loves Park, IL, USA.

Aqua-Aerobic Systems Inc (2013) Privately communicated, Loves Park, IL, USA.

Ardern E. and Lockett W. T. (1914). Experiments on the oxidation of sewage without the aid of filters. *J. Soc. Chem. Ind.*, **33**, 523–539.

ATV (1995). ATV-Merkblatt ATV-M 770: *Behandlung und Verwertung von Reststoffen aus Schlacht- und Fleischverarbeitungsbetrieben* (Treatment and recovery of wastes from slaughterhouses and meat processing plants). Abwassertechnische Vereinigung e.V., Hennef, Germany.

Avnimelech Y. (1999). Carbon nitrogen ratio as a control element in aquaculture systems. *Aquaculture*, **176**(3–4), 227–235.

Avnimelech Y. and Taw N. (2012). Chapter 17: Field experience. In: Biofloc Technology – A Practical Guide Book. Y. Avnimelech (ed.), The World Aquaculture Society, Baton Rouge, LA, USA, 231–252.

Berends D. H. J. G., Salem S., van der Roest H. F. and van Loosdrecht M. C. M. (2005). Boosting nitrification with the BABE technology. *Water Sci. Technol.*, **52**(4), 63–70.

Bischofsberger W., Dichtl N., Rosenwinkel K.-H., Seyfried C.-F. and Böhnke B. (2005). *Anaerobtechnik* (Anaerobic technology), 2nd edn., Springer Verlag, Berlin, Germany.

Böhnke B. (1977). *Das Adsorptionsbelebungsverfahren* (The adsorption activated sludge process). *Korresp. Abwasser*, **24**(2), 33–42.

Boon N., Top E. M., Verstraete W. and Siciliano S. D. (2003). Bioaugmentation as a tool to protect the structure and function of an activated-sludge microbial community against a 3-chloroaniline shock load. *Appl. Environ. Microbiol.*, **69**(3), 1511–1520.

Bramucci, M. and Nagarajan V. (2006). Bacterial communities in industrial wastewater bioreactors. *Current Opinion Microbiol.*, **9**(3), 275–278.

Burgess J. E., Quarmby J. and Stephenson T. (1999). Role of micronutrients in activated sludge-based biotreatment of industrial effluents. *Biotechnol. Adv.*, **17**(1), 49–70.

Cao Y. S., Ang C. M., Raajeevan K. S., Kiran A. K., Lai K. C., Ng S. W., Zulkifli I. and Wah Y. L. (2006). Analysis of phosphorus removal and anaerobic selector performance in a full-scale activated sludge process in Singapore. *Water Sci. Technol.*, **54**(8), 237–246.

Chen J., Zhan P., Koopman B., Fang G. and Shi Y. (2012). Bioaugmentation with *Gordonia* strain JW8 in treatment of pulp and paper wastewater. *Clean Technol. Environ. Policy*, **14**(5), 899–904.

Cornel P. and Krause S. (2006). Membrane bioreactors in industrial wastewater treatment-European experiences, examples and trends. *Water Sci. Technol.*, **53**(3), 37–44.

Crab R., Defoirdt T., Bossier P. and Verstraete W. (2012). Biofloc technology in aquaculture: Beneficial effects and future challenges. *Aquaculture*, **356**, 351–356.

De Schryver P., Boon N., Verstraete W. and Bossier P. (2012). Chapter 15: The biology and biotechnology behind bioflocs, In: Biofloc Technology – A Practical Guide Book. Y. Avnimelech (ed.), The World Aquaculture Society, Baton Rouge, LA, USA, pp. 199–215.

Desloover J., De Clippeleir H., Boeckx P., Du Laing G., Colsen J., Verstraete W. and Vlaeminck S. E. (2011). Floc-based sequential partial nitritation and anammox at full scale with contrasting N_2O emissions. *Water Res.*, **45**(9), 2811–2821.

Diering B. (1980). *Einstufige oder zweistufige biologische Klärwerke?* (One or two-step biological wastewater treatment plants?). *Korresp. Abwasser*, **2,** 163–166.

Duine A. and Kunst S. (2002). Control of bulking sludge caused by Type 021N and type 0961 in an industrial wastewater treatment plant with an aerobic selector. *Water Sci. Technol.*, **46**(1–2), 39–33.

DWA (2004). *Aerobe Biofilmverfahren in der Industrieabwasserreinigung: Definitionen, Verfahrenstechniken, Einsatzgebiete, Bemessungshinweise* (Aerobic biofilm processes in industrial wastewater treatment) Deutsche Vereinigung für Wasserwirtschaft, Abwasser und Abfall e.V., Hennef, Germany.

DWA (2007). *Treatment of Industrial Wastewater and Process Water with Membrane Processes and Membrane Bioreactor Technology (DWA-Topics).* Deutsche Vereinigung für Wasserwirtschaft, Abwasser und Abfall e.V., Hennef, Germany.

DWA (2008). *Behandlung von Industrie- und Gewerbeabwasser* (Treatment of industrial wastewater). Deutsche Vereinigung für Wasserwirtschaft, Abwasser und Abfall e.V., Hennef, Germany.

DWA (2010). Merkblatt DWA-M 732: *Abwasser aus Brauereien* (Wastewater from Breweries). Deutsche Vereinigung für Wasserwirtschaft, Abwasser und Abfall e.V., Hennef, Germany.

DWA (2011). Merkblatt DWA-M 708: *Abwasser aus der Milchverarbeitung* (Wastewater from Dairy Industry). Deutsche Vereinigung für Wasserwirtschaft, Abwasser und Abfall e.V., Hennef, Germany.

FAO (2004). State of World Fisheries and Aquaculture. FAO, Rome, Italy.

Fitzke B. and Wienands H. (2008). *Abwasser aus Petrochemie- und Raffineriebetrieben.* (Wastewater from petrochemical plants and refineries). *Wasserwirtschaft Wassertechnik*, **9**, 16–19.

Fux C. and Siegrist H. (2004). Nitrogen removal from sludge digester liquids by nitrification/ denitrification or partial nitritation/anammox: environmental and economical considerations. *Water Sci. Technol.*, **50**(10), 19–26.

Görner K. and Hübner K. (2001). *Gewässerschutz und Abwasserbehandlung* (Water conservation and Wastewater treatment). Springer Verlag, Berlin, Germany.

Head M. A. and Oleszkiewicz J. A. (2004). Bioaugmentation for nitrification at cold temperatures. *Water Res.*, **38**(3), 523–530.

Henkel J., Lemac M., Wagner M. and Cornel P. (2009). Oxygen transfer in membrane bioreactors treating synthetic greywater. *Water Res.*, **43**(6), 1711–1719.

Hines D. A., Bailey M., Ousby J. C. and Roesler F. C. (1975). The ICI deep shaft aeration process for effluent treatment. *Inst. Chem. Eng. Symp.Series*, **41**, D1–D10.

Hitachi Puranto Kenetsu Company, *Firmenprospekt über das Deep-Shaft-Verfahren bei der Ohtsu Itagami Cardboard AG* (Company brochure on the deep shaft process at Ohtsu Itagami Cardboard AG).

Imhoff, K. (1955). *Das zweistufige Belebungsverfahren für Abwasser* (The two-step activated sludge process for wastewater treatment). *gwf Wasser Abwasser*, **96**, 43–45.

Jäppelt W. and Neumann H. (1985). *Verarbeitung tierischer Produkte – Schlacht- und Fleischverarbeitungsbetriebe* (Processing of animal products – Slaughterhouses and meat processing plants). In: Lehr- und Handbuch der Abwassertechnik, Band V, Chapter 6.1, ATV Abwassertechnische Vereinigung, Hennef, Germany, 320–382.

Hobus I., Kolisch G., Oldenburg M., Dlabacs C., Metzen P. and Oidtmann T. (2004). *Prozessoptimierung durch integrierte Modellierung einer großen kommunalen Kläranlage* (Process optimization by integrated modelling of a large municipal wastewater treatment plant) in Hamburger Berichte zur Siedlungswasserwirtschaft – 16. Fachtagung: Norddeutsche Tagung für Abwasserwirtschaft und Gewässerschutz, 45, 83–92.

Hrudy, S. E. (1981) Activated sludge response to emulsified lipid loading. *Water Res.*, **15**, 361–373.

Jogdand S. N. (2006). Environmental Biotechnology, 3rd version. Himalaya Publishing House, Mumbai, India.

Judd S. (2011). The MBR Book – Principles and Applications of Membrane Bioreactors in Water and Wastewater Treatment. Butterworth-Heinemann, Oxford, UK.

Keays G. M., Marchand W., Marstaller T. and GaydeMontella R. (1996). Activated sludge pilot studies at Stone-Consolidated Corporation, Kenora, Ontario. *Pulp & Paper-Canada,* **97**(4), 34–37.

Kehr D. (1964). *Die Behandlung gemischter häuslicher und industrieller Abwässer, Bericht über Versuche zu ihrer Reinigung* (Treatment of mixtures of municipal and industrial wastewater, Report about experiments for their purification) in Veröffentlichungen des Instituts für Siedlungswasserwirtschaft und Abfalltechnik, Volume 15. Hannover, Germany.

Keppel Seghers Belgium NV, Willebroek, Belgium. (2013) Privately communicated.

Kilgore H. D. and Sawyer C. N. (1953). Nutritional requirements in the biological stabilization of industrial wastes. *Sewage Industr. Wastes,* **25**(5), 596–604.

Klinzmann G. and Stammann D. (1981). *Biologische Abwasserreinigung mit technisch reinem Sauerstoff – Die Kläranlage der KNOLL AG im Werk Minden* (Biological wastewater treatment using pure oxygen – the treatment plant of Knoll A.G. at Minden). *gwf Wasser Abwasser,* **122**, 258–262.

Kuai L. P. and Verstraete W. (1998). Ammonium removal by the oxygen-limited autotrophic nitrification-denitrification system. *Appl. Environ. Microbiol.,* **64**(11), 4500–4506.

Lackner S., Gilbert E. M., Vlaeminck S. E., Joss A., Horn H. and van Loosdrecht M. C. M. (2014). Full-scale Partial Nitritation/Anammox Experiences – an Application Survey. *Water Res,* doi: 10.1016/j.watres.2014.02.032

Lom T. and Fedderson C. C. (1981). Practical experience with a novel method of treatment of brewery wastewaters. Proc. 18th Congress of the European Brewery Conv., Copenhagen, Denmark, 121–128.

Longdong J. and Rosenwinkel K.-H. (2007). *Industrieabwasserbehandlung* (Treatment of industrial wastewater). Weiterbildendes Studium Wasser und Umwelt, B.U.W., in cooperation with DWA (Deutsche Vereinigung für Wassserwirtschaft, Wasser und Abwasser e.V.).

Mace S. and Mata-Alvarez J. (2002). Utilization of SBR technology for wastewater treatment: An overview. *Industr. Eng. Chem. Res.,* **41**(23), 5539–5553.

Marzorati M., Wittebolle L., Boon N., Daffonchio D. and Verstraete W. (2008). How to get more out of molecular fingerprints: practical tools for microbial ecology. *Environ. Microbiol.,* **10**(6), 1571–1581.

Meidl J. A. (1999). Industrial Wastewaters for Direct Discharge or Reuse. *Jubilee XX Conference of Science and Technology,* Katowice, Poland.

Melcer H., Roeder M., McKinnon, J. and McGuffin R. (2003). Full-scale aerobic selector evaluation at a recycle fiber mill. In: Industrial Wastewater and Best Available Treatment Technologies: Performance, Reliability, and Economics, DEStech Publications Lancaster, PA, USA.

Munirathinam K. (2003). Effective biological treatment technology for the treatment of high-strength pulp mill wastewater – A case study. In: Industrial Wastewater and Best Available Treatment Technologies: Performance, Reliability, and Economics, DEStech Publications, Lancaster, PA, USA.

Nejidat Z., Khaimov A., Ronen Z. and Brenner A. (2006). Enhanced nitrification in industrial wastewater after augmentation by soil nitrifying enrichments. *Intl J. Environ. Technol. Management,* **6**(5), 489–496.

Oehme C. (1981). New possibilities for oxidative purification of wastewater. *Umwelt*, **6**, 470–476.

Prendl L. and Kroiss H. (1998). Bulking sludge prevention by an aerobic selector. *Water Sci. Technol.* **38**(8–9), 19–27.

Rosenwinkel K.-H. (1984). *Beitrag zur Frage der wirtschaftlichen Reduzierung von Abwassermengen und Schmutzfrachten aus Brauereien und Fruchtsaftfabriken* (Contribution concerning an economic reduction of volume and load of wasterwater from breweries and fruit juice producing industry) in Veröffentlichungen des Instituts für Siedlungswasserwirtschaft und Abfalltechnik, 56. Hannover, Germany.

Rosenwinkel K.-H. (1994). *Betriebserfahrungen mit Misch- und Ausgleichsbecken - Auswirkungen auf den Betrieb kommunaler Anlagen* (Practical experience with mixing and equalizing tannks – Impacts on operation of municipals WWTPs) in Veröfffentlichungen des Instituts für Siedlungswasserwirtschaft und Abfalltechnik, 87. Hannover, Germany.

Rosenwinkel K.-H. (1996). *Ursachen und Bekämpfung von Schwimm- und Blähschlamm-bildung bei der biologischen Abwasserreinigung* (Origin and methods for forestalling bulking and floating sludge in biological waste water treatment plants). *Wasser und Boden*, **48**(5), 38–52.

Rosenwinkel K.-H. (1999). *ATV – Seminar - Bemessung von Belebungsanlagen: Bemessungshinweise für Belebungsanlagen mit hohen organischen Industrieabwasseranteilen* (Instructions for the design of activated sludge plants with a high proportion organically polluted industrial wastewater), Essen, Germany.

Rosenwinkel K.-H. and Brinkmeyer J. (2004). *Verfahrensentwicklung in der Brauereiabwasserbehandlung* (Process development concerning wastewater treatment from breweries), 8. Hannoversche Industrieabwassertagung – Brauerei-Seminar in Veröffentlichungen des Instituts für Siedlungswasserwirtschaft und Abfalltechnik, 132, Hannover, Germany.

Rosenwinkel K.-H., Austermann-Haun U. and Meyer H. (2005). Industrial wastewater sources and treatment strategies. In: Environmental Biotechnology: Concepts and Applications, H.-J. Jördening, and J. Winter (eds), Wiley, Weinheim, Germany, 49–78.

Rosenzweig M. D. and Ushio S. (1974). Protein from methanol. *Chem. Eng.*, **81**, 62–63.

Rudolph K.-U. (2013). Integrated Industrial Wastewater Management – Experience from Federal republic of Germany. International conference on 'Integrated Wastewater Management in Mekong Delta Industrial Zones (IWWM – Mekong), October 11, 2013. Can Tho City, Vietnam.

Rüffer H. (1995). *Hefefabriken* (Yeast factories), In: Lehr- und Handbuch der Abwassertechnik, Band V, Chapter 7.3., ATV Abwassertechnische Vereinigung, Hennef, Germany, 549–565.

Schmitt-Schmieder V. (2001). *Vergleich der Leistungsfähigkeit von Biofilmverfahren bei Sanierung bzw. Erweiterung von kommunalen Kläranlagen* (Comparison of performance of biofilm processes by rehabilitation and expansion of municipal waste water treatment plants). Dissertation Universität Karlsruhe, Schriftenreihe des Institut für Siedlungswasserwirtschaft und Wassergütewirtschaft, 104, Universität Karlsruhe, Germany.

Schulze-Rettmer R. and Marzinkowski J. M. (2001). *Textilindustrie* (Textile industry). In: ATV Handbuch Industrieabwasser – Dientsleistungs- und Veredelungsindustrie,

ATV-DVWK, K. H. Rosenwinkel and H. Meyer, (eds), Ernst & Sohn Verlag, Berlin, Germany, pp. 227–233.

Seyfried C. F. (1969). *Erfahrungen mit Brauereikläranlagen* (Experience with wastewater treatment plants for breweries). *Münchner Beiträge zur Abwasser-, Fischerei-, und Flußbiologie,* **16**, 258–267.

Seyfried C. F. (1978). *Biologische Abwasserreinigung nach dem ICI-Tiefstrombelüftungsverfahren (Deep-Shaft) – Erfahrungen auf der Kläranlage Emlichheim* (The ICI-Deep-Shaft-Process for biological wastewater treatment – Experiences at the WWTP Emlichheim). *Gewässerschutz – Wasser – Abwasser (GWA)*, **25**, 535–544.

Stephenson T., Judd S., Jefferson B. and Brindle K. (2000). Membrane Bioreactors for Wastewater Treatment. IWA Publishing, London, UK.

Tilley D. F. (2011). Aerobic Wastewater Treatment Processes: History and Development. IWA-Publishing, London, UK.

Tokutomi T., Yamauchi H., Nishimura S., Yoda M. and Abma W. (2011). Application of the nitritation and anammox process into inorganic nitrogenous wastewater from semiconductor factory. *J. Environ. Eng., Amer. Soc. Civil Eng.*, **137**(2), 146–154.

Van den Hende S., Vervaeren H., Desmet S. and Boon N. (2011). Bioflocculation of microalgae and bacteria combined with flue gas to improve sewage treatment. *New Biotechnol.*, **29**(1), 23–31.

Van Limbergen H., Top E. M. and Verstraete W. (1998). Bioaugmentation in activated sludge: current features and future perspectives. *Appl. Microbiol. Biotechnol.*, **50**(1), 16–23.

Verstraete W., de Caveye P. V. and Diamantis V. (2009). Maximum use of resources present in domestic 'used water'. *Biores. Technol.*, **100**(23), 5537–5545.

Verstraete W. and Vlaeminck S. E. (2011). ZeroWasteWater: short-cycling of wastewater resources for sustainable cities of the future. *Internl J. Sustainable Dev. World Ecol.*, **18**(3), 253–264.

Vlaeminck S. E., De Clippeleir H. and Verstraete W. (2012). Microbial resource management of one-stage partial nitritation/anammox. *Microbial Biotechnol.*, **5**(3), 433–448.

von der Emde W. (1964). *50 Jahre Schlammbelebungsverfahren – Die Geschichte des Belebungsverfahrens* (50 Years Activated Sludge Treatment – History of the Activated Sludge Process). *gwf Wasser Abwasser*, **105**(28), 755–760.

von der Emde W., Fleckseder H., Huber L. and Viehl K. (1973). *Zellstoffabwässer – Anfall und Reinigung* (Wastewater from pulp industries – Amount and treatment). *Wiener Mitteilungen Wasser-Abwasser-Gewässer*, **13**, Vienna, Austria.

Wagner F. (2004). *Membranbelebungsanlagen mit externer Membrananordnung zur Reinigung hochbelasteter Industriabwässer* (Membrane bioreactors with externally arranged membranes for treatment of highly loaded industrial wastewaters). ATV-DVWK Membrantage, Kassel, Germany.

Watanabe K., Teramoto M. and Harayama S. (2002). Stable augmentation of activated sludge with foreign catabolic genes harboured by an indigenous dominant bacterium. *Environ. Microbiol.*, **4**(10), 577–583.

WEF and ASCE (2005). Biological nutrient removal (BNR) operation in wastewater treatment plants - WEF Manual of Practice, 29 – *ASCE/EWRI Manual and Report on Engineering Practice*, 109. McGraw-Hill, NY.

Wett B., Hell M., Nyhuis G., Puempel T., Takacs I. and Murthy S. (2010). Syntrophy of aerobic and anaerobic ammonia oxidisers. *Water Sci. Technol.*, **61**(8), 1915–1922.

Wittebolle L., Van Vooren N., Verstraete W. and Boon N. (2009a). High reproducibility of ammonia-oxidizing bacterial communities in parallel sequential batch reactors. *J. Appl. Microbiol.*, **107**(2), 385–394.

Wittebolle L., Verstraete W. and Boon N. (2009b). The inoculum effect on the ammonia-oxidizing bacterial communities in parallel sequential batch reactors. *Water Res.*, **43**(17), 4149–4158.

Wu W., Timpany P. and Dawson B. (2001). Simulation and applications of a novel modified SBR system for biological nutrient removal. *Water Sci. Technol.*, **43**(3), 215–222.

Wuhrmann K. (1958). *Grundlagen der Planung biologischer Reinigungsanlagen für organisch verunreinigtes Industrieabwasser* (Planning of biological treatment plants to treat organically polluted industrial wastewater). *Schweiz. Zeit. Hydrol.*, **20**(2), 255–283.

Zhang F. G., Liu J. X. and Sui J. (2007). Sludge concentration dynamic distribution and its impact on the performance of UNITANK. *J. Environ. Sci.-China*, **19**(2), 141–147.

Chapter 18

Planning and design

Burkhard Teichgräber (Germany)

Over the past 100 years the AS process has become the most common system for biological treatment worldwide (Bott *et al.*, 2011; de la Espino *et al.*, 2011). The fundamental processes of aeration and mixing followed by phase separation and recycling of the AS can be designed with many technical options and process parameters. The process is highly flexible and allows cost- and energy-effective solutions.

Von der Emde (1957) modified the term sludge age, first used by Gould (1953), and developed the SRT concept which now serves as a key design parameter. Nutrient removal can also be provided by specific designs and layouts. Dynamic modeling, developed in the 1980s, expanded our ability to interpret the design and performance of existing plants. However, even today, steady state approaches are still widely used for design purposes.

18.1 BIOLOGICAL PROCESS AND TRANSLATION INTO DESIGN PARAMETERS

18.1.1 The SRT concept

The simplest way to construct a bioreactor is to provide an SRT long enough to develop the desired microorganisms. The 'once-through' reactor is only appropriate for highly concentrated substrates and is used in some anaerobic treatment processes. Decoupling HRT and microbial growth can be achieved by fixing microorganisms on carriers or recycling them from the effluent of the bioreactor such as in the AS system. The SRT depends on the sludge concentration in the reactor, X_{SS}, (determined by the properties of the secondary clarifier) and the sludge production (determined by the amount of sludge wasting and biomass loss in the effluent).

$$\text{SRT} = \frac{V * X_{\text{SS}}}{Q_{\text{ES}} * X_{\text{ES}} + Q * X_{\text{Q}}}$$

Von der Emde (1957) formulated the relationship between organic load, sludge production and efficiency of BOD_5 and COD removal as follows:

$$B_{V,\text{BOD}_5} = \frac{Q * C_{\text{BOD}}}{V} \quad \text{or} \quad B_{X,\text{BOD}_5} = \frac{Q * C_{\text{BOD}}}{V * X_{\text{SS}}}$$

These loading parameters were later revised to replace BOD_5 with COD for expressing the organic load.

18.2 NITROGEN REMOVAL

Very early in the development of biological wastewater treatment engineers discovered the sequence of organic matter oxidation and the oxidation of ammonia to nitrate (Dunbar, 1912). Downing and Hopwood (1964) applied growth rate constants from *Nitrosomonas* spp. (isolated from the River Thames) to the SRT concept in AS plants. Nitrification can take place when the SRT significantly exceeds the reciprocal maximum growth rate of *Nitrosomonas* spp.

$$\text{SRT} > 1/\mu_{\text{max}}$$

$$\mu_{\text{a,max}} = 0.47 \text{ d}^{-1}$$

Since the maximum growth rate for *Nitrobacter* spp. at 15°C is reported as $\mu_{\text{max}} = 0.79\,\text{d}^{-1}$ (Downing & Hopwood, 1964)

Under steady state conditions the *Nitrosomonas* spp. growth rate limits the rate of complete nitrification. During start-up phases there can be a short period of nitrite production without nitration until a sufficient number of *Nitrosomonas* spp. have developed.

Sludge production depends on organic matter and N loads. Since the sludge yield produced by oxidizing organic matter (Y_h) is significantly higher than the yield of autotrophic biomass (Y_a) such as nitrifying bacteria, and the organic load much higher than N load, the sludge production attributable to nitrifiers can be neglected when calculating the gross sludge production (Table 18.1). This means that the organic load determines the ability of an AS plant to nitrify.

Table 18.1 Comparison of excess sludge production from BOD_5 and N.

Parameter	Heterotrophic biomass	Autotrophic biomass
Y	0.55 g TSS/g BOD_5	0.15 g TSS/g N
c_{Influent}	200 g/m³	50 g/m³
X_{BM}	110	7.5

18.3 PHOSPHORUS REMOVAL

P is an essential nutrient for AS anabolism. WAS solids contain about 1% P on a dry solids basis. When AS is operated in an anaerobic/aerobic mode in the presence of VFAs it can incorporate up to 5% P (Barnard *et al.*, 2011).

Chemical P removal can be achieved by precipitation. This process takes place within milliseconds. Reaction times and reactor volume are determined by mixing efficiency. It can be carried out during primary sedimentation (pre-precipitation) or in the AS stage (simultaneous precipitation) without any additional reactors. It influences the AS process by decreasing the influent organic load (pre-precipitation) or by increasing the WAS production (simultaneous precipitation).

18.4 PROCESS MODIFICATIONS

The AS process is based on retaining suspended biomass in the process (aeration basin) for longer than the wastewater being treated. Key process parameters such as SRT are independent of the phase separation (secondary clarifier) step which retains the biomass in the system. Figure 18.1 shows the main process modifications.

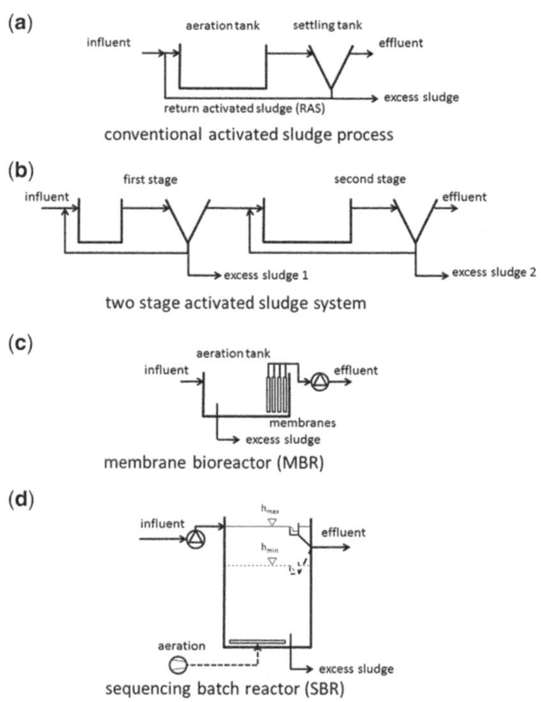

Figure 18.1 AS process modifications.

The conventional AS process (Figure 18.1a) is widely used. By controlling SRT, either a highly loaded carbonaceous removal process with comparatively low energy consumption or a low loaded nitrification process can be obtained in a single sludge system.

Both processes also can be divided in a two-sludge system (Figure 18.1b) which combines low energy consumption and complete nitrification together with high COD removal.

When membranes are used for phase separation (Figure 18.1c), an effluent virtually free of TSS is produced. Moreover the phase separation process becomes independent of sludge settling properties. MLSS concentrations of about 15 g/L are possible and this results in low aeration basin volumes. The head loss through the membranes and the requirement for cross flow aeration to clean them creates higher energy demands.

Ardern and Lockett (1914) based their investigations on a fill-and-draw system in bottles. This system at a technical scale is called a sequencing batch reactor (SBR) (Figure 18.1d) and is frequently used for industrial applications and high strength wastewaters.

Combinations of AS with fixed film biological treatment systems are well known. Two stage systems use trickling filters either as a 1st stage (roughing filter) or as a 2nd stage (e.g. nitrification filter). In RBC systems, fixed film biomass and suspended biomass perform biological treatment simultaneously in the same reactor. If carriers (e.g. sponges, plastic media) are added to the AS system to support fixed film growth they must be screened out of the mixed liquor as it leaves the aeration tank.

18.5 CONFIGURATION

18.5.1 Nitrogen removal

Process configuration for nitrification is simple, since carbonaceous removal and nitrification can occur simultaneously. Most carbonaceous material oxidizers can utilize nitrate in the absence of dissolved oxygen (DO). The aim of denitrification process configurations is to provide a zone with adequate DO to nitrify and a zone containing nitrate that has no DO but has a high respiration rate for denitrification. The higher the SRT is, the lower the respiration rate becomes that can be used for denitrification purposes.

Pre-denitrification (Figure 18.2a) achieves high denitrification rates but cannot achieve very low nitrate concentrations due to limitations in internal recycle flow rates.

Simultaneous and intermittent nitrification/denitrification (18.2c) requires more sophisticated control systems but they are very flexible and allow for low effluent nitrate concentrations. Because they have low denitrification rates a larger reactor volume is needed.

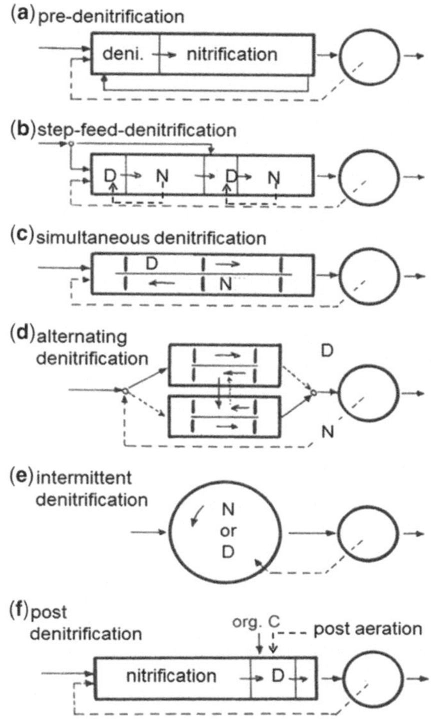

Figure 18.2 Configurations for denitrification.

Step-feed nitrification/denitrification (Figure 18.2b) requires 10–15% less total reactor volume than pre-denitrification because the MLSS concentration in the 1st step-feed compartments is higher than in a complete-mix aeration basin. For high nitrate removal rates, additional recycle streams, at least to the 1st step-feed compartment are necessary.

External carbon sources are needed for post denitrification in single (18.2f) or separate sludge systems.

For high N loads, external carbon sources can be dosed into the denitrification tanks of all denitrification configurations.

18.5.2 EBPR

EBPR requires an initial anaerobic phase in which the AS is mixed with influent. In non-nitrifying systems this can be achieved by mixing, but not aerating (RAS) with influent (Figure 18.3a). For nitrifying systems the RAS can contain nitrate so that the system (shown in Figure 18.3b) has difficulty in meeting very low effluent P concentrations. Several configurations such as those in Figures 18.3c–18.3f have been developed to protect the initial anaerobic zone from nitrate.

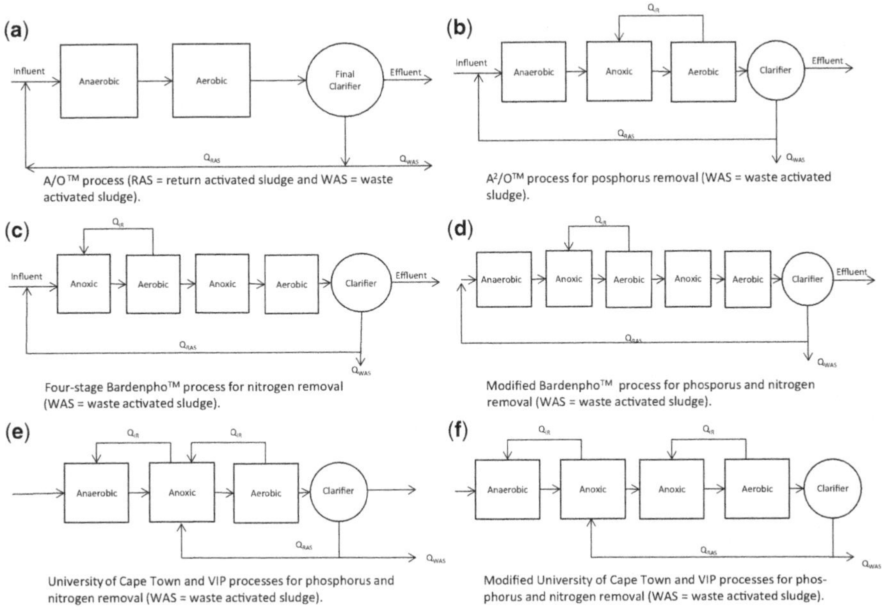

Figure 18.3 Mainstream EBPR configurations (from WEF, 2002).

The only sidestream EBPR configuration (the PhoStrip^{TM}) process used lime precipitation on a side stream and produced a chemical sludge from which P could be recovered (Figure 18.4).

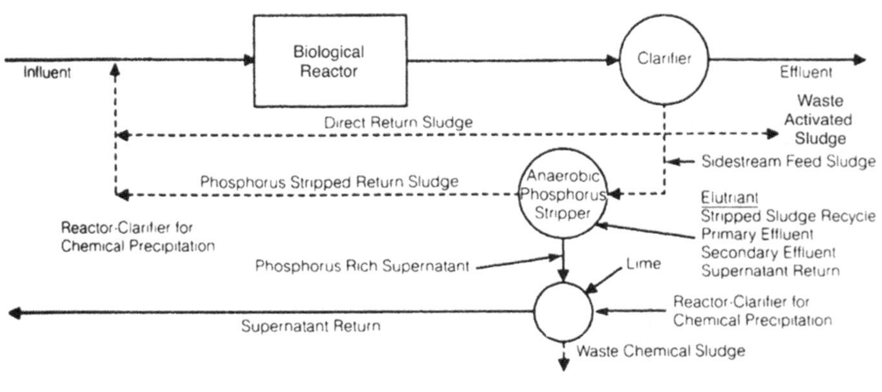

Figure 18.4 Sidestream EBPR, configuration process (Phostrip^{TM}) (WEF, 2002).

18.6 DESIGN PROCEDURES

Von der Emde (1957) described the relationship between AS BOD_5 load, SRT and BOD_5 removal efficiency shown in Figure 18.5.

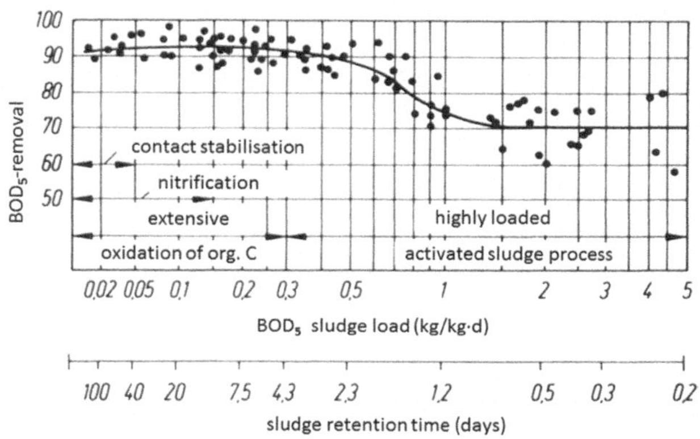

Figure 18.5 BOD_5 removal efficiency as a function of SRT and BOD_5 sludge loading (ATV, 1997).

BOD_5 has mostly been replaced by COD for load assessment and effluent quality measurement. However the basic principles illustrated in Figure 18.5 are still valid and the relationship is applied using COD/2 instead of BOD_5.

All modern design procedures use the aerobic SRT as a key parameter for the AS system. The maximum growth rate of *Nitrosomonas* spp $\mu_{a,max} = 0.47\ d^{-1}$ (15°C)) indicates when there is an SRT at which there is a balance between wash out and accumulation of nitrifiers in the AS system. The procedures for calculating the required aerobic SRT for a given temperature, environmental conditions and effluent requirements and the additional anoxic SRT required for denitrification differ widely from country to country. All procedures contain assumptions on load data acquisition, safety factors and the method of effluent monitoring. Kayser (1991) reports a comparison of design approaches.

Many design approaches for EBPR have been published (Scheer, 1994). For practical design purposes, an anaerobic tank HRT of between 0.5 and 2 h is used. To minimize reactor volumes it is common to install flexible (swing) zones that can be operated under either anoxic or aerobic conditions (for nitrification/denitrification) or under anaerobic, anoxic or oxic conditions (for EBPR).

All design procedures require an assumption of flow (wastewater + RAS), SVI and MLSS concentration.

Since the development of AS models, planners need to decide whether to use steady state or dynamic modeling for design purposes. Several design procedures partially combine these approaches by using nitrification or denitrification rates derived from Downing and Hopwood (1964), pilot experiments or other literature.

18.6.1 USA

The second edition of the USEPA design manual for N control (USEPA, 1993) presents a steady state design procedure with some dynamic elements. The steps are:

(1) Calculate aerobic SRT using Monod kinetics and a design factor (DF) to $1/\mu_{a,\max}$
(2) Use the DF in combination with WAS sludge production to calculate the nitrification volume.
(3) Calculate the N balance.
(4) Compare the resulting nitrification rate with reference data.
(5) If necessary change the DF (Repeat Step 1).
(6) Estimate anoxic volume required for denitrification.
(7) Calculate total SRT and organic load.
(8) Calculate denitrification rates.
(9) Calculate the required denitrification volume.
(10) Compare to the value from Step 6 and if necessary iterate until convergence.

The effect of DO concentration on nitrification rate can be introduced using an additional Michaelis–Menten term and a check on alkalinity can be performed applying an alkalinity mass balance (Metcalf & Eddy, 2003).

18.6.2 Germany A 131

In Germany, design guidelines for wastewater management are developed by the German Water Association, formerly ATV and DVWK. Guideline A 131 (ATV-DVWK, 2000) for the design of a single stage AS sludge plant first included N removal in 1991 using a method based on the work of Kayser (1983). The current version was issued in 2000 and is currently under revision by the specialized group (DWA KA 6) on aerobic biological treatment.

The amended guidelines use a steady state approach based on SRT. In spite of the current discussion in Germany concerning the environmental impacts of COD, a COD balance is used to describe the main biological reactions because of the good experience with this parameter for more than twenty years of dynamic modeling. (DWA has initiated the development of an alternative COD analysis that does not use mercuric sulfate and potassium dichromate.)

The new A 131 guideline will define the data set necessary for design; the corresponding guideline 'DWA-A 198 Determination of Loads' presents the procedures for data collection and harmonization with sewer design.

The design procedure (Table 18.2) calculates the required aerobic SRT using the maximum growth rates of nitrifiers and applies a safety factor. Iterations are necessary to match the oxygen demand by respiration and the oxygen supply from nitrate for sizing the denitrification portion of the total biomass. This biomass can be used for further design of various single sludge process modifications (conventional AS, MBR, SBR).

Table 18.2 Design procedure proposed by DWA KA 6.

Step	Procedure
Nitrogen load	Determination of nitrogen load (influent, reject water from sludge treatment
Aerobic SRT	$t_{TS,aer} = PF * 1.6 \cdot \frac{1}{\mu_{A,max}}$ [d]
	$\mu_{A,max} = 0.47 \cdot 1.103^{(T-15)}$ [1/d]; (PF = 1.4-2.4)
estimation $t_{TS,anoxisch}$ bzw. V_D/V_{BB}	$\frac{V_D}{V_{BB}}$ (estimation 0.2-0.6)
Total SRT $t_{TS,BEM}$	$t_{TS,Bem} = t_{TS,aerob} \cdot \frac{1}{1 - V_D/V_{BB}}$
Calculation of excess sludge production	$X_{CSB,BM} = C_{CSB,abb,ZB} \cdot Y_{CSB,abb} \cdot \frac{1}{1+b \cdot t_{TS} \cdot F_T} + C_{CSB,dos} \cdot Y_{CSB,dos} \cdot \frac{1}{1+b \cdot t_{TS} \cdot F_T}$ $X_{CSB,inert,BM} = 0.2 \cdot X_{CSB,BM} \cdot b \cdot t_{TS} \cdot F_T$
Yield	$Y_{CSB,abb} = 0.67$ kg CSB_{BM}/kg CSB_{abb}, $Y_{CSB,dos} = 0.42$ kg CSB_{BM}/kg CSB_{Meth}, $Y_{CSB,dos} = 0.45$ kg CSB_{BM}/kg CSB_{Ess}
calculation of nitrate for denitrification	$S_{NO3,D} = C_{N,ZB} - S_{orgN,AN} - S_{NH4,AN} - S_{NO3,AN} - X_{orgN,BM}$
	• $S_{orgN,AN} = 2$ mg/l • $S_{NH4,AN} \approx 0$ mg/l • $S_{NO3-N,AN} = 0.8$ bis $0.6\ S_{anorgN,ÜW}$ ($S_{anorgN,ÜW}$ = effluent requirement) • $X_{orgN,BM} = 0.086 \cdot X_{CSB,BM}$
O_2-consumption for carbonaceous removal: - total; - portion OV_C readily biodegradable COD, separately for pre- and intermitt.-Denitrification - O2-consumption in denitrification zone	$OV_C = C_{CSB,abb,ZB} + C_{CSB,dos} - X_{CSB,BM} - X_{CSB,inert,BM}$
	1) $OV_{C,fast,vorg} = f_{CSB} \cdot C_{CSB,abb,ZB} \cdot (1 - Y_{CSB,abb}) + C_{CSB,dos} \cdot (1 - Y_{CSB,dos})$ 2) $OV_{C,fast,int.} = C_{CSB,dos} \cdot (1 - Y_{CSB,dos})$, only when dosing solely in denitrification phase
	1) $OV_{C,D} = 0.75 \cdot [OV_{C,fast,vorg.} + (OV_C - OV_{C,fast,vorg.}) \cdot (V_D/V_{BB})^{0,68}]$ pre-denitrification **2) $OV_{C,D} = 0,75 \cdot [OV_{C,fast,int.} + (OV_C - OV_{C,fast,int.}) \cdot (V_D/V_{BB})]$ intermittend denitrification with ext. C-source solely in deni-phase** **3) $OV_{C,D} = 0,75 \cdot OV_C \cdot V_D/V_{BB}$, simultaneous denitrification**
	• factor for nitrate respiration = 0.75; Enhanced respiration rate in the denitrification zone of pre-denitrification = $(V_D/V_{BB})^{0.68}$ • $f_{CSB} = C_{CSB,l.a.,ZB} / C_{CSB,abb,ZB}$ (portion of readily biodegradable COD in domestic wastewater 0.15-0.25; other factors must be proven by designer)
Comparison O_2-consumption und NO_3-O-supply	$x = \frac{OV_{C,D}}{2.86 \cdot S_{NO3,D}}$

x > 1	x = 1	x < 1
V_D/V_{BB} ↓ resp. $C_{CSB,dos}$ ↓	**design values**	V_D/V_{BB} ↑ resp. $C_{CSB,dos}$ ↑
$\Sigma ÜS_d = ÜS_{d,C} + ÜS_{d,C,Dos} + ÜS_{d,Bio-P} + ÜS_{d,P,Fäll}$		
TS_{BB} from sec. clarifier design [kg/m³] (MLSS)		
$V_{BB} = M_{TS} / TS_{BB} = \Sigma ÜS_d \cdot t_{TS} / TS_{BB}$ [m³]		

All design formulae can be expanded to accommodate dynamic modeling. The steps are:

(1) Determine the N load including sludge reject and other recycled process streams.
(2) Calculate the aerobic SRT and apply a safety factor of between 1.4 and 2.4.
(3) Estimate the denitrification tank volume.
(4) Calculate the excess sludge production including that generated from the addition of external carbon sources.
(5) Determine the N to be denitrified from a nitrate balance.
(6) Determine the oxygen demand due to carbonaceous removal.

(7) Compare the results of Steps 5 and 6; if not equal then return to Step 3 and iterate until convergence.
(8) The resulting total biomass (based on aerobic and anoxic SRT) can be used for all AS process modifications.

From the nitrate balance, this procedure gives an effluent nitrate concentration; the median effluent ammonia concentration can be kept below 1 mg NH_4-N/L while its peak concentrations will vary between 3 and 10 mg NH_4-N/L. Effluent total COD concentrations are largely determined by TSS and refractory substance concentrations (mostly humic acids). Removal of the biodegradable COD is virtually complete in an AS system with nitrification.

For German conditions, design temperatures of 10 or 12°C are typically used but the design procedure is applicable between 8 and 20°C. The investigations of Orth *et al.* (2010) showed that below about 10°C the temperature coefficient was higher than the design guideline value of $1.103^{(15-T)}$ (Figure 18.6).

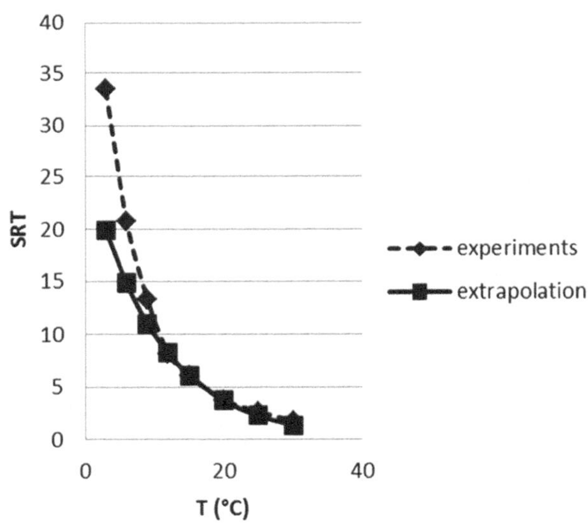

Figure 18.6 Comparison of experimental values of the limiting SRT for nitrification (Orth *et al.*, 2010) with design guideline values using a temperature coefficient of $1.103^{(15-T)}$.

18.6.3 Japan

A completely different method for determining nitrification volume is used in Japan (based on Ichimura *et al.* (1994), quoted in Scheer (1998)). A large database from existing WWTPs is used to derive the SRT needed to achieve a certain level of nitrification. The regression line in Figure 18.7 is drawn to represent the conditions that will provide 80% nitrification efficiency. Sludge production and denitrification are calculated similarly to the USEPA and German approaches described above.

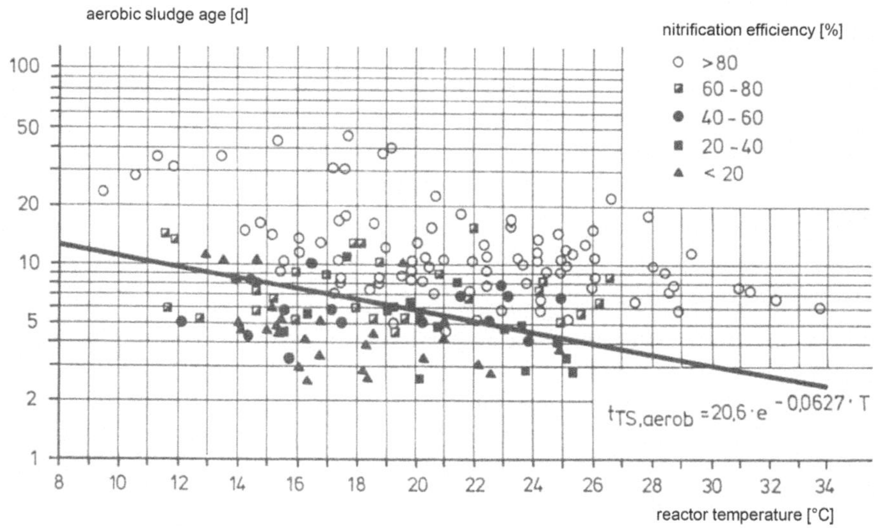

Figure 18.7 Japanese design criteria for nitrification (Ichimura *et al.*, 1994, quoted in Scheer, 1998).

18.6.4 Pilot tests and modeling

The design and layout of AS plants has been based on pilot plant data for many years. Von der Emde (1957) developed a set of design criteria (SRT, volumetric and organic loadings, respiration rates, and energy consumption) which have allowed the scale-up of laboratory and pilot test data to full scale plants.

Dynamic modeling (Daigger, 2011) has become a useful tool for understanding the numerous processes that can take place in the AS process, for configuring control loops, aeration devices, and so on and for predicting reactions to shock loads. The prediction of the effect of small variations of conditions on effluent concentrations with dynamic modeling is not yet possible. Determination of the numerous model coefficients is the most crucial and costly part of dynamic modeling.

Existing design procedures provide sufficient information for plant layout for treating domestic wastewater with standard AS technologies without further investigations. More sophisticated configurations can be designed using dynamic modeling. More challenging wastewaters or environmental conditions (industrial loads, high or low temperatures, stringent effluent requirements) still require the operation of a pilot plant for at least 6–12 months (depending on seasonal variations). Pilot plant test results also allow the determination/verification of dynamic model parameters.

18.7 ECOLOGICAL FOOTPRINT

18.7.1 Space requirements

The AS system allows a wide variety of layouts. When space is limited, a highly integrated layout is possible for conventional AS plants whose space requirement is not much higher than for MBR or biofilter plants (Teichgräber, 1991). The footprint of AS plants is comparatively small compared to natural systems such as aerated lagoons or constructed wetlands.

18.7.2 Emissions

Noise emissions from AS treatment systems mainly arise from aeration and mixing. Hoods on motors combined with fine bubble aeration produce very low noise levels.

Highly loaded AS plants may produce odors. These can be controlled with covers and off-gas treatment, which result in additional energy consumption. Nitrifying AS plants only emit low odor levels since they produce low amounts of odorous materials in large volumes of air.

18.7.3 Carbon footprint

IPCC (2006) states 'Carbon dioxide (CO_2) emissions from wastewater are not considered in the IPCC Guidelines because these are of biogenic origin and should not be included in national total emissions.' Because CO_2 emissions from energy production from fossil sources must be counted in a carbon footprint it is important to minimize energy demand when designing process modifications.

'Wastewater can be a source of methane (CH_4) when treated or disposed anaerobically. It can also be a source of nitrous oxide (N_2O) emissions' (IPCC, 2006). Becker *et al.* (2012) investigated the carbon footprint of wastewater treatment plants. They found contributions of 32% from CO_2 emissions resulting from the actual energy intake and 28% CO_{2e} (CO_2 equivalents) from the CH_4 and N_2O emissions of the wastewater treatment including sludge processing. Denitrification processes emitted 0.5 kg/$CO_{2,e}$/(PE * a) methane and EBPR and nitrification processes emitted nitrous oxide in amounts of 0.1 kg $CO_{2,e}$/(PE * a) and 0.9 kg $CO_{2,e}$/(PE * a) respectively.

Differences in $CO_{2,e}$ emissions from different process modifications or configurations have not yet been determined.

Future investment decisions will also need to account for CO_2 emission from energy consumption as well as direct emissions of CH_4 and N_2O from biological treatment processes. Tools are available for these determinations (Guo *et al.*, 2012).

18.8 SUSTAINABILITY

AS systems are suitable for both very small and very large WWTPs. More than half of the Earth's population currently lives in urban areas and this fraction is increasing. For such densely populated areas a conventional sanitation system with separate sewers, a properly designed AS treatment system with optimized sludge treatment and usage is preferable compared to a new sanitation system with decentralized treatment. Even though decentralized systems may have some ecological advantages, the long term cost/benefit and the operation, maintenance and management aspects make conventional systems preferable. Furthermore the 'social dimension' makes central systems also more attractive because people are happy when they can flush their wastes down the drain rather than having to deal with them where they live.

18.9 CONCLUSIONS

Because of its high efficiency and cost effectiveness AS technology is the standard method for treating wastewater biologically all over the world. For domestic wastewater and environmental conditions, standard design procedures are available. More difficult tasks can be solved using dynamic modeling and pilot tests. The ecological footprint is modest and it enables sustainable solutions. Therefore this technology should maintain its leading role in wastewater treatment for a many years to come (perhaps even 100 more!).

18.10 REFERENCES

Ardern E. and Lockett W. T. (1914). Experiments on the oxidation of sewage without the aid of filters. *J. Soc. Chem. Ind.*, **33**(10), 523–539.

ATV (1997). ATV-Handbuch Biologische und weitergehende Abwasserreinigung. Berlin: Ernst, 849 pp, 4. Aufl.

ATV-DVWK (2000). Guideline A 131E: Dimensioning of Single-Stage Activated Sludge Plants. Hennef. DWA, p. 54.

Barnard J., Houweling D., Analla H. and Steichen M. (2011). Saving phosphorus removal at the Henderson NV plant. Proc. 11th IWA Specialized Conference on Design, Operation and Economics of Large Wastewater Treatment Plants, September 4–8, Budapest, Hungary, 117–124.

Becker A., Düputell D., Gärtner A., Hirschberger, R. and M. Oberdörfer. (2012). Emissionen klimarelevanter Gase aus Kläranlagen. *Immissionsschutz*, No. 4, 182, 188.

Bott C. B., Parker D., Jimenez J., Miller M. and Neethling J. B. (2011). WEF/WERF study of BNR plants achieving very low N and P limits: evaluation of technology performance and process reliability. Proc. 11th IWA Specialized Conference on Design, Operation and Economics of Large Wastewater Treatment Plants, September 8, Budapest, Hungary, pp. 141–148.

Daigger G. T. (2011). A practitioner's perspective on the uses and future – developments for wastewater treatment modeling. *Water Sci. Technol.*, **63**(3), 516–526.

de la Espino O. E., Sandino J. and Mendoza H. (2011). Implementing the World's Largest Wastewater Treatment Project to Date: The Atotonilco WWTP. Proc. 11th IWA Specialized Conf. Design, Operation and Economics of Large Waste Water Treatment Plants, September 4–8. Budapest, Hungary, 43–49.

Downing A. and Hopwood A. (1964). Some observations on the kinetics of nitrifying activated-s plants. *Schweizerische Z. Hydrol.*, **26**, 271–288.

Dunbar W. P. (1912). Leitfaden für die Abwasserreinigung. München, Oldenbourg, 643 pp, 2. Aufl.

Guo L., Porro J., Sharma K. R., Amerlinck Y., Benedetti L., Nopens I., Shaw A., Van Hulle S. W. H., Yuan Z. and P. A. Vanrolleghem (2012). Towards a benchmarking tool for minimizing wastewater-utility greenhouse gas footprints. *Water. Sci. Technol.,* **66**(11), 2483–2495.

Gould R. H. (1953). Sewage aeration practice in New York City. *Proc. Amer. Soc. Civil. Eng.,* 307-1 – 307-11. *Wasser,* **94**(12), 343–347.

IPCC (2006). In: 2006 IPCC Guidelines for National Greenhouse Gas Inventories, National Greenhouse Gas Inventories Programme. H. S. Eggleston, L. Buendia, K. Miwa, T. Ngara and K. Tanabe (eds), IGES, Japan.

Kayser R. (1983). Ein Ansatz zur Bemessung einstufiger Belebungsanlagen für Nitrifikation-Denitrifikation. *GWF-Wasser-Abwasser,* **124**(9), 419–427.

Kayser R. (1991). Design for nitrogen removal and guarantees for aeration. Veröffentlichungen des Instituts für Siedlungswasserwirtschaft der TU Braunschweig, 50E.

Metcalf & Eddy, Inc. (2003). Wastewater Engineering-Treatment and Reuse, 4th edn., New York: McGraw Hill, p. 1819.

Orth H., Grube S. and Tserashchuk, M. (2010). Bemessungs- und Betriebsparameter des Belebungsverfahrens in Abhängigkeit von der Abwassertemperatur und stofflichen Eigenschaften. In: Final presentation BMBF Verbundprojekt 'Exportorientierte Forschung und Entwicklung auf dem Gebiet Abwasser' June 10. Bochum: Lehrstuhl für Siedlungswasserwirtschaft und Umwelttechnik, Ruhr-Universität Bochum, pp. 11–22.

Scheer H. (1994). Vermehrte Biologische Phosphorelimination – Bemessung und Modellierung in Theorie und Praxis. Veröff. Des Instituts für Siedlungswasserwirtschaft der Universität Hannover, **88**, 276.

Scheer H. (1998). Bemessung von Kläranlagen auf Stickstoff- und Phosphorelimination im internationalen Vergleich. Veröff. Des Instituts für Siedlungswasserwirtschaft und Abfalltechnik der Universität Hannover, **102**.

Teichgräber B. (1991). Decentralisation of the Emscher sewerage system. Planning of the Bottrop biological sewage treatment plant. *Water Sci. Technol.*, **25**(4–5), 51–58.

USEPA (1993). Process Design Manual for Nitrogen Control. Technomic Publishing Company, Lancaster, PA, USA. p. 311.

von der Emde W. (1957). *Beitrag zu Versuchen zur Abwasserreinigung mit belebtem Schlamm.* Veröff. Des Instituts für Siedlungswasserwirtschaft der TH Hannover, Bd. 1. Hannover: Eigenverlag, p. 154.

WEF (2002). Activated Sludge – Manual of Practice No. OM-9, 2nd edn. Water Environment Federation, Alexandria, VA, p. 211.

Chapter 19

Activated sludge process economics

Norbert Jardin (Germany), Julian Sandino (USA)

19.1 INTRODUCTION

The purpose of this chapter is to present an overview of the economic aspects of the AS process. The term economics is interpreted by the authors mainly on a cost basis that considers both construction and operating components. To provide a comparative basis for the cost discussion, most economic information will be given in relative terms rather than as absolute values to take account of the significant differences in design, construction and operation of AS plants in different countries. These differences are mainly due to different effluent requirements, for example nutrient removal or carbon removal only, and differing cost situations for construction, energy, personnel and material costs.

This chapter has been structured so that, as a starting point, the major cost elements of the AS process are described to provide an overview of the relative cost importance in comparison to other unit processes in wastewater and sludge treatment. Because nowadays cost control is one of the most important tasks of every plant operator, emphasis is given to appropriate methods of its implementation, that is benchmarking, energy auditing or cost reporting. After 100 years of using the AS process the question still remains as to how cost effective this process is in comparison to other biological methods of wastewater treatment. Therefore, some indications will be given in this respect to show that the AS process is still a comparatively favorable wastewater treatment process. Finally, mechanisms will be described for decreasing cost in all phases of the implementation of an AS plant from its design to its operation.

19.2 AS PROCESS COST ELEMENTS

19.2.1 Total costs of wastewater treatment

Numerous European benchmarking projects have shown that the total costs of wastewater treatment are in the range of 35–45 €/(PE, y) (EWA, 2009; ÖWAV, 2010, 2011; EBC, 2012) for a nutrient removal level that fulfils the effluent requirements of the European Directive on Urban Wastewater Treatment (EU, 1991) for N and P removal for all plants above 10,000 PE. As expected, larger plants have lower specific costs. Capital costs are 50–60% of the total costs.

19.2.2 Differentiation of cost elements

Figure 19.1 shows the differentiation between the capital and operating unit process costs for wastewater treatment plants with separate anaerobic sludge stabilization determined by benchmarking projects in Germany, Austria and Switzerland (EWA, 2009; Bertzbach, 2012; Möller, 2012a, b). It clearly shows that the biological treatment – which in these countries is mainly the AS process – largely determines the capital costs while only 24% of the operating costs are due to biological treatment.

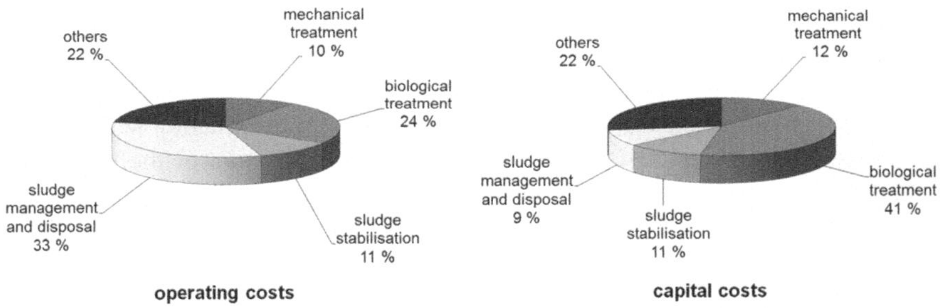

Figure 19.1 Differentiation of operating and capital costs based on unit processes for wastewater treatment plants in Austria and Germany (ÖWAV, 2010, 2011; EWA, 2009; Bode & Grünebaum, 2000; Ruhrverband, 2013).

Figure 19.2 presents a summary of capital and energy cost information for a large scale (2.6 million PE) AS-based wastewater treatment facility that began operation in Santiago, Chile in 2001 (Sandino & Steichen, 1999). This figure shows that the cost of the AS process (aeration basins, clarifiers and RAS pump station) was approximately 25% of the total capital cost of the facility. In the operating cost, the cost of energy for the AS process (by far the largest cost component for operating an AS process) accounts for 16% of the total yearly costs including sludge disposal costs.

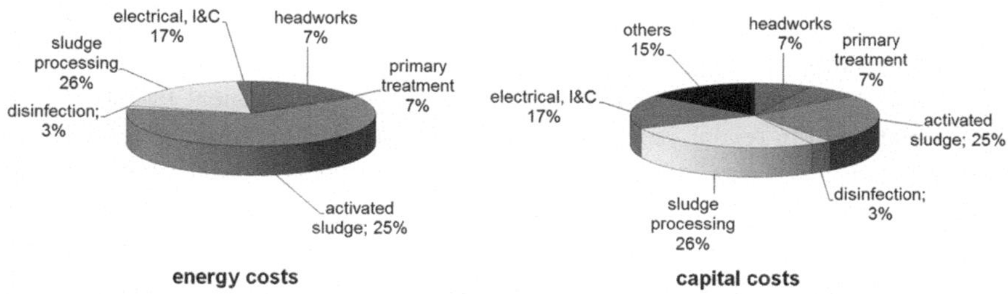

Figure 19.2 Energy and capital costs for a large-scale (2.6 million PE) AS WWTP in Santiago, Chile (Sandino & Steichen, 1999).

19.2.3 AS operating costs

In Western Europe, wastewater treatment plant operating costs are largely determined by personnel and sludge disposal costs. Figure 19.3 shows the results of regular cost reporting in Austria and within Ruhrverband. According to benchmarking projects in Austria, total operating costs vary between 10 €/(PE, y) for larger plants (50,000 to 100,000 PE) and 17 €/(PE, y) for smaller plants (20,000 to 50,000 PE) (ÖWAV, 2010, 2011).

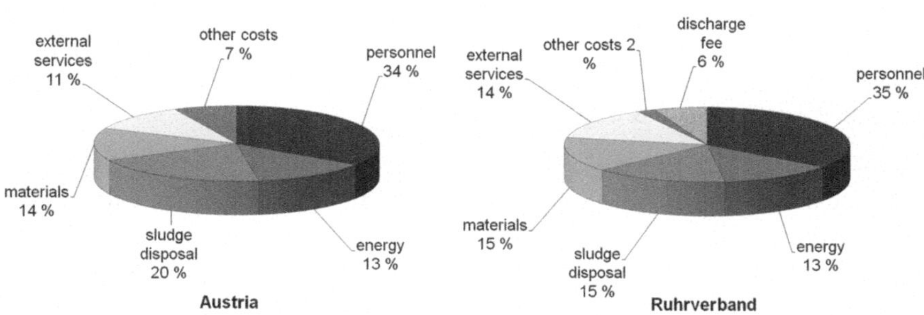

Figure 19.3 Differentiation of operating costs on wastewater treatment plants based on benchmarking results from Austria and Ruhrverband (Lindtner *et al.*, 2008; ÖWAV, 2010, 2011; Ruhrverband, 2013).

The cost of external energy consumption varies substantially depending on the plant size with an average for a large number of different plant sizes of some 11% to 15% (Lindtner *et al.*, 2008; EWA, 2009; ÖWAV, 2010, 2011; Ruhrverband, 2013). Generally, it can be stated that with increasing plant size, the fraction of the total operating costs that are for energy decreases due to the use of digester biogas combined heat power (CHP) schemes. These can lead to up to 70 to 80%, and in extraordinary cases, 100%, energy sufficiency, meaning that these plants can be operated without consumption of external electrical energy and without adding external carbon sources to the anaerobic digesters to increase their co-generation potential (Sandino *et al.*, 2013).

In Latin America, where several new large AS-based treatment facilities have been constructed in recent years, personnel costs are a much lower percentage of the total operating costs, while sludge disposal and energy costs typically tend to be higher. Figure 19.4 presents a summary of the operational costs for the 2.6 million PE facility in Santiago, Chile referenced above (Sandino & Steichen, 1999). In this plant, approximately 27% of the yearly operating costs are associated with electrical energy consumption and 16% of the total costs are for the energy demand of the AS process.

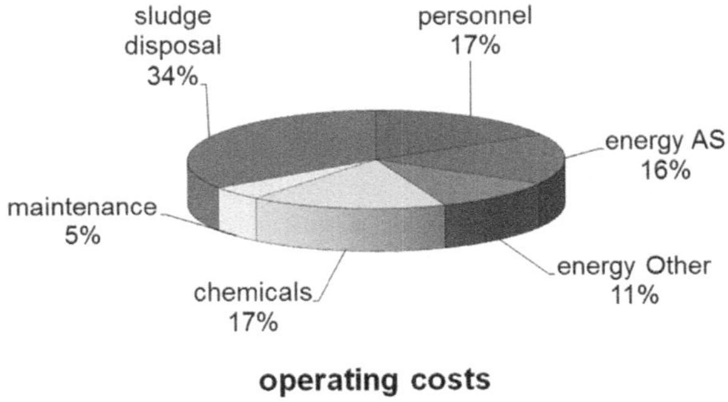

Figure 19.4 Operating costs for a large-scale (2.6 million PE) AS WWTP in Santiago, Chile (Sandino & Steichen, 1999).

The main operating cost components of an AS system are those associated with the energy demand for aeration, mixing and pumping, and typically by far the largest of these is aeration. Configuration of the biological process with denitrification capabilities will result in oxygen demand credits, but these will be somewhat offset by the energy required to mix the unaerated zones and by internal recycle pumping. Figure 19.5 (based on regularly performed energy audits for the Ruhrverband AS wastewater treatment plants (100,000–200,000 PE)) shows the distribution of energy consumption for the various uses of electrical energy compared to the benchmarks derived for treatment plants operating under optimum conditions. It is clear that under optimum conditions, aeration is the main consumer of energy. Nevertheless, several energy audits have shown that a substantial amount of energy can be consumed by RAS pumping, AS tank mixing and internal recirculation. As a consequence of the energy audit of plant A (capacity 320,000 PE), which had an extremely high energy consumption for mixing and pumping (RAS and internal recirculation), pre-denitrification was recently changed to three-step-feed-denitrification to avoid internal recirculation and reducing RAS pumping. Life cycle cost estimates confirmed the financial benefits of these modifications.

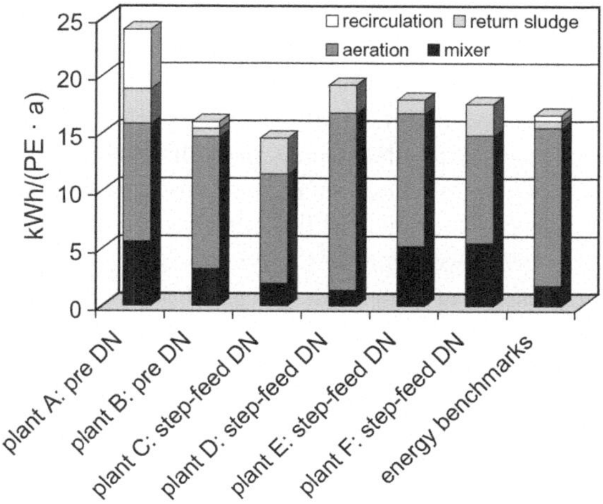

Figure 19.5 Energy consumption for AS operation (Ruhrverband, 2013).

Energy costs for pumping (internal recirculation, RAS) can represent a major part of the total energy costs for operating an AS system that is not well-designed or inefficiently operated. Since pumping life cycle costs are dominated (>90%) by energy costs it is always worthwhile to pay close attention to pumps and to renew or service them whenever an energy audit or a cost analysis reveals increased operating costs. The same applies to mixing, although the energy cost of well-designed mixers should be substantially lower than other AS process elements, practical experience and energy audits often identify mixing devices as major energy consumers. The maintenance costs of mixers should always be considered when comparing alternative AS process layouts.

19.3 COMPARISION OF AS COSTS TO OTHER WASTEWATER TREATMENT PROCESSES

19.3.1 Trickling filters

Non-submerged biofilm processes such as trickling filters are among the oldest methods for biological wastewater treatment and are still used worldwide from small to very large scales. Usually no external aeration energy is needed in these non-submerged systems other than that necessary to elevate the influent to the filter distribution system; this provides them with an advantage over AS with respect to the external energy consumption. Furthermore, sludge settling and dewatering behavior are more favourable and their operation is simpler than for AS. However trickling filters cannot easily be integrated into treatment schemes for enhanced N and P removal because they have limited ability to denitrify.

Where only carbon removal and nitrification are the goals of wastewater treatment, trickling filters (and RBCs for small plants) alone or in combination with anaerobic process can have economic benefits compared to AS. Zahid (2007) performed a comparative economic analysis of a trickling filter and a single stage extended aeration AS system with separate anaerobic stabilization for a treatment goal of 'carbon removal only'. Based on the unit costs for the treatment of 1 m^3 of wastewater, the yearly costs for the trickling filter system including investment and operating cost were 18% lower than for the AS process.

In warmer climates and at high wastewater carbon and nutrient concentrations a two-stage process with an anaerobic 1st stage (UASB) followed by a 2nd stage trickling filter (Elmitwalli *et al.*, 2003; Almeida *et al.*, 2009) or other biofilm system (e.g. downflow hanging sponge (Machdar *et al.*, 2000)), can be economically compared to an AS system. Indeed these two-stage systems have particularly low energy, sludge disposal and maintenance costs for similar carbon removal and nitrification efficiencies (Tandukar *et al.*, 2005, 2007; Sato *et al.*, 2007; Jordão & Volschan, 2004). However, N removal is usually less efficient than in AS systems.

19.3.2 Biofiltration

As an alternative to conventional AS systems, different aerated biofiltration technologies (e.g. BAF) have been developed over the last 50 years and have been used especially when there are severe space limitations. A direct comparison of operating costs was conducted at the WWTP Frederikshavn, Denmark, where an AS system is operated in parallel to a BAF unit achieving nearly identical treatment performance for nutrient and carbon removals. The electrical energy consumption of the BAF was 2.5 times higher than the AS system, giving the AS process a 25% cost advantage with respect to total operating costs (Hansen *et al.*, 2007).

19.3.3 Membrane bioreactor systems

Membrane bioreactor (MBR) technology for wastewater treatment uses the principles of the AS process but instead of a secondary clarifier a compact filtration system is used for separating sludge and treated wastewater. This allows the maintenance of a higher biomass concentration in the biological reactor and makes the process more resilient to unfavourable sludge characteristics, that is high SVIs, dispersed bacteria. Furthermore, membrane bioreactor systems usually achieve a higher effluent quality with regard to TSS and provide disinfection. For a direct comparison with a conventional AS system with secondary clarifiers, subsequent filtration and disinfection has to be considered economically when a decision has to be made about which process scheme is more cost effective under the site circumstances under consideration.

As a consequence of the higher TSS concentration in the biological reactor, MBRs require a much lower footprint than a conventional AS. However the energy and operating costs for membrane fouling and clogging control are considerably higher. Furthermore, especially when operating at high MLSS concentrations the aeration alpha factor can decrease substantially (Henkel *et al.,* 2011). The energy demand of full-scale MBR systems has been reduced in recent years but is still in the range of 0.7–1 kWh/m^3 (Lesjean *et al.,* 2011; Le-Clech, 2010) with some further potential for reduction.

The prices of membranes have also been decreasing over the recent decades making MBR systems more cost competitive compared to conventional AS systems with tertiary filtration and disinfection. A comparison between the MBR plant Nordkanal (80,000 PE) built and operated by Erftverband, Germany showed that the cost differences between these two treatment techniques were highly dependent on membrane lifetime and energy costs. The break-even cost occurred when a membrane life of 7–8 year was assumed for the MBR system and it was also assumed that the effluent requirements for plants were the same and included the requirement to disinfect (Brepols *et al.,* 2010). Considering the recent substantial increase of prices for electricity in Germany it can be assumed that the cost difference between AS systems and MBRs has further increased.

Due to the high effluent quality of MBR systems regarding pathogens, they are increasingly being considered as an alternative to AS in water reclamation/reuse schemes, especially in arid areas (Bolzonella *et al.,* 2010).

19.4 COST CONTROL IN PLANNING, CONSTRUCTION AND OPERATION

19.4.1 Planning

Long-term practical experience has shown that the greatest potential for cost optimization is usually found in the early phases of project planning. It is in these phases that the main conceptual decisions are made and all of the design data are collected. Together these largely determine not only the size of the wastewater treatment plant but also the design of the wastewater collection system and stormwater management (in combined sewer systems).

Ten critical factors responsible for the overall cost have been identified (Bode, 2007; Jardin, 2013):

(1) Considering the interactions in the urban drainage system between wastewater collection, wastewater treatment, stormwater management and especially the ecological situation in the receiving water find the optimum project concept. Such integrated planning approaches (described in the DWA German Rules and Standards, ((DWA, 2006) or used by Ruhrverband (Jardin, 2008)) have substantial cost optimization potential for the entire urban drainage system.

(2) Pay special attention to comprehensive data collection. The cost of additional data collection campaigns that will improve the design database for an urban wastewater system, usually pays for itself in the identification of the most cost-effective solution.

(3) Besides investigating the most cost-effective process scheme for a particular wastewater treatment plant, also consider system-wide solutions, for example centralization of several wastewater treatment plants.
(4) Critically evaluate the utility of existing structures when expanding an existing AS plant. However, especially in regions with high energy costs it can be more cost-effective to construct new well-designed biological reactors with the proper geometry for mixing and aeration rather than using existing structures that mostly have inappropriate geometry for high efficiency mixing and aeration. Always make the final decision on the most cost-effective process layout on the basis of a life-cycle assessment.
(5) Continuously train and qualify the project engineers in the areas of process design and techniques for performing cost-analysis and life-cycle-assessments.
(6) Develop a sense of proportion when deciding on materials quality, amount of redundancy, operability and architectural treatment. Make use of process automation and control to reduce operational and process monitoring costs.
(7) Critically discuss the required level of plant performance and effluent quality with the authorities. For example: during the implementation of the Water Framework Directive In Europe (EU, 2000) it was often shown that the most cost-effective reductions in receiving water nutrient concentrations could be achieved by methods other than through wastewater treatment plant modification.
(8) Manage the entire project to provide as short a project schedule as possible. Employ tight project management from start to finish. Especially for large projects, a separate project management team is most often superior compared to a single team responsible for planning, process design and project management.
(9) Select the most appropriate form of tendering and enforce the interests of the owner and operator during the construction phase.
(10) Determine the right time for financing.

19.4.2 Operation

19.4.2.1 Benchmarking

Nowadays, benchmarking is a universally established method for assessing performance quality although the term is still widely used in the context of cost-comparison. In a broader sense, benchmarking assesses performance using five criteria: customer service, quality, safety, sustainability and structure/technology. The objective of benchmarking – as defined by the IWA specialist group on benchmarking (Cabrera *et al.*, 2011) – is mainly to determine the potential of improving performance and identifying practical measures for realizing this potential.

In Germany, benchmarking has been used for more than fifteen years to access the performance of wastewater treatment (Bertzbach *et al.*, 2012). Two levels of benchmarking have been established among wastewater treatment plant operators and benchmarking methods have been standardized. Corporate benchmarking deals with assessment of the general level of performance and focuses on the performance of the entire utility; process benchmarking looks in detail at the performance of wastewater treatment unit processes. Typically, corporate benchmarking is used by most utilities as the entry point to process benchmarking (Möller *et al.*, 2012a, b). By comparing corporate level benchmarking unit total costs (€/PE, y) for wastewater treatment or sludge disposal with best practice unit costs and by differentiating the cost elements (personnel, energy, sludge disposal, materials, etc.) it will become fairly clear whether a more detailed analysis is needed to improve performance. Process benchmarking is then conducted to analyze the detailed performance of the unit processes. Figure 19.6 shows the effect on cost development for wastewater management including urban drainage based on the results from corporate benchmarking and for wastewater treatment in particular based on the results from process benchmarking in relation to the inflation rate. These projects demonstrate that benchmarking can help a utility to identify measures for improvement and that the implementation of these measures pays off in terms of economic performance.

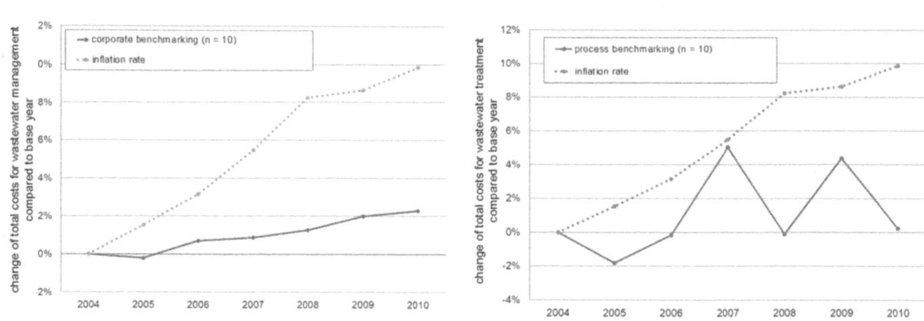

Figure 19.6 Cost developments for wastewater management (including urban drainage) and wastewater treatment among benchmarking participants in Germany (Bertzbach *et al.*, 2012).

From experience in the process benchmarking of numerous (mostly AS) wastewater treatment plants in Germany and Austria, the following key success factors can be identified:

- Compare nationwide operating figures to check the plant performance.
- Be open-minded and willing to learn from others.
- Make benchmarking part of the corporate strategy.
- Pay attention to data collection and establish QA/QC for all data.

- Use a standardized procedure to collect and analyze data.
- Involve plant operating personnel in all stages of process benchmarking for their plant.
- Find potential methods of optimization.
- Embed the identification of measures in the operational and investment strategy for the plant.
- Critically review the success of the implemented measures.

Today, process benchmarking is used worldwide as a method for identifying and implementing wastewater treatment plant performance improvements. Successful examples can be found in Austria (Lindtner *et al.*, 2008), Canada (Main *et al.*, 2006), Sweden (Balmér & Hellström, 2012), many European countries (EBC, 2012; Berbeka *et al.*, 2012) and many other parts of the world (Benedetti *et al.*, 2006; Mizuta & Shimada, 2010; Yang *et al.*, 2010; Krampe, 2013; Science Applications International Corporation, 2006).

19.4.2.2 Energy auditing and energy benchmarking

Energy consumption and approaches for assessing the energy efficiency by energy audits and energy benchmarking of AS systems have been addressed in Chapter 12. It is well proved by the experience of WWTP operators that these assessments gave valuable indications on how to improve the energy efficiency of the AS system and the whole treatment plant when considering the interactions between biological treatment and other unit processes.

From an operations perspective it is important to point out that besides the regular, mostly externally performed auditing, it is of equal importance that a continuous monitoring of the energy situation of the whole wastewater treatment plant is established. This helps to identify changes of the energy situation in the plant at a very early point and gives valuable indications when to initiate a thorough analysis in the form of a qualified auditing. Energy monitoring approaches which should be established today in most plants include:

- Regular energy check with the calculation of specific energy consumption indicators, for example specific consumption electricity for aeration: $e_{\mathrm{aer}} = E_{\mathrm{aer}}/\mathrm{PE}_{\mathrm{COD}}$.
- Yearly reporting, and communication to the operating personnel, on the energy situation in the plant by an energy audit that includes all relevant energy consumption and production information (loadings, flows, energy consumption, biogas production, etc.).
- Setting strategic targets within the organization regarding energy consumption and production, for example increase the number of CHP systems, improve the efficiency of digester gas usage in the plants or increase the amount of co-digestion.
- Make the identification and implementation of energy efficiency measure part of the utility's incentive programme.

19.5 OPTIONS FOR DECREASING TREATMENT COSTS

19.5.1 Alternative treatment concepts

19.5.1.1 Hybrid® process

A more general option to decrease the total costs of wastewater treatment by using the AS system arises when plant has to be modified to meet more stringent effluent requirements. In such cases experience from Austria has shown that it is worthwhile to investigate whether it is possible to change the plant layout from a conventional one-stage to a two-stage Hybrid® process with a high load 1st stage and a low load 2nd stage (Figure 19.7) (Winkler *et al.*, 2008). This process is a modification of the well-known AB process (Böhnke, 1979) with (1) a sludge recirculation line that feeds part of the AS from the 1st stage to the 2nd stage to enhance denitrification (SC1 in Figure 19.7), and (2) an AS recirculation line from the 2nd stage to the 1st stage to enhance nitrification in the 1st stage (SC2 in Figure 19.7). The main advantages of this process scheme are a lower overall AS tank volume, better sludge settling characteristics (lower SVI) and a higher gas production in the digester due to the higher excess sludge production compared to a one-stage AS plant.

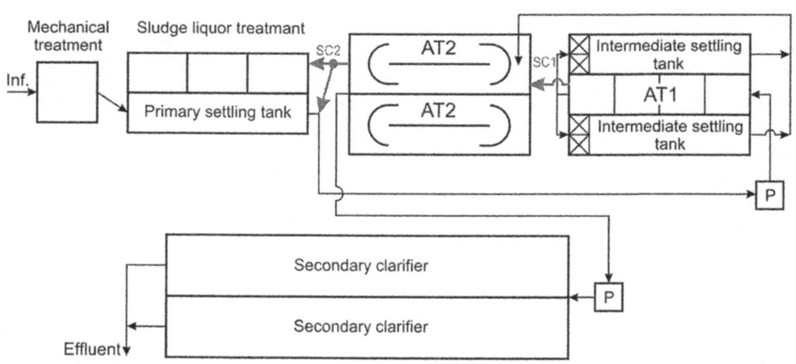

Figure 19.7 New layout of the Knittelfeld WWTP as a 2-stage HYBRID® process AS plant (Winkler *et al.*, 2008).

19.5.1.2 Lamella settling

During the expansion of wastewater treatment plants to meet more stringent effluent requirements, the main approach when dealing with AS systems is to enlarge the mass of AS in the reactor, in the simplest way by increasing the tank volume. Another approach is to increase the MLSS concentration in the tank by a form of 'pre-sedimentation' already in the AS tank, which can be achieved by lamella settlers integrated in an existing AS tank. This option can be economically superior to the enlargement of tank volume under appropriate conditions and especially under severe space limitations on the plant site.

A successful example of this kind of plant extension has been realized on the Finnentrop WWTP (Jardin *et al.*, 2008). The overall expansion consists of a reconstruction of the former primary clarifier into a primary settling tank with a short retention time and using the remaining volume for pre-denitrification purposes. Four lamella separators were installed in the existing carousel-type AS tank as shown in Figure 19.8. The existing secondary settling tanks could be further used if the lamella separators assure a limit solids loading to the secondary settling tank. To control the solids concentration in the AS tank and to avoid an overloading of the secondary settling tank, a newly developed bypass strategy was used. With a controlled mixing of direct effluent from the lamella separators and the content of the AS tank, the solids concentration of the inflow to the secondary settling tank could be maintained at a value of 2.2 kg/m^3. Lamella separation does not appear to have any significant influence on the sludge characteristics. The overall removal of nutrients and organic carbon was excellent after the optimization of the lamella operational strategy.

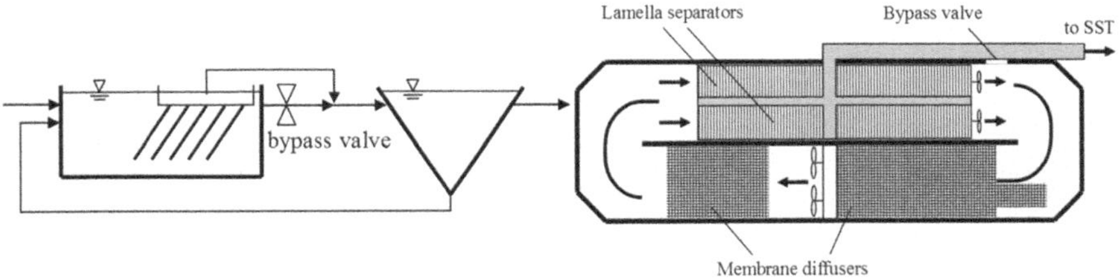

Figure 19.8 Schematic diagram of the bypass valve and schematic layout of the AS tank equipped with four lamella separators and membrane diffusers (Jardin *et al.*, 2008).

19.5.1.3 IFAS processes

Integrated fixed film AS (IFAS) is a hybrid process in which biofilm support media is incorporated into AS basins, typically in applications requiring some level of N control (either nitrification or nitrification–denitrification). The process details are described in Chapter 15. It achieves wastewater treatment by combining suspended and attached growth processes (thus the term 'hybrid') by encouraging development of slow-growing autotrophic nitrifying organisms in the media, thus allowing the operation at much lower mixed liquor suspended solids (MLSS) concentrations. This combination typically results in significantly smaller bioreactor volume as compared to a conventional AS process, in particular in cold weather applications. The IFAS process can utilize either (1) fixed bed media, or (2) floating carrier media, which could be either plastic or sponges. The media differs in specific surface areas and capability of supporting biofilm growth. They also have different requirements for media retention and mixing to keep the media suspended when used in unaerated zones. Fixed bed media is held in place with frames and do not need in-basin screens to retain the media. Plastic carrier media and sponge type media require screens at the end of each zone or a multi-stage cell configuration to retain media.

The cost-effectiveness of an IFAS configuration alternative over a conventional AS process is for the most part determined by the relative cost of providing the media and retention systems of the former and the provision of additional bioreactor volume for the latter. The outcome of this comparison, however, is ultimately dependent on the site specific conditions of the project being considered. These conditions include space availability, existing hydraulic profile, basin configuration, subsurface conditions (which could make providing an appropriate foundation for additional volume very expensive), cost of concrete, and the need to cover and provide odor control for the biological reactor, amongst others. In terms of operating costs, energy requirements are essentially the same for both solutions, yet the inclusion of an eventual replacement of media as part of the life cycle cost evaluation of an IFAS solution needs to be considered.

19.5.1.4 Nereda®-process

It has been shown some fifteen years ago that beside a granulation in anaerobic systems also aerobic microorganisms are able to form stable granules under optimized conditions for the granulation. These early findings (Beun *et al.*, 1999) have led to the development of a technology nowadays named as Nereda® in which an aerobic granular sludge is used for biological wastewater treatment as an alternative for the conventional AS process. The Nereda®-process is operated as a sequencing batch reactor with granular sludge (GSBR). The granules (>0.2 mm) are formed due to too high a substrate gradient at the beginning of the treatment cycle and the use of high hydraulic stress to select mainly granule forming bacteria. To realize this shear stress on the granules, GSBR-systems can be operated, for example with high air upflow velocities. Finally, the settling phase is comparatively short with a substantially higher surface volume loading rate (up to 10 to 15 m^3/m^2, h) (de Bruin *et al.*, 2004). To support stable granule formation it is advisable to start-up GSBR-systems at warmer temperatures (Kreuk *et al.*, 2005). More details on the Nereda®-process are presented in Chapter 20.

As a consequence of the granular formation and the operation as a sequencing batch reactor, no separate secondary clarifier is needed. Furthermore, due to the usually low SVI and the high settling velocity of the granules, the space requirements for such a plant are much lower (75% smaller footprint) compared to a conventional AS plant. Finally, the operation as a sequencing batch reactor with continuous aeration in which nitrification and denitrification are taking place simultaneously in the granules, no mixers or pumps for recirculation are required which results in energy savings of some 25 to 35%.

Recent operational results of the first full-scale plants with this process show stable results with regard to N and P removal (mainly by EBPR) (Giesen *et al.*, 2012).

Nevertheless, it is still unclear if the effluent quality concerning suspended solids can sustainably meet very stringent requirements with, for example values below 10 mg/L SS as an average (Li *et al.*, 2008). Or, in other words, it will be proven in the near future if this kind of process can work without post-filtration in cases where low effluent TSS levels are required from the plant.

19.5.2 Reducing capital costs

19.5.2.1 Chemically enhanced primary treatment

Incorporation or upgrading existing primary treatment with chemical addition to enhance performance has been demonstrated to reduce the capital costs of the AS process (Chagnon & Harleman, 2005; Melcer *et al.*, 2010; Sandino & Steichen, 1999). Chemically enhanced primary treatment (CEPT) consists of the addition of a coagulant (typically iron or alum salts) as well as a flocculation aid (polymers) to improve removal efficiencies of suspended and colloidal organics present in raw wastewater. This in turn reduces the loading to a subsequent AS process, resulting in decreased aeration basin volume and oxygen demand requirements. Incorporation of CEPT can also reduce the size of primary clarifiers, as these units can be operated at higher hydraulic loading rates given the improvement in settleability of coagulated and flocculated suspended solids. Addition of a metal salt also results in the precipitation of P, which could be used as an effective tool in applications requiring removal of this nutrient. Care should be taken, however, in making sure that sufficient P and alkalinity are left for subsequent biological treatment stages, and that proper consideration is given to the additional sludge processing needs that will result from increased solids removals and the generation of an additional chemical sludge fraction.

Table 19.1 presents a summary of the impact in terms of facilities requirements of incorporating CEPT as part of a large-scale wastewater treatment facility that was originally sized for carbonaceous treatment only (Sandino & Steichen, 1999). As it can be observed, significant savings in primary clarifier facilities, aeration basin volume and aeration supply capacity can be achieved by incorporating CEPT. Increases in sludge processing (primarily in sludge stabilization and ultimate disposal requirements) will somewhat offset the capital cost savings of this upgrade.

When aiming at denitrification in the AS system, extended removal of carbonaceous substances via CEPT can lead to situations where only limited amounts of carbon remain after primary treatment, therefore, limiting the degree of N removal. In these cases, care has to be taken in implementing any enhancement to the primary treatment process.

19.5.2.2 Options for influencing SVI

One of the most decisive factors determining the tank volume of both the AS tank and the secondary clarifier is the settleability propensity of the biomass as reflected by the SVI. Therefore, different approaches have been implemented in the past and proved in full scale which might be able to influence the sludge properties. One of these options is the application of selectors to avoid bulking in AS plants (Parker *et al.*, 2004). These selectors can either be used as anoxic or anaerobic systems with high F/M ratio or as aerobic selectors with aeration. Several studies in Europe and the US focusing on the effect of anaerobic versus anoxic selectors have shown that in general an improvement of sludge volume could be achieved, although the effect was highly variable depending on the type of selector and the loading situation of the AS plant (with or without N removal). In general terms, incorporation of unaerated zones (both anaerobic and anoxic) as part of a biological nutrient removal configuration is considered as a measure to improve SVI values (IWA, 2011). Entrained gasses (primarily CO_2 and N_2) can also adversely affect sludge settleability. Common methods of degasification include inducing a rapid change of hydraulic pressure or the stripping of fine bubbles of these gasses using coarse-bubble aeration. A more recent development for this purpose is the incorporation of a vacuum chamber between the biological reactor and the gravity clarification stage of the AS process to remove dissolved gasses (Maciejewski *et al.*, 2010).

Table 19.1 Example of impact of incorporating CEPT in the facility requirements of a large-scale WWTP.

Unit process	Conventional primary treatment and AS[1]	Chemically enhanced primary treatment and AS[1]
Primary clarification Surface area	1.0	0.5
Aeration basin volume	1.0	0.7
Aeration system capacity	1.0	0.7
Anaerobic sludge Stabilization volume	1.0	1.17

[1]Expressed as relative fractions.
Source: Adapted from Sandino and Steichen (1999).

19.5.3 Reducing (external) energy consumption

As has been shown in Figures 19.4 and 19.5 the (external) energy consumption of the AS stage is a decisive factor determining overall operating costs of the biological wastewater treatment. The energy related aspects of the AS process have been discussed in detail by Kroiss and Cao in Chapter 12 'Energy Considerations'. In addition to this, practical measures in order to reduce the energy consumption of AS plants based on the experience of operators are presented in the following. Due to the fact that the major part of energy usage for the process is related to oxygen supply, the main emphasis is given to strategies and full scale implementations in order to increase energy efficiency of the aeration system or the related process control.

When planning a new, or expanding an existing, AS plant one of the most decisive factors determining the future operating costs is the selection of the most appropriate aeration system. Although some general trends have been manifested in identifying the most cost efficient system, that is fine bubble diffusers, high efficiency blowers and medium to high tank depth (~5–7 m), the final decision should always be based on a detailed life cycle cost calculation considering all aspects of tank geometry, process selection for nutrient removal, mixing devices, wastewater quality, and so on. When performing such life cycle assessments, the aeration system consisting of blowers, piping and diffusers has to be considered as a whole, taking into account the interactions between the subsystems and – if necessary – compare an optimized fine bubble diffuser system with a surface aeration system.

If only nitrification is required as the target for wastewater treatment it should be considered to implement an additional anoxic for denitrification in order to reduce the oxygen consumption for the whole process.

Although strong emphasis is given to the selection of a high energy efficient aeration system, practical experience based on regular energy audits has shown that the optimization of mixing and recirculation is of equal importance. In this context, a critical evaluation of process layouts for N removal involving a substantial recirculation of nitrified wastewater into the anoxic zone (e.g. pre-denitrification), especially in cases where the delivery head for pumping the wastewater is comparatively high, should be mandatory. In these cases, a step feed process or alternating denitrification can offer a substantial improvement of energy efficiency.

The practical experiences operating AS plants aiming at enhanced nutrient removal have shown that there is still no 'best' fine bubble aeration system available. Depending on the local conditions (tank geometry, wastewater composition, process scheme for N removal, etc.) fine bubble diffusers using EPDM, PU or ceramic can be of equal energy efficiency. When implementing a new aeration system in very large plants, full scale tests with different aeration systems can be very helpful in identifying the most cost-efficient solution for the plant under consideration. Experiences by Berliner Wasserbetriebe revealed that under the process conditions of one of their AS plants, ceramic diffusers are more cost effective compared to membrane diffusers on the basis of total yearly costs (Libra *et al.*, 2005). Although the specific aeration efficiency (αSAE) was nearly the same for the ceramic and for the membrane diffusers, the comparatively short reinvesting intervals for membrane diffusers resulted in considerably higher yearly costs.

It is well known that the oxygen transfer efficiency is – beside other factors – depended largely on the SRT in the AS system (Henkel *et al.*, 2011). Therefore, when comparing different AS systems layouts (carbon removal only, nitrification only, nitrification/denitrification, extended aeration with simultaneous sludge stabilization) a thorough lifecycle cost analysis should be performed in order to identify to most cost effective option. Especially in cases of high electrical energy costs an N/DN system can easily outcompete process layouts aiming at carbon removal or nitrification only due to lower overall sludge production and higher aeration efficiency (Rosso & Stenstrom, 2005). Furthermore, recent investigations have shown that due to the strong dependency of oxygen transfer efficiency (α-factor) from sludge age, the expected disadvantage of plants with extended aeration (simultaneous aerobic stabilization) and, therefore, long sludge ages with regard to energy consumption for aeration might be substantially lower compared to conventional N/DN plants with primary sedimentation (Henkel *et al.*, 2011).

In addition, also the implementation of appropriate control strategies for aeration on plants with enhanced N removal can effectively reduce operating costs of the AS system (Sahlmann *et al.* 2004; Stare *et al.*, 2007).

19.5.4 Reducing sludge processing and disposal costs

The sludge processing and ultimate disposal costs in AS plants are a substantial part of the overall operating costs. Potential energy recovery from sludge is also affected by the quantity and quality of the sludge generated from the AS process. In recent years, several technologies and process innovations have been proposed to reduce or eliminate WAS and/or render more amenable to yield its embedded energy as part of sludge-to-energy schemes. Some of these technologies affect the rate of biodegradation of WAS, while others affect the extent of biodegradation. Some of these processes are modifications or additions to the AS process itself (such as adding ozone to the recycle stream), while others are applications to the WAS (such as sonication, chemical induced hydrolysis, thermal hydrolysis, electropulse energy, pressure release amongst others) usually after thickening and before subsequent processing such as anaerobic digestion. The challenge in applying these technologies is that their ultimate performance is dependent on the actual characteristics of the AS process, and this is determined by raw wastewater characteristics, the AS process configuration adopted, and the actual operational conditions. This normally requires the use of extensive pilot testing to demonstrate the actual cost-effectiveness of incorporating these modifications to an existing facility (Sandino & Whitlock, 2010; Dhar *et al.*, 2012).

Therefore, any additional measures or changes in the process conditions leading to a decrease in the overall sludge production would be beneficial with regard to the cost situation. In this context, considerable efforts have been invested in strategies to reduce excess sludge production of an AS system. One of the options in this regard is the continuous dosing of ozone in a range of 0.03–0.05 g O_3/g TSS (Chu *et al.*, 2009). Ozone can be added to the RAS line causing cell lysis of some parts of the sludge and promoting the so-called cryptic growth (Mason *et al.*, 1986). The flocs are broken down into fine, dispersed particles which are subject to further mineralization in the AS process (Chu *et al.*, 2009). The amount of sludge reduction varies greatly depending on wastewater characteristics, for example amount of inert solids, and the process conditions in the biological process and might reach under optimal conditions a state in which no further excess sludge is produced (Yasui & Shibata, 1998; Sakai *et al.*, 1997). Boehler & Siegrist (2007) performed an economic evaluation for a Swiss wastewater treatment (35,000 PE) and clearly showed that the cost benefit for sludge reduction is considerably lower than the costs associated with the sludge-ozonation due to investment costs, operating costs (mainly energy and oxygen) and the additional loss of biogas, which leads to net costs of more than 1 €/(PE, y) (Boehler & Siegrist, 2007).

19.5.5 Reducing personnel cost

Development of BNR configurations have increased the operational complexity of the AS process, however advancements in ICA (instrumentation, control and automation) have come a long way in facilitating the operations (and thus reducing the personnel requirements) of these advanced processes (Chapter 13). It should be noted that challenges still remain associated with the ability to keep these advanced ICA systems properly calibrated and operational, creating now the demand for the highly skilled personnel required to maintain them. Another consequence of the increased level of complexity in AS applications is related to the increase in maintenance requirements for additional pieces of equipment, such as mixers and odor control facilities for unaerated zones, and pump systems for internal flow recirculation. All of these additional demands have resulted in an increased reliance on preventative maintenance programs as an essential component in the management of wastewater treatment facilities.

19.6 CONCLUSIONS

The historic and continuing reliance on the AS process in its many variances since the introduction of the concept 100 years ago for secondary treatment and nutrient removal applications, speaks loudly to the favorable economics of this biological treatment process. Nevertheless, the cost driving factors for this process have been changed considerably in recent decades, that is costs for external energy, costs for sludge management and disposal and personnel costs. Therefore, a thorough planning strategy aiming at identifying the most cost efficient process for biological wastewater treatment is nowadays mandatory when designing a new AS plant. The main basis for the decision on the most cost effective process should be based on a detailed life cycle analysis whenever possible. This is in particular due to the selection of the most appropriate process layout for the AS system and the selection of the oxygen supply system. There is no 'best' process and no 'best' aeration system on the market, therefore, a thorough evaluation and cost analysis considering the actual conditions of the plant is an absolute prerequisite for the identification of the most cost effective solution. Moreover, the interactions between aeration, mixing and recirculation of sludge and wastewater have to be taken into account.

Although several alternative biological wastewater treatment concepts have been developed in the past and were successfully applied in full scale, the AS process still offers significant advantages in terms of energy efficiency and the achievable degree of nutrient removal.

The increasing pressure on nearly every wastewater treatment plant operator with regard to the cost situation of the plant makes a thorough and continuous monitoring of the economics mandatory. For this purpose, cost controlling on different levels has now been applied worldwide. Besides a continuous benchmarking on the utility level, a detailed process benchmarking is nowadays used by numerous operators to identify subsequent cost saving potential in all unit processes of the plant. In addition, regular energy audits and systematic evaluation of the long term energy efficiency complement the continuous cost controlling approaches.

To improve the economics, several new technological approaches have been introduced making use of the basic principles of the AS process but adding new elements to save construction costs for the AS tank. Some examples in this context are:

- increasing the suspended solids concentration, for example hybrid systems or lamella settlers incorporated into the AS tank;
- modifying the biological characteristics in such a way that a granulation of the AS organisms occur with the consequence that a secondary clarifier is not any longer needed;
- improving the overall energy efficiency of the whole plant by using a two-stage biological process layout or by increasing the biogas production when applying disintegration technologies to the WAS;
- reducing the organic loading in the influent of the AS stage and, thereby, saving tank volume by increased elimination of carbonaceous substances and solids during chemically enhanced primary clarification.

The future will show if the conceptually new method of N removal via deammonification will make its way into mainstream treatment after this process has been successfully applied for N elimination of high strength wastewaters concerning N (Wett *et al.*, 2009), for example process water from dewatering after anaerobic sludge stabilization.

But in addition to being able to influence the cost situation of an AS system as early as in the planning phase, a continuous monitoring and optimization of the process economics should be performed during full scale operation. The major focus during operation should be on the energy consumption of the aeration system, mixing and recirculation of sludge and wastewater streams, and the personnel resources needed for maintenance and monitoring of the AS process.

19.7 REFERENCES

Balmér P. and Hellström D. (2012). Performance indicators for wastewater treatment plants. *Water Sci. Technol.*, **65**(7), 1304–1310.

Beun J. J., Hendriks A., van Loosdrecht, M. C .M., Morgenroth E., Wilderer P. A. and Heijnen, J. J. (1999). Aerobic granulation in a sequencing batch reactor. *Water Res.*, **33**(10), 2283–2290.

Benedetti L., Bixio D. and Vanrolleghem P. A (2006). Benchmarking of WWTP design by assessing costs, effluent quality and process variability. *Water Sci. Technol.*, **54**(10), 95–102.

Berbeka K., Czajkowski M., and Markowska A. (2012). Municipal wastewater treatment in Poland – efficiency, costs and returns to scale. *Water Sci.Technol.*, **66**(2), 394–401.

Bertzbach F., Franz T. and Möller K. (2012). How to achieve and prove performance improvement – 15 years of experience in German wastewater benchmarking. *Water Sci. Technol.*, **65**(4), 661–668.

Bode H. (2007). *Einflussfaktoren auf Investitions- und Betriebskosten von Abwasseranlagen* (Factors influencing the investment and operating costs of wastewater facilities). DWA Wasserwirtschaftskurs M/3, 1-1–1-32.

Bode H. and Grünebaum T. (2000). The cost of municipal sewage treatment – structure, origin, minimization – methods of fair cost comparison and allocation. *Water Sci. Technol.*, **41**(9), 289–298.

Boehler M. and Siegrist H. (2007). Potential of activated sludge ozonation. *Water Sci. Technol.,* **55**(12), 181–187.

Böhnke B. (1979). Installation for the treatment of waste water by the activated sludge process. United Kingdom Patent 2018736-A.

Bolzonella D., Fatone F., Di Fabio S., and Cecchi F. (2010). Application of membrane bioreactor technology for wastewater treatment and reuse in the Mediterranean region: Focusing on removal efficiency of non-conventional pollutants. *J. Environ. Management*, **91**(12), 2424–2431.

Brepols Ch., Schäfer H. and Engelhardt N. (2010). Considerations on the design and financial feasibility of full-scale membrane bioreactors for municipal applications. *Water Sci. Technol.*, **61**(10), 2461–2468.

Cabrera E. J., Dane P., Haskins S. and Theuretzbacher-Fritz H. (2011). Benchmarking Water Services. Guiding water utilities to excellence. IWA Publishing (Manual of best practice), London, UK.

Chagnon F. J .F. and Harleman D. R. F. (2005). Chemically Enhanced Primary Treatment of Wastewater. Wiley & Sons, Inc. Chichester, UK.

Chu L., Yan, S., Xing X-H., Sun X. and Jurcik B. (2009). Progress and perspectives of sludge ozonation as a powerful pretreatment method for minimization of excess sludge production. *Water Res.,* **43**(7), 1811–1822.

de, Almeida P. G. Chernicharo C. A. L. and Souza, C. L. (2009). Development of compact UASB/trickling filter systems for treating domestic wastewater in small communities in Brazil. *Water Sci. Technol.*, **59**(7), 1431–1439.

de, Kreuk M. K., Pronk M. and van Loosdrecht M. C. M. (2005). Formation of granules and conversion processes in an aerobic granular sludge reactor at moderate and low temperatures. *Water Res.*, **39**(18), 4476–4484.

De Bruin L. M. M., de Kreuk, M. K. van der Roest H. F. R., Uijterlinde C. and van Loosdrecht M. C. M. (2004). Aerobic granular sludge technology: an alternative to activated sludge? *Water Sci. Technol.,* **49**(11–12), 1–7.

Dhar B. R., Nakhla G. and Ray M. B. (2012). Techno-economic evaluation of ultrasound and thermal pretreatments for enhanced anaerobic digestion of municipal waste activated sludge. *Waste Management*, **32**(3), 542–549.

DWA, German Association for Water, Wastewater and Waste. (2006). DWA-A 100E: Guidelines of Integrated Urban Drainage (IUD). Hennef, Germany. ISBN: 978-3-941089-71-6.

Elmitwalli T. A., van Lier J., Zeeman G. and Lettinga G. (2003). Treatment of domestic sewage at low temperature in a two-anaerobic step system followed by a trickling filter. *Water Sci. Technol.*, **48**(11–12), 199–206.

EBC (2012). Learning from International Best Practices. European Benchmarking Co-operation, www.waterbenchmark.org (accessed 10 March 2013)

EU (1991). European Union Council directive 91/271/EEC of 21, May 1991 concerning urban wastewater treatment. Official Journal of the European Communities, L 135/40-L 135/52.

EU (2000). European Union Council directive 2000/60/EC of October 23, 2000 establishing a framework for Community action in the field of water policy. Official Journal of the European Communities, L 327/1-L 327/72.

EWA (2009). International workshop 'European wastewater benchmarking reporting – discussion of current activities and chances of cross-national comparisons'. European Water Association, Vienna, Austria. www.ewaonline.de

Giesen A., van Loosdrecht M. C .M. and Niermans R. (2012). Aerobic granular biomass: the new standard for domestic and industrial wastewater treatment? *Water21*, April, 28–30.

Hansen R., Thøgersen T. and Rogalla F. (2007). Comparing cost and process performance of activated sludge (AS) and biological aerated filters (BAF) over ten years of full sale operation. *Water Sci. Technol.*, **55**(8–9), 99–106.

Henkel J., Cornel P. and Wagner M. (2011). Oxygen transfer in activated sludge – new insights and potentials for cost saving. *Water Sci. Technol.*, **63**(12), 3034–3038.

IWA (2011). Biological Wastewater Treatment, 3rd edn; C.P. L. Grady Jr., G. T. Daigger, G. T., Love, N. G., Filipe, C. CRC Press, Boca Raton, FL, USA.

Jardin N. (2008). *Integrale Niederschlagswasserbehandlung im Einzugsgebiet der Ruhr – Erste Erfahrungen mit der ganzheitlichen Optimierung am Beispiel des Einzugsgebietes der Kläranlage Wenden* (Integrated planning of urban drainage systems – first experiences with the integrated optimisation by using the example of the WWTP Wenden). Schriftenreihe Siedlungswasserwirtschaft der Ruhr-Uni Bochum, Band 56, 165–184.

Jardin N. (2013). *Einflussfaktoren auf Investitions- und Betriebskosten von Abwasseranlagen* (Factors influencing the investment and operating costs of wastewater facilities). DWA Wasserwirtschafts-Kurse 'Kostenanalyse und Kostensteuerung in der Abwasserwirtschaft', Hennef, 7–37. ISBN: 978-3-942964-92-0.

Jardin N., Rath L., Schönfeld A. and Grünebaum T. (2008). Cost-effective upgrading of a biological wastewater treatment plant by using lamella separators with bypass operation. *Water Sci. Technol.*, **57**(10), 1619–1625.

Jordão E. P. and Volschan I. (2004). Cost-effective solutions for sewage treatment in developing countries – the case of Brazil. *Water Sci. Technol.*, **50**(7), 237–242.

Krampe J. (2013). Energy benchmarking of South Australian WWTPs. *Water Sci. Technol.*, **67**(9), 2059–2066.

Le-Clech P. (2010). Membrane bioreactors and their uses in wastewater treatments. *Appl. Microbiol. Biotechnol.*, **88**(6), 1253–1260.

Lesjean B., Tazi-Pain A., Thaure D., Moeslang H. and Buisson H. (2011). Ten persistent myths and the realities of membrane bioreactor technology for municipal applications. *Water Sci. Technol.*, **63**(1), 32–39.

Li Z. H., Kuba T., Kusuda T. and Wang X. C. (2008). A comparative study on aerobic granular sludge and effluent suspended solids in a sequence batch reactor. *Environ. Eng. Sci.*, **25**(4), 577–584.

Libra J. A., Sahlmann C., Schuchardt A., Handschag J., Wiesmann U. and Gnirss R. (2005). Evaluation of ceramic and membrane diffusers under operating conditions with the dynamic offgas method. *Water Environ. Res.*, **77**(5), 447–454.

Lindtner S., Schaar H. and Kroiss H. (2008). Benchmarking of large municipal wastewater treatment plants treating over 100,000 PE in Austria. *Water Sci. Technol.*, **57**(10), 1487–1493.

Machdar I., Sekiguchi Y., Sumino H., Ohashi A. and Harada H. (2000). Combination of a UASB reactor and a curtain type DHS (downflow hanging sponge) reactor as a cost-effective sewage treatment system for developing countries. *Water Sci. Technol.*, **42**(3-4), 83–88.

Maciejewski M., Oleszkiewicz J. A., Golcz A. and Nazar A. (2010). Degasification of mixed liquor improves settling and biological nutrient removal. *Water Practice Technol.*, WPT **5**(1), doi:10.2166/wpt.2010.008.

Main D., Ng L. and North A. (2006). The Canadian National Water and Wastewater Benchmarking Initiative. Using process to drive improvement: strategic management of water in urban areas. *Water Sci. Technol. Water Supply*, **6**(5), 111–121.

Mason C. A., Hamer G. and Bryers J. D. (1986). The death and lysis of microorganisms in environmental processes. *FEMS Microbiol. Letters*, **39**(4), 373–401.

Melcer H., Ciolli M., Lilienthal R., Ott G., Land G., Dawson D., Klein A. and Wightman D. (2010). Bringing CEPT technology into the 21st century. Proc. 83rd Ann. WEFTEC, New Orleans, LA, October 2–6; Water Environment Federation; Alexandria, VA, USA.

Mizuta K. and Shimada M. (2010). Benchmarking energy consumption in municipal wastewater treatment plants in Japan. *Water Sci.Technol.*, **62**(10), 2256–2262.

Möller K., Bertzbach F., Nothhaft S., Waidelich P. and Schulz A. (2012a). *Benchmarking in der Abwasserbeseitigung – eine Bestandsaufnahme Teil 1: Ziele und Ergebnisse des Benchmarkings.* (Benchmarking in Wastewater Disposal – a Stocktaking Exercise, Part I: Benchmarking Objectives and Outcomes). *Korrespondenz Abwasser, Abfall,* **59**(8), 730–737.

Möller K., Bertzbach F., Nothhaft S., Waidelich P. and Schulz A. (2012b). *Benchmarking in der Abwasserbeseitigung – eine Bestandsaufnahme Teil 2: Erfolgsfaktoren des Benchmarkings.* (Benchmarking in Wastewater Disposal – a Stocktaking Exercise, Part II: Benchmarking Success Factors). *Korrespondenz Abwasser, Abfall,* **59**(9), 829–835.

ÖWAV Österreichischer Wasser- und Abfallwirtschaftsverband. (2010). *Benchmarking für Kläranlagen, Geschäftsjahr* 2009 (Benchmarking for wastewater treatment plants – 2009). www.benchmarking.at.

ÖWAV Österreichischer Wasser– und Abfallwirtschaftsverband (2011). Benchmarking für Kläranlagen, Geschäftsjahr 2010 (Benchmarking for wastewater treatment plants – 2010). www.benchmarking.at

Parker D., Appleton R., Bratby J. and Melcer H. (2004). North American performance experience with anoxic and anaerobic selectors for activated sludge bulking control. *Water Sci. Technol.*, **50**(7), 221–228.

Rosso D. and Stenstrom M. K. (2005). Comparative economic analysis of the impacts of mean cell retention time and denitrification on aeration systems. *Water Res.*, **39**(16), 3773–3780.

Ruhrverband. (2013). Internal cost reporting. Essen Germany.

Sahlmann C., Libra J. A., Schuchardt A., Wiesmann U. and Gnirss R. (2004). A control strategy for reducing aeration costs during low loading periods. *Water Sci. Technol.*, **50**(7), 61–68.

Sakai Y., Fukase T., Yasui H. and Shibata M. (1997). An activated sludge process without excess sludge production. *Water Sci. Technol.*, **36**(11), 163–170.

Sandino J. and Steichen M. (1999). Chemically enhanced primary treatment (CEPT): a viable alternative for sanitary solutions in Latin America, Proc. 72nd WEFTEC, New Orleans, LA, Water Environment Federation, Alexandria, VA, USA.

Sandino J. and Whitlock D. (2010). Evaluation of Processes to Reduce Activated Sludge Solids Generation and Disposal, *WERF*, Alexandria, VA, USA.

Sandino J., Nielsen P. H., Whitlock D., Johnson T., Leth M., Nielsen M. and Jensen T. (2013). Achieving significant GHG reductions and positive net energy balance at a BNR treatment facility. Proc. IWA Leading Edge Technology Conf. Bordeaux, France, IWA, London, UK.

Sato N., Okubo T., Onodera T., Agrawal L. K., Ohashi A. and Harada H. (2007). Economic evaluation of sewage treatment processes in India. *J. Environ.*, **84**(4), 447–460.

Science Applications International Corporation (2006). Water and Wastewater Energy Best Practice Guidebook, Wisconsin Focus on Energy, Madison, WI, USA.

Stare A., Vrečko D., Hvala N. and Strmčnik S. (2007). Comparison of control strategies for nitrogen removal in an activated sludge process in terms of operating costs: a simulation study. *Water Res.,* **41**(9), 2004–2014.

Tandukar M., Uemura S., Machdar I., Ohashi A. and Harada A. (2005). A low-cost municipal sewage treatment system with a combination of UASB and the 'fourth-generation' downflow hanging sponge reactors. *Water Sci. Technol.*, **52**(1–2), 323–329.

Tandukar M., Ohashi A. and Harada, H. (2007). Performance comparison of a pilot-scale UASB and DHS system and activated sludge process for the treatment of municipal wastewater. *Water Res.*, **41**(12), 2697–2705.

Wett B., Jardin N. and Katehis D. (2009). Environmental technologies to remove nitrogen from high-strength wastewaters. In: Environmental Technologies to Treat Nitrogen Pollution, Francisco J. Cervantes (ed.), IWA Publishing, London, pp. 115–137.

Winkler S., Gasser M., Schättle W., Kremmel D., Kletzmayr P. and Matsché N. (2008). Upgrading of wastewater treatment plants for nutrient removal under optimal use of existing structures. *Water Sci. Technol.*, **57**(9), 1437–1443.

Yang L., Zeng S., Chen J., He M. and Yang W. (2010). Operational energy performance assessment system of municipal wastewater treatment plants. *Water Sci. Technol.*, **62**(6), 1361–1370.

Yasui H. and Shibata M. (1994). An innovative approach to reduce excess sludge production in the activated sludge process. *Water Sci. Technol.*, **30**(9), 11–20.

Zahid W. M. (2007). Cost Analysis of Trickling-Filtration and Activated-Sludge Plants for the Treatment of Municipal Wastewater. In: Proc.7th Saudi Engineering Conference, Riyadh, Saudi Arabia, December 2–5, 67–81.

Chapter 20

The next 100 years

Mark van Loosdrecht (The Netherlands), Harry Seah (Singapore), Yuen Long Wah (Singapore), Yeshi Cao (Singapore)

20.1 WASTEWATER TREATMENT: A HISTORY OF PROCESS INTENSIFICATION

20.1.1 History

Water has always been a major concern for growing cities. Even in ancient times large cities relied on the importation of safe fresh water by channels/pipelines and also used this water to flush wastes from them. Without a safe water infrastructure, waterborne diseases and pollution prevent the growth and prosperity of a society. Even after 2000 years, the water infrastructure from Roman times is visible in the form of aqueducts, ancient sewers and storage facilities such as the Basilica Cistern in Istanbul, Turkey. Much of this infrastructure could even be used nowadays with the Cloaca Maxima in Rome being an iconic example. Basic water infrastructure in cities has hardly undergone significant changes over the centuries because of the absolute necessity for preventing diseases in large human populations and the simplicity, durability, reliability and relatively low cost of the water infrastructure. Thus, in general, the water industry has been a slowly changing industry when compared to, for example, the electronics industry. In the last century, however, water treatment has developed rapidly as a result of the increasing pressure of growing populations and this will likely be a strong driver for further developments in the coming decades.

Water treatment and especially wastewater treatment are logical results of the industrial revolution in the early 19th century that led to an increased growth of cities and resulted in the discharge of large amounts of wastewater. Moreover, the availability of artificial fertilizer has allowed a rapid and exponential growth of human population since then. These have resulted in a still increasing pressure on fresh water resources. The outbreaks of major cholera and typhoid epidemics in the mid-19th century in industrialized cities catalyzed the search to combat these diseases by better water supply and sanitation and led, 100 years ago, to the introduction of the AS process.

20.1.2 Wastewater treatment a history of process intensification

When small populations discharge their organic wastewater into rivers the natural microbial population is capable of removing the pollutants and incorporating them into the natural cycling of elements. Effectively, the biological processes used in modern-day treatment facilities are the same as those occurring in nature. The natural cleansing capacity is still reflected in BOD_5-based regulations. The 5 d period for this test was chosen because it gave BOD results that most closely matched those from the already-existing chemically based oxygen demand method (Anon, 1912). Limiting BOD_5 discharge was aimed at keeping these rivers and estuaries well-oxygenated.

The rates of natural processes in rivers and lakes are limited by mass transfer rates (especially of oxygen) and by the quantity of bacteria present. The development of biological wastewater treatment has been mainly an intensification of these natural processes achieved by improvement of mass transfer processes and of increased biomass retention, resulting in higher biomass concentrations. These improvements started with imitating the natural process in rivers by taking the stones covered with slime layers or biofilms from the stream bed and piling them into a reactor over which the wastewater was trickled. In this way the treatment was intensified and the oxygen transfer was enhanced. Trickling filters and other types of biofilm reactor systems are still designed and constructed as a secondary wastewater treatment process. The advantage of easy operation makes them especially suitable for 'low tech' or decentralized applications. The mass transfer limitation of oxygen from the liquid to the biofilm surface (maximum of approx. 10 $gO_2/m^2 \cdot d$ for typical specific surface areas of 100–300 m^2/m^3) is the main limitation on minimizing the reactor volume. This gives volumetric capacities of approximately 1–3 kg $BOD_5/m^3 \cdot d$) (Nicolella *et al.*, 2000).

Another route for process intensification was to apply forced aeration so as to introduce sufficient oxygen into the water in a reactor that contained suspended biomass. Initially (and nowadays at locations with sufficient available area) this was based on the simple aeration of polluted water in artificial lakes or ponds. When it became clear that the sludge being produced in these processes, or the slime layer in trickling filter systems, could enhance the process rate, the retention of this 'activated sludge' was a logical step … and with the presentation of the first paper by Ardern and Lockett in 1914, the AS process was born. This process depended on flocculation of the bacteria grown on the BOD in wastewater with non-biodegradable wastewater particulates in a clarifier and settling under the influence of gravity. This process has proven so reliable, versatile and economical that, even after a century of research, it is still the state-of-the-art biological wastewater treatment technology.

The search for improved wastewater treatment has led to the discovery of new types of microorganisms (such as methanogenic bacteria, anaerobic ammonium oxidizing bacteria and polyphosphate accumulating organisms) but genetically engineered organisms have never been introduced successfully in environmental engineering. The main field of wastewater treatment has therefore been the improvement of the process engineering aspects combined with an increasing understanding of microbial ecology. We foresee that the interaction between environmental engineers and microbiologists will remain a major driver for developments in the coming decades; albeit that for systems producing high-purity effluents it can be expected that more physical/chemical methods will be introduced for post-treatment. However since currently less than 1% of the microbial life on our planet has been discovered, there should be plenty of room left for improving biological processes and discovering interesting new microbiological activities.

20.1.3 General developments for future WWTPs

The development of biological wastewater treatment has been driven mainly by a search for more compact, economical and simpler processes and reactors combined with a steadily improving effluent quality. Newly developed processes and treatment methods generally tend to lead to stricter regulations. Treatment requirements were originally focused on minimizing diseases and nuisances such as odors in channels and rivers. Gradually requirements increased to prevent eutrophication with discharge controls for nutrients (N and P) especially for 'sensitive areas' in Europe (European Community, 1991) and the USA (USFR, 1999). Now sustainability requirements (energy efficiency, reuse of water, and resource efficiency) are slowly gaining importance.

In recent years, criticism of the current municipal wastewater treatment plants and sanitation systems has arisen due to their consumption of energy (carbon footprint), emission of greenhouse gasses (GHGs) (CO_2, N_2O and CH_4) and their production of residual solids. It has been argued that the requirements for improved wastewater treatment performance are often accompanied by more negative side aspects. A typical case is the use of external or internal carbon sources for denitrification that can have a net negative energy impact and result in more GHG emissions (van Loosdrecht *et al.*, 1997). The balance between environmental impact and effluent quality will become increasingly scrutinized as effluent standards become more and more stringent. It has been argued that some of the important issues with current wastewater treatment plants are related to the current centralized sanitation system due to its dilution through transportation of pollutants from the sources to the treatment plants. In essence, the scale (level of centralization) of a system should be an outcome of a system optimization. Increasing population growth and resource depletion generate a need to divert materials (water, energy, nutrients etc.) from the current urban water/wastewater treatment chain so that there is a better integration with other aspects of urban metabolism (Guest *et al.*, 2009; WERF, 2009; STOWA, 2010).

The future of wastewater treatment will likely follow the trend of the past century: process intensification combined with introduction of newly discovered microbial processes, governed by increased population pressure and urbanization, which drives a paradigm shift to general resource efficiency in combination with the protection of public health and the environment.

20.1.4 Performance criteria for future municipal wastewater treatment plants

The performance indicators of current wastewater treatment plants are largely limited to those of the liquid effluent quality and treatment costs. The performance indicators of future municipal wastewater treatment plants will cover liquid, solids (residuals), air, energy and chemicals (WERF, 2009; Guest *et al.*, 2009; Hering *et al.*, 2011).

20.1.4.1 Water

Depending on the local hydrological situation, tiered reuse criteria, based on end user needs such as potable, agricultural, industrial usage, and so on, will be imposed. This trend is already seen in developments such as NEWater in Singapore (Tan, 2004). New standards such as lower effluent nutrient and metal concentrations will be introduced. There will be more micro-pollutants on discharge standards lists. Significant changes may occur due to the ability of using the rapidly developing genetic methods to set standards for specific pathogens instead of using indicator organisms. Thermal energy in the effluent will be considered for municipal wastewater heat energy recovery or for influencing local natural ecology.

20.1.4.2 Solids

Solids include dewatered or dried sludge and ash after incineration. Similarly to water, future requirements for solids will depend on the nature of the end uses, for example agricultural, industrial, disposal. It can be expected that more types of micro-pollutants and pathogens will be regulated in solids than currently. Specific solids production (e.g., kg solids produced/kg COD removed) will be included as a plant performance indicator. With the trend towards producing more resources from the wastewater, the composition of the solids produced will be specified in greater detail.

20.1.4.3 Air

Odor emissions are often well controlled. Reduction of GHG emissions from future municipal wastewater treatment plants will be mandatory. The legal requirements on GHG emission (equivalent GHG/PE, d or equivalent GHG/m^3) will be promulgated, similarly to today's effluent 'discharge' standards. Also, it can be expected that aerosols will be subjected to regulation.

20.1.4.4 Energy

High energy efficiency in municipal wastewater treatment plants is currently achievable using today's best available practices and experience. In future plants, energy efficiency will be improved substantially; but energy consumption will increase due to the additional energy required for micro-pollutant removal and effluent disinfection, etc. This depends on the development of new processes, particularly for N conversion. However, it can be expected that energy consumption (kWh/PE; kwh/kgCOD; kWh/m^3) and energy efficiency (%) will be regulated.

20.1.4.5 Chemicals

Chemical consumption for uses such as thickening, dewatering and external carbon sources for nutrient removal, will be restricted in municipal wastewater treatment plants. Parameters such as g chemical/PE and g chemical/m^3 wastewater and soon on. will be adopted as performance indicators.

Below, we try to sketch some of the expected future developments of the AS process itself:

- Process intensification
- Improved effluent quality – potential new microorganisms
- Towards energy neutrality or production
- Resource recovery
- Integration of functionalities in urban areas.

20.2 PROCESS INTENSIFICATION

20.2.1 Improving SVI or granular sludge

AS processes are effectively limited by the settling capacity of the final clarifier. The sludge flocs have a limited settling velocity because they have a bulk density close to water and an open, porous structure. This can be further hampered by the growth of filamentous bacteria in the AS. High SVIs limit sludge loading rate as well as the biomass content of the aeration tank leading to treatment systems with long aeration times. Sludge morphology and filamentous bacteria have been an ongoing research topic over the last century. Chudoba *et al.* (1973) pointed out the need for plug flow conditions to limit growth of filamentous bacteria, and with the selector, paved the way to smaller systems. Initially, such gains have been relatively small and have mainly focused on preventing detrimental conditions caused by sludge bulking. Recently, however, granular sludge (with very low SVIs) has made its entry into aerobic wastewater treatment, making it possible to integrate the clarification step into the aeration basin. This leads to treatment plants with a footprint that is at least 75% lower, and in future will allow better integration of wastewater treatment in urban environments.

Granular sludge has already been used for several decades in anaerobic treatment systems (Lettinga *et al.*, 1980). With a better understanding of floc and biofilm morphogenesis (Martins *et al.*, 2004; van Loosdrecht *et al.*, 1995) it has become possible to specifically select for granular biomass for all kind of microorganisms. The main objective is to select for slow growing microbial populations, and counteract rapid growth by selecting for granular sludge retention (wash out of flocculent biomass) or higher shear rates. Steep diffusion gradients in the outer layers of the granules are essential. For slow growing bacterial systems (methanogenic, nitrifying and anammox) granular sludge forms almost spontaneously. For faster growing bacteria (fermentative or aerobic heterotrophs) special measures are needed.

Granular sludge is a combination of biofilm and flocculent sludge processes. Essentially the granules are biofilm systems with gradients in solute concentrations and microbial populations over the depth of the granule. Because of the very high surface area of the sludge granules, these systems are not limited by mass transfer as in typical biofilm systems. Moreover, the process operation is similar to AS systems, making for simpler reactor construction and process control. In aerobic granular sludge the aerobic and anoxic zones are integrated in the granule and effectively only one reactor compartment is needed for full nutrient removal. This eliminates the need for recycle pumps and sludge return pumps so that both construction costs and energy demand are reduced.

It is interesting to note that in the development of AS systems sequencing batch reactors were replaced by continuously operated reactors. This was the case with the original system (Ardern & Lockett, 1914) as well as the oxidation ditch from (Pasveer, 1962). The first aerobic granular sludge process – the Nereda technology (van Loosdrecht, 2011; Inocencio *et al.*, 2013) – is also based on a SBR concept in which several traditional disadvantages have been overcome. There is no variable volume (no need for decanters, efficient aeration) and the flow to and from the process can be constant. The batchwise operation allows for much better process control and optimization and produces a better effluent than continuously operated completely mixed systems. This should make these systems more adaptable to more stringent effluent standards in the future. Also granular sludge should be advantageous for novel applications of slow growing bacteria such as anammox bacteria due to the ease with which it can be operated at long SRTs and high biomass concentrations. Indeed currently the most commonly used nutrient removal side stream processes are based on developing granular sludge to retain the anammox bacteria.

20.2.2 Hybrid biological processes

Currently, there are two main categories of biological treatment processes, namely, suspended growth processes (AS) and immobilized biomass or biofilm processes (e.g. trickling filters (TFs), rotating biological contactors (RBCs), fixed-film activated sludge (IFAS), moving bed biofilm reactors (MBBRs), and airlift reactors, etc.), In reality, experimental and modeling simulation results show that in conventional TFs and RBCs, the suspended biomass plays an important role that is comparable to, or even more essential, than the biofilm for COD biodegradation and nutrient removal, even at low MLSS concentrations (Cao & Alaerts, 1996). Recent studies have shown that, even in the IFAS process, where the specific surface area of the carrier materials is high, the suspended biomass plays an essential role in nitrification (Sen *et al.*, 2010; Bott *et al.*, 2011). Even in granular sludge processes Winkler *et al.* (2011) found that GAOs were enriched in the smaller granules, while PAOs were present in larger granules. Hybrid biofilm processes might provide additional selective options for microbial population control. The combination of slowly growing bacteria in biofilms with rapidly growing bacteria in suspended growth is an interesting future direction for designing more compact integrated processes.

20.3 IMPROVED EFFLUENT QUALITY

20.3.1 Exploring natural diversity

The development of wastewater treatment has been accompanied by the discovery of many new microbial conversion processes such as nitrification, polyphosphate accumulation in anaerobic/aerobic cyclic processes and anaerobic ammonium oxidation. With only a small fraction of the biodiversity explored to date, there should be many more opportunities for discoveries that could be interesting for wastewater treatment. Examples already exist: for example N-recovery using bacteria that can accumulate nitrate in their cells (Jorgensen *et al.*, 1999), application of methane oxidizing bacteria for denitrification (Raghoebarshing *et al.*, 2006), sulfur-cycle associated PAOs (Wu *et al.*, 2013) and fungi with very powerful hydrolyzing activity (Fujii *et al.*, 2013) for sludge minimization. In the coming decades the further investigation of microbial diversity will provide even more potential for incorporating new microbial processes in wastewater treatment plants allowing not only a better effluent quality but also resource recovery and energy optimization.

Microbial ecology has become an indispensable tool for process engineers to understand the cause–response relationships and selective mechanisms in biological wastewater treatment processes that involve multiple-substrates and multiple populations. Molecular methods such as FISH and q-PCR and metagenomic methods provide powerful tools for species identification and quantification, which are essential for understanding microbial community interactions. Its development and application have moved rapidly in recent years (van Loosdrecht *et al.*, 2013). Further standardization of the procedure together with additional experience will lead to a better understanding of the complex ecological system and the interrelationships between biological processes and microbial community structure. This will lead to better control of the microbial treatment systems and provide improved effluent quality.

20.3.2 Emerging pollutants

Effluent standards have the tendency to become stricter with time. Where public health and ecological diversity are of concern, discharge control of emerging (micro) pollutants including endocrine disrupting compounds (EDCs), pharmaceutically active compounds (PhACs), personal care products, household chemicals and refractory dissolved organic N (rDON) have been on the agenda. Research and development to tackle these issues has been carried out in Europe (NEPTUNE, 2010). Recently the removal of micropollutants has assumed importance and the first legal requirements for their removal are in the process of being implemented (Swiss Federal Government, 2013). Other emerging contaminants such as nanoparticles will have to be addressed in future. Besides source control, optimization of the treatment process is needed to prevent the need for relatively expensive physical/chemical treatment of the effluent.

20.3.3 Optimize process design

Besides effluent polishing, better treatment in the AS process will always be more advantageous. Improved effluent quality can be obtained by appropriate process design because smarter integration of reactor design and reaction kinetics will lead to better effluent quality. Most micropollutants are present in low concentrations, and are removed at a first order rate. This will give an additional advantage in the future to plug flow or SBR processes. Because emerging pollutant removals will likely require some physical/chemical processes it is possible that inert soluble COD will interfere with removals and will therefore increase in importance. The use of computational fluid dynamics simulations will allow for better designs. GHG emissions such as methane and N_2O seem to depend on complex interactions between microbial conversions and process design. Improved process design and process control will likely be introduced to minimize such emissions as well as to improve the overall process performance.

20.4 ENERGY NEUTRALITY/MINIMUM CLIMATE IMPACT

Human societies, especially those in densely populated urban areas, need to increase their sustainability. Minimizing energy and resource needs are top priority; and energy and resource recovery are a close second. GHG emissions need to be minimized. With the extensive use of biogas in wastewater treatment, CH_4

emissions are a potential concern (Daelmans *et al.*, 2012). In the last century, energy consumption was a point of attention. Anaerobic processes minimizing aeration energy requirements and producing energy rich biogas have always been around the corner in wastewater treatment (PUB, 2013). Clarigesters, Dortmund tanks and anaerobic sludge digestion where introduced at about the same time as the AS process. They were originally used for energy generation but nowadays anaerobic digesters are mainly constructed for sludge reduction, with energy production as just an extra benefit. Modern forms of anaerobic treatment by UASB or integrated concepts such as the Paques-UBOX system (Paques, Balk, The Netherlands) are currently attracting interest for direct anaerobic wastewater treatment in high temperature climates; however, currently this is also driven primarily by costs rather than by the requirement for energy conservation. Although energy minimization/energy generation is a major driver for current developments it should be recognized that making wastewater treatment independent of external energy supply can be an additional driver for development, especially in countries with unreliable electricity supplies for plant operation.

20.4.1 Energy consumption and recovery

A first priority should always be the minimization of energy needs for aeration, mixing and pumping, which are the largest energy consumers in a wastewater treatment plant. Membrane diffusers and on-line sensors based on blower control can save up to 20% of the aeration energy. SBRs have the advantage that processes are scheduled in time in the same reactor and not in space. This minimizes the need for water/sludge recycle pumps. This will become increasingly important as effluent limits for N discharge become more stringent and higher recirculation rates between continuous flow aerated and anoxic tanks are required. Development of simultaneous nitrification/denitrification processes will minimize pumping requirements. Developments in the field of aerobic granular sludge (with or without anammox) should also reduce pumping requirements. Pumps, aerators, and mixing equipment are generally positioned on the basis of constructability. Better placement and design (aided by developments in CFD technology) could further contribute to energy minimization. Energy consumption can be further reduced by optimizing the biological processes (see below).

Energy recovery from wastewater treatment has received special attention due to the high energy prices caused by a politically based energy shortage in the western world. Anaerobic wastewater treatment by UASB and follow-up technology (such as IC (Benev, Yiixing, China) and EGSB (Veolia Water) reactors) were successful developments from this period; these processes however are currently used mainly for industrial wastewater treatment. For municipal treatment, collection of the organic carbon in sludge combined with sludge digestion evolved as a more successful strategy. Direct anaerobic municipal wastewater treatment has found low applicability not only due to the long SRT required for methanogens, but because a relatively large fraction (10–40%, Meda *et al.*, 2010) of the methane produced is dissolved and is lost in the effluent. Recovery, requiring energy, is essential because of the GHG potential of the CH_4. The large amounts of TSS present in municipal wastewaters, make the sludge production in UASB treatment and the conventional activated sludge AS process similar (Cao, 2011).

Energy production can be enhanced by directing more COD from the influent to the anaerobic digesters and by optimizing the digestion process. Upstream COD harvesting (or pre-concentration) is a process that concentrates COD (and a significant fraction of the N and P) from wastewater prior to biological treatment. The pre-concentration can be by biological, physical and chemical means or by a combination of these (PUB, 2013). Physical/chemical processes can be used to concentrate the particulate/colloidal COD from the wastewater. For soluble COD, conversion into biomass in a highly loaded AS (SRT on the order of several hours) is needed, after which the biomass can be removed with the particulate matter. The two-stage (A/B process) technology developed by Bohnke (1981) in the 1970s seems to be a logical approach. This A/B process was popular for a decade after its introduction but increasing demands for N removal led to a decreased application of this process since the BOD removed in the A stage could not be used for denitrification in the B stage. The introduction of anammox technology for fully autotrophic N removal will overcome this limitation.

Despite the application of the A/B technology in dozens of large treatment plants, information on its design criteria and mechanism is lacking in the open literature. The A-stage is an AS process with a short SRT (typically <0.5 d) and HRT (<1 h). Produced biopolymers and added chemicals result in extensive BOD/COD removal. The excess sludge from the A-stage can contain up to 60% of influent COD load (Wett *et al.*, 2007), while a fraction of the influent COD (10–30%) is oxidized (Roest *et al.*, 2012). The reasons to adopt such a unit in the treatment process are: (i) to direct more COD to anaerobic digestion to increase biogas and energy generation; and (ii) to reduce the oxygen demand of the B stage, thus saving aeration energy. Indeed, high efficiency COD accumulation is not only the key for achieving energy neutrality, but also might be an essential step in resource recovery from wastewater.

20.4.2 Mainstream anammox

The anammox process (anaerobic ammonium oxidation; also termed deammonification) is an innovative N removal process that eliminates the carbon requirements for denitrification. This process has been successfully applied worldwide on the recycle streams from anaerobic sludge digestion units as well as for post-treatment of industrial anaerobic wastewater treatment processes. Jetten *et al.* (1997) were the first to propose the use of this technology in mainstream wastewater treatment in combination with COD removal in the A-stage of an A/B process for increasing the sustainability of municipal wastewater treatment. Currently efforts have shifted to mainstream applications since they provide a chance to achieve energy neutrality (Siegrist *et al.*, 2008). Significant progress has been made and application of this technology seems feasible within the coming decade. A pilot-scale investigation using granular sludge at the Dokhaven WWTP, Rotterdam, The Netherlands has achieved a treatment capacity of 0.2 kgNH_4-N/m^3 · d at 20°C (Lotti *et al.*, 2013) without anammox sludge augmentation. Full-scale demonstrations and investigations, in which the mainstream process is augmented by transferring anammox bacteria from the side-stream processes, at the Strass WWTP, Austria and the Glarnerland WWTP, Switzerland, show promising results (Wett *et al.*, 2012). Cao *et al.*, (2013) have recently shown significant autotrophic N removal in one of the 200,000 m^3/d trains of the Changi, Singapore Water Reclamation Plant. Several challenges still need to be addressed. Important among these are the stability of nitritation, the repression of nitrite oxidizing bacteria (NOB), and balancing of the aerobic and anoxic ammonium oxidation activities. Recent findings (Winkler *et al.*, 2012) that show that, at low temperature, anammox bacteria can use BOD for the conversion of nitrate into their substrates nitrite and ammonium, further expands the potential for optimum design of anammox-based treatment technology.

20.4.3 Energy recovery from low temperature and thermal treatment

Heat is the major energy component of the urban water cycle. There is almost 10 times more heat energy in wastewater than the energy that is needed for treating the wastewater. Heat optimization in urban environments integrated with water management is therefore a major goal for energy efficiency. Heat pumps provide a method for recovering this heat. Full-scale extraction of heat from municipal wastewater has already been applied in some places (Steinherr, 2010). Furthermore the waste heat from biogas combustion might be better integrated into the energy needs for cities. Further research areas include developing technologies for efficient recovery of thermal energy from wastewater and side streams, identifying users of low grade heat, and understanding the energy and cost models to make these concepts economically feasible.

20.5 RESOURCE RECOVERY

20.5.1 Water

Treated wastewater is used for augmentation of ornamental water in cities, industry, agriculture, natural water bodies and can be a reliable source for water supply. A typical example is Singapore NEWater, which is consistently of high quality and well within the requirements of the USEPA and WHO standards for drinking water. PUB hence embarked on a program for the large scale production and supply of NEWater directly to the industrial and commercial sectors for direct non-potable use (Tan, 2004). Effluent from a conventional secondary treatment plant is further treated by ultrafiltration (UF) followed by reverse osmosis (RO) and ultraviolet (UV) disinfection. It has been demonstrated that this treatment process produces water that is even more suitable as a raw water source for many industrial applications than 'virgin' drinking water. Based on successful demonstration of the technology, PUB has moved aggressively to construct several full-scale NEWater plants to serve industry and to supplement public water supply. The current production capacity of NEWater is 50% of the secondary effluent generated in Singapore, and will eventually be expanded to 80% of the secondary effluent. The next phase in process development involves replacing the conventional biological secondary treatment and ultrafiltration steps with a membrane bioreactor (MBR).

20.5.2 P and N recovery

In the last decade, resource recovery from wastewater has been mainly related to energy recovery and water reuse. From a sustainable society point of view, reuse of chemical compounds is more desirable. The complex mixture of compounds in wastewater has thus far prevented serious recovery of materials from wastewater. Traditionally, sludge has been used in agriculture as nutrient and soil conditioner. In general, this has its limits due to transport distances and to the presence of contaminants in the sludge. Also the increased efforts towards achieving nutrient efficiency in agriculture to protect the natural water systems does not align with use of wastewater sludge as fertilizer. New routes will have to be developed.

N and P forms are the major nutrients in wastewater. P can be partly recovered from wastewater due to its formation of the sparingly soluble solids, struvite and apatite. In general, recovery from a solution that is more concentrated than municipal wastewater is preferred, for example the anaerobic zones of EBPR or in anaerobic digester effluents. Struvite formation integrated with EBPR also allows the recovery of magnesium, ammonium and potassium from wastewater. In North America there are currently six prototype plants using Ostara technology (Ostara, Vancouver, BC, Canada) to recover struvite and, in Europe, there are full-scale plants using Paques technology for struvite recovery (Abma *et al.*, 2010) and Crystalactor technology to produce $Ca_3(PO_4)_2$ crystals (Britton *et al.*, 2009). It seems that for phosphate, but also for other compounds, recovery is not really a technical challenge, but a problem of developing a value chain where the product is produced in small amounts at many wastewater treatment plants, while the market is currently a bulk market based on a few suppliers/mines. Better integration of technology and market development will be essential for resource recovery.

For N there are fewer options for recovery. In essence, N is not a limiting compound in society, and biological and chemical N fixation can generate sufficient ammonium. It is more a matter of energy. Current chemical ammonium recovery methods require more energy than the modern Haber–Bosch processes and are therefore not effective. Nevertheless, smart solutions could be helpful, for example, ammonium recovery during sludge drying (Winkler *et al.*, 2013).

20.5.3 Organics

A real challenge lies in recovery of organic compounds from wastewater. This might partially be based on separation technology. A recent report on cellulose recovery is a good example of the potential of looking at wastewater from a resource recovery perspective. Hundreds of influent and modeling studies have overlooked the fact that toilet paper is a major fraction in raw municipal wastewater; the realization of this from a recovery point of view showed potential for wastewater treatment optimization (Ruiken *et al.*, 2013). Also, waste sludge may contain interesting compounds for recovery such as alginates (Lin *et al.*, 2013), bioplastics (Tabakatake *et al.*, 2002) polyelectrolytes and surfactants. Recognizing this, could lead to the discovery of routes for maximizing the production of such compounds. Finally, the chemical or biological conversion of the complex mixture of organic carbon compounds in wastewater to a few chemicals that can be recovered could be a promising development. Gasification or production of volatile fatty acids would then be a first step. Smart recovery technology could recover these VFAs directly or convert them into chemical building blocks such as poly-β-hydroxybutyrate.

20.6 INTEGRATION OF FUNCTIONALITIES

Service functions in urban environments have traditionally been the domain of individual departments (e.g. wastewater, drinking water, surface water, energy, solid waste, town planning and transport). In order to improve sustainability, minimize energy and costs and recover resources a stronger integration is needed. The techniques and methods developed in the industrial ecology field could form the basis for such development. It should however be recognized that the complexity of 'urban ecosystems' is at least an order of magnitude larger then than for industrial ecosystems.

20.6.1 Water supply and wastewater in an integrated urban water cycle

Wastewater reuse links the boundary between wastewater treatment and water supply. As previously mentioned, in the NEWater program in Singapore, half of secondary effluent is further purified to produce potable quality water for various uses. Conventional AS plants were upgraded to accommodate N removal for this purpose. These uses may mean that additional wastewater characteristics (e.g., hardness) may need to be considered and could lead to additional plant design and performance criteria for future AS processes and wastewater treatment plants. Other alternative water sources will also inevitably be a part of new urban water systems. These might be related to rainwater harvesting, seawater usage for flushing and more innovative sources such as water harvesting from the off-gasses of power plants by membrane processes. These methods will impact the wastewater composition and the flow at the wastewater treatment plant. For example, the use of seawater for toilet flushing in Hong Kong has led to wastewater with high sulfate concentrations. This can lead to increased sulfide production during anaerobic digestion. Such changes should be seen as an opportunity for innovations in process design. Thus in Hong Kong, the design of a wastewater treatment plant based on sulfate reduction has become feasible (Lu *et al.*, 2011).

20.6.2 Water and energy

Because water has a large heat capacity it takes a lot of energy to heat it up and it can store a large amount of heat. Thus the amount of energy a family uses to heat up water for showering and washing is about 10 times more than the total energy used to produce and treat the water for the same household (Hofman *et al.*, 2010). Biogas production from the BOD in wastewater can make treatment plants energy neutral, but much larger energy effects will be associated with smart heat exchange systems in urban areas. However, it should be noted that in colder climates the recovery of heat energy from the sewer network will decrease the wastewater temperature and could significantly impact treatment processes. Therefore integration of water and energy with respect to climate regulation in dense urban areas will be an important development that will require collaboration between many groups from the water, urban planning and construction sectors.

20.6.3 Centralized *vs.* De-centralized systems

The relative merits of centralized and decentralized wastewater treatment have been widely debated. One typical advantage of decentralized systems is the ability to recover P and energy from black water by source separation; grey water would then be treated in a centralized plant (Zeeman *et al.*, 2007). Another advantage would be to treat micropollutants in a decentralized system to avoid the 'dilution factor' of a centralized system. For an example, the pharmafilter hospital wastewater treatment process (www.pharmafilter.com) shows that wastewater treatment can be economical for hospitals and at the same time remove a significant fraction of pharmaceuticals currently discharged to the sewer system. The operational gains in the hospital can easily compensate for the costs of an extensive wastewater treatment system. Combination of such decentralized facilities at micropollutant hot spots is an interesting option for optimizing the AS process. With greater integration of water systems in urban systems it is likely that more decentralized treatment systems will be developed. While smaller treatment units generally are less economical than larger scale systems (even when including the cost of sewer construction for centralized systems), it should be noted that even for small scale treatment processes a large fraction of the centralized system sewer network will still be needed. Therefore case-by-case feasibility and economic evaluation is needed for such developments.

With the topics mentioned above in mind, the segregation of domestic and industrial areas will likely be a condition required for implementing integration. Singapore's NEWater is an example of this because wastewater segregation was achieved before the NEWater program was developed. This allowed the attainment of NEWater after membrane treatment of secondary effluent. Integrating land and urban sanitation planning is an essential factor in for the integration of utility functionality.

20.7 CONCLUDING REMARKS

Over the last century, the AS process has been shown to be a versatile concept, adaptable to changes in societal demands with respect to effluent standards or energy use. In the coming century, the AS process is likely to remain the workhorse of wastewater treatment. However, further improvements (such as the recently introduced aerobic granular sludge process) will lead to intensified processes requiring less space and energy. Societal demands will also convert wastewater treatment facilities into resource recovery units, initially for water and energy but, in the near future, also for heat and chemicals.

20.8 REFERENCES

Abma W. R., Driessen W., Haarhuis R. and van Loosdrecht M. C. M. (2010). Upgrading of sewage treatment plant by sustainable and cost-effective separate treatment of industrial wastewater. *Water Sci. Technol.*, **61**(7), 1715–1722.

Anon (1912). Standards and Tests for Sewage and Sewage Effluents Discharging into Rivers and streams. Royal Commission on Sewage Disposal, 8th rpt. HMSO, London, UK, pp. 2–3.

Ardern E. and Lockett W. T. (1914). Experiments on the oxidation of sewage without the aid of filters. *J. Soc. Chem. Ind.*, **33**(10), 523–539.

Baur R. (2011). Results of the first full year of operation of North America's first full scale nutrient removal facility, *WEF-IWA Nutrient Recovery and Management*, January 9–12, Miami, FL, USA.

Böhnke B. (1981). *Leistngssteigerung und Energiekostensenkung durch Einsatz Mehrstufiger Biologischer verfahren* (Performance improvement and energy minimisation by application of multistage biological treatment). *Wissenschaft und Umwelt,* **4**, 247–256.

Bott C., Waltrip D., Baumler R., Rutherford B., Jones R., Regmi P., Schafran G., Thomas W., Pinto A. and McQuarrie J. (2011). Evaluation of IFAS AOB and NOB kinetics and mass transfer. Workshop A: Backstage with (bio)film's hottest stars: carbon footprint, O & M costs and nutrient removal reliability, *WEF-IWA Nutrient Recovery and Management*, January 9–12, 2011, Miami, FL, USA.

Britton A., Prasad R., Balzer B. and Cubbage L. (2009). Pilot testing and economic evaluation of struvite recovery from dewatering centrate at HRSD's Nansemond WWTP. In:

International Conference on Nutrient Recovery from Wastewater Streams, K. Ashley, D. Mavinic and F. Koch (eds), IWA Publishing, London, UK, ISBN: 9781843392323.

Cao Y. S. (2011). COD, nitrogen conversion and mass flow in coupled UASB-Activated sludge process for municipal wastewater treatment in warm climates. In: Y. S. Cao (ed.) Mass Flow and Energy Efficiency of Municipal Sewage Treatment Plants, IWA Publishing, London, p. 132, ISBN: 9781843393825.

Cao Y. S. and Alaerts G. J. (1996). A model for oxygen consumption in aerobic heterotrophic biodegradation in dual-phase drainage systems. *Wat. Res.*, **30**(4), 1010–1022.

Cao Y. S., Kwok B. H., Yong W. H., Chua S. C., Wah Y. L. and Yahya A. G. (2013). The main stream autotrophic nitrogen removal in the largest full scale activated sludge process in Singapore: process analysis. WEF/IWA Nutrient Removal and Recovery 2013: Trends in Resource Recovery and Use, July 28–31, Vancouver, BC, Canada.

Chudoba J., Grau P., Ottova V. (1973). Control of activated sludge filamentous bulking II. Selection of microorganisms by means of a selector. *Water Res.*, **7**, 1389–1398.

Daelmans M. R. J., van Voorthuizen E. M., van Dongen U. G. J. M., Volcke E. I. P., and van Loosdrecht M. C. M. (2012). Methane emission during municipal wastewater treatment. *Water Res.*, **46**(11), 3657–3670.

European Community (1991). Urban Wastewater Treatment Directive (91/271/EEC) of May 21, 1991.

Fujii K., Kai, Y., Matsunobu S., Sato H. and Mikami A. (2013). Isolation of digested sludge-assimilating fungal strains and their potential applications, *J. Appl. Microbiol.*, **115**, 718–726.

Guest J. S., Skerlos S. J., Barnard J., Beck M. B., Daigger G. T., Hilger H., Jackson S. J., Karvazy K., Kelly L., Hering J. G., Hoehn E., Klinke A., Maurer M., Peter A., Reichert P., Robinson C., Schirmer K., Schirmer M., Stamm C. and Wehrli B. (2011). Moving targets, long-lived infrastructure, and increasing needs for integration and adaptation in water management: an illustration from Switzerland. *Environ. Sci. Technol.*, **46**, 112–118.

Hofman J., Hofman-Caris R., Nederlof M., Frijns J. and van Loosdrecht M. C. M. (2010). Water and energy as inseparable twins for sustainable solutions. *Water Sci. Technol.*, **63**, 88–92.

Inocencio P., Coelho F., van Loosdrecht M. C. M. and Giesen A. (2013). The future of sewage treatment: Nereda technology exceeds high expectations. *Water21*, 28–29.

Jetten M. S. M., Horn S. J. and van Loosdrecht M. C. M. (1997). Towards a more sustainable municipal wastewater treatment system. *Water Sci. Technol.*, **35**(9), 171–180.

Jørgensen B. B. and Gallardo V. A. (1999). Thioploca spp: filamentous sulfur bacteria with nitrate vacuoles (1999) *FEMS Microbiol. Ecol.*, **28**(4), 301–313.

Lettinga G., van Velsen A. F. M., Hobma S. W., de Zeeuw W. J. and Klapwijk A. (1980). Use of upflow sludge blanket (USB) reactor concept for biological wastewater treatment. *Biotechol. Bioeng.*, **22**(4), 699–734.

Lin Y. M., Sharma P. K. and van Loosdrecht M. C. M. (2013). The chemical and mechanical differences between alginate-like exopolysaccharides isolated from aerobic flocculent sludge and aerobic granular sludge. *Water Res.*, **47**(1), 57–65.

Lotti T., Kleerebezema R., Kartal B., de Kreuk M. K., van Erp Taalman Kip, C., Kruit J., Hendrickx T. L. G. and van Loosdrecht M. C. M. (2013). Pilot scale evaluation of anammox based main stream nitrogen removal from municipal wastewater. WEF/IWA Nutrient Removal and Recovery 2013: Trends in Resource Recovery and Use, July 28–31, Vancouver, BC, Canada.

Lu H., Wu D., T. W., Tang D. T. W., Chen G. H. van Loosdrecht M. C. M. and Ekama, G. (2011). Pilot scale evaluation of SANI® process for sludge minimization and greenhouse gas reduction in saline sewage treatment. *Water Sci. Technol.,* **63**, 2149–2154.

Martins A. M. P., Pagilla K., Heijnen J. J. and van Loosdrecht M. C. M. (2004). Filamentous bulking sludge – a critical review. *Water Res.,* **38**, 793–817.

Meda A., Cornel P. and Henkel J. (2010). Wastewater as a source of energy – can wastewater treatment plants be operated energetically self-sufficient? IWA Leading-Edge Conference on Water and Wastewater Technologies, June 2–4, Phoenix, AZ, USA.

Nicolella C., van Loosdrecht M. C. M. and Heijnen J. J. (2000). Particle-based biofilm reactor technology. *Trends in BioTechnol.,* **18**(7), 312–320.

NEPTUNE (2010). New sustainable concepts and processes for optimization and upgrading municipal wastewater and sludge treatment, Work Package 2, Novel Technologies.

Paques (2013). BIOPAQ®UBOX system http://en.paques.nl/pageid=210/BIOPAQ%C2%AEUBOX.html (accessed 10 July 2013)

Pasveer, A. (1962). The oxidation ditch. *Environ. Health,* **4**, 245–257.

PUB (2013). A road map: toward energy self-sufficient, Internal report.

Quadros S., Rosa M. J., Alegre H. and Silva C. (2009). A performance indicators system for urban wastewater treatment plants, IWA, PI09, March 12–13, Amsterdam, The Netherlands.

Raghoebarsing A. A., Pol A., van de Pas-Schoonen K. T., Smolders, A. J. P., Ettwig K. F., Rijpstra W. I. C., Schouten S., Damsté J. S. S., den Camp H. J. M. O., Jetten M. S. M. and Strous, M. (2006). A microbial consortium couples anaerobic methane oxidation to denitrification. *Nature,* **440**, 918–921.

Roest K., Daamen B., de Graaff M. S., Hartog L., Zandvoort M. H., Uijterlinde C. A., Dilven S., van Lier J. B and van Loosdrecht M. C. M. (2012). Energy production from wastewater – dynamic filtration of activated sludge. IWA World Water Congress. September 15–20, Busan, Korea.

Ruiken C. J., Breuer G., Klaversma E., Santiago T. and van Loosdrecht M. C. M. (2013). Sieving wastewater – cellulose recovery, economic and energy evaluation. *Water Res.,* **47**(1), 43–48.

Sen D., Lodhi A., Randall C., Gold L., Brandt K., Copithorn R., Pehrson R. and Chandran K. (2010). Improving our understanding of the differences between fixed and moving bed media IFAS systems for design, operations and for real time control of plants (in Aquifas+) to simultaneously enhance nutrient removal and minimize GHG emissions. *WEFTEC 2010.* October 2–6, New Orleans, LA, USA.

Siegrist H., Salzgeber D., Eugster J. and Jos, A. (2008). Anammox brings WWTP closer to energy autarky due to increased biogas production and reduced aeration energy for N-removal. *Water Sci. Technol.,* **57**(3), 383–388.

Steinherr A. (2010). Heat Recovery from Wastewater, IFAT, 6–19 October 2010, Straubing, Germany.

STOWA (2010). NEWs: The Dutch Roadmap for the WWTP of 2030. Utrecht, The Netherlands.

Strous M., Fuerst J. A., Kramer E. H. M., Logemann S., Muyzer G., van de Pas-Schoonen K. T., Webb, R., Kuenen J. G. and Jetten M. S. M. (1999). Missing lithotroph identified as new planctomycete. *Nature,* **400**, 446–449.

Swiss Federal Government (2013). *Botschaft zur Änderung des Gewässerschutzgesetzes* (Announcement on a change of the Swiss water protection act). *Verursachergerechte Finanzierung der Elimination von Spurenstoffen im Abwasser* (application of the polluter pays principle to the elimination of micropollutants from municipal waste water). http://www.admin.ch/opc/de/federal-gazette/2013/5549.pdf (accessed 10 July 2013)

Tabakatake H., Satoh H., Mino T. and Matsuo T. (2002). PHA (polyhydroxyalkanoate) production potential of activated sludge treating wastewater. *Water Sci. Technol.*, **45**(12),119–126.

Tan G. P. (2004). After the NEWater, there are new challenges. In: Water for All, Public Utilities Board, Singapore, Annual Report, 2004, pp. 18–21.

USFR (1999). CFR 40 Part 130 – Water Quality Planning and Management. www.epa.gov/fedrgstr/EPA-WATER/1999/October/Day-01/w25307.htm/ (accessed 22 June 2013).

van Loosdrecht M. C. M., Eikelboom C., Gjaltema A., Mulder A., Tijhuis L. and Heijnen J. J. (1995). Biofilm structures. *Water Sci. Technol.*, **32**(8), 35–45.

van Loosdrecht M. C. M., Kuba T. van Veldhuizen H. M. Brandse F. A. and Heijnen J. J. (1997). Environmental impacts of nutrient removal processes: case study. *J. Environ. Eng.*, **123**, 33–40.

van Loosdrecht M. C. M. (2011). Granular sludge for nutrient removal. Workshop: Backstage with (Bio) Film's Hottest Stars: Carbon Footprint, O&M Cost, and Nutrient Removal Reliability. WEF-IWA Nutrient Recovery and Management, January 9–12, Miami, FL, USA.

van Loosdrecht M. C. M., Keller J., Nielsen P. H., Lopez-Vazquez C. M. and Brdjanovic D. (2013). *Experimental Methods in Wastewater Treatment.* M. C. M. van Loosdrecht, J. Keller, P. H. Nielsen, C. M. Lopez-Vazquez and D. Brdjanovic (eds), IWA Publishing, London, p. 300, ISBN: 9781780404745.

WERF (2009). Technology Roadmap for Sustainable Wastewater Plants in a Carbon-Constrained World (draft version). WERF Workshop, May 20–21, Chicago, IL, USA.

Wett B., Buchauer K. and Frimml C. (2007). Energy self-sufficiency as a feasible concept for wastewater treatment systems, Leading-Edge Conference, June 4–6, Singapore.

Wett B., Podmirseg S. M., Hell M., Nyhuis G., Bott C. and Murthy S. (2012). Expanding DEMON sidestream deammonification technology towards mainstream application. IWA Conference on Autotrophic Nitrogen Removal: From Research to Applications, June 29, 2012. Milan, Italy.

Winkler M. K. H., Bassin J. P., Kleerebezem R., de Bruin L. M. M., van den Brand T. P. H., van Loosdrecht M. C. M. (2011). Selective sludge removal in a segregated aerobic granular biomass system as a strategy to control PAO-GAO competition at high temperatures. *Water Res.,* **45**, 3291–3299.

Winkler M. K. H., Kleerebezem R., van Loosdrecht M. C. M. (2012). Integration of anammox into the aerobic granular sludge process for main stream wastewater treatment at ambient temperatures. *Water Res.*, **46**(1), 136–144.

Winkler M. K., Bennenbroek M. H., Horstink F. H., van Loosdrecht M. C. M. and van de Pol G. J. (2013). The biodrying concept: an innovative technology creating energy from sewage sludge. *Bioresource Technol.,* **147**, 124–129.

Wu D., Ekama G. A., Lu H., Chui H.-K., Liu W.-T., Brdjanovic D., van Loosdrecht M. C. M. and Chen G.-H. (2013). A new biological phosphorus removal process in association with sulfur cycle. *Water Res.*, **47**(9), 3057–3069.

Zeeman G., Katarzyna K., Brendo M. and Flip K. (2007). Full Scale Demonstration of Vacuum Collection, Transport & Treatment of Black Water. Advanced Sanitation Conference, Aachen, Germany.

IWA Publishing's authorised EU representative for General Product Safety Regulations is Diane D'Arras, 15 rue Duret, 75116 Paris, France, e-mail: safety@iwap.co.uk.

Printed and bound by CPI Group (UK) Ltd, Croydon, CR0 4YY
12/05/2026
02108864-0012